INORGANIC POLLUTANTS IN WATER

INORGANIC POLLUTANTS IN WATER

Edited by

POOJA DEVI
CSIR-Central Scientific Instruments Organisation, Chandigarh, India

PARDEEP SINGH
Department of Environment Sciences, University of Delhi, New Delhi, India

SUSHIL KUMAR KANSAL
Dr. S. S. Bhatnagar University Institute of Chemical Engineering and Technology,
Panjab University, Chandigarh, India

ELSEVIER

Elsevier
Radarweg 29, PO Box 211, 1000 AE Amsterdam, Netherlands
The Boulevard, Langford Lane, Kidlington, Oxford OX5 1GB, United Kingdom
50 Hampshire Street, 5th Floor, Cambridge, MA 02139, United States

Notices
Knowledge and best practice in this field are constantly changing. As new research and experience broaden our understanding, changes in research methods, professional practices, or medical treatment may become necessary.

Practitioners and researchers must always rely on their own experience and knowledge in evaluating and using any information, methods, compounds, or experiments described herein. In using such information or methods they should be mindful of their own safety and the safety of others, including parties for whom they have a professional responsibility.

To the fullest extent of the law, neither the Publisher nor the authors, contributors, or editors, assume any liability for any injury and/or damage to persons or property as a matter of products liability, negligence or otherwise, or from any use or operation of any methods, products, instructions, or ideas contained in the material herein.

British Library Cataloguing-in-Publication Data
A catalogue record for this book is available from the British Library

Library of Congress Cataloging-in-Publication Data
A catalog record for this book is available from the Library of Congress

ISBN: 978-0-12-818965-8

For Information on all Elsevier publications
visit our website at https://www.elsevier.com/books-and-journals

Publisher: Candice Janco
Acquisitions Editor: Louisa Munro
Editorial Project Manager: Sara Valentino
Production Project Manager: Vignesh Tamil
Cover Designer: Matthew Limbert

Typeset by MPS Limited, Chennai, India

Working together
to grow libraries in
developing countries

www.elsevier.com • www.bookaid.org

Contents

List of contributors

Raj Mohan Balakrishnan Department of Chemical Engineering, National Institute of Technology Karnataka Surathkal, Surathkal, India

Amit Bansiwal CSIR-National Environmental Engineering Research Institute, Nagpur, India

Shefali Bhardwaj School of Engineering, Indian Institute of Technology Mandi, Mandi, Himachal Pradesh, India

Preetismita Borah CSIR-Central Scientific Instruments Organisation, Chandigarh, India

Rohit Chauhan Department of Chemical Engineering, Indian Institute of Technology Roorkee, Roorkee, India

Rashmi Dahake CSIR-National Environmental Engineering Research Institute, Nagpur, India

Pooja Devi CSIR-Central Scientific Instruments Organisation, Chandigarh, India

Mohsen Jahanshahi Department of Chemical Engineering, Babol Noshirvani University of Technology, Babol, Iran

Sushil Kumar Kansal Dr. S. S. Bhatnagar University Institute of Chemical Engineering and Technology, Panjab University, Chandigarh, India

Harminder Kaur Punjab Engineering College (Deemed to be University), Chandigarh, India

Manish Kumar CSIR-Central Scientific Instruments Organisation, Chandigarh, India

Pradeep Kumar Department of Chemical Engineering & Technology, IIT (BHU), Varanasi, India

Arindam Malakar University of Nebraska–Lincoln, Lincoln, NE, United States

Vishnu Manirethan Department of Chemical Engineering, National Institute of Technology Karnataka Surathkal, Surathkal, India

Vishal Mishra School of Biochemical Engineering, IIT (BHU) Varanasi, Varanasi, India

Yuvraj Singh Negi Department of Polymer and Process Engineering, Indian Institute of Technology (IIT Roorkee), Roorkee, India

Bhagavan Nvs Department of Physics, Govt Degree College (Men), Srikakulam, India

Majid Peyravi Department of Chemical Engineering, Babol Noshirvani University of Technology, Babol, Iran

Prachi Rajput Academy of Scientific and Innovative Research, New Delhi, India; CSIR-Central Scientific Instruments Organisation, Chandigarh, India

Kirpa Ram Institute of Environment and Sustainable Development, Banaras Hindu University, Varanasi, India

Manish Ranjan Department of Civil Engineering, Indian Institute of Technology (BHU), Varanasi, India

Jatinder Kumar Ratan Department of Chemical Engineering, Dr. B. R. Ambedkar National Institute of Technology Jalandhar, Jalandhar, India

Keyur Raval   Department of Chemical Engineering, National Institute of Technology Karnataka Surathkal, Surathkal, India

P.L. Saranya   Department of Physics, Govt College for Women (Autonomous), Srikakulam, India

Chhavi Sharma   Department of Polymer and Process Engineering, Indian Institute of Technology (IIT Roorkee), Roorkee, India

Dericks Praise Shukla   School of Engineering, Indian Institute of Technology Mandi, Mandi, Himachal Pradesh, India

Nancy Sidana   CSIR-Central Scientific Instruments Organisation, Chandigarh, India; Punjab Engineering College (Deemed to be University), Chandigarh, India

Jyoti Singh   School of Biochemical Engineering, IIT (BHU) Varanasi, Varanasi, India

Pardeep Singh   Department of Environmental Science, PGDAV College, University of Delhi, New Delhi, India

Surinder Singh   Dr. S. S. Bhatnagar University Institute of Chemical Engineering and Technology, Panjab University, Chandigarh, India

R.K. Sinha   CSIR-Central Scientific Instruments Organisation, Chandigarh, India

Daniel D. Snow   University of Nebraska–Lincoln, Lincoln, NE, United States

Richa Soni   School of Engineering, Indian Institute of Technology, Mandi, India

Arun Lal Srivastav   Chitkara University School of Engineering and Technology, Chitkara University, Himachal Pradesh India

Vimal Chandra Srivastava   Department of Chemical Engineering, Indian Institute of Technology Roorkee, Roorkee, India

Houshmand Tourani   Department of Chemical Engineering, Babol Noshirvani University of Technology, Babol, Iran

Priyanka Uddandarao   Department of Chemical Engineering, National Institute of Technology Karnataka Surathkal, Surathkal, India

Priyanshu Verma   Department of Chemical and Biochemical Engineering, Indian Institute of Technology Patna, Patna, India

Kailas L. Wasewar   Advance Separation and Analytical Laboratory (ASAL), Department of Chemical Engineering, Visvesvaraya National Institute of Technology (VNIT), Maharashtra, India

Deepak Yadav   Chemical Engineering Department, HBTU, Kanpur, India

Priyanka Yadav   School of Biochemical Engineering, IIT (BHU) Varanasi, Varanasi, India

Preface

Water is a vital element for the existence of life on Earth. According to a report by the World Water Council, about 1.2 billion people worldwide do not have access to safe drinking water. Furthermore, the distribution and availability of water are declining globally with predicted intense impact between 2020 and 2050. This is because of increased global population, industrialization, irrigation, and thus overexploitation, which has led to water-stressed situation around the globe. Besides, increased contamination of naturally occurring water resources, ground and surface water, is further making them unsuitable for consumption. Thus it is important to undertake the best water management practices for its usage, conservation, and safety in terms of quality. Water contamination is generally caused by geogenic and anthropogenic activities, leading to accumulation of variety of foreign pollutants into water bodies. These pollutants of concern are majorly categorized as organic, inorganic, and biological contaminants.

Inorganic pollutants are a class of nonbiodegradable chemical elements, which when consumed above a permissible limit may lead to several acute and chronic diseases. These mainly include heavy metals (lead, copper, chromium, arsenic, selenium, nickel, mercury, cadmium, nitrate, phosphate, etc.). The origin of these pollutants in ground, surface, and river water is mainly associated with unregulated release of chemicals in water resources and imbalanced geogenic leaching due to environmental and humans' interventions. When consumed above a permissible limit, they could cause damage to vital organs of the body and in severe case cancer leading to death. The clinical symptoms are generally observed as discoloration of skin, hair, nail, etc.; vomiting; nausea; headache; mental retardness; etc. Also, due to their nonbiodegradable nature, they get accumulate into a living system through water as well as food chain. It is thus important to monitor these inorganic pollutants in water resources and develop sustainable remedial technologies for the same.

This book presents the current state of art of these pollutants in water resources, policies for their regulation, monitoring technologies based on various transduction techniques, and remedial solution. The futuristic technologies for their monitoring based upon paper microfluidics, colorimetric reagent kits, etc., are discussed. Likewise, the performance of various remedial systems and their adaptability for their removal is presented. This book is handy information for the researchers, scientists, engineers, environmentalists, who are working toward this domain in aspects of their regulation, monitoring, assessment, and remediation.

Acknowledgment

We thank the reviewers for their useful suggestion on the chapters to improve the quality of the book. Sincere thanks goes to Director, CSIO, Chandigarh, for his kind permission and encouragement to work in this book project, which is the need of the hour.

1

Inorganic water pollutants

Arun Lal Srivastav[1] and Manish Ranjan[2]

[1]Chitkara University School of Engineering and Technology, Chitkara University, Himachal Pradesh India [2]Department of Civil Engineering, Indian Institute of Technology (BHU), Varanasi, India

1.1 Water: a natural resource and a basic need

Water is a vital natural resource necessary to ensure the sustainment of the living systems on the earth, as without it, life cannot exist. Moreover, it has the most important role in the development of any society (CPCB, 2008). Of the Earth's total area, 70.9% is covered by water, that is, hydrosphere, as it contains most of the living organisms of the ecosystem. Out of this 70.9% of the Earth's area, around 96.5% is water present as salty water (e.g., oceans, seas, estuaries), whereas 1.7% water is present as underground water, as well as glaciers/ice caps of the polar areas and Greenland. Only 0.001% is present in the vaporized form as clouds (Gleick, 1993). Freshwater is an essential commodity for the sustenance of living entities on the Earth (Sajid et al., 2018). According to an estimate of the United Nations, the quantity of water present on the Earth is sufficient to cover a 3000 m deep layer (MOWR, 2012). Water quality may be categorized into physical, chemical, and biological, and the people need pure and safe water to use for drinking, as well as domestic purposes (CPCB, 2008).

Cosmopolitan economic growth is also water dependent because of its properties of being a universal solvent, coolant for factories, etc. Since the last few decades, the awareness and treatment for providing safe drinking water to the human community have relatively got better throughout the world. As per the statistics of world water research agencies, around 1 billion people still lacking the availability of pure drinking water, and researches says that this situation will become worse by the year 2025 (Kulshreshtha, 1998). The quality of water may be affected by several natural, as well as human activities. Naturally, the composition of water varies widely in accordance with the local geological conditions. World Health Organization (WHO) has suggested some international guidelines for safe drinking water quality to the humans, which includes the maximum desirable concentrations of chemicals (inorganic, organic) and biological contaminations (WHO, 2004).

Both WHO and UNICEF have reported that >2.2 million of the populations around the world expire every year because of the consumption of contaminated drinking water, as well as poor cleanliness, only in developing countries (Azizullah et al., 2011) and ~60% newly born babies are also under life-threatening conditions due to contagious and parasitic diseases (Ullah et al., 2009). Overuse of inorganic fertilizers/pesticides in agriculture has also elevated the levels of inorganic contamination of soil, groundwater, and surface water, which is not good for human health (Volk et al., 2009; Kundu et al., 2009).

1.2 Water pollution due to inorganic chemicals

Aquatic reservoirs of the world are getting polluted by several types of pollutants, such as hydrocarbons, insecticides, pharmaceutics, cosmetics, chemical causes, hormonal imbalance, as well as toxic heavy metals (Chowdhury et al., 2016; Huber et al., 2016; Grandclement et al., 2017; Kim et al., 2018). Groundwater is the often used source of drinking water throughout the rural, as well as urban worlds, also including India (95% rural and 30%–40% urban Indian population). Since the last few decades, heavy industrialization, as well as mismanaged agricultural activities, have added toxic pollutants (e.g., inorganic chemicals, heavy metals, radionuclide microorganisms, and synthesized organic reactants) to the groundwater (Velizarov et al., 2004).

Commonly found inorganic contaminants of water include arsenic, fluoride, iron, nitrate, heavy metals, etc., and their presence at more than permissible levels degrades water potability for living organisms.

Out of these contaminants, arsenic, fluoride, and iron are having a geogenic origin, whereas nitrates, phosphates, heavy metals are added by anthropogenic behaviors like poor sewage systems, mismanaged agricultural practices, industrial discharges, etc. Modernization of the societies, heavy industrial growth, urban expansion, and huge growth in the human population are the major causes of corrupting groundwater quality (Srivastav, 2013).

1.3 Major inorganic water pollutants

Examples of the inorganic pollutants are heavy metals, halides, oxyanions and cations, and radioactive materials (Mohan et al., 2014). Inorganic pollutants do not degrade easily and persist longer in aqueous systems and cause further deterioration. High concentrations of metals (especially heavy metals) and other toxicants, such as fluoride and nitrate have been found beyond the threshold limit in the groundwater of many parts of the world, including India, and making it unfit for drinking (Srivastav et al., 2013). Arsenic, copper, chromium, lead, mercury, nickel, and zinc are the most studied heavy metals in the wastewater generated by various industries (Ahluwalia and Goyal, 2007; Fu and Wang, 2011).

1.3.1 Water contamination by heavy metals

Wastewater of industries may have huge concentrations of toxic heavy metals along with other pollutants as well, which may damage the environment and living entity of any ecosystem (Zheng et al., 2013; Ariffin et al., 2017; Ali et al., 2019). Some metals which have $>4 \pm 1$ g/cm^3 elemental density are considered as heavy metals; for example, Cd, Cr, Cu, Co, Hg, Ni, Pb, Sn, and Zn (Ali et al., 2019).

Generally, wastewaters from mines, smelters, sewage, battery industries, dyes, alloys, and electronic factories are the source of toxic heavy metals, such as As, Cd, Cr, Cu, Hg, Pb, and Zn. Heavy metals can contaminate the water both from natural or anthropogenic sources. The natural source includes volcanoes, erosion of soil, and disintegration of rocks, etc., whereas the petroleum combustion, mineral extraction, landfilling, urban water discharge, mining activities, industrial discharge, agriculture, metal refining, manufacturing of printed circuit boards, coloring dyes, etc., are among the human activities, which are responsible for heavy metal contamination of water (Baldwin and Marshall, 1999; Barakat, 2011; Akpor et al., 2014; Harvey et al., 2015).

1.3.1.1 Health problems due to heavy metals concentration in water

Heavy metals are very lethal, sometimes carcinogenic, and can also create big problems in the health of any kind of living creatures. If the concentration of heavy metals exceeds the level prescribed by WHO, it will create toxic effects on the soil and aquatic systems (Ali et al., 2019). Heavy metals are well known for higher reactivity, rapid complexation, as well as biochemical processes (Mohammed et al., 2011; Salem et al., 2000). Moreover, these heavy metals get circulated among all the living systems of any ecosystem through food chains (Ali et al., 2019).

Some common heavy metals and health problems, along with their allowable maximum contaminant level (MCL) prescribed by USEPA, are compiled in Table 1.1.

1.3.1.2 Removal of heavy metals from water

Water treatment can be achieved by using many technologies like chemical precipitation and oxidation, reverse osmosis, adsorption, electrodialysis, etc. (Ali et al., 2007, 2011; Chen et al., 2018). However, comparatively, adsorption is better than the other ones because of its easy operation, low cost, practicability, and also because it can remove several contaminants from water (Chowdhury and Balasubramanian, 2014; Ali et al., 2019; Park et al., 2019). Further, it does not produce any type of secondary contamination in treated water (Dubey et al., 2009; Santhosh et al., 2016; Ersan et al., 2017). Moreover, the

TABLE 1.1 Common heavy metals, standard, and health problems.

S. no.	Heavy metal	Source	MCL (mg/L) by USEPA	Health problems
1.	Arsenic	Natural: Geogenic	0.05	Skin diseases, carcinogenicity
		Anthropogenic: Industrial discharge		
2.	Cadmium	Natural: Geogenic	0.005	Kidney diseases
		Anthropogenic: metallurgical process, discarded batteries		
3.	Chromium	Natural: Geogenic	0.1	Allergy and skin inflammation
		Anthropogenic: Steel factories, paper and pulp mills		
4.	Copper	Natural: Geogenic	1.3	Problems to the stomach, liver, kidney
		Anthropogenic: Chemicals used for preservation of woods, corrosion of pipes		
5.	Lead	Natural: Geogenic	0.015	Retards physical and mental growth, kidney disorders, and high blood pressure
		Anthropogenic: corrosion of pipes		
6.	Mercury	Natural: Geogenic	0.002	Kidney and spinal card disorders
		Anthropogenic: refineries discharge		

water purification effectiveness of either technique relies on the nature of the contaminants present in it (Ren et al., 2018; Kim et al., 2018). Latest research shows that graphene oxide is being used widely in the elimination of various inorganic (including heavy metals) (Dong et al., 2014; Hu et al., 2017), as well as organic pollutants (Chen and Chen, 2015; Ersan et al., 2016; Jiao et al., 2017) from water because of containing peculiar surface reactive groups, higher surface area, etc. (Zhou et al., 2012; Kim et al., 2018).

1.3.2 Water contamination by arsenic

Arsenic contamination in water has been considered as one the most perilous threat for human beings, as it can cause several types of serious illnesses. It is widely documented by the world's researchers, as compiled in Table 1.2.

1.3.2.1 Sources of arsenic contamination in water

Both natural and human-generated sources of arsenic in water have been observed, such as volcanic activities and weathering of arsenic bearing rocks; for example, realgar (AsS), orpiment (As_2S_3), arsenopyrite (FeAsS), and lollingite ($FeAs_2$) are natural sources. Whereas, pesticide application, burning of arsenic-containing substances, industrial

TABLE 1.2   Arsenic contamination in groundwater: global and Indian scenario.

S. no.	Country	References
1.	Indian states (Assam, Bihar, Chhattisgarh, West Bengal, and Uttar Pradesh)	Ahmad (2001); Ng et al. (2003); Brinkel et al. (2009); CGWB (2010a,b); Chen et al. (2014)
2.	Mongolia, China	Guo et al. (2006); Wade et al. (2009); Xia et al. (2009); Chen et al. (2014)
3.	Bangladesh	Brinkel et al. (2009); Hassan et al. (2014)
4.	Hungary	Sugar et al. (2013)
5.	Croatia	Habuda-Stanić et al. (2007)
6.	Central Austria, New Zealand, North Afghanistan, and North Mali and Zambia	Amini et al. (2008)
7.	Thailand and Taiwan	Ning (2002)
8.	Argentina, Chile, El Salvador, Mexico, Nicaragua, and Peru	Armienta and Segovia (2008)
9.	Vietnam	Agusa et al. (2014)

wastewater discharge, mine works, mechanization of arsenic compounds come under anthropogenic causes (Ning, 2002).

1.3.2.2 Adverse health effects of high arsenic concentration in water

Out of 64 districts of Bangladesh, 59 are suffering from the serious problem of arsenic contamination in groundwater (Hassan et al., 2014).

A variety of skin disorders along with other venomous effects of arsenic (e.g., melanosis, keratosis, gangrene, cancer) are very common in arsenic contaminated areas. The children are more at risk, as compared to the adults. The poor socioeconomic population is found to be much affected by arsenic skin lesions. Ghosh and Singh (2009) reported some common problems in the Indian Territory during 1983–2006, which includes different ailments of skin, eyes, weight loss, loss of hunger, laziness, limited physical activities, respiratory diseases, coughing, gastric problems, vomiting, indigestion, flavor disorders, stomach pain, liver and spleen diseases, anemia, etc.

1.3.2.3 Removal of arsenic from aqueous solutions

Coagulation, flocculation, and filtration are among the most common traditional techniques to remove arsenic from aqueous solutions. However, there are some issues still unresolved related to the consistency, protection, and residual eliminations of the contaminants from the water using these techniques (Jekel, 1994). Mondal et al. (2006) have published a review on the recent advancements of laboratory-based approaches for the elimination of arsenic from water.

Zaw and Emett (2002) and Leupin and Hug (2005) reported arsenate removals from aqueous solutions using the oxidation/precipitation approach. Gomes et al. (2007) have

TABLE 1.3 Best available techniques for the removal of arsenic (As^{5-}) from aqueous solution (EPA, 2005).

S. no.	Techniques	As^{5-} removal (%)
1.	Reverse osmosis	>95
2.	Ion exchange	95
3.	Coagulation or filtration	95
4.	Adsorption	90
5.	Electrodialysis	85

described electrocoagulation/coprecipitation of As(III) and As(V). EPA (2005) has documented some of the best techniques for the removal of arsenate from water, as given in Table 1.3.

1.3.3 Water contamination by nitrate

Eutrophication of surface water reservoirs and staid human health problems are among the major problems occurring due to the high soluble property of nitrate in water (Bhatnagar and Sillanpää, 2011). The nitrogen cycle is a complex type of nutrient cycling as it involves a variety of biotics, as well as abiotic alterations from seven valence states ($+5$ to -3). The nitrogen compounds may be found as both inorganic (e.g., ammonium, nitrite, and nitrate) and organic nitrogen nature in water, which are considered as vital for life support systems on the earth (Vymazal, 2007), whereas ammonia, dinitrogen, nitric oxide, and nitrous oxide are present in gaseous forms (Bialowiec et al., 2012). Nitrate is a problem primarily in groundwater, and it has shown negative health impacts on the biological systems because excess nitrate levels in groundwater for the methemoglobinemia among infants also can be a reason for gastric, as well as intestinal carcinogenicity. Indiscriminate application of inorganic fertilizers during agriculture is the main culprit of nitrate contamination of groundwater (Bouchard et al., 1992; Rao and Puttanna, 2000; Bhatnagar and Sillanpää, 2011). Apart from this, animal husbandry, septic tanks, ambiance dumping, as well as industrial wastewater release, are the probable sources of nitrate in groundwater (Aelion and Conte, 2004). Excess nitrate from agricultural fields may enter into the surface waters through overflowing, and in groundwater, through seeping (Limbrick, 2003). However, Vinten and Dunn (2001) observed that animal manures are also the main source of elevated nitrate levels in the groundwater. Apart from these, it is cited that sewage pipe leakage, inappropriate wastewater treatment of the effluents, overdosing of inorganic fertilizers, wastes generated by animal farms adds considerable amounts of nitrate in water. Central Pollution Control Board (CPCB) of India has released a report on groundwater quality of 27 metropolitan cities. Many parts of Andhra Pradesh (Hyderabad, Vishakhapatnam), Bihar (Patna), Delhi, Gujrat (Ahmedabad, Rajkot, Surat, Vadodara), Haryana (Faridabad), Jharkhand (Dhanbad, Jamshedpur), Karnataka

(Bangalore), Kerala (Kochi), Rajasthan (Jaipur), Madhya Pradesh (Bhopal, Indore, Jabalpur), Maharashtra (Nagpur, Nasik, Pune, Mumbai), Punjab (Amritsar), Uttar Pradesh (Kanpur, Allahabad, Varanasi) and West Bengal (Asansol, Kolkata) have been found to have high nitrate concentrations (CPCB, 2008). According to WHO (2007), nitrate contamination can be present in groundwater, even for more than 10 years.

1.3.3.1 Nitrate contamination: global and Indian scenario

The problem is prevalent in many parts of Europe, including Great Britain, France, Germany, and Switzerland, several parts of the United States, and Israel (Elyanow and Persechino, 2005). In the European territory, the nitrate concentrations in groundwater exceeded more than the prescribed WHO guidelines for drinking purposes (50 mgNO^{3-}/L) (WHO, 1993). Similarly, China and the United States are also suffering from high nitrate contamination of groundwater (Laegreid et al., 1999), Spain (Mesa et al., 2002). Apart from these, Bulgaria (Gatseva and Argirova, 2008), Senegal (Sall and Vanclooster, 2009), and Italy (Ghiglieri et al., 2009) are also at risk regarding nitrate contamination in groundwater. Several researchers (Kazmi and Khan, 2005; Naeem et al., 2007; Farooqi et al., 2007; Tahir and Rasheed, 2008) have also reported drinking water contamination due to nitrate in the megacities of Pakistan, such as Islamabad, Kasur, Lahore, and Rawalpindi. In Iran, the Hamadan area also reported a level of more than 50 mg/L in 37% of the water of 311 wells used for drinking purpose (Jalali, 2005); in Turkey, 45% of the well water samples and the adjoining topsoils had 108 mgN/L of nitrate concentration (Sönmez et al., 2007), which is even greater than two times of the WHO prescribed guidelines. In Ireland, nitrate concentration depends upon the various factors, including spatial and temporal changes of groundwater, denitrification capacity of subsoils, recharging magnitude of groundwater, and physicochemical properties of soil textures (Fenton et al., 2011). Similarly, according to CGWB (2010a,b), 21 Indian states are suffering with high nitrate contamination (above permissible limit), which are Andhra Pradesh, Bihar, Chhattisgarh, Delhi, Goa, Gujarat, Haryana, Himachal Pradesh, Jammu and Kashmir, Jharkhand, Karnataka, Kerala, Maharashtra, Madhya Pradesh, Odisha, Punjab, Rajasthan, Tamil Nadu, Uttar Pradesh, Uttarakhand, and West Bengal.

1.3.3.2 Sources of nitrate in water

Elevated nitrate levels in groundwater are mainly due to human-induced factors (indiscriminate use of inorganic fertilizers in agriculture) and wastes from animal farms. In regions where agricultural activities are highly intensive, nitrate concentrations in groundwater are usually above its permissible level in drinking water. Overuse of agrochemicals (fertilizers, pesticides) has increased the threat of groundwater contamination (Wang et al., 2009; Barrabes and Sa, 2011; Bhatnagar and Sillanpää, 2011). Nitrate contamination of groundwater occurs due to several causes, such as nitrogenous fertilizers in agriculture, animal wastes generated from farms, municipal solid wastes, landfill sites, septic tanks, soil organic materials (Trevisan et al., 2000; Suthar et al., 2009), and household and industrial wastewater release (Wang et al., 2009; Barrabes and Sa, 2011).

1.3.3.3 *Adverse human health effects of nitrate contamination in water*

Excess concentration of nitrate in water can cause several problems, as the lethality of nitrate on human beings can be seen as the prevalence of methemoglobinemia and cancers (WHO, 2007; Rios et al., 2013) because these human health disorders due to the exceeded nitrate concentration in drinking water have attracted the attention of the world's researchers (Ghafari et al., 2009).

1.3.3.4 *Removal techniques of nitrate from water*

Among the common techniques of nitrate removal, biological denitrification, and ion exchange methods have been suggested by the WHO, while ion exchange, reverse osmosis, and electrodialysis are the recognized techniques from the United States Environmental Protection Agency (USEPA, 2009) as Best Available Technologies (BAT). Similarly, common methods for the elimination of fluoride from aqueous solutions are adsorption, ion exchange process, precipitation−coagulation, reverse osmosis, electrolytic defluoridation, etc.

1.3.4 Fluoride contamination in groundwater

Groundwater contamination by fluoride has been identified as a very severe cosmopolitan problem (Amini et al., 2008). Among most abundant anions, fluoride is one, present widely in groundwater, and creates problems in the supply of safe potable water. Fluorine has the highest electronegativity, as well as reactivity across the periodic table. It is the high reactivity because of which fluorine cannot remain stable in nature in the elemental state. Two forms of fluoride are reported, either as inorganic fluorides (including the free anion F) or as organic compounds bearing fluoride elements. In a study (Jagtap et al., 2012), it has been found that organic fluoride occurrence in the environmental systems is relatively less than the inorganic fluoride compounds. It is mostly found in water due to natural causes, and almost every part of the world is suffering from the contamination of fluoride in water, especially drinking water sources. In India, 201 districts are having fluoride pollution problem affecting millions of public health (Chakraborti et al., 2011). The influence of contamination of fluoride through water consumption can be beneficial or detrimental for humans because it depends on the level of fluoride concentration present in feeding water, and a very small range of fluoride concentration present in drinking water is beneficial for human health. According to Mahramanlioglu et al. (2002), a small amount of fluoride intake rate is beneficial to prevent dental problems, especially in children. However, its elevated levels can be the reason of creating teeth decay, crippling, as well as skeletal fluorosis (WHO, 2008).

1.3.4.1 *Fluoride contamination: global and Indian scenario*

Higher fluoride concentration levels are observed in the groundwater of several countries, such as Kenya (Gaciri and Davies, 1993), Northern and Central Poland (Czarnowski et al., 1996), India (Ayoob and Gupta, 2006; Mohapatra et al., 2012), China (Wang and Huang, 1995), Tanzania (Mjengera and Mkongo, 2003), Mexico (Diaz-Barriga et al., 1997), and Argentina (Kruse and Ainchil, 2003). According to Zhang et al. (2013), there are similar reported results of excessive concentration of fluoride in water/wastewater of countries such as China, India, Africa, and Mexico. More than 20 nations (including both developed and developing world) are found to be affected by fluorosis problems (Ayoob and Gupta, 2006), including Algeria, Argentina, Australia, Canada, China, Egypt, India, Iran, Iraq, Japan, Jordan, Kenya, Libya, Morocco, New Zealand, Saudi Arabia, South Africa, Sri Lanka, Syria, Tanzania, Thailand, Turkey, the United States, etc. (Mameri et al., 1998), and Pakistan (DevBrahman et al., 2013). Montoya et al. (2012) observed that all three major continents like American, Asian, as well as African, are having severe problems of fluoride contamination in groundwater as much as 30 mg/L due to only natural sources.

Seventeen Indian states, especially Andhra Pradesh, Gujarat, Madhya Pradesh, Rajasthan, Tamil Nadu, and Uttar Pradesh (Ayoob and Gupta, 2006) have been affected by fluorosis since it was diagnosed for the first time in Nellore district of Andhra Pradesh in 1937 (Shortt, 1937). It indicates that fluorosis is one of the most disturbing community health harms to the people of India (Jagtap et al., 2012). Mohapatra et al. (2012) reported that in Odisha state, Balasore, Bolangir, Cuttack, Dhenkanal, Kalahandi, Nayagarh, Phulbani, Puri, Sambalpur, and Sundergarh districts are suffering from this problem of highly contaminated fluoride levels in groundwater. Moreover, the situation in India is worse regarding the intake of fluoride contaminated water as it stands among the 25 countries of the world where fluoride generated diseases are more common among the public (Islam et al., 2011).

1.3.4.2 Sources of fluoride in water

Dissolution of minerals of the rocks and soils in the groundwater is the major source of fluoride there. The typical source of fluoride in water is fluoride compound bearing rocks, and thus, only the groundwater is at more risk than the surface water reservoirs. Leaching of fluoride occurs when water percolates from fluoride-containing rocks to the underground water. The rocks rich in fluoride include Fluorspar, Cryolite, and Fluorapatite (Mohapatra et al., 2009; Jagtap et al., 2012). In the human body, the important sources of fluoride are air, cosmetics, drugs, foodstuffs, and drinking water. Drinking water is a major source of fluoride ingestion, and it accounts for approximately around 60% of the total fluoride intake (Jagtap et al., 2012). Its minute concentrations can promote the health wellness for both human, as well as animals (Mourabet et al., 2012), as it helps in the

TABLE 1.4 Fluoride contamination and human health disorders (Mohapatra et al., 2009).

Fluoride concentration (mg/L)	Effects on human beings
<0.5	Dental carries
0.5–1.5	Optimal dental health
0.5–4.0	Dental fluorosis
4.0–10.0	Dental and skeletal fluorosis
>10	Crippling fluorosis

mineralization of bones and formation of teeth enamels, whereas, excessive consumption can cause crippling and skeletal fluorosis in humans (Chen et al., 2011; Mourabet et al., 2012).

1.3.4.3 Adverse human health effects of fluoride contamination in water

Both lower and higher fluoride concentration in drinking water causes serious health disorders among humans, and it depends upon the concentration of fluoride present in potable water as given in Table 1.4.

1.3.4.4 Removal of fluoride from aqueous solutions

Several methods, including reverse osmosis (Sourirajan and Matsurra, 1972), nanofiltration (Simons, 1993), ion exchange (Popat et al., 1994; Sundaram et al., 2008), precipitation (Sujana et al., 1998), electrodialysis (Kabay et al., 2008), ultrafiltration (Guo and Chen, 2005) and adsorption (Tor et al., 2009) have been used in the removal of fluoride from aqueous solutions.

1.4 Concluding remarks and future scope

Water contamination, primarily due to the presence of inorganic pollutants, has become a serious problem to society. The concentration of inorganic pollutants in the aqueous medium could be reduced via the adopting of some treatment techniques, as discussed in previous sections. However, these techniques may have some demerits as well, including cost, operation, and maintenance, the need for electricity, and also the safe disposal of exhausted materials containing toxic elements. Therefore it is the need of the time to research and develop technically advanced, energy-efficient (or electricity less), as well as environmentally friendly water purification systems. In this regard, phytoremediation can be a better option for the elimination of inorganic pollutants from the watery systems because it has been acknowledged as a green technique. Moreover, another sustainable and relatively inexpensive material, that is, biochar may also be derived using waste residues either from agricultural wastes or from forest residues for the treatment of water for a variety of contaminants.

References

Aelion, C., Conte, B., 2004. Susceptibility of residential wells to VOC and nitrate contamination. Environ. Sci. Technol. 38, 1648–1653.

Agusa, T., KimTrang, P.T., Lan, V.M., Anh, D.H., Tanabe, S., Viet, P.H., et al., 2014. Human exposure to arsenic from drinking water in Vietnam. Sci. Total Environ. 488–489, 562–569.

Ahluwalia, S.S., Goyal, D., 2007. Microbial and plant derived biomass for removal of heavy metals from wastewater. Bioresour. Technol. 98, 2243–2257.

Ahmad, K., 2001. Report highlights widespread arsenic contamination in Bangladesh. Lancet 358, 133.

Akpor, O.B., Ohiobor, G.O., Olaolu, T.D., 2014. Heavy metal pollutants in wastewater effluents, sources, effect and remediation. Adv. Biosci. Bioeng. 2, 37–43.

Ali, I., Gupta, V.K., Aboul-Enein, H.Y., 2007. Role of racemization in optically active drugs development. Chirality 19, 453–463.

Ali, I., Khan, T.A., Asim, M., 2011. Removal of arsenic from water by electrocoagulation and electrodialysis techniques. Sep. Purif. Technol. 240, 25–42.

Ali, I., Basheer, A.A., Mbianda, X.Y., Burakov, A., Galunin, E., Burakova, I., et al., 2019. Graphene based adsorbents for remediation of noxious pollutants from wastewater. Environ. Int. 127, 160–180.

Amini, M., Abbaspour, K.C., Berg, M., Winkel, L., Hug, J.S., Hoehn, E., et al., 2008. Statistical modeling of global geogenic arsenic contamination in groundwater. Environ. Sci. Technol. 42, 3669–3675.

Ariffin, N., Abdullah, M.M.A.B., Mohd Arif Zainol, M.R.R., Murshed, M.F., Zain, H., Faris, M.A., et al. 2017. Review on adsorption of heavy metal in wastewater by using geopolymer. MATEC Web Conference. vol. 97, p. 01023.

Armienta, M.A., Segovia, N., 2008. Arsenic and fluoride in the groundwater of Mexico. Environ. Geochem. Health 30, 345–353.

Ayoob, S., Gupta, A.K., 2006. Fluoride in drinking water: A review on the status and stress effects. Crit. Rev. Environ. Sci. Technol. 36, 433–487.

Azizullah, A., Khattak, M.N.K., Richter, P., Häder, D.P., 2011. Water pollution in Pakistan and its impact on public health - a review. Environ. Intern. 37, 479–497.

Baldwin, D.R., Marshall, W.J., 1999. Heavy metal poisoning and its laboratory investigation. Ann. Clin. Biochem. 36, 267–300.

Barakat, M.A., 2011. New trends in removing heavy metals from industrial wastewater. Arabian J. Chem. 4, 361–377.

Barrabes, N., Sa, J., 2011. Review: Catalytic nitrate removal from water, past, present and future perspectives. Appl. Catal. B: Environ. 104, 1–5.

Bhatnagar, A., Sillanpää, M., 2011. A review of emerging adsorbents for nitrate removal from water. Chem. Eng. J. 168, 493–504.

Bialowiec, A., Davies, L., Albuquerque, A., Randerson, P.F., 2012. The influence of plants on nitrogen removal from landfill leachate in discontinuous batch shallow constructed wetland with recirculating subsurface horizontal flow. Ecol. Eng. 40, 44–52.

Bouchard, D.C., Williams, M.K., Surampalli, R.Y., 1992. Nitrate combination of groundwater sources and potential health effects. J. Am. Med. Assoc. 7, 85–90.

Brinkel, J., Khan, M.H., Kraemer, A., 2009. A systematic review of arsenic exposure and its social and mental health effects with special reference to Bangladesh. Int. J. Environ. Res. Public. Health 6, 1609–1619.

Central Pollution Control Board (CPCB), 2008. Status of Groundwater Quality in India –Part II. CPCB, Ministry of Environment and Forest. Government of India, New Delhi.

CGWB, 2010a. A report of Central Groundwater Board 2010. Central Groundwater Board, Ministry of Water Resources, Government of India.

CGWB, 2010b. Status of Groundwater Quality in India –Part II 2008. Central Pollution Control Board, Ministry of Environment and Forest, Government of India.

Chakraborti, D., Das, B., Murrill, M.T., 2011. Examining India's groundwater quality management. Environ. Sci. Technol. 45, 27–33.

Chen, X.X., Chen, B.L., 2015. Macroscopic and spectroscopic investigations of the adsorption of nitroaromatic compounds on graphene oxide, reduced grapheme oxide, and graphene nanosheets. Environ. Sci. Technol. 49 (10), 6181–6189.

Chen, L., Wang, T.J., Wu, H.X., Jin, Y., Zhang, Y., Dou, X.M., 2011. Optimization of a Fe–Al–Ce nano-adsorbent granulation process that used spray coating in a fluidized bed for fluoride removal from drinking water. Powder Technol. 206, 291–296.

Chen, M.L., Ma, L.Y., Chen, X.W., 2014. Review: New procedures for arsenic speciation: A review. Talanta 125, 78–86.

Chen, Y., Liang, W., Li, Y., Wu, Y., Chen, Y., Xiao, W., et al., 2018. Modification, application and reaction mechanisms of nano-sized iron sulfide particles for pollutant removal from soil and water: A review. Chem. Eng. J. Available from: https://doi.org/10.1016/j.cej.2018.12.175.

Chowdhury, S., Balasubramanian, R., 2014. Recent advances in the use of graphene family nanoadsorbents for removal of toxic pollutants from wastewater. Adv. Colloid. Interface. Sci. 204, 35–56.

Chowdhury, S., Mazumder, M.A.J., Al-Attas, O., Husain, T., 2016. Heavy metals in drinking water: occurrences, implications, and future needs in developing countries. Sci. Total Environ. 569, 476–488.

Czarnowski, W., Wrzesniowska, K., Krechniak, J., 1996. Fluoride in drinking water and human urine in Northern and Central Poland. Sci. Total Environ. 191, 177–184.

DevBrahman, K., GulKazi, T., Afridi, H., Naseem, S., Arain, S.S., Wadhwa, S.K., et al., 2013. Simultaneously evaluate the toxic levels of fluoride and arsenic species in undergroundwater of Tharparkar and possible contaminant sources: A multivariate study. Ecotoxicol. Environ. Saf. 89, 95–107.

Diaz-Barriga, F., Navarro-Quezada, A., Grijalva, M., Grimaldo, M., Loyola Rodriguez, J.P., Ortiz, M.D., 1997. Endemic fluorosis in Mexico. Fluoride 30, 233–239.

Dong, Z.H., Wang, D., Liu, X., Pei, X.F., Chen, L.W., Jin, J., 2014. Bio-inspired surface functionalization of graphene oxide for the adsorption of organic dyes and heavy metal ions with a superhigh capacity. J. Mater. Chem. A 2 (14), 5034–5040.

Dubey, S.P., Gopal, K., Bersillon, J.L., 2009. Utility of adsorbents in the purification of drinking water, a review of characterization, efficiency and safety evaluation of various adsorbents. J. Environ. Biol. 30, 327–332.

EPA, 2005. Arsenic Removal From Drinking Water by Ion Exchange and Activated Alumina Plants 2005. United States Environmental Protection. EPA/600/R-00/ 088.

Elyanow, D. and Persechino, J., 2005. Advances in Nitrate Removal, GE Water and Process Technology, Technical Paper.

Ersan, G., Kaya, Y., Apul, O.G., Karanfil, T., 2016. Adsorption of organic contaminants by graphene nanosheets, carbon nanotubes and granular activated carbons under natural organic matter preloading conditions. Sci. Total Environ. 565, 811–817.

Ersan, G., Apul, O.G., Perreault, F., Karanfil, T., 2017. Adsorption of organic contaminants by graphene nanosheets, a review. Water Res. 126, 385–398.

Farooqi, A., Masuda, H., Firdous, N., 2007. Toxic fluoride and arsenic contaminated groundwater in the Lahore and Kasur districts, Punjab, Pakistan and possible contaminant sources. Environ. Pollut. 145, 839–849.

Fenton, O., Healy, M.G., Henry, T., Khalil, M.I., Grant, J., Baily, A., et al., 2011. Exploring the relationship between groundwater geochemical factors and denitrification potentials on a dairy farm in south east Ireland. Ecol. Eng. 37, 1304–1313.

Fu, F., Wang, Q., 2011. Removal of Heavy Metal Ions from Wastewaters: A Review. J. Environ. Manage. 92, 407–418.

Gaciri, S.J., Davies, T.C., 1993. The occurrence and geochemistry of fluoride in some natural waters of Kenya. J. Hydrol. 143, 395–412.

Gatseva, P.D., Argirova, M.D., 2008. High-nitrate levels in drinking water may be a risk factor for thyroid dysfunction in children and pregnant women living in rural Bulgarian areas. Int. J. Hyg. Environ. Health 211, 555–559.

Ghafari, S., Hasan, M., Aroua, M.K., 2009. Nitrate remediation in a novel up flow bioelectrochemical reactor (UBER) using palm shell activated carbon as cathodematerial. Electrochim. Acta 54, 4164–4171.

Ghiglieri, G., Barbieri, G., Vernier, A., Carletti, A., Demurtas, N., Pinna, R., et al., 2009. Potential risks of nitrate pollution in aquifers from agricultural practices in the Nurra region, northwestern Sardinia, Italy. J. Hydrol. 379, 339–350.

Ghosh, N.C., Singh, R.D., 2009. 'Groundwater Arsenic Contamination in India: Vulnerability and Scope for Remedy', Special Session on Groundwater in the 60th IEC and 5th Asian Regional Conference of ICID Held during 6–11 December at Vigyan Bhawan, New Delhi.

Gleick, P.H., 1993. Water in Crisis: A Guide to the World's Freshwater Resources, vol. 13. Oxford University Press.

Gomes, J.A., Daida, P., Kesmez, M., Weir, M., Moreno, H., Parga, J.R., et al., 2007. Arsenic removal by electrocoagulation using combined Al-Fe electrode system and characterization of products. J. Hazard. Mater. 139 (2), 220–231.

Grandclement, C., Seyssiecq, I., Piram, A., Wong-Wah-Chung, P., Vanot, G., Tiliacos, N., et al., 2017. From the conventional biological wastewater treatment to hybrid processes, the evaluation of organic micropollutant removal: a review. Water Res. 111, 297–317.

Guo, X., Chen, F., 2005. Removal of arsenic by bead cellulose loaded with iron oxyhydroxide from groundwater. Environ. Sci. Technol. 39 (17), 6808–6818.

Guo, X., Fujino, Y., Ye, X., Liu, J., Yoshimura, T., 2006. Association between multi-level inorganic arsenic exposure from drinking water and skin lesion in China. Int. J. Environ. Res. Public. Health 3 (3), 262–267.

Habuda-Stanić, M., Kuleš, M., Kalajdzić, B., Romic, Z., 2007. Quality of groundwater in eastern Croatia. The problem of arsenic pollution. Desalination 210, 157–162.

Harvey, P.J., Handley, H.K., Taylor, M.P., 2015. Identification of the sources of metal (lead) contamination in drinking waters in north-eastern Tasmania using lead isotopic compositions. Environ. Sci. Pollut. Res. 22, 12276–12288.

Hassan, K.M., Fukushi, K., Turikuzzaman, K., Moniruzzaman, S.M., 2014. Effects of using arsenic–iron sludge wastes in brick making. Waste Manage. 34, 1072–1078.

Hu, B.W., Hu, Q.Y., Li, X., Pan, H., Tang, X.P., Chen, C.G., et al., 2017. Rapid and highly efficient removal of Eu (III) from aqueous solutions using graphene oxide. J. Mol. Liq. 229, 6–14.

Huber, M., Welker, A., Helmreich, B., 2016. Critical review of heavy metal pollution of traffic area runoff: occurrence, influencing factors, and partitioning. Sci. Total Environ. 541, 895–919.

Islam, M., Mishra, P.C., Patel, R., 2011. Fluoride adsorption from aqueous solution by a hybrid thorium phosphate composite. Chem. Eng. J. 166, 978–985.

Jagtap, S., Yenkie, M.K., Labhsetwar, N., Rayalu, S., 2012. Fluoride in drinking water and defluoridation of water. Chem. Rev. (ACS) 112 (4), 2454–2466.

Jalali, M., 2005. Nitrates leaching from agricultural land in Hamadan Western Iran. Agric. Ecosyst. Environ. 110, 210–218.

Jekel, M., 1994. In: Nriagu, J.O. (Ed.), Arsenic in the Environment, Part I, Cycling and Characterization. Wiley, New York, pp. 119–132.

Jiao, X., Zhang, L.Y., Qiu, Y.S., Guan, J.F., 2017. Comparison of the adsorption of cationic blue onto graphene oxides prepared from natural graphites with different graphitization degrees. Colloids Surf A: Physicochem. Eng. Aspects 529, 292–301.

Kabay, N., Ara, O., Samatya, S., Yuksel, U., Yuksela, M., 2008. Separation of fluoride from aqueous solution by electrodialysis: effect of process parameters and other ionic species. J. Hazard. Mater. 153 (1–2), 107–113.

Kazmi, S.S., Khan, S.A., 2005. Level of nitrate and nitrite contents in drinking water of selected samples received at AFPGMI, Rawalpindi. Pak. J. Physiol. 1, 1–2.

Kim, S., Park, C.M., Jang, M., Son, A., Her, N., Yu, M., et al., 2018. Aqueous removal of inorganic and organic contaminants by graphene-based nanoadsorbents: a review. Chemosphere 212, 1104–1124.

Kruse, E., Ainchil, J., 2003. Fluoride variations in groundwater of an area in Buenos Aires Province Argentina. Environ. Geol. 44, 86–89.

Kulshreshtha, S.N., 1998. A global outlook for water resources to the year 2025. Water Resour. Manage. 12 (3), 167–184.

Kundu, M.C., Mandal, B., Hazra, G.C., 2009. Nitrate and fluoride contamination in groundwater of an intensively managed agroecosystem: a functional relationship. Sci. Total Environ. 407, 2771–2782.

Laegreid, M., Bockman, O.C., Kaarstad, O., 1999. Agriculture, Fertilizers and the Environment. CABI Publishing, Norsk Hydro ASA, Porsgrunn, Norway, p. 294.

Leupin, O.X., Hug, S.J., 2005. Oxidation and removal of arsenic(III) from aerated groundwater by filtration through sand and zero-valent iron. Water Res. 39 (9), 1729–1740.

Limbrick, K., 2003. Baseline nitrate concentration in groundwater of the chalk in South Dorset, UK. Sci. Total Environ. 314–316, 89–98.

Mahramanlioglu, M., Kizilcikli, I., Bicer, I.O., 2002. Sorption of fluoride from aqueous solution by acid treated spent bleaching earth. J. Fluorine Chem. 115, 41–47.

Mameri, N., Yeddou, A.R., Lounici, H., Grib, H., Belhocine, D., Bariou, B., 1998. Defluoridation of septentrional Sahara water of North Africa by electrocoagulation process using bipolar aluminium electrodes. Water Res. 32 (5), 1604–1610.

Mesa, J.M.C., Armendariz, C.R., Torre, A.H.D.L., 2002. Nitrate intake from drinking water on Tenerife island (Spain). Sci. Total Environ. 302, 85–92.

Ministry of Water Resources (MOWR), 2012. New Delhi, Government of India. <http://mowr.gov.in/writwerreaddata/linkimages>.

Mjengera, H., Mkongo, G., 2003. Appropriate deflouridation technology for use in flourotic areas in Tanzania. Phys. Chem. Earth 28, 1097–1104. 2003.

Mohammed, A.S., Kapri, A., Goel, R., 2011. Biomanagement of metal-contaminated soils. Heavy Metal Pollution, Source, Impact, and Remedies. Springer, Dordrecht, pp. 1–28.

Mohan, D., Sarswat, A., Ok, Y.S., Pittman Jr, C.U., 2014. Organic and inorganic contaminants removal from water with biochar, a renewable, low cost and sustainable adsorbent – a critical review. Bioresour. Technol. 160, 191–202.

Mohapatra, M., Anand, S., Mishra, B.K., Giles, D.E., Singh, P., 2009. Review of fluoride removal from drinking water. J. Environ. Manage. 91, 67–77. 2009.

Mohapatra, M., Hariprasad, D., Mohapatra, L., Anand, S., Mishra, B.K., 2012. Mg-doped nano errihydrite—a new adsorbent for fluoride removal from aqueous solutions. Appl. Surf. Sci. 258, 4228–4236.

Mondal, P., Majumdar, C.B., Mohanty, B., 2006. Laboratory based approaches for arsenic remediation from contaminated water: recent development. J. Hazard. Mater. 137 (1), 464–479.

Montoya, V.H., Ramírez-Montoya, L.A., Bonilla-Petriciolet, A., Montes-Morán, M.A., 2012. Optimizing the removal of fluoride from water using new carbons obtained by modification of nut shell with a calcium solution from egg shell. Biochem. Eng. J. 62, 1–7.

Mourabet, M., Rhilassi, E.I.A., Boujaady, E.I.H., Bennani-Ziatni, M., Hamri, E.I.R., Taitai, R., 2012. Removal of fluoride from aqueous solution by adsorption on apatitic tricalcium phosphate using Box–Behnken design and desirability function. Appl. Surf. Sci. 258 (10), 4402–4410.

Naeem, M., Khan, K., Rehman, S., Iqbal, J., 2007. Environmental assessment of groundwater quality of Lahore area, Punjab, Pakistan. J. Appl. Sci. 7, 41–46.

Ng, J.C., Wang, J.P., Shraim, A., 2003. A global health problem caused by arsenic from natural resources. Chemosphere 52, 1353–1359.

Ning, R.Y., 2002. Arsenic removal by reverse osmosis. Desalination 143, 237–241.

Park, C.M., Kim, Y.M., Kim, K.H., Wang, D., Su, C., Yoon, Y., 2019. Potential utility of graphene-based nano spinel ferrites as adsorbent and photocatalyst for removing organic/inorganic contaminants from aqueous solutions: a mini review. Chemosphere. Available from: https://doi.org/10.1016/j.chemosphere.2019.01.063.

Popat, K.M., Anand, P.S., Dasare, B.D., 1994. Selective removal of fluoride ions from water by the aluminium form of the aminomethylphosphonic acid-type ion exchanger. React. Polym. 23 (1), 23–32.

Rao, E.V.S.P., Puttanna, K., 2000. Nitrates, agriculture and environment. Curr. Sci. 79, 1163–1168.

Ren, X.Y., Zeng, G.M., Tang, L., Wang, J.J., Wan, J., Feng, H.P., et al., 2018. Effect of exogenous carbonaceous materials on the bioavailability of organic pollutants and their ecological risks. Soil. Biol. Biochem. 116, 70–81.

Rios, J.F., Ye, M., Wang, L., Lee, P.Z., Davis, H., Hicks, R., 2013. ArcNLET: a GIS-based software to simulate groundwater nitrate load from septic systems to surface water bodies. Comput. Geosci. 52, 108–116.

Sajid, M., Nazal, M.K., Ihsanullah, Baig, N., Osman, A.M., 2018. Review-removal of heavy metals and organic pollutants from water using dendritic polymers based adsorbents: a critical review. Sep. Purif. Tech 191, 400–442.

Salem, H.M., Eweida, E.A., Farag, A., 2000. Heavy Metals in Drinking Water and Their Environmental Impact on Human Health. ICEHM, pp. 542–556.

Sall, M., Vanclooster, M., 2009. Assessing the well water pollution problem by nitrates in the small scale farming systems of the Niayes region, Senegal, Agri. Water Manage. 96, 1360–1368.

Santhosh, C., Velmurugan, V., Jacob, G., Jeong, S.K., Grace, A.N., Bhatnagar, A., 2016. Role of nanomaterials in water treatment applications, a review. Chem. Eng. J. 306, 1116–1137.

Shortt, W.E., 1937. Endemic fluorosis in Nellore District, South India. Ind. Med. Gazette 72, 396.

Simons, R., 1993. Trace element removal from ash dam waters by nanofiltration and diffusion dialysis. Desalination 89 (3), 325–341.

Sönmez, I., Kaplan, M., Sönmez, S., 2007. Investigation of seasonal changes in nitrate contents of soil and irrigation waters in greenhouse located in Antalya-Demre region. Asian J. Chem. vol.19, 5639–5646.

Sourirajan, S., Matsurra, T., 1972. Studies on reverse osmosis for water pollution control. Water Res. 6 (9), 1073–1086.

Srivastav, A.L., 2013. Development of Inorganic Adsorptive Media for the Removal of Nitrate and Fluoride From Water (Thesis). Indian Institute of Technology (BHU), Varanasi (India).

Srivastav, A.L., Madhav, S., Sharma, Y.C., Singh, P.K., 2013. Adsorptive removal of arsenic, fluoride and nitrate from aqueous solutions: a Brief Review. J. Water Res. 135, 237–256.

Sugar, E., Tatar, E., Zaray, G., Mihucz, V.G., 2013. Field separation-based speciation analysis of inorganic arsenic in public well water in Hungary. Microchem. J. 107, 131–135.

Sujana, M.G., Takhur, R.S., Rao, S.B., 1998. Removal of fluoride using aqueous solutions using alum sludge. J. Colloid Interface Sci. 206 (1), 94–101.

Sundaram, C.S., Viswanathan, N., Meenakshi, S., 2008. Defluoridation chemistry of synthetic hydroxyapatite at nano scale: equilibrium and kinetic studies. J. Hazard. Mater. 155 (1−2), 206−215.

Suthar, S., Bishoni, P., Singh, S., Mutiyer, P.K., Nema, A.K., Patil, N.S., 2009. Nitrate contamination in groundwater of some rural areas of Rajasthan, India. J. Hazard. Mater. 171, 189−199.

Tahir, M.A., Rasheed, H., 2008. Distribution of nitrate in the water resources of Pakistan. Afr. J. Environ. Sci. Technol. 2, 397−403.

Tor, N., Danaoglu, G., Arslan, Y., Cengeloglu, 2009. Removal of fluoride from water by using granular red mud: batch and column studies. J. Hazard. Mater. 164 (1), 271−278.

Trevisan, M., Padovani, L., Capri, E., 2000. Nonpoint-source agricultural hazard index: a case study of the province of Cremona, Italy. J. Environ. Manage. 26 (5), 577−584.

Ullah, R., Malik, R.N., Qadir, A., 2009. Assessment of groundwater contamination in an industrial city, Sialkot, Pakistan. Afr. J. Environ. Sci. Technol. 3, 429−446.

United States Environmental Protection Agency, 2009. Drinking Water Contaminants, <http://water.epa.gov/drink/contaminants/index.cfm>.

Velizarov, S., Crespo, J.G., Reis, M.A., 2004. Removal of inorganic anions from drinking water supplies by membrane bio/processes. Rev. Environ. Sci. Bio/Technol. 3, 361−380.

Vinten, A., Dunn, S., 2001. Assessing the effects of land use on temporal change in well water quality in a designated nitrate vulnerable zone. Sci. Total Environ. 265, 253−268.

Volk, M., Liersch, S., Schmidt, G., 2009. Towards the implementation of the European water framework directive? Lessons learned from water quality simulations in an agricultural watershed. Land Use Policy 26, 580−588.

Vymazal, J., 2007. Removal of nutrients in various types of constructed wetlands. Sci. Total Environ. 380, 48−65.

Wade, T.J., Xia, Y., Wu, K., Li, Y., Ning, Z., Le, C., et al., 2009. Increased mortality associated with well-water arsenic exposure in Inner Mongolia, China. Int. J. Environ. Res. Public. Health 6, 1107−1123.

Wang, L.F.M., Huang, J.Z., 1995. Outline of control practice of endemic fluorosis in China. Soc. Sci. Med. 41, 1191−1195.

Wang, Q.H., Feng, C.P., Zhao, Y.X., Hao, C.B., 2009. Denitrification of nitrate contaminated groundwater with a fiber-based biofilm reactor. Bioresour. Technol. 100, 2223−2227.

WHO, 1993. Guidelines for Drinking Water Quality, second ed. World Health Organization, Geneva, 1993, Recommendations.

WHO, 2004. Guidelines for Drinking-Water Quality, third ed. World Health Organization, Geneva, pp. 301−303, 2004.

WHO, 2007. Nitrate and Nitrite in Drinking-Water. World Health Organization, Geneva, 2007, Background Document for Development of WHO Guidelines for Drinking-water Quality.

WHO, 2008. Guidelines for Drinking-Water Quality, vol. 1. World Health Organization, Geneva, 2008, third ed. Recommendations. Incorporating 1st and 2nd Addenda.

Xia, Y., Wade, T.J., Wu, K., Li, Y., Ning, Z., Le, X.C., et al., 2009. Well water arsenic exposure, arsenic induced skin-lesions and self-reported morbidity in Inner Mongolia. Int. J. Environ. Res. Public. Health 6, 1010−1025.

Zaw, M., Emett, M.T., 2002. Arsenic removal from water using advanced oxidation processes. Toxicol. Lett. 133 (1), 113−118.

Zhang, T., Li, Q., Xiao, H., Mei, Z., Lu, H., Zhou, Y., 2013. Enhanced fluoride removal from water by non-thermal plasma modified CeO_2/Mg−Fe layered double hydroxides. Appl. Clay Sci. 72, 117−123.

Zheng, Ch, Zhao, L., Zhou, X., Fu, Z., Li, A., 2013. Treatment technologies for organic wastewater. Water Treat. 11, 249−286.

Zhou, X., Yi, H.H., Tang, X.L., Deng, H., Liu, H.Y., 2012. Thermodynamics for the adsorption of SO_2, NO and CO_2 from flue gas on activated carbon fiber. Chem. Eng. J. 200, 399−404.

Further reading

Cui, J., Shi, J., Jiang, G., Jing, C., 2013. Arsenic levels and speciation from ingestion exposures to biomarkers in Shanxi, China: implications for human health. Environ. Sci. Technol. 47 (10), 5419−5424.

Gao, Q., Xu, J., Bu, X.H., 2019. Review: recent advances about metal−organic frameworks in the removal of pollutants from wastewater. Coord. Chem. Rev. 378, 17−31.

2

Types of inorganic pollutants: metals/metalloids, acids, and organic forms

Preetismita Borah, Manish Kumar and Pooja Devi

CSIR-Central Scientific Instruments Organisation, Chandigarh, India

OUTLINE

2.1 Introduction

Water pollution is a great issue among humans and the ecosystem, caused largely due to rapid industrialization and globalization. It is expected to become worse in the near future. Several organic and inorganic pollutants have been found in water causing environmental deterioration, which are found to be toxic and carcinogenic in nature (Ali and Aboul-Enein, 2006). Most organic and nonbiodegradable metal ions are persistent in the environment for a long time. The present chapter focuses on the inorganic water pollutants/chemical elements.

Currently, the presence of inorganic contaminants in water is a major environmental issue. Water today has been largely contaminated by infectious pathogens and carcinogenic organic and inorganic chemicals. Over the last century, the increase in the population as well as natural activities has increased the quantity of waste and also produced large amounts of chemicals, such as pharmaceuticals and hormones. Despite the introduction of various analytical methods, the discharge of such contaminants is still increasing. The emerging contaminants have now gained remarkable attention to health professionals, engineers, and environmental scientists. The majority of water pollutants are generated from human activity, natural activities such as volcanic eruptions, and anthropogenic activities like sewage, agricultural (fertilizers, pesticides), industrial chemical wastes, mining, and so on (Moore, 2012). In addition to the organic matter, inorganic compounds and other heavy metal ions also contaminate the water, and these compounds are nonbiodegradable and cannot be removed easily from the environment. These inorganic pollutants comprise numerous mineral acids, different inorganic compounds, metals, cyanides, metals complexes, organic metal complexes, and so on. The accumulation of heavy metal ions has a negative effect on aquatic life of the earth system and causes diverse health effects on the liver, kidney, nervous system, circulatory system, blood, gastrointestinal system, bones, and skin (Gadd, 2010). Thus inorganic pollution introduces many inorganic chemicals into the environment, which are destructive to the flora and fauna of the ecosystem. Typically, inorganic pollutants are the compounds of inorganic by-products arising due to radiant energy and noise, heat, or light. Generally, inorganic pollutants include arsenic, cadmium, lead, mercury, chromium, aluminum, nitrates, nitrites, and fluorides. Most of them have long tenacity and resistance to degradation. Heavy metals and metalloids are the main constituents of inorganic pollutants, which are associated with the processes like mining and fossil fuel burning, municipal solid waste, industrial waste, and fertilizers (Halim et al., 2003; Rao and Kashifuddin, 2014).

Furthermore, heavy metals are carcinogenic in nature and cause serious issues to the environment or living system owing to their nondegenerated, stability and accumulative nature (Hu, 2002). Although few heavy metals play crucial role in the human metabolism, they can harm the organs if the intake is in excess (Uddin, 2017). Because of the high solubility in water, they can enter living organisms through food chain and cause several health problems (Speight, 2014). Generally, inorganic pollution arises due to failure of natural environment to decompose the anthropogenic pollutants and due to insufficient knowledge on disposal of chemicals. Table 2.1 shows the examples of some inorganic pollutants and their sources.

2.2 Inorganic pollutants in water and their sources

Water pollution occurs in various bodies like streams, ponds, lakes, rivers, oceans, aquifers, and groundwater when inorganic pollutant/chemicals are directly or indirectly released into the water matrix, which affects the aquatic life. Primary anthropogenic sources of water pollution include municipal sewage, agricultural fertilizers/pesticides,

TABLE 2.1 Examples of inorganic pollutants.

Pollutant	Sources
Sulfur dioxide	Volcanoes
Nitrogen dioxide	Volcanoes
Carbon monoxide	Transports
Ammonia	Industry
Hydrogen sulfide	Industry, anaerobic fermentation
Hydrogen chloride, HCl	Industrial fumes
Hydrogen fluoride	Industry
Ammonium salts	Farms, factories
Nitrate salts	Farms, factories
Nitrite salts	Farms, factories
Lead salts	Heavy industry
Mercury salts	Industry
Zinc salts	Industry

industrial chemical wastes, petroleum products, waste solvents, mine drainage, and so on. Once chemical compounds are released into the surface water and the groundwater through discharge processes or through atmospheric precipitation, they are transported within the water cycle. The inorganic pollutants also undertake various physiochemical and biological transformations. Thus inorganic water pollutants encompass several chemicals, waste, metals, or nonmetals. In addition, thermal pollution occurs from power plants and industries imbalance the inorganic chemicals in river, lakes, and other water sources. Due to globalization and greenhouse effects, the temperature of water has been increasing day by day, therefore decreasing the quantity of oxygen dissolved in water affecting the flora and fauna of aquatic life system (Speight, 2017).

Contamination of the environment by inorganic chemicals such as heavy metals, frequent acid rain due to heavy industrialization, chemical waste, and chemical fertilizers is increasing day by day and their effects on environment can be observed by both physical and chemical properties of the compounds (Speight and Islam, 2016). For example, the dangerous pollutants arise from different man-made mobile sources like vehicles (car, buses, and trucks) and stationary sources like industries, refineries, and power plants as well building materials. On the other hand, fossil fuels like coal, natural gas and gasoline, diesel fuel, and fuel oil are the major sources of inorganic pollutants that are responsible for acid rain (Speight and Islam, 2016; Fawell, 1993). The inorganic chemicals that results from the combustion of fossil fuels are carbon monoxide, carbon dioxide, nitrogen, particulate matter, and heavy metals. Because of the global warming and greenhouse gases, several natural phenomena like melting of ice, an increase in sea level and severe weather events such as hurricanes, tornadoes, heat waves, floods, and so on are occurs (Our Planet, 1996).

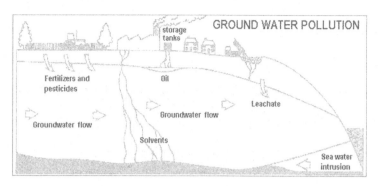

FIGURE 2.1 Sources of inorganic pollutant (Our Planet, 1996). http://www.gdrc.org/uem/water/water-pollution.html

Further, drinking water contains large amount of inorganic contaminant and these are occurs through some natural and man-made activities (Fergusson, 1990). While many of the chemicals such as calcium, sodium, iron, and manganese are regulated and may be indigenous to the water body and may be deposited as minerals in water. Although, numerous inorganic pollutants cannot be diminished by living organisms and exist in the ecosphere for long time. Since inorganic pollutants demolish ecology and the habitat and become a global problem. To control the inorganic pollutants in water, we need better understanding of their chemical/physical characteristics, their occurrence, transport pathways, ultimate fate, detection methods, health effects and treatment potential. The present chapter proposes an overview of the types and different forms of inorganic water pollutant. Fig. 2.1 depicts the different sources of water pollutant (Bradl, 2002).

2.3 Types of inorganic water pollutants: physiochemical properties

Inorganic pollutants consist of both major and minor constituents where about twelve elements considered as major components, 28 elements are minor elements and thirteen elements of less common minor elements. The types of inorganic pollutants from which water is contaminated are categorized as (1) acids, which often occur due to industrial discharges, most specifically sulfur dioxide from coal-burning power plants; (2) base like ammonia from food industry; (3) chemical waste, obtained from industrial process and agricultural; (4) heavy metals, which arises from motor vehicles mine drainage; and (5) silt or sediment-fine grained powdery materials, mostly occurs through construction sites, land clearing sites, and so on. Table 2.2 describes different types of inorganic pollutants and their sources and activities and all these chemicals are originated from minerals not from livening organism.

The following *inorganic contaminants* have numerous health effects or physiochemical properties:

i. Arsenic (As)

Arsenic enters the environment by the fossil fuel combustion and when the disposal of fly ash and coal cleaning waste is leached into the groundwater (Herawati et al., 2000). Arsenic, however, is recognized as an essential but hazardous in excess amount of intake (Salomons et al., 1995). It can cause ulceration and chronic

TABLE 2.2 Types of inorganic chemical pollutants and sources.

Inorganic pollutants	Examples	Activity
Acidity, pyrites, sulfur	Deep or open cast mines, crushing	Mineral, extraction/processing coal
All metals and associated anions	Ferrous and nonferrous, deep, or open cast, crushing, waste lagoons	Metals
Sulfides, sulfates	Asbestos, chalk, china clay, gypsum	Other minerals
Fluoride, sulfur dioxide, As, Cd, Cr, Cu, cyanide, Pb, Hg, Ni, sulfate, Zn	Battery, electronics, and pigments manufacture, electroplating	Industries brick works, chemical works
Sulfur, cyanides	Coal gasification	Gasworks
Metal wastes, iron, and steel slag	Blast furnaces	Iron and steel
Pb, Zn, Cu, Hg, Cd, Ni, acidity	Lead, zinc, copper smelters	Nonferrous smelting/processing
Lead compounds	Petroleum production	Oil refineries
Fuel ash, fly ash, various radionuclides	Coal, gas, nuclear plant	Power generation
Acidity, asbestos, Pb, Ni, Cd	Batteries, cars, domestic appliances	Recycling/scrap
Metals, phosphate, chromium	Various processes	Sewage treatment
Various, fuels	Munitions, anticorrosion or antifouling chemicals, preservatives	Tanneries, military bases air/army/navy bases
Acidity, alkalinity, antifouling compounds, metal wastes	Runoff, workshops, dredged sediments, dumped waste	Airports, canals, waterways, docks
Metal wastes, asbestos	Railways, borders, sidings/yards	Railways
Pb, Cd, Zn	Borders, central reservation, runoff areas	Roads
Zn, Cu	Oil pipelines, cables and towers, substations, transformers	Energy supply pipelines, electricity supply network
Pb		Oil/petroleum storage

diseases such as bladder/kidney, lung, and skin cancer (Finkelman et al., 2001). Besides, it damages peripheral nerves and blood vessels. For example, in the Gizhou Province of China, chronic arsenic poisoning occurs as a result of drying chili pepper over open burning stoves where arsenic rich coal with 35,000 ppm were used to heat and cook (Clarke and Sloss, 1992; Emsley, 1989). Chili peppers, which itself contain 1 ppm arsenic, after drying it can reach up to 500 ppm. As a result, almost 3000 people are manifesting the symptoms of arsenic poisoning such as hyperpigmentation, Bowen's disease, and cell carcinoma.

ii. Zinc

Zinc is commonly used as a corrosive metal for resistant coatings, in alloys, in dry-cell batteries, paints, plastics, wood preservatives, rubber, cosmetics, and so on. The waste from the related industries may pollute the water. Zinc contaminant waste also exists in different process like metal production, worn rubber tires on vehicles and combustion of coal. Its consumption above standard limit may cause various health problems such as nausea, stomach cramps, lungs and the body's temperature control system (Finkelman et al., 2002).

iii. Cadmium (Cd)

Cadmium is toxic and carcinogenic in nature and can cause several health issues such as cardiovascular disease, fibrosis of the lung, renal injury, and so on (Swaine and Goodarzi, 1995). In the 1960s, in Japan, the metal cadmium was the cause of itai-itai disease, which is a painful bone disease. In that incidence, the cadmium arises from a source of zinc—lead mine and accumulated in the downstream rice fields and the river water. Apparently, in general, high cadmium levels can cause diseases such as decreased growth, kidney damage, cardiac enlargement, hypertension cancer, and so forth.

iv. Chromium (Cr)

Chromium is an essential trace element and its chromates are toxic and carcinogenic (Herawati et al., 2000). When chromium is ingested by humans, it is found to be dangerous for liver and spleen (Finkelman et al., 2001).

v. Hydrogen sulfide

Hydrogen sulfide naturally exists as gas in the environment. It is generated from inorganic waste of electric power and oil—gas extraction operations, pig farms, sewage treatment plants, and other confined animal feeding operations. Polluted water from this chemical may cause collapse, hemorrhage, headaches, coma, blurred vision, dizziness, and nausea.

vi. Fluorine (F)

Fluorine is also an essential element and mainly used to protect enamel of teeth and excess of fluoride is hazardous (Finkelman et al., 2001). Fluorosis, caused by excess of fluorine and causes numerous forms of skeletal damage like osteosclerosis, less movement of joints, and outward manifestations like knock-knees and spinal curvature (Clarke and Sloss, 1992). Fluorosis, also causes severe bone deformation in children. In Guizhou Province, China, more than 10 million people were affected by fluorosis due to burning of corn over high fluorine coal.

vii. Manganese (Mn)

Manganese may cause several health problems as a result of combustion of fossil and leaching of ash (Finkelman et al., 2001). However, it is considered an essential nutrient, nontoxic but causes respiratory problems (Geiger and Cooper, 2010).

viii. Lead (Pb)

Lead commonly exists in ore with zinc, silver and with copper and isolated together with these metals. Metallic lead generally occurs in nature and it can be accumulated in body by inhalation of air and ingestion of this metal in to food, water, and soil matrixes. Generally, it absorbs in the body through blood, soft tissues and bones (EPA (United States Environmental Protection Agency), 1998). It has diverse health issues such as kidneys damages, liver, and nervous system. Over consumption to lead causes numerous neurological disorders like mental retardation, and behavioral disorders. The metal also damages the central nervous system of embryo and young children. It is also a high-risk factor for high blood pressure and heart disease.

ix. Selenium (Se)

Selenium is also recognized as an essential element and toxic and carcinogenic when accumulate in excess (Finkelman et al., 2001). It has several adverse impacts on human health such as gastrointestinal disturbance, anemia, liver, and spleen damage. A normal symptom of selenium poisoning is loss of hair and nail. In southwest of China, selenosis was reported, which is occurs due to use of selenium contain carbonaceous shales (stone coals) for home heating and cooking purposes (Emsley, 1989).

x. Mercury (Hg)

Mercury is the very harmful when it is used in high level amount. It is existed in various water pollutants agents like scale gold mining, primary production of nonferrous metals and fossil fuel burning. Cement production, contaminated sites, consumer products waste and chloro-alkali industry also possess this element in small percentages (Finkelman et al., 2001).

Presently, environmental chemist performing nanotechnology to eliminating the mercury from the polluted water. Apart from the pathogens in sewage, mercury is only contaminant accumulated to the sea which caused death in mankind. In 1956 Japan Minamata disaster was caused only because of mercury. This was happened due to the release of industrial wastewater from the Chisso Corporation's chemical factory which releases methylmercury to the food chain and then accumulated in shellfish or fish in the local market which was used by the local population and caused Minamata disease. Mercury has lots of health problem such as progressive and irreversible brain damage, hearing and speech damage, a perturbed vision, muscle weakness, paralysis, coma, and death. The United States Environmental Protection Agency (EPA) stated that Mercury can develop side effects such as loss of peripheral vision, deteriorated movement coordination, and weakened muscles, impairment of speech and hearing.

Thus the abovementioned inorganic water contaminants are commonly found in water and required to be aware of them. By applying preventive methods can make the ecosystem and aquatic life healthy.

2.3.1 Metals/metalloids

Metals contamination in water is becoming a serious issue of concern globally. Naturally occurring metals/heavy metals have a high density and atomic weight and are five times greater than that of water. Due to wide application in industry, domestic, medical, agricultural, and technological applications, they are widely distributed in the ecosystem and have shown potential impact on human health and the atmosphere. Toxicity of the metals mainly depends on numerous factors such as dose and route of exposure, also age, gender, nutritional status of individuals. Owing to their high extent of toxicity, cadmium, chromium arsenic, mercury, and lead, are of human health significance. Due to increase in concentrations levels of heavy metal ions in food, humans are more prone to serious health issues. These toxic metals damage the human organ at very low concentration.

In the bygone era, it has been found that the metals are contaminants in the environment and affects the ecological as well as the human health. Besides human exposure have increases drastically due to the increase use of numerous agricultural, industrial, pharmaceutical domestic effluents, and technological applications and owing to this the sources of heavy metal ions increases gradually (He et al., 2005). The prominent sources such as mining, foundries and metal-based factories are mainly responsible for the environmental pollution (Shallari et al., 1998; Nriagu, 1989). However, some of the heavy metals are naturally originating and readily found in the earth's surface due to different human activities. The metal can contaminate the environment in several ways such as corrosion of metal, soil erosion, leaching of heavy metals, and evaporation of metal from surface water to groundwater and to soil matrixes. Besides, several natural circumstances like weathering and volcanic eruptions have remarkable contribution to metal or heavy metal pollution (Pacyna, 1996). However, industrial sources such as refineries, power plants, nuclear power stations, plastics, textiles, wood preservation, and paper processing plants are responsible for heavy metal pollutant (Shallari et al., 1998). Some heavy metals such as zinc, copper, selenium, chromium, cobalt, iodine, manganese, molybdenum, aluminum and nickel are considered as trace elements and are also necessary for biological activities and needed for healthy life and these are called essential metals (Pacyna, 1996). Biological factors like species characteristics, tropic interactions, and physiological properties also play an important role and each metal has specific physiochemical mechanism. However, heavy metals have some toxic and carcinogenic mechanistic aspects. Thus pollution caused by metals or heavy metals, like cadmium, lead, and mercury, are significantly hazardous. Earlier lead was used in gasoline as an antiknocking additive but in recent times it strictly banned in many countries (Arruti et al., 2010). The battery companies are still using mercury and cadmium in batteries. Table 2.3 describes the metals and biological function associated with them.

Arsenic is considered as toxic to both plant and animals and most remarkable water pollutant metalloid. The actual arsenic poisoning can occur when ingestion is more than 100 mg. Besides, chronic poisoning can occur due to ingestion of small amounts of arsenic over a long period. Arsenic is carcinogenic in nature and causes several health problems such as respiratory cancers, chronic diseases, bladder/kidney cancers. Its toxicity depends on its valency and trivalent arsenic is more toxic than pentavalent arsenic. Generally, arsenic exists in the Earth surface at an average of 2—5 ppm. However, all the marine pathogens contain arsenic in the stable pentavalent form. The combustion of fossil fuels, especially combustion of coal, produces huge amount of arsenic into the environment and reaches natural waters sources. Another major source of arsenic is mine dumps. Normally, arsenic is found to be present with phosphate minerals and gets into the atmosphere with phosphorus complexes. Before World War II, the pesticides used contained toxic arsenic complex such as lead arsenate, sodium arsenite, and Paris green (Fig. 2.2) (Garbarino et al., 1995).

2.3.2 Acids

Acids are used in different industrial applications and chemical research laboratories. Generally, they are low hazardous chemicals because these can be easily neutralized in

TABLE 2.3 Metals and their physiochemical properties.

Metal	Biological function
Sodium	Balance the plasma membrane
Potassium	Balance the plasma membrane
Calcium	Muscle functioning and bone tissue synthesis
Magnesium	Central ion in chlorophyll heme group
Zinc	As a cofactor in many enzymes, for example, carboxypeptidase and liver alcohol dehydrogenase
Iron	Vitamin C updates and found in hemoglobin
Copper	Cytochrome oxidase
Selenium	Activate the antioxidant enzyme such as glutathione peroxidase
Cobalt	Use in synthesis of vitamin B12
	Mining wastes
	Inorganic

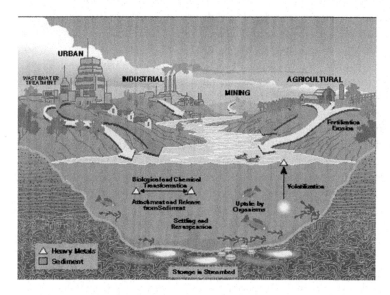

FIGURE 2.2 Sources of metals/heavy metals (Garbarino et al., 1995). https://pubs.usgs.gov/circ/circ1133/heavy-metals.html.

the environment, but if excess amount is spread, it will cause bad impact on environment and human health. Due to human interest and increase in industrialization, the emission of nitrogen oxides (NO, NO_2) and sulfur dioxide from automobile and power is increased in a tremendous way. And these gases are converted into nitric acid and sulfuric acid in the atmosphere, which changes the pH of natural water and results in acid rain or snow.

2.3.3 Organic forms

Organic contaminants easily enter the environment through various man-made activities like the use of numerous synthetic pesticides and detergents, food additives, pharmaceuticals, insecticides, paints, synthetic fibers, waste solvents, and some volatile organic compounds. All these compounds are hazardous and resistant to microbial degradation. In fact, if concentration of any of these is imbalanced, it can cause water unsuitable for use. Since the 1930s, polychlorinated biphenyls, which are complex mixtures of chlorobiphenyls, are used as an organic form of pollutants in the industries. They are fat-soluble compounds and can readily enter the environment and within the cells. Due to their high stability to chemical reagents and persistence in nature, they are readily introduced into the ecosystem. The interaction of organic compounds with metals has ubiquitous importance to determine the role of metals ions in aquatic life. There are two forms metal−organic interactions in aquatic system such as metal−organic ligand complexation and organometallic compound formation. The metal−organic ligand complexation is applicable to natural water and wastewater, where the species reversibly dissociates to metal ions and organic complex species as a hydrogen ion concentration. On the other hand, in organometallic compounds, the metal is bound to organic molecule by a carbon atom and do not dissociate in water even at lower pH or higher dilution.

Hazardous organic compounds are chemicals having carbon as the main constituent. They are generally intact in the environment for long durations and finally have been found to accumulate in aquatic life and sediments. The organic chemicals easily pollute the drinking water supplies and can cause long-term health problems to humans. So, many toxic organic chemicals are the main cause of water pollution nationwide. These pollutants include a very large number of different organic substances such as organic solvents, pesticides, dioxins, polychlorinated biphenyls, furans, and other nitrogen-containing derivatives. The most typical sources of the pollutants include wood preservatives, antifreeze, dry-cleaning materials, cleansers, and different other chemical products. Improper disposal of industrial and household wastes and runoff of pesticides are considered to the two significant sources of hazardous organic compounds in water. Excessive use of insecticides, herbicides, fungicides, and rodenticides can result in toxic compounds carried by storm water runoff from agricultural lands, construction sites, and residential lawns to receiving waters.

In addition, one of the significant volatile organic compounds is methyl tert-butyl ether (MTBE). Earlier MTBE was commonly used as an air-cleaning gas ingredient but nowadays it is banned because; if once it is accumulated in water it talks long time to remove from contaminated water. This form of organic compound causes several health problems such as leukemia, thyroid glands, and kidneys lymphoma and tumors in the testicles.

2.4 Status of water resources contamination with inorganic water pollutants

Concerning human health and ecosystem, numerous acts/laws have been introduced by the government to preserve the quality of groundwater and other surface water. Transboundary is one of the significant problems with water pollution. There are many rivers and seas that are part of more than one country. If one country is affected by the factory waste, it can cause problems in the neighboring countries. Environmental law such as UN Convention on the Law of the Sea, the London Dumping Convention, has provided restriction to industries and factories. Also, MARPOL International Convention controls the pollutant originated from ships.

Clean Water Act and Safe Drinking Water Act has contributed various pure water resource prevention schemes such as Wellhead Protection Program, Sole Source Aquifer Program, and Source Water Assessment Program. These programs monitor the quality of water with minimum health issues. The Clean Water Act controls the protection of surface as well as groundwater setting the standard for allowing the pollutants supply to the water sources.

The polluted water can be purified by several standard methods such as osmosis, coagulation, filtrations, modular anaerobic system, precipitations, microbial fuel cell, and advanced oxidation process. The above pathogens cannot be applied everywhere so monitoring of water is possible by easily detecting the fecal contamination of spotting contamination. If fecal contaminations are present in higher concentrations, then pathogens would be there and water is called hazardous for use. The common indicators of fecal contamination in water are total coliforms, *Escherichia coli*, fecal coliforms, fecal streptococci/enterococci, coliphage, and *Clostridium perfringens*. Various pathogens must be controlled through multiple barrier approach to prevent waterborne diseases. Fish as living organism can be applied for detecting the water pollutants. Behavior changes and growth of water living inform about the polluting aspects. There are computer-generated programs to watch hazards in the standard water at laboratories. The data are analyzed to determine the water pollutants.

There are various analytical techniques available to detect the water contamination in the polluted water such as membrane filtration, advanced oxidation, ultraviolet (UV) irradiation, ion exchange, and biological filtration. Fluorometry and fluorescence-based techniques are utilized for the detection of hydrocarbons in the oil and gas industry and for the detection of different algae in the water. UV-visible light has been introduced to determine the total organic carbon and measure turbidity transform because of water contamination. In the present scenario, smart phone sensor-based application uses DNA-magnetic particle technology to identify certain bacteria in the water bodies. Moreover, according to the WHO, almost 1.6 million people die every year due to diarrheal diseases, which occurs mainly due to the lack of safe drinking water as well as sanitation. Very recently, researchers are coming up with a solution to this perennial problem by using nanotechnology in filtration for water purification. The technology basically diminishes microbes, bacteria, and other particulate matters.

The application of abovementioned techniques is costly and contains limitations, so many new approaches are being developed for groundwater remediation like subsurface barriers, pump and treat, soil vapor extraction, bioremediation to control the flowing of polluted water in the specified area.

2.5 Health effects of inorganic water pollutants

Polluted water has diverse health effects in human as well as in aquatic life. According to a survey by WHO (last updated in 2019), more than 2.1 billion people are living with without safe drinking water, and around 2.3 billion people do not have hygienic sanitation system. As per the report, it is found that more than 525,000 children died due to diarrhea as a result of sewage disposal. The World Health Organization also calculated approximately 135 million people will die by 2020 due to water-related diseases (Kabata-Pendias and Pendias, 2001).

Several inorganic contaminants exist in water, which are present as a lower concentration as compared to the major components. Some of them attain great importance because of diverse health effect or associated with different health problem. The emissions of trace elements are responsible for severe health problems to human and these pollutants are accumulated in our bodies through food chain. Here we are discussing the health impacts of various metals/heavy metals and trace elements. Among all inorganic pollutants, heavy metals have attracted great attention of environmental chemists due to their toxicity. Usually they are present in trace amounts in natural waters, but some of them are very toxic even at very low concentration (Hamelink et al., 1994). Metals such as arsenic (As), lead (Pb), cadmium (Cd), nickel (Ni), mercury (Hg), chromium (Cr), cobalt (Co), zinc (Zn), and selenium (Se) are highly toxic even in low quantity. Moreover, pollutants such as heavy metals, dyes, and other organic pollutants are known as carcinogens whereas hormones, pharmaceuticals, and cosmetics and personal care product wastes are called endocrine disruptive chemicals. In the present scenario, the quantity of heavy metals is increasing due to the direct discharge of metal-containing effluents into the water sources. Humans consume these metals in their food and water. Toxic heavy metals have severe health effects on various organs of human such as mild eye, nose and skin irritations, headache, stomach ache, diarrhea, hematemesis, vomiting, cirrhosis, necrosis, low blood pressure, hypertension, and gastrointestinal distress (Verma and Schneider, 2017).

It is very interesting that some heavy metals are also recognized as essential elements like cobalt, copper, iron, manganese, vanadium, and zinc, which are essential to the body in little quantities for various biochemical systems. Arsenic consumption through polluted water causes cancer of the lungs, liver, and bladder. Cadmium contaminated water may harm kidney and lungs, and cause bone fragility.

Most specifically, the consumption of lead has very negative effect. It may damage the brain and kidneys. A small amount of lead may disturb the children in the learning process, and cause memory loss, response functions, and make children aggressive (Sun et al., 2017). Pregnant women can get miscarriage due to the high consumption of lead use and it affects sperm production in men. Mercury is also known as a global pollutant as it is widely used in many purposes, so it has wide side effects on health. Mercury enters the body through blood vessels and exits through urinal excretion and scat. It has several side effects such as loss of peripheral vision, deteriorated movement coordination, weakened muscles, and impairment of speech and hearing.

However, pharmaceuticals have used numerous chemical samples from ppb to ppm levels and the waste of these samples cause water contamination and affect the human health and other lives (Archer et al., 2017). Water contaminated by Pharmaceuticals may decrease the eggs in women as well sperms in men (Akanyeti et al., 2017; Ng et al., 2017).

Several traditional as well as modern methods of remediation of contaminants from water resources are available. One of the most traditional approaches is the filtration of water through a column of bed and bank material. Apart from filtration, numerous natural processes, such as leaching, hydrolysis, precipitation, oxidation, and reduction, also help in the remediation of water. One of the most important environmental challenges is caused by emerging organic pollutants. Several techniques are available to remove the emerging organic contaminants (EOCs) from water such as adsorption, ultrafiltration, electrodialysis, Fenton reaction, photocatalysis, and so on. One of the significant phenomena used for the remediation of water is the photocatalysis. In this process, two sources of energy, solar energy and artificial UV light, are widely utilized in water treatment.

Further, the fluoride remediation can be easily carried out by using three strategies, namely, precipitation- and coagulation-based remediation techniques, membrane-based filtration techniques, and adsorption-based remediation techniques.

In the present scenario of water remediation studies, recent technologies like remote sensing and geographic information systems (GIS) are widely used. Remote sensing-based techniques and GIS have been found to detect the water quality at low cost and in short time. GIS can easily detect the water quality parameter like chlorophyll-a, algae bloom, turbidity, suspended sediments, and mineral content in water bodies with high accuracy and low cost.

However, nanotechnology (nanoparticles) is a recent advancement in the water remediation and extensive studies are ongoing for wastewater treatment. Several carbon-based nanocomposites used for water remediation are carbon$-$ZnO, graphene$-$SiO$_2$/Cu$_2$O, graphdiyne $-$ ZnO, carbon nanoparticles$-$gold, platinum nanoparticle, carbon$-$TiO$_2$ nanotubes, carbon aerogel$-$TiO$_2$, graphene$-$SiO$_2$ nanoplatelets, multiwalled carbon nanotubes$-$metal-doped ZnO nanohybrid, multiwall carbon nanotubes$-$TiO$_2$-SiO$_2$, carbon nanofibers$-$Ag-TiO$_2$, carbon nanotube-Ag$_3$PO$_4$ in pickering emulsions, carbon$-$nitrogen-doped TiO$_2$-SiO$_2$, carbon-Ag-TiO$_2$, and so on.

2.6 Conclusions

The chapter briefly summarizes the sources of inorganic water pollutants, their classification, and negative health impact, with a focus on physiochemical properties of the inorganic chemicals. Besides, a trend toward using biological and chemical processes for inorganic pollutant degradation is covered and discussed in brief. Further several newer techniques for water remediation are also discussed in brief.

Acknowledgment

The sincere support and encouragement by Director, CSIR-Central Scientific Instruments Organisation, Chandigarh, is acknowledged.

References

Akanyeti, I., Kraft, A., Ferrari, M., 2017. Hybrid polystyrene nanoparticle-ultrafiltration system for hormone removal from water. J. Water Process Eng. 17, 102–109.

Ali, I., Aboul-Enein, H.Y., 2006. Instrumental Methods in Metal Ion Speciation. CRC Press, Boca Raton, FL.

Archer, E., Petrie, B., Hordern, B.K., Wolfaardt, G.M., 2017. The fate of pharmaceuticals and personal care products (PPCPs), endocrine disrupting contaminants (EDCs), metabolites and illicit drugs in a WWTW and environmental waters. Chemosphere 174, 437–446.

Arruti, A., Fernández-Olmo, I., Irabien, A., 2010. Evaluation of the contribution of local sources to trace metals levels in urban PM2.5 and PM10 in the Cantabria region (Northern Spain). J. Environ. Monit. 12 (7), 1451−1458.

Bradl, H. (Ed.), 2002. Heavy Metals in the Environment: Origin, Interaction and Remediation, vol. 6. Academic Press, New York.

Clarke, L.E., Sloss, L.L., 1992. Trace Elements-Emissions From Coal Combustion and Gasification. IEA Coal Research, London.

Emsley, J., 1989. The Elements. Clarendon Press, Oxford.

EPA (United States Environmental Protection Agency), 1998. Latest Findings on National AIR QUality: (1997) Status and Trends. Office of Air Quality Planning and Standards, Washington, DC.

Fawell, J.K., 1993. The impact of inorganic chemicals on water quality and health. Ann. Ist Super Sanita. 29 (2), 293−03.

Fergusson, J.E., 1990. The Heavy Elements: Chemistry, Environmental Impact and Health Effects. Pergamon Press, Oxford.

Finkelman, R.B., Skinner, H.C.W., Plumlee, G.S., Bunnell, J.E., 2001. Medical Geology. Geosciences and Human Health, American Geological Institute, Alexandria, VA, pp. 20−23.

Finkelman, R.B., Orem, W., Castranova, V., Tatu, C.A., Belkin, H.E., Zheng, B., 2002. Health impacts of coal and coal use: possible solution, Int. J. Coal Geol., Vol. 50. Elsevier Science, pp. 425−443.

Gadd, G.M., 2010. Metals: minerals and microbes: geomicrobiology and bioremediation. Microbiology 156, 609−643.

Garbarino, J.R., Hayes, H.C., Roth, D.A., Antweiler, R.C., Brinton, T.I., Taylor, H.E., 1995. Contaminants in the Mississippi River: Heavy Metals in the Mississippi River. U.S. Geological Survey Circular, Reston, VA, p. 1133.

Geiger, A., Cooper, J., 2010. Overview of Airborne Metals Regulations, Exposure Limits, Health Effects, and Contemporary Research. Cooper Environmental Services LLC.

Halim, M., Conte, P., Piccolo, A., 2003. Potential availability of heavy metals to phytoextraction from contaminated soils induced by exogenous humic substances. Chemosphere 52, 265−275.

Hamelink, J.L., Landrum, P.F., Harold, B.L., William, B.H. (Eds.), 1994. Bioavailability: Physical, Chemical, and Biological Interactions. CRC Press, Taylor & Francis Group, Boca Raton, FL.

He, Z.L., Yang, X.E., Stoffella, P.J., 2005. Trace elements in agroecosystems and impacts on the environment. J. Trace Elem. Med. Biol. 19 (2−3), 125−140.

Herawati, N., Suzuki, S., Hayashi, K., Rivai, I.F., Koyoma, H., 2000. Cadmium, copper and zinc levels in rice and soil of Japan, Indonesia and China by soil type. Bull. Environ. Contam. Toxicol. 64, 33−39.

Hu, H., 2002. Human Health and Heavy Metals Life Support: The Environment and Human Health, 65. MIT Press, Cambridge, MA.

Kabata-Pendias, A., Pendias, H., 2001. Trace Metals in Soils and Plants, second ed. CRC Press, Taylor & Francis Group, Boca Raton, FL.

Moore, J.W., 2012. Inorganic Contaminants of Surface Water: Research and Monitoring Priorities. Springer, New York.

Ng, C.K., Bope, C.D., Nalaparaju, A., Cheng, Y., Lu, L., Wang, R., et al., 2017. Concentrating synthetic estrogen 17a-ethinylestradiol using microporous polyethersulfone hollow fibre membranes: experimental exploration and molecular simulation. Chem. Eng. J. 314, 80−87.

Nriagu, J.O., 1989. A global assessment of natural sources of atmospheric trace metals. Nature 338, 47−49.

Our Planet, Vol. 8, No. 3, 1996., http://www.gdrc.org/uem/water/water-pollution.html.

Pacyna, J.M., 1996. Monitoring and assessment of metal contaminants in the air. In: Chang, L.W., Magos, L., Suzuli, T. (Eds.), Toxicology of Metals. CRC Press, Taylor & Francis Group, Boca Raton, FL, pp. 9−28.

Rao, R.A.K., Kashifuddin, M., 2014. Kinetics and isotherm studies of Cd (II) adsorption from aqueous solution utilizing seeds of bottlebrush plant (*Callistemon chisholmii*). Appl. Water Sci. 4, 371−383.

Salomons, W., Forstner, U., Mader, P., 1995. Heavy Metals: Problems and Solutions. Springer-Verlag, Berlin, Germany.

Shallari, S., Schwartz, C., Hasko, A., Morel, J.L., 1998. Heavy metals in soils and plants of serpentine and industrial sites of Albania. Sci. Total Environ. 192 (09), 133−142.

Speight, J.G., 2014. The Chemistry and Technology of Petroleum, fifth ed. CRC Press, Taylor & Francis Group, Boca Raton, FL.

Speight, J.G., 2017. Handbook of Petroleum Refining. CRC Press, Taylor & Francis Group, Boca Raton, FL.

Speight, J.G., Islam, M.R., 2016. Peak Energy-Myth or Reality. Scrivener, Salem, MA.

Sun, B., Zhang, X., Yin, Y., Sun, H., Ge, H., Li, W., 2017. Effects of sulforaphane and vitamin E on cognitive disorder and oxidative damage in lead-exposed mice hippocampus at lactation. J. Trace Elem. Med. Bio. 44, 88–92.

Swaine, D.J., Goodarzi, F., 1995. Environmental Aspects of Trace Elements in Coal. Kluwer Academic Publishers, Dordrecht, The Netherlands.

Uddin, M.K., 2017. A review on the adsorption of heavy metals by clay minerals, with special focus on the past decade. Chem. Eng. J. 308, 438–462.

Verma, M., Schneider, J.S., 2017. Strain specific effects of low level lead exposure on associative learning and memory in rats. Neurotox. 62, 186–191.

Further reading

Dada, A.O., Adekola, F.A., Odebunmi, E.O., 2015. Kinetics and equilibrium models for sorption of cu (II) onto a novel manganese nano-adsorbent. J. Disper. Sci. Tech. 37 (1), 119–133.

World Health Organization (WHO): Diarrhoeal disease, Fact sheet Number 330, May 2017.

Priority and emerging pollutants in water

Manish Kumar, Preetismita Borah and Pooja Devi

CSIR-Central Scientific Instruments Organisation, Chandigarh, India

3.1 Introduction: inorganic water pollutants

Water pollution is a crucial problem of global concern faced by the current generation. It is well known that earth is covered with 78% of water and also body contains approximately 70% water. The increased industrialization and urbanization is adding up high level of various pollutants in water resources due to their unregulated wastewater discharge into the river and thus surface and groundwater. Inorganic pollutants are the nonbiodegradable toxic elements, which when present in trace amount can lead to adverse health effects (Carson, 2002). These pollutants are mainly originated from naturally occurring activities and anthropogenic contribution through industrialization, urbanization, uncontrolled discharge of sewerage, and so on. They are found to impact metabolic activities of variety of systems such as flora, fauna, and humans. They are naturally known to occur in earth's crust; however the concern of their presence is raised owing to the negative health impacts associated with them on their long-term exposure. Their consumption leads to potential damages to ecological system and environment. It is therefore important to regulate activities causing their elevated level in water sources. Also, technological interventions are required for their monitoring as well as remediation (Sauve and Desrosiers, 2014). Although contamination with inorganic pollutants is a global problem, however, certain countries including India, Peru, China, and Russia, are facing this problem to a large extent. In general water pollutants are categorized into following categories: Organic Pollutants and Inorganic Pollutants. In the first category, pollutants such as herbicides, insecticides, petrochemicals, persistent organic pollutants are originated from agricultural and industrial activities (Landers et al., 2012). Similarly, detergent, chemical wastes from cosmetic industries, chlorinated solvents, hydrocarbons, and many more compounds from research laboratories also add up to the pollutants in water. While in later, urban and industrial activities add their level into water channels such as river and ocean (Lombi and Hamon, 2005). Listed below are various inorganic water pollutants generally found in water bodies:

1. Ammonia

    Ammonia release in water channels is generally from food processing and agricultural industrial waste. Its excess quantity generates toxicity in the body (Bleibel and Al-Osaimi, 2012). When mixed with bleach, it further produces toxic by-products in the form of gaseous compounds.

2. Arsenic

    Arsenic (As, atomic number 33) is a commonly found inorganic water pollutant and originates from industrial waste such as from processing of glass, pigments, textiles, paper, metal adhesives, and wood preservatives. It has been used earlier in pesticides, pharmaceuticals, and feed additives. It is a well-known group 1 carcinogen by International Agency for Research on Cancer.

3. Barium

    Barium (Ba) has features and effects similar to Arsenic. When it is consumed in an excess amount, it may cause cardiac abnormalities, tremors, weakness, shortness of breath, and paralysis. It is originated from the wastes of metal industries. The permissible standard range of barium in water is 0.23 and 2.5 mg/L for water consumption (Fawell and Mascarenhas, 2004).

4. Chloride

    Chloride is found in the waste of salt storage, sewage effluent, manure, industrial residue, and gas drilling. The high consumption of chloride-contaminated water may cause poisoning beyond a permissible limit of 250 mg/L.

5. Chromium

    Chromium (Cr) as waste is discharged from dyes, paints, and leather tanning industries (Resende et al., 2014). The standard permissible limit of total chromium is drinking water is 0.1 mg/L. It also exists in two oxidation states, wherein Chromium (VI) is known for its carcinogenic effects.

6. Copper

Copper is released in water resources from corrosion of water plumbing units. Although it is an essential element for metabolic activities when consumed as 0.3 mg/L per day, its excess consumption can cause stomach cramps and intestinal diseases (Vargas et al., 2017).

7. Mercury

Mercury is a very harmful and toxic inorganic element event at ultra-trace level. It is released in water bodies from various activities such as scale gold mining, primary production of nonferrous metals, and fossil fuel burning. Cement production, consumer products waste, and chloro-alkali industry also possess this element in small percentages (Sundseth et al., 2017).

The US Environmental Protection Agency (EPA) revealed that consumption of mercury can lead to side effects such as loss of peripheral vision, deteriorated movement coordination, weakened muscles, and impairment of speech and hearing.

8. Uranium

Uranium is a rare element on the earth. It is found in water-polluting sources due to the mining or refining of nuclear fuels or radioactive waste. It could have negative health impacts on kidney, liver, heart, brain, and other organ systems (Dawe and Ferguson, 2003). The negative effects of uranium can be resolved by regulating the nuclear mining or refining sites.

9. Zinc

Zinc is used widely for corrosion-resistant coatings, in alloys, in dry-cell batteries, paints, dyes, plastics, wood preservatives, rubber, and cosmetics. The related industries may pollute the water from their acidic wastes. Waste having zinc contaminant is also existed in metal production processes, waste incineration, rubber tires, and industrial combustion of coal. Its consumption above standard limit may cause various health problems such as nausea and stomach cramps, lungs and the body's temperature control system (Geiger and Cooper, 2010).

3.2 Emerging water pollutants

Emerging water pollutants are elements or their acids, which are not generally found in water sources and hence not much concern was given on them by scientific community. But with observed increasing ecological disturbance and improved scientific methods, they have been analyzed for their origin and mechanism of health impacts. These emergent water pollutants can make changes in human behavior, landscape, water resources, and demography, due to developing technologies, microbial adaptation, climate change, increased travel, and so on. This category of pollutants includes pharmaceuticals, algal toxins, microorganisms, and several other chemicals. To prevent these contaminants from the water, it is required to understand their features, generation,

TABLE 3.1   Emerging water contaminants, effects and sources.

Contaminant	Adverse effects	Source
Endocrine-disrupting compounds and personal care products	Endocrine system disruption	Wastewater
Pharmaceuticals	Unknown	Wastewater
Antibiotic resistance genes	Pathogen resistance to antibiotics	Human and animal antibiotics
Cyanotoxins	Liver and nervous system damage	Blue-green algae
Human parasites, bacteria, viruses	Infections	Wastewater
Zoonotic parasites, bacteria, viruses pathogenic to humans	Infections	Animal waste
N-nitrosodimethylamine	Carcinogenesis	by-product of wastewater chlorination
Perchlorate	Uncertain	Rocket and missile propellant
1,4-Dioxane	Carcinogenesis	Stabilizer of solvents
Methyltertbutylether	Toxicity	Fuel oxygenate
Alkylphenolpolyethoxylates	Toxicity	Degradation of surfactants
Fluorinated alkyl surfactants Toxicity Industrial processes	Toxicity	Industrial processes
Polybrominated diphenyl ethers	Toxicity	Flame retardants
Benzotriazoles	Toxicity	Anticorrosives, wastewater
Naphthenic acids	Toxicity	Crude oil, wastewater
New, chiral, and transformed pesticides	Toxicity, carcinogenesis	Wastewater
Disinfection by-products	Toxicity	Chlorination by-products of emerging contaminants

transportation, side effects on human health and environments, analysis approaches, and resolving techniques. Table 3.1 lists out emerging contaminants with their effects and sources mentioned:

3.3 Priority water pollutants

The pollutants are known as a group of chemical toxic pollutants as per EPA regulations. The priority pollutant list suggests that toxic pollutants are more applicable practically for assigning purpose of EPA through Clean Water Act. Some toxic pollutants have open-ended sets like chlorinated benzenes which remain hundreds of elements. Agency for Toxic Substances and Disease Registry (ATSDR) concerned about health issues of human beings and included all health departments widely to identify the pollutants from the water and environment. ASTDR is an agency who takes responsibility for health issues, provides demanding information, and prevents the harmful and toxic compounds. The ATSDR recognize those chemicals which are industrial wastes, known as, minimal risk levels (MRLs). MRL is a standard level for the chemicals quantity which may be taken by us as food, drink, or breathe without any harmful risk on health. MRLs have been prepared regarding health domain other than cancer (ATSDR, 1993).

MRL may be taken above standard level if it is not harmful for our health otherwise avoid the MRLs in higher amount. The ATSDR is associated with EPA for zonal and national level to trace the MRLs of toxic substances. MRLs are formed in three time zones as exposing duration: (1) acute: 1—14 days, (2) intermediate: 15—364 days, and (3) chronic: above 364 days. The criterion for MRLs calculation includes:

- How much duration, chemical is exposed by people?
- The amount of the chemical and its health issues.
- What is the age (infant or adult/old) of the person at exposing time?
- Data are either from human or animal. Sometimes data are not concerned with human exposure so animal studies are taken because health issues are equal in human and animals.
- Quality of data from the human and animal with same health effects.

Uncertain parameters may affect the calculation of MRLs due to the dependency on type and quality of data available. Uncertain parameters are helpful to recognize the differences between human beings and animals by health issues. This factor also provides the information about health effects on sensitive population. The priority water pollutants along with their MRLs as per ASTDR list are mentioned below in Table 3.2.

TABLE 3.2 Agency for Toxic Substances and Disease Registry: minimal risk levels (MRLs), August 2018.

Sr. No.	Name	MRL in oral route	CAS number
1	ACENAPHTHENE	0.6 mg/kg/day	83-32-9
2	ACETONE	2 mg/kg/day	67-64-1
3	ACROLEIN	0.004 mg/kg/day	107-02-8
4	ACRYLAMIDE	0.01 mg/kg/day	79-06-1
5	ACRYLONITRILE	0.1 mg/kg/day	107-13-1
6	ALDRIN	0.002 mg/kg/day	309-00-2
7	ALUMINUM	1.0 mg/kg/day	7429-90-5
8	AMERICIUM		7440-35-9
9	AMMONIA		7664-41-7
10	ANTHRACENE	10 mg/kg/day	120-12-7
11	ANTIMONY	1 mg/kg/day	7440-36-0
12	ARSENIC	0.005 mg/kg/day	7440-38-2

(Continued)

TABLE 3.2   (Continued)

Sr. No.	Name	MRL in oral route	CAS number
13	ATRAZINE	0.01 mg/kg/day	1912-24-9
14	BARIUM, SOLUBLE SALTS	0.2 mg/kg/day	7440-39-3
15	BENZENE	0.0005 mg/kg/day	71-43-2
16	BERYLLIUM	0.002 mg/kg/day	7440-41-7
17	BIS(2-CHLOROETHYL)ETHER (BCEE)		111-44-4
18	BIS(CHLOROMETHYL)ETHER (BCME)		542-88-1
19	BORON	0.2 mg/kg/day	7440-42-8
20	BROMODICHLOROMETHANE	0.1 mg/kg/day	75-27-4
21	BROMOFORM	0.7 mg/kg/day	75-25-2
22	BROMOMETHANE		74-83-9
23	1-BROMOPROPANE	0.2 mg/kg/day	106-94-5
24	2-BUTOXYETHANOL (ETHYLENE GLYCOL MONOBUTYL ETHER)	0.4 mg/kg/day	111-76-2
25	CADMIUM	0.0005 mg/kg/day	7440-43-9
26	CARBON DISULFIDE	0.01 mg/kg/day	75-15-0
27	CARBON TETRACHLORIDE	0.02 mg/kg/day	56-23-5
28	CESIUM		7440-46-2
29	CHLORDANE	0.001 mg/kg/day	57-74-9
30	CHLORDECONE	0.01 mg/kg/day	143-50-0
31	CHLORFENVINPHOS		470-90-6
32	CHLORINE	0.002 mg/kg/day	7782-50-5
33	CHLORINE DIOXIDE		10049-04-4
34	CHLORITE	0.1 mg/kg/day	7758-19-2
35	CHLOROBENZENE	0.4 mg/kg/day	108-90-7
36	CHLOROETHANE		75-00-3
37	CHLOROFORM	0.3 mg/kg/day	67-66-3
38	CHLOROMETHANE		74-87-3
39	4-CHLOROPHENOL	0.01 mg/kg/day	106-48-9
40	CHLORPYRIFOS	0.003 mg/kg/day	2921-88-2
41	CHROMIUM(III) SOLUBLE PARTICULATES		16065-83-1
42	CHROMIUM(III) INSOL. PARTICULATES		16065-83-1

(*Continued*)

TABLE 3.2 (Continued)

Sr. No.	Name	MRL in oral route	CAS number
43	CHROMIUM(VI)	0.005 mg/kg/day	18540-29-9
44	CHROMIUM(VI), AEROSOL MISTS		18540-29-9
45	CHROMIUM(VI), PARTICULATES		18540-29-9
46	COBALT	0.01 mg/kg/day	7440-48-4
47	COPPER	0.01 mg/kg/day	7440-50-8
48	CRESOLS	0.1 mg/kg/day	1319-77-3
49	CYANIDE, SODIUM	0.05 mg/kg/day	143-33-9
50	CYHALOTHRIN	0.01 mg/kg/day	68085-85-8
51	CYPERMETHRIN	0.02 mg/kg/day	52315-07-8
52	DDT, P,P'-	0.0005 mg/kg/day	50-29-3
53	DEET	1 mg/kg/day	134-62-3
54	DIAZINON	0.006 mg/kg/day	333-41-5
55	DIBROMOCHLOROMETHANE	0.1 mg/kg/day	124-48-1
56	1,2-DIBROMO-3-CHLOROPROPANE	0.002 mg/kg/day	96-12-8
57	DI-N-BUTYL PHTHALATE	0.5 mg/kg/day	84-74-2
58	1,2-DICHLOROBENZENE	0.7 mg/kg/day	95-50-1
59	1,3-DICHLOROBENZENE	0.4 mg/kg/day	541-73-1
60	1,4-DICHLOROBENZENE	0.07 mg/kg/day	106-46-7
61	1,2-DICHLOROETHANE	0.2 mg/kg/day	107-06-2
62	1,1-DICHLOROETHENE	0.009 mg/kg/day	75-35-4
63	1,2-DICHLOROETHENE, CIS-	1 mg/kg/day	156-59-2
64	1,2-DICHLOROETHENE, TRANS-	0.2 mg/kg/day	156-60-5
65	2,4-DICHLOROPHENOL	0.003 mg/kg/day	120-83-2
66	2,4-DICHLOROPHENOXYACETIC ACID(2,4-D)	0.009 mg/kg/day	94-75-7
67	1,2-DICHLOROPROPANE	0.1 mg/kg/day	78-87-5
68	1,3-DICHLOROPROPENE	0.04 mg/kg/day	542-75-6
69	2,3-DICHLOROPROPENE		78-88-6
70	DICHLORVOS	0.004 mg/kg/day	62-73-7
71	DIELDRIN	0.0001 mg/kg/day	60-57-1
72	DI(2-ETHYLHEXYL)PHTHALATE	0.1 mg/kg/day	117-81-7
73	DIETHYL PHTHALATE	7 mg/kg/day	84-66-2
74	DIISOPROPYL METHYLPHOSPHONATE(DIMP)	0.8 mg/kg/day	1445-75-6

(Continued)

TABLE 3.2   (Continued)

Sr. No.	Name	MRL in oral route	CAS number
75	DIMETHYLARSINIC ACID (DMA)	0.02 mg/kg/day	75-60-5
76	1,1-DIMETHYLHYDRAZINE		57-14-7
77	1,2-DIMETHYLHYDRAZINE	0.0008 mg/kg/day	540-73-8
78	1,3-DINITROBENZENE	0.08 mg/kg/day	99-65-0
79	4,6-DINITRO-O-CRESOL	0.004 mg/kg/day	534-52-1
80	2,4-DINITROPHENOL	0.01 mg/kg/day	51-28-5
81	2,3-DINITROTOLUENE	0.09 mg/kg/day	602-01-7
82	2,4-DINITROTOLUENE	0.05 mg/kg/day	121-14-2
83	2,5-DINITROTOLUENE	0.007 mg/kg/day	619-15-8
84	2,6-DINITROTOLUENE	0.09 mg/kg/day	606-20-2
85	3,4-DINITROTOLUENE	0.03 mg/kg/day	610-39-9
86	3,5-DINITROTOLUENE	0.03 mg/kg/day	618-85-9
87	DI-N-OCTYL PHTHALATE	3 mg/kg/day	117-84-0
88	1,4-DIOXANE	5 mg/kg/day	123-91-1
89	DISULFOTON	0.001 mg/kg/day	298-04-4
90	ENDOSULFAN	0.007 mg/kg/day	115-29-7
91	ENDRIN	0.002 mg/kg/day	72-20-8
92	ETHION	0.002 mg/kg/day	563-12-2
93	ETHYLBENZENE	0.4 mg/kg/day	100-41-4
94	ETHYLENE GLYCOL	0.8 mg/kg/day	107-21-1
95	ETHYLENE OXIDE		75-21-8
96	FLUORANTHENE	0.4 mg/kg/day	206-44-0
97	FLUORENE	0.4 mg/kg/day	86-73-7
98	FLUORIDE, SODIUM	0.05 mg/kg/day	7681-49-4
99	FLUORINE		7782-41-4
100	FORMALDEHYDE	0.3 mg/kg/day	50-00-0
101	FUEL OIL NO.2		68476-30-2
102	GLUTARALDEHYDE	0.1 mg/kg/day	111-30-8
103	GUTHION (AZINPHOS-METHYL)	0.01 mg/kg/day	86-50-0
104	HEPTACHLOR	0.0006 mg/kg/day	76-44-8
105	HEXACHLOROBENZENE	0.008 mg/kg/day	118-74-1
106	HEXACHLOROBUTADIENE	0.0002 mg/kg/day	87-68-3

(Continued)

Inorganic Pollutants in Water

TABLE 3.2 (Continued)

Sr. No.	Name	MRL in oral route	CAS number
107	ALPHA-HEXACHLOROCYCLOHEXANE	0.008 mg/kg/day	319-84-6
108	BETA-HEXACHLOROCYCLOHEXANE	0.05 mg/kg/day	319-85-7
109	GAMMA-HEXACHLOROCYCLOHEXANE	0.003 mg/kg/day	58-89-9
110	HEXACHLOROCYCLOPENTADIENE	0.1 mg/kg/day	77-47-4
111	HEXACHLOROETHANE	1 mg/kg/day	67-72-1
112	HEXAMETHYLENE DIISOCYANATE		822-06-0
113	N-HEXANE		110-54-3
114	2-HEXANONE	0.05 mg/kg/day	591-78-6
115	HMX (CYCLOTETRAMETHYLENETETRANITRAMINE)	0.1 mg/kg/day	2691-41-0
116	HYDRAZINE		302-01-2
117	HYDROGEN FLUORIDE		7664-39-3
118	HYDROGEN SULFIDE		7783-06-4
119	IODIDE	0.01 mg/kg/day	7553-56-2
120	IONIZING RADIATION, N.O.S.		HZ1800-45-T
121	ISOPHORONE	3 mg/kg/day	78-59-1
122	JP-4		50815-00-4
123	JP-5		8008-20-6
124	JP-7		HZ0600-22-T
125	JP-8	3 mg/kg/day	8008-20-6
126	KEROSENE		8008-20-6
127	MALATHION	0.02 mg/kg/day	121-75-5
128	MANGANESE, RESPIRABLE		7439-96-5
129	MERCURIC CHLORIDE	0.007 mg/kg/day	7487-94-7
130	MERCURY		7439-97-6
131	METHOXYCHLOR	0.005 mg/kg/day	72-43-5
132	4,4'-METHYLENEBIS(2-CHLOROANILINE)(MBOCA)	0.003 mg/kg/day	101-14-4
133	METHYLENE CHLORIDE	0.2 mg/kg/day	75-09-2
134	4,4'-METHYLENEDIANILINE	0.2 mg/kg/day	101-77-9
135	METHYLENEDIPHENYL DIISOCYANATE(MDI), polymeric		101-68-8
136	METHYLMERCURY	0.0003 mg/kg/day	22967-92-6
137	1-METHYLNAPHTHALENE	0.07 mg/kg/day	90-12-0
138	2-METHYLNAPHTHALENE	0.04 mg/kg/day	91-57-6

(Continued)

TABLE 3.2   (Continued)

Sr. No.	Name	MRL in oral route	CAS number
139	METHYL PARATHION	0.0007 mg/kg/day	298-00-0
140	METHYL-T-BUTYL ETHER	0.4 mg/kg/day	1634-04-4
141	MIREX	0.0008 mg/kg/day	2385-85-5
142	MOLYBDENUM	0.05 mg/kg/day	7439-98-7
143	MONOMETHYLARSONIC ACID(MMA)	0.1 mg/kg/day	124-58-3
144	NAPHTHALENE	0.6 mg/kg/day	91-20-3
145	Nitrate	4 mg/kg/day	14797-55-8
146	Nitrite	0.1 mg/kg/day	14797-65-0
147	N-NITROSODI-N-PROPYLAMINE	0.095 mg/kg/day	621-64-7
148	NICKEL		7440-02-0
149	PARATHION	0.009 mg/kg/day	56-38-2
150	2,3,4,7,8-PENTACHLORODIBENZOFURAN	0.001 ug/kg/day	57117-31-4
151	PENTACHLOROPHENOL	0.005 mg/kg/day	87-86-5
152	PERCHLORATES	0.0007 mg/kg/day	7778-74-7
153	PERMETHRIN	0.3 mg/kg/day	52645-53-1
154	PHENOL	1 mg/kg/day	108-95-2
155	PHOSPHORUS, WHITE	0.0002 mg/kg/day	7723-14-0
156	POLYBROMINATED BIPHENYLS (PBBs)	0.01 mg/kg/day	36355-01-8
157	POLYBROMINATED DIPHENYL ETHERS(PBDEs), LOWER BROMINATED	0.00006 mg/kg/day	32536-52-0 60348-60-9 5436-43-1
158	PBDEs, DECABROMINATED	0.01 mg/kg/day	1163-19-5
159	PERFLUOROHEXANE SULFONIC ACID(PFHxS)	0.00002 mg/kg/day	355-46-4
160	PERFLUORONONANOIC ACID(PFNA)	0.000003 mg/kg/day	375-95-1
161	PERFLUOROOCTANOIC ACID (PFOA)	0.000003 mg/kg/day	355-67-1
162	PERFLUOROOCTANE SULFONIC ACID(PFOS)	0.000002 mg/kg/day	1763-23-1
163	POLYCHLORINATED BIPHENYLS (PCBs)(Aroclor 1254)	0.03 ug/kg/day	11097-69-1
164	PROPYLENE GLYCOL		57-55-6
165	PROPYLENE GLYCOL DINITRATE		6423-43-4
166	RDX (Cyclonite)	0.2 mg/kg/day	121-82-4
167	REFRACTORY CERAMIC FIBERS		HZ0900-26-T
168	SELENIUM	0.005 mg/kg/day	7782-49-2
169	STRONTIUM	2 mg/kg/day	7440-24-6

(Continued)

Inorganic Pollutants in Water

TABLE 3.2   (Continued)

Sr. No.	Name	MRL in oral route	CAS number
170	STYRENE	0.1 mg/kg/day	100-42-5
171	SULFUR DIOXIDE		7446-09-5
172	SULFUR MUSTARD	0.5 ug/kg/day	505-60-2
173	2,3,7,8-TETRACHLORODIBENZO-P-DIOXIN	0.0002 ug/kg/day	1746-01-6
174	1,1,2,2-TETRACHLOROETHANE	0.5 mg/kg/day	79-34-5
175	TETRACHLOROETHYLENE	0.008 mg/kg/day	127-18-4
176	TIN, INORGANIC	0.3 mg/kg/day	7440-31-5
177	TIN, DIBUTYL-, DICHLORIDE	0.005 mg/kg/day	683-18-1
178	TIN, TRIBUTYL-, OXIDE	0.0003 mg/kg/day	56-35-9
179	TITANIUM TETRACHLORIDE		7550-45-0
180	TOLUENE	0.8 mg/kg/day	108-88-3
181	TOLUENE DIISOCYANATE (TDI)		26471-62-5
180	TOXAPHENE	0.05 mg/kg/day	8001-35-2
183	TRIBUTYL PHOSPHATE(TnBP)	1.1 mg/kg/day	126-73-8
184	1,2,4-TRICHLOROBENZENE	0.1 mg/kg/day	120-82-1
185	1,1,1-TRICHLOROETHANE	20 mg/kg/day	71-55-6
186	1,1,2-TRICHLOROETHANE	0.3 mg/kg/day	79-00-5
187	TRICHLOROETHYLENE	0.0005 mg/kg/day	79-01-6
189	1,2,3-TRICHLOROPROPANE	0.06 mg/kg/day	96-18-4
190	TRICRESYL PHOSPHATE(TCP)	0.04 mg/kg/day	1330-78-5
191	2,4,6-TRINITROTOLUENE	0.0005 mg/kg/day	118-96-7
192	TRIS(2-BUTOXYETHYL) PHOSPHATE(TBEP)	4.8 mg/kg/day	78-51-3
193	TRIS(2-CHLOROETHYL) PHOSPHATE(TCEP)	0.6 mg/kg/day	115-96-8
194	TRIS(1,3-DICHLORO-2-PROPYL)PHOSPHATE (TDCP)	0.05 mg/kg/day	13674-87-8
195	URANIUM, SOLUBLE SALTS	0.002 mg/kg/day	7440-61-1
196	URANIUM, INSOLUBLE COMPOUNDS		7440-61-1
197	VANADIUM	0.01 mg/kg/day	7440-62-2
198	VINYL ACETATE		108-05-4
199	VINYL CHLORIDE	0.003 mg/kg/day	75-01-4
200	XYLENES, MIXED	1 mg/kg/day	1330-20-7
201	ZINC	0.3 mg/kg/day	7440-66-6

MRLs in draft toxicological profiles are provisional. See bottom of the table for more information.

Inorganic Pollutants in Water

3.4 Sources of emerging inorganic water pollutants

Emerging inorganic water pollutants are artificial substances or man-made chemicals such as pesticides, care products, pharmaceuticals, and cosmetics. These pollutants are utilized widely and show the status of present life style (Thomaidis et al., 2012). These inorganic pollutants and acids or microorganisms are produced by waste and wastewaters through industries, agriculture or municipal activities. Acidic pollutants are normally produced by the reduction of organic substances (Sørensen et al., 2007) and also through the waste of pharmaceuticals (hospitals, medical facilities) residue in the natural environment. Most of reports are not satisfied with the availability of toxicity data for these inorganic pollutants, which are generated through residue, wastage, or polluted water (Murray et al., 2010; Verlicchi et al., 2013; Stuart et al., 2012).

The contaminations are originated from geogenic or anthropogenic sources. The groundwater is a combination of the composition of water that enters the subsurface and kinetically controlled reactions with the aquifer matrix for a long duration, which increase the ion concentrations in the groundwater (Appelo and Postma, 2005). The groundwater may be modified through various ways like anthropogenic impacts, using nitrate as fertilizer and waste of industries. Geogenic groundwater pollution means that groundwater which contains all natural substances in various concentrations with harmful health effects. Fluoride and arsenic are those geogenic contaminations, which extremely affect the groundwater widely (Eawag, 2015). Salinity, iron, manganese, uranium, radon, and chromium are also found in higher amount in groundwater, which are again assigned to geogenic activities.

The second type of source is anthropogenic, which arises due to municipal, industrial, and agricultural activities. Anthropogenic contamination source is classified in two categories: direct and indirect impacts. Direct anthropogenic contaminations are those that directly affect the groundwater such as nitrate, phosphate, salinity, addition from agricultural activity; salinity, heavy metals, etc., addition from industrial, sewage and improper waste disposal activities. Moreover, some anthropogenic activities transform the geochemical conditions such as dewatering in lignite mines or acid mine drainage (Kohfahl, 2004). Arsenic has amazing nature that it increases in concentration under reducing conditions because of groundwater abstraction for water supply, irrigation, geothermal power plants or mining activities (Guidelines for Drinking-water Quality, 1997). So, the observed concentration distribution (f_{obs}) is explained by the sum of the natural (f_{nat}) and the influenced (f_{infl}) component, which are two statistical distribution functions as shown in Fig. 3.1 (Wendland et al., 2003):

$$f_{obs}(c) = f_{nat}(c) + f_{infl}(c) \tag{3.1}$$

Agricultural activities are main sources/causes of anthropogenic pollution. When pesticides, fertilizers, herbicides, and animal wastes are applied in higher amount, then anthropogenic contaminations are developed in ground as well as in groundwater. Industrial outcomes are the second most source of anthropogenic pollution. The industrial residues are thrown without proper treatment leading to groundwater pollution. Along with agricultural activities and industrial effluent, the wastes of transportation, manufacturing,

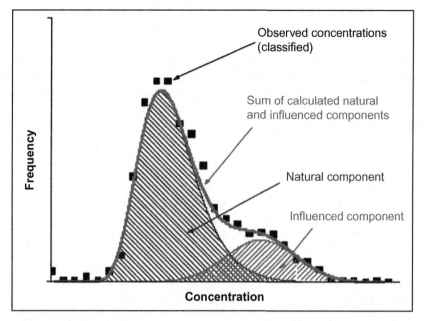

FIGURE 3.1 Pattern to understand the basic approach of separating the natural and influenced component from groundwater concentration (Wendland et al., 2003).

processing, and constructions industries are other sources of groundwater contamination. Earlier, it was also noticed that irregular disposal of sewerage waste became very harmful for groundwater. The storage tanks which contain fuels/oils, acids, various solvents, and chemicals may have leaks due to corrosion, defects, lack of installation, or mechanical failure in the supply pipes and fittings. So that mining fuel and nonfuel minerals creates various causes for groundwater contamination. Wastewater from residential area is also a source of several types of contaminants such as bacteria, viruses, nitrates, and organic compounds. In a recent scenario, injection wells are utilized for house wastewater disposal as septic systems.

3.5 Health effects of inorganic water pollutants

Polluted water has adverse effects onto environment as well as human health. Statistical results confirmed that 1 billion people are directly affected from toxic polluted water every year as they get illness and possess various health problems. The population residing near industrial areas is in much risky zone proportionally than others. These pollutants are mainly reported to cause carcinogenic effects (Adeogun et al., 2016; Popa and Petrus, 2017). Toxic heavy metal effects have several health effects such as skin irritations, headache, stomach ache, diarrhea, vomiting, cirrhosis, necrosis, low blood pressure, hypertension, and gastrointestinal distress (Dada et al., 2015). It is very surprising that some heavy metals are also known as essential elements such as cobalt, copper, iron, manganese, vanadium, and zinc, which are necessary to be consumed in small quantity in body for various biochemical systems. While some other heavy metals such as lead, cadmium, arsenic, and mercury are called foreign in the body due to their negative health effects. Arsenic consumption through polluted water can cause cancer of the lungs, liver, and bladder. Cadmium-contaminated water may harm kidney, lungs, and bone fragility. Lead consumption has negative health effects in terms of damage to brain, kidneys, and reproductive systems. The little amount of lead may disturb learning process in children and also could cause memory loss (Verma and Schneider, 2017; Sun et al., 2017).

Mercury is a global toxic pollutant. It exits inorganic and inorganic forms, wherein organic forms include methyl mercury and dimethyl mercury (Liu et al., 2017). Mercury enters in the body through blood and is excreted through urinal excretion. It could exist in urine for about 60 days (Li et al., 2015). Several mentally hazardous can crustaceans' exposure to metals and instant metabolic systems transformations. Heavy metal consumption in extreme concentration may disturb hunger for food that could further affect body weight loss and also decrease reproduction in adults as well as larvae growth (Zhang et al., 2017) (Fig. 3.2).

Organic pollutants have been presented in the groundwater in several ranges of toxicity. These are very harmful for aquatic organisms, plants, and humans through the wastes of dyes, pharmaceuticals, personal care products, petroleum pollutants, and so on. Dyes are utilized in various applications in liquid form for textile, leather, tanning, food, paper, and so on, to color these products. Dyes obstruct sun rays insertion into water lives and decrease mixed oxygen so that the photosynthetic organism and other lives may die within the aquatic environment (Jung et al., 2015). Human beings are also affected through dye toxicity by eating vegetables and fish as bioaccumulate dyes. Colored paper towels are applied to dry hands and for cooking food that is also another reach of dyes to human (Oplatowska et al., 2011). As a precaution, it is important to eliminate the dyes from wastewater because they are carcinogenic and mutagenic. Pharmaceutical pollutants have also been found in ppb to ppm levels in hospitals and medical area. The waste of these samples makes water contamination and affects human health and other lives (Archer et al., 2017).

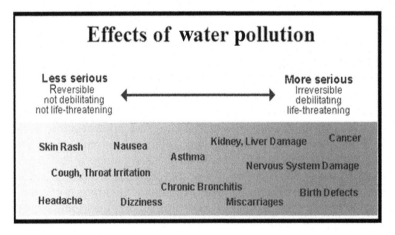

FIGURE 3.2 Impact on human health by water contamination (https://www.quora.com/What-is-the-impact-of-environmental-issues-on-our-health-and-social-lives, 2019).

They could cause both acute and chronic toxicity to aquatic organisms leading to increased cancer risk and other health problems (Kar and Roy, 2010; Aguirre-Martínez et al., 2016). Pharmaceutical-contaminated water consumption may decrease the eggs in females and as well sperms in men (Akanyeti et al., 2017; Ng et al., 2017).

3.6 Future outlook

Various health-related regulations support to preserve the quality of groundwater and other water resources. The Safe Drinking Water Act (SDWA) has been entrenched three pure water resource prevention schemes: (1) Wellhead Protection Program, (2) Sole Source Aquifer Program, and (3) Source Water Assessment Program. These programs ensure the quality of water with minimum health risks. The Federal Insecticide, Fungicide, and Rodenticide Act (FIFRA) monitors pesticide utilizations. The Clean Water Act controls the protection of groundwater as well as surface water by setting the standard for allowing the pollutants' supply to the water sources. The Comprehensive Environmental Response, Compensation, and Liability Act (CERCLA) allows the government to eliminate toxic substances from polluted and unsafe waste which may damage human health and living things systems. CERCLA have a "community right-to-know" provision. The Resource Conservation and Recovery Act (RCRA) regulates treatment, storage, and disposal of unsafe wastes. The Toxic Substances Control Act (TSCA) controls manufactured chemicals.

Water treatment measurement may be classified into ex situ and in situ techniques. Above-mentioned techniques are costly and contain limitations, so many new approaches are developed to groundwater remediation such as subsurface barriers, pump and treat, soil vapor extraction, and bioremediation to control the flow of polluted water in the specified area. Though several remediation technologies based on reverse osmosis, coagulation, filtrations, modular anaerobic system, precipitations, microbial fuel cell, and advanced oxidation process have been devised for the removal of various organic and inorganic water pollutants. Several analytical methods could be applied to detect the contaminations in water. Fluorimeter and fluorescence-based determination are utilized for the detection of hydrocarbons in the oil and gas industry and for the detection of algae in water. UV-visible spectroscopy can be used to determine the total organic carbon and measure turbidity in contaminated water. Nowadays, smart phone sensor—based app uses DNA-magnetic particle technology to detect certain bacteria in the water bodies. Satellite imaging—based sensors provide the assessment of water quality. The multispectral and hyperspectral sensors detect the behavior changes in ecological and hydrological systems accurately using specific indicators and then provide information about water quality and toxic substances in it.

3.7 Conclusion

It is well known that water is a necessary requirement for human beings and other living things for existence. In the present scenario, water resources are getting contaminated through several anthropogenic activities leading to their impact on the ecosystem. A brief summary of these priority and emerging pollutants has been provided in this chapter along with the discussion on their health effects. For preventive measures, it is important to treat wastewater before its discharge to water bodies. Also, strict regulatory actions should be exercised by the government and EPAs. These policies should be based on facts and performance and focused on the objectives as well as goals. It will definitely be a good step to ensure safe and pure water.

Acknowledgment

The sincere support and encouragement by Director, CSIR-Central Scientific Instruments Organisation, Chandigarh, is acknowledged.

References

Adeogun, A.O., Ibor, O.R., Adeduntan, S.D., Arukwe, A., 2016. Intersex and alterations in the reproductive development of cichlid, Tilapia Guineensis, from a municipal domestic water supply lake (Eleyele) in south western Nigeria. Sci. Total Environ. 541, 372–382.

Aguirre-Martínez, G.V., Okello, C., Salamanca, M.J., Garrido, C., Del Valls, T.A., Martín-Díaz, M.L., 2016. Is the step-wise tiered approach for ERA of pharmaceuticals useful for the assessment of cancer therapeutic drugs present in marine environment? Environ. Res. 144, 43–59.

Akanyeti, I., Kraft, A., Ferrari, M., 2017. Hybrid polystyrene nanoparticle-ultrafiltration system for hormone removal from water. J. Water Process Eng. 17, 102–109.

Appelo, C.A.J., Postma, D., 2005. Geochemistry, groundwater and pollution, second ed Balkema, Rotterdam, pp. 241–309.

Archer, E., Petrie, B., Hordern, B.K., Wolfaardt, G.M., 2017. The fate of pharmaceuticals and personal care products (PPCPs), endocrine disrupting contaminants (EDCs), metabolites and illicit drugs in a WWTW and environmental waters. Chemosphere 174, 437–446.

ATSDR, 1993. Cancer policy framework. US Department of Health and Human Services, Atlanta, GA.

Bleibel, W., Al-Osaimi, A.M.S., 2012. Hepatic Encephalopathy. Saudi J. Gastroenterol. 18 (5), 301–309.

Carson, R., 2002. Silent Spring - 40th Anniversary Edition. Boston, MA, Mariner.

Dada, A.O., Adekola, F.A., Odebunmi, E.O., 2015. Kinetics and equilibrium models for sorption of cu (II) onto a novel manganese nano-adsorbent. J. Dispersion Sci. Technol. 37 (1), 119–133.

Dawe, R.S., Ferguson, J., 2003. Environmental effects and skin disease. Br. Med. Bull 68, 129–142.

Eawag, 2015. Geogenic Contamination Handbook – Addressing Arsenic and Fluoride in drinking water. Johnson, C.A., Bertzler, A. (Eds.), Swiss Federal Institute of Aquatic Sci. Tech. (Eawag), Dubendorf, Switzerland.

Fawell, J.K., Mascarenhas, R., 2004. Barium in Drinking Water. WHO.

Geiger, A., Cooper, J., 2010. Overview of Airborne Metals Regulations, Exposure Limits, Health Effects, and Contemporary Research. Cooper Environmental Services LLC.

Guidelines for Drinking-water Quality, 1997. Third ed., World Health Organization, Geneva, Switzerland.

https://www.quora.com/What-is-the-impact-of-environmental-issues-on-our-health-and-social-lives May 9, 2019 at 10.05 AM.

Jung, C., Son, A., Her, N., Zoh, K., Cho, J., Yoon, Y., 2015. Removal of endocrine disrupting compounds, pharmaceuticals, and personal care products in water using carbon nanotubes: a review. J. Ind. Eng. Chem. 27, 1–11.

Kar, S., Roy, K., 2010. First report on interspecies quantitative correlation of ecotoxicity of pharmaceuticals. Chemosphere 81, 738–747.

Kohfahl,C., 2004. The Influence of Water Table Oscillations on Pyrite Weathering and Acidification in Open Pit Lignite Mines; Column Studies and Modelling of Hydro Geochemical and Hydraulic Processes in the Lohsa Storage System (Ph.D. thesis). Berlin, Germany, 111 pp; ISBN 3-89825-774-6.

Landers, T.F., Cohen, B., Wittum, T.E., Larson, E.L., 2012. A review of antibiotic use in Food animals: perspective, policy, and potential. Public Health Rep. 127 (1), 4–22.

Li, P., Du, B., Chan, M.C., Feng, X., 2015. Human inorganic mercury exposure, renal effects and possible pathways in Wanshan mercury mining area, China. Environ. Res. 140, 198–204.

Liu, Z., Wang, L., Xu, J., Ding, S., Feng, X., Xiao, H., 2017. Effects of different concentrations of mercury on accumulation of mercury by five plant species. Ecol. Eng. 106, 273–278.

Lombi, E., Hamon, R.E., 2005. Encyclopedia of Soils in the Environment.

Murray, K.E., Thomas, S.M., Bodour, A.A., 2010. Prioritizing research for trace pollutants and emerging contaminants in the freshwater environment. Environ. Pollut. 58, 3462–3471.

Ng, C.K., Bope, C.D., Nalaparaju, A., Cheng, Y., Lu, L., Wang, R., et al., 2017. Concentrating synthetic estrogen 17a-ethinylestradiol using microporous polyether sulfone hollow fibre membranes: experimental exploration and molecular simulation. Chem. Eng. J. 314, 80–87.

Oplatowska, M., Donnelly, R.F., Majithiya, R.J., Kennedy, D.G., Elliot, C.T., 2011. The potential for human exposure, direct and indirect, to the suspected carcinogenic triphenylmethane dye brilliant green from green paper towels. Food Chem. Toxicol. 49, 1870–1876.

Popa, C., Petrus, M., 2017. Heavy metals impact at plants using photoacoustic spectroscopy technology with tuneable CO2 laser in the quantification of gaseous molecules. Microchem. J. 134, 390–399.

Resende, J.E., Gonçalves, M.A., Oliveira, L.C.A., da Cunha, E.F.F., Ramalho, T.C., 2014. Use of ethylenediamine tetraacetic acid as a scavenger for chromium from "Wet Blue" leather waste: thermodynamic and kinetics parameters. J. Chem. 754526.

Sauve, S., Desrosiers, M., 2014. A review of what is an emergent contaminant? Chem. Cent. J. 8, 15.

Sørensen, S.R., Holtze, M.S., Simonsen, A., Aamand, J., 2007. Degradation and mineralization of nanomolar concentrations of the herbicide dichlobenil and its persistent metabolite 2,6-dichlorobenzamide by *Aminobacter* spp. isolated from Dichlobenil-treated soils. Appl. Environ. Microbiol. 73, 399–406.

Stuart, M., Lapworth, D., Crane, E., Hart, A., 2012. Review of risk from potential emerging contaminants in UK ground waters. Sci. Total Environ. 446, 1–21.

Sun, B., Zhang, X., Yin, Y., Sun, H., Ge, H., Li, W., 2017. Effects of sulforaphane and vitamin E on cognitive disorder and oxidative damage in lead-exposed mice hippocampus at lactation. J. Trace Elem. Med. Biol. 44, 88–92.

Sundseth, K., Pacyna, J.M., Pacyna, E.G., Pirrone, N., Thorne, R.J., 2017. Global sources and pathways of mercury in the context of human health. Int. J. Environ. Res. Public Health 14 (1), 105.

Thomaidis, N.S., Asimakopoulos, A.G., Bletsou, A.A., 2012. Emerging contaminants: a tutorial mini-review. Global NEST J. 14, 72–79.

Vargas, I.T., Fischer, D.A., Alsina, M.A., Pavissich, J.P., Pastén, P.A., Pizarro, G.E., 2017. Copper corrosion and biocorrosion events in premise plumbing. Mater 10 (9), 1036.

Verlicchi, P., Galletti, A., Aukidy, M.A., 2013. Hospital wastewaters: quali-quantitative characterization and for strategies for their treatment and disposal. Wastewater reuse and management. Springer, Dordrecht, Germany, pp. 225–252.

Verma, M., Schneider, J.S., 2017. Strain specific effects of low level lead exposure on associative learning and memory in rats. Neurotoxicology 62, 186–191.

Wendland, F., Hannappel, S., Kunkel, R., Schenk, R., Voigt, H.J., Wolter, R., 2003. A procedure to define natural groundwater conditions of groundwater bodies in Germany. In: Diffuse Pollution Conference, Dublin.

Zhang, C., Yua, K., Li, F., Xiang, J., 2017. Acute toxic effects of zinc and mercury on survival, standard metabolism, and metal accumulation in juvenile ridge tail white prawn, Exopalaemon Carinicauda. Ecotoxicol. Environ. Saf. 145, 549–556.

Further reading

Xagoraraki, I., Kuo, D., 2008. Water pollution: emerging contaminants associated with drinking water. Int. Encycl. Public Health 6, 539–550.

4

Policy and regulatory framework for inorganic contaminants

Shefali Bhardwaj and Dericks Praise Shukla

School of Engineering, Indian Institute of Technology Mandi, Mandi, Himachal Pradesh, India

OUTLINE

4.1 Introduction

A nation is considered to be water-deficient if the availability of water per capita drops below 1700 m^3. In India, this number is 1000 m^3, making our country "water-deficient." According to the Ministry of Environment and Forest (MOEF, 2009), more than 70% of surface water and an increasing percentage of groundwater in India are contaminated with toxic, organic, and inorganic pollutants. Recently, the major concerns in India are the accumulation of inorganic pollutants and hazardous metals such as arsenic (As), mercury (Hg), cadmium (Cd), lead (Pb), nickel (Ni), zinc (Zn), copper (Cu), chromium (Cr), nitrate (NO$_3$), etc. The main problem with these inorganic pollutants is their nonbiodegradability and bioaccumulation. They enter into living organisms through water, air, or food. More than 200 districts in India are facing the problem of heavy metal contamination in groundwater such as As, Fe, Pb, etc. (Murthy and Kumar, 2011). Rivers of India are the main sinks of pollution, as polluted effluents from industries and various treatment plants are ultimately disposed into them. Recently, Central Water Commission (CWC) has reported the contamination of 42 rivers with at least two heavy metals (CWC, 2018). Ganga, the sacred river of India, is polluted with Cr, Ni, Pb, Fe, and Cu. Even the lakes of India are facing the problem of inorganic contamination. From Dal Lake in Kashmir to Bellandur Lake in Bangalore, various contaminations have been reported as a result of the discharging of industrial effluents in lakes (Nusrath et al., 2012). Groundwater that accounts for 80% of drinking water in India is also affected by both geogenic and anthropogenic sources of inorganic pollutants. Nineteen Indian states such as Punjab, Haryana, Gujarat, Telangana, Rajasthan, etc. are facing the problem of heavy metal pollution. The lack of monitoring and improper regulation has worsened the condition of water resources. As a result of the growth of industrialization, mining, and urbanization, various harmful pollutants have been discharged into rivers and streams that ultimately affect groundwater through seepage. The generation of wastewater is 38,254 million L/day (MLD) in India out of which 2647 MLD remains untreated (CPCB, 2008). According to the Central Pollution Control Board (CPCB), Class 1 cities generate 35,558 MLD of wastewater out of which only 32% is being treated. Major states that produce wastewater are Delhi, Uttar Pradesh, Gujarat, Maharashtra, and West Bengal.

4.2 Drinking water quality standard (focus on inorganic contaminants)

Human beings are always confused with the question of what constitutes water. Development in science and technology has tried to characterize both the physical and chemical nature. Standards and limits are the quantitative limits for every contaminant present in water. For various types of contaminants, upper and lower limits are prescribed

TABLE 4.1 Drinking water quality standards of BIS (2012), WHO (2011), and USEPA (2009) for various heavy metals.

Heavy metal	BIS (2012) (acceptable limit; ppb)	BIS (2012) (permissible limit; ppb)	WHO (maximum acceptable; ppb)	USEPA (maximum acceptable; ppb)
Arsenic	10	50	10	10
Lead	10	No relaxation	10	15
Mercury	1	No relaxation	1	2
Cadmium	3	No relaxation	3	5
Chromium	50	No relaxation	50	100
Nickel	20	No relaxation	70	100
Zinc	5000	15,000	Ngl	5000
Copper	50	1500	2000	1300
Iron	300	No relaxation	–	300

that usually represent the value of the contaminant that will not affect human health if taken over a lifetime. Under the Water Act 1974, CPCB lays down standards and State Pollution Control Board (SPCB) is responsible for the implementation of it. In India, Bureau of Indian standard (BIS) is the main body for the development of the standard. It specifies various drinking water quality standards. BIS standards apply to various water resources, supplied by authorities in the country. Different standards are laid down for different parameters such as color, odor, pH, TDS, hardness, sulfate, nitrate, chlorine, bromine, arsenic, copper, cyanide, mercury, zinc, coliform bacteria, etc. Drinking water standards also formed by World Health Organization (WHO) are considered to be international standards for drinking water. Table 4.1 shows different standards for common heavy metals found in water as laid down by BIS, WHO, and USEPA.

4.3 Main sources of inorganic pollutants

Water gets contaminated due to both natural and anthropogenic-related processes. Natural process that includes leaching of natural minerals, interaction of rock with water, groundwater movement, and mineralization (Shukla et al., 2010; Bhardwaj and Shukla, 2019), while anthropogenic activities include agriculture usages of fertilizers and pesticides, sewage discharge, industrial discharge (slurry water and drain water), and mining activities (Dubey et al., 2012; Usham et al., 2018; Bhardwaj and Shukla, 2019).

4.3.1 Geogenic sources

Heavy metals are naturally present in Earth's crust and upon weathering are released in groundwater. In aquifers, groundwater moves very slowly, and as it is in contact with rocks, it gets contaminated with heavy metals. In these aquifers, the geological materials are made up of minerals that are dissolved in groundwater slowly, thus contaminating the groundwater upon coming in contact with oxygen, or dissolution due to changing redox conditions.

4.3.2 Anthropogenic sources of heavy metals

Heavy metals appear in various Indian rivers, streams, and wells as a result of human or anthropogenic activities. Metal pollution due to anthropogenic sources is an emerging issue, having serious environmental implications (Rees and Wackernagel, 2008; Blackman and Baumol, 2008). One of the most toxic inorganic pollutants, mercury, is released into the atmosphere because of various operations such as refining and mining of mercury, using mercury-based pesticides, and agriculture runoff. Arsenic is released into the atmosphere through fungicides and pesticides containing arsenic, mining, and thermal power plants. Chromium finds its way in the atmosphere through chemical and metallurgical industries and cement units. Similarly, copper is also released through steel and iron industries, wood burning, mine tailings, and fly ash discharge. Lead is released into the environment through automobile emission, burning of coal and oil, and smoking.

The main anthropogenic sources for growing inorganic pollutants in the country are industries, population growth, and waste and sewage disposal in water bodies. Various key industries in the country like mining, thermal power plant, plating, chemical, and metallurgical industries are causing continuous contamination of water resources (Dubey et al., 2012; Usham et al., 2018; Bhardwaj and Shukla, 2019). Improper regulation and inadequate treating facilities are the key reasons for contaminated bodies. Inorganic pollutants, especially heavy metals, not only pollute water bodies but also get accumulated in soils and plants and further get bioaccumulated in the human body (Bahadir et al., 2007). Some important sources of inorganic pollutants are discussed below.

4.3.3 Mining industries

As India is flourished with a rich variety of mineral resources, the mining industry is a key industry of the country. Acid mine drainage that is formed after the oxidation of sulfide and pyrite ores can degrade the quality of water by dissolution of toxic elements in tailing and carry them to both surface- and groundwater. Water runoff from mine tailings and coal washeries can contaminate surface- and groundwater by carrying loads of heavy metals (Finkelman, 2007). Leaching from these mines is the major source of groundwater contamination of heavy metals such as arsenic and mercury (Usham et al., 2018; Bhardwaj and Shukla, 2019).

4.3.4 Coal-based thermal power plants

Burning of coal in coal-based thermal power plants is the major source of pollution of water bodies (Dubey et al., 2012; Usham et al., 2018; Bhardwaj and Shukla, 2019). This results in the formation of fly ash and bottom ash. These ashes that contain many harmful heavy metals like arsenic and mercury further are deposited in nearby water bodies with very less utilization (Bhattacharjee and Kandpal, 2002). Bottom ash accumulated in ash ponds can further contaminate groundwater through leaching (Bhardwaj and Shukla, 2019).

4.3.5 Domestic wastewater effluents

Domestic effluents also carry large amounts of trace elements such as Pb, Zn, Cu, Cd, Mn, Cr, and Ni. These metals cannot be removed by effluent treatment plants and are deposited in the form of sludge. Wastewater from household and industrial sludge is the major cause of heavy metal pollution such as Cr, Cu, Fe, Pb, and Hg (Csuros and Csuros, 2002).

4.3.6 Runoff storm water

These days, storm water coming from urbanized areas is also growing as a major source of heavy metal pollution in water bodies. Urban runoff water is having a varying composition of heavy metals in it depending on factors such as the construction of roads, usage of land, as well as watershed physical characteristics (Csuros and Csuros, 2002).

For a proper assessment of water quality, there is a need for adequate monitoring stations. India faces a problem of inadequate infrastructure such as stations and institutions for monitoring. According to CPCB (2009), 32% of monitoring stations monitor water resource monthly, 28% monitor half-yearly, while 38.64% monitor quarterly, thus indicating the need to increase monitoring frequency. Various regulatory agencies in India like Ministry of Environment, Forest and Climate change (MOEFCC), Central Ground Water Board (CGWB), CWC, CPCB, SPCB, etc. have taken measures for subsiding and abating pollution in water resources. They frame different policies and programs for the control and regulation of pollutants in both groundwater and surface water. In nutshell, management and regulation of pollution in India pass through many agencies, institutes, and bureaucrats. Furthermore, many national- and state-level policies that are framed for the regulation of pollutants have succeeded on many fronts but have also failed in some areas. Hence, there is a need for proper regulation, management, and monitoring of inorganic contaminants in water resources of the country.

4.4 Water policies of India

Water policies are imperative for the sustainable use of water resources for both economic and human development. It is a monologue of various guidelines and standards. The key element in any policy is its flexibility, meaning that a water policy should be adaptable to suit in every condition. This section will elaborate on various steps taken after independence for water management and will address the existing water policies.

After the Bengal famine of 1943, the security of food became a major concern, and in view of its development, water resources were given priority. Even after four decades of independence, there were no serious attempts for the formulations of water policy and guidelines. National Water Resource Council (NWRC) was constituted in 1983. The National Water Board (NWB) supported the NWRC and together they formed the national water policy (NWP) for the first time. The first NWP was adopted in September 1987 with an emphasis on domestic water supply, security of groundwater, and monitoring of water quality (NWP, 1987). The progress of NWP is required to be reported to the NWRC. To fulfill their purpose, the government constituted the NWB in 1990. One of the most important works of NWB is to review and check the implementation of NWP and suggests suitable measures for the same. After the implementation of NWP (1987), the emergence of new challenges in the water resource sector led to the revision of NWP and a new water policy was adopted in 2002 (NWP, 2002). The impact of climate change, population growth, and water resource delineation further led to the revision of NWP (2002) in 2010. The NWRC adopted the NWP in 2012 and released it during India Water week in 2013 (NWP, 2012). Some of the major highlights and of each NWP are mentioned below.

4.4.1 National Water Policy (1987)

The main objective of NWP (1987) is to define water as a national asset. It envisaged that the management and development of national resources should be an issue of the entire nation. Various elements were discussed in it, such as information system, maximizing the availability of resources, planning of projects, modernization, maintenance and safety of structures, development of groundwater, water rates, various water allocation priorities, farmer and voluntary agency participation, management of flood and drought, and the quality of water. Surprisingly enough, the phrase "water quality" or "pollution" rarely occurs in the whole document. Under the heading need for water policy, the following statement was made "improvement in existing strategies and the innovations of new techniques will be required to abolish the pollution of surface and groundwater" *(sic)*.

4.4.1.1 Limitations

The main limitations of water policy of 1987 are its predetermined priorities and problems in its legislative and regulatory framework. Water rights are not clearly defined and have many uncertainties. In regard to pollution control, there is no legal provision for water pollution control and partial monitoring. There is no adequate application of water qualities and its regulations; furthermore, there are no penalties defined for polluting the water resource. More focus was given to the monitoring of surface- and groundwater rather on paying attention to water quality. Water is regarded as one of the most important entities in development and planning; however, the 1987 policy lacks in discussing about the water quality and pollution.

4.4.2 National Water Policy (2002)

Many new implementation and changes were adopted in the next NWP (2002). The main emphasis was given to Integrated Water Resource Management and the management of river basins. It reemphasizes that various new water-related institutions and the present ones must to be reoriented and reorganized. The policy stresses on both the quality and preservation of the environment. A new participatory approach to water resource management was included and further encouraged the participation of various government-based agencies, local bodies, stakeholders and consumers, and, especially, women. The policy envisaged that for better management of water resources, the knowledge needed to be enhanced in different directions. Hence, various remote sensing techniques were regarded as the new area for strengthening research efforts. The policy foresaw the need to resettle and rehabilitate the communities affected by water-related projects. Prime consideration was given to the ecological balance and as well as the quality of the environment. Both the central and state governments were given rights and were held equally responsible for preventing overexploitation of water. For the first time, the emphasis was given to water quality such as:

1. Regular monitoring of surface- and groundwater quality.
2. Treatments of effluents to acceptable standards prior to discharge into rivers and streams.
3. A new principle of polluter pays was followed for the management of polluted water.

4.4.2.1 Limitations

Major limitations of water policy 2002 are the same as that of water policy 1987 such as the brief mentioning of private sector engagement and participation. Problems related to river basins continue to remain. There were no proper water rights. The policy of 2002 does not focus on the sharing of information through the information system. Various water markets continued to be unauthorized and still decide the value of water. There were no major reformation in water standards and no emphasis on inorganic pollutants.

4.4.3 National Water Policy (2012)

The draft of NWP 2012 included a water framework law that helped in the establishment of river basin authorities and empowered them with appropriate powers to plan and manage the water resources (NWP, 2012). The policy brought various concerns regarding the management of water resources in the country such as:

1. For the first time, the safe drinking water was discussed and many problems related to inadequate safe potable water were addressed.
2. Rapid urbanization, industrialization, and poor management of water resources have made several parts of India water-stressed.
3. Groundwater is exploited inequitably, which led to its overexploitation in different areas.
4. Various environmental and health hazards are increasing in the country as the result of the discharging of effluents in water streams.
5. In many parts of the country, rivers are heavily polluted that disturb the natural ecology.
6. Major attention was given to sanitation and hygiene including concerns of improper sanitation and lack of treatment of sewage, leading to pollution of water resources.

4.4.3.1 Principles of National Water Policy, 2012

1. Both the water quality and quantity are connected, and hence, their management has to be done in an integrated manner.
2. Safe drinking water and sanitation are preemptive needs.
3. Climate change impacts must be included in water-related management decisions.

4. A common integrated practice should be included for the governance and management of water resources considering regional, state, and national context.
5. To fulfill the objective of social justice and equality, informed transparent decisions and good governance are crucial.
6. For planning, the river basin is considered as a basic hydrological unit.

4.4.3.2 *Other aspects of National Water Policy 2012*

4.4.3.2.1 Conservation of river corridors and water bodies

Strict regulations are imposed on the new urban settlements or any development activities that can affect or pollute the water bodies. Stringent actions for the person responsible for pollution should be taken in addition to a third-party inspection to avoid the pollution of water bodies. Groundwater pollution has to be avoidable at any cost, as its cleaning is difficult. For the first time, proper regulations of the fertilizers, chemicals, and industrial discharge are considered.

4.4.3.2.2 Water supply and sanitation

There has to be proper sewage treatment facilities and water quality needs to be improved for domestic water supply. Rainwater harvesting should be encouraged in the urban areas, and industrial sectors should be monitored to avoid the contamination of groundwater. New subsidies and incentives need to be enforced for the recovery of effluents and pollutants from industries.

4.4.3.2.3 Limitations of national Water Policy 2012

Various provisions of NWP (2012) have been discussed and criticized by many persons. They argue that firm steps should be taken for the rebalancing of recharge and extraction of groundwater. While the draft specifically stated water as a community right but at the same time, it envisages the need to treat water as an economic good. Water allocation priorities should be clear, which means at no cost, there should be allocations of water for industries at the expense of agriculture.

4.4.4 Comparison of all national water policies

Elements	1987	2003	2012
Maximizing availability	Water resources in the country have to be brought within the category of utilizable resources. Important part of water resource development is recycling and reusing of water. In water-short areas, water is available by transferring from other river basins	Various nonconventional as well as traditional methods should be used for water resource development. Traditional methods include rainwater harvesting	Reviewing and scientific assessment of water resource availability has to be done in an interval of 5 years. Some new strategies are made that can increase water resources available like direct use of rainwater and avoiding evapotranspiration

(Continued)

(Continued)

Elements	1987	2003	2012
Groundwater development	Regulations to avoid exploitation of groundwater resource and avoidance of groundwater exploitation near coasts. Main stress is given to reassessment of groundwater periodically and on scientific basis	Overexploitation of groundwater should be prevented by central and state governments Development and implementation of new groundwater recharge projects	Decline in groundwater level and quality should be halted by using new technologies, efficiently. Use of water and management of aquifers by community are encouraged
Water allocation	Water allocation priorities were given to drinking water, irrigation, hydropower, navigation, and industrial purpose. With reference to an area or any region, these priorities can be modified	Agriculture-based industries, nonagro industries, ecology, and navigation were given priorities, and again, these priorities can be modified depending on a specific area	Highest priority was given to human beings and ecology. Water allocation priorities were not clear. It also talks about giving priorities to drinking water. It focuses on the use of water as an economic good
Water rates	Water rates must be such that as to bring the value of scarcity to the users. Rates must be adequate to fulfill the need for annual maintenance and charges of operation. Small farmers were considered, and it was recommended to rationalize water rates with regard to them	Operations and maintenance charges were included in water rates. They are considered as important and valuable while creating alternative water resources. Subsidies on water rates are encouraged for poor society	It was elaborated that every state should have a mechanism for the incorporation of water tariff system and to fix water charge criteria. Water pricing will result in less water wastage
Water quality	It envisages regular monitoring of both surface- and groundwater resources. Phased programs encourage improving the quality of water. Government has given responsibility to avail potable water to all	Effluents from industries or trade effluents must be treated to permissible standards and then they can be discharged into rivers. A new principle of polluter pays was formulated for maintaining the quality of water	On the basis of polluter pays principle, heavy penalty should be imposed on the person of industries who discharges effluents in river or streams. Policy focuses on the management of different sources of water and their auditing. A third-party inspection was involved. Groundwater cleaning was considered a difficult task; hence, conservation of quality and improvements are considered important

(Continued)

(Continued)

Elements	1987	2003	2012
Conservation of water	Conservation of water resources and further promotion of them through various regulation, education, and awareness programs. To avail water in water-stressed region of country, conservation is necessary	Conservation of resources is recommended by abating pollution and minimizing the loss of resources. For this, distinct measures are encouraged such as rehabilitating the existing system (tanks), modernization, recycling, and reusing of industrial effluents. Some new techniques like drip and sprinkle are promoted	Communities are recommended and encouraged to use water as per local availability of it. For efficient use of water, water footprints and auditing are promoted. A national-level institution arrangement is necessary for efficient use of water and its regulation
Information system	Well-developed information system was given prime consideration for planning of water resources. With a wide network of data banks and databases, a national standardized system information system must be established. Despite the availability of data related to water and its usage, the information system also incorporated data regarding the demand of future generations	Various standards were adopted such as standard for coding, data processing, collection procedure, and its classification. Free exchanges of data among states are encouraged for the management of water resources	Data related to water or hydrological data must be considered in public province. For the collection of data, a national water information center is to be established. It is the responsibility of them to collect data periodically from the entire country and maintain them on Geographic Information System in a transparent manner
Water zoning	Agricultural, urban, and industrial developments must be planned with due regard to limitations obtruded by the arrangement of water availability. All the economic activities and regulation of them should be escorted in accordance with water zoning	No additional features	No additional features

4.5 Water Regulatory Framework of India

The constitution of India gives right to the states for dealing with matters related to water, even though water has become a national importance. Due to this, there have been many inter-state water disputes leading to poor regulation of water resources. Different states adopt different policies for water utilization; hence, the rights for the water of different river basins falling in separate states become very difficult. Hence, the Central Government of India formed two ministries to discuss and chalk out the regulatory framework.

4.5.1 Ministry of the Environment, Forest, and Climate Change

MOEFCC came into existence after the Bhopal gas explosion on September 25, 1985. Being an administrative structure, its main role is to plan, coordinate, and promote the implementation of environment-related policies and programs. The main concern of the MOEFCC is to protect the country's natural resources. Its main objectives are to conserve the forest, flora, and fauna of the country, to prevent deforestation, to protect the environment, to encourage environmental awareness, and to prevent pollution. For the first time in the year 1992, MOEFCC laid down a policy statement to prevent pollution in the country. The policy stresses the need for a positive attitude in people to abate pollution. According to it, a comprehensive approach is encouraged for the integration of environmental and economic features in development planning. To reduce industrial pollution, new technological inputs are recommended. It strongly envisages the incorporation of the environment in all decision-making process. Some basic fundamental principles of the policy are:

1. Prevention of pollution at source.
2. To adopt available technology.
3. Polluter pays and participation of public.

4.5.2 Ministry of Water Resource

Water resources in our country are mainly managed by the Ministry of Water Resource (MoWR), River Development, and Ganga Rejuvenation under the Central Government. The ministry was formed in 1985, following the division of the Ministry of Irrigation and Power. Observing the deteriorating condition of river Ganga, the ministry was renamed the Ministry of Water Resources, River Development, and Ganga Rejuvenation.

Important functions of the Ministry are (MoWR, RD & GR, 2016) as follows:

1. Management of the water resources in an indiscriminate manner.
2. Regulation of water resources with the help of various policies and guidelines.
3. Giving support and infrastructure for the development of water resources.
4. Help in providing financial assistance to various water-related projects.
5. Responsible for the overall development of groundwater, its monitoring, and formulation of various policies to prevent overexploitation of groundwater.
6. Prevent pollution and to rejuvenate the river, Ganga.

4.5.2.1 National Water Framework Bill (2016)

The union MoWR released a comprehensive and uniform legal framework for the management of water resources in our country. As shown in a flowchart in Fig. 4.1, it addresses the principle of protection, conservation, management, and regulation of water resources. Some important headings of the policies are as follows.

4.5.2.1.1 Standard for water quality and water footprints

Under Chapter III on basic principles, the focus was given to water quality and standards. It envisages that for every kind of water, there have to be national water quality standard. Both water quality and quantity are interlinked with each other; hence; they need to be managed in an integrated manner with the implementation of some broader

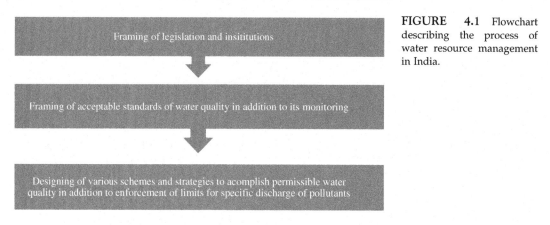

FIGURE 4.1 Flowchart describing the process of water resource management in India.

management perspectives like new economic schemes and penalties. Encouragement should be given to recycle, reduce, and reuse the water resources.

4.5.2.1.2 Preservation of water quality

In Chapter IV, The Planning of Water Security, the phrase "preservation of water quality" appears for the first time. It further discusses various approaches for the prevention of contamination of water resources. Subjected to the stipulations of the Water Act of 1974 and Environment Protection Act (EPA) of 1986, various perspectives for the prevention of pollution are as follows:

1. To reuse and recycle water and enforce it.
2. To minimize waste in all types of water use.
3. To minimize the source of pollution (nonpoint).
4. To ensure that materials that do not meet the quality standards should not be allowed to invade the water resources.
5. Water quality of various resources in the country such as river, groundwater, aquifers, etc. should be protected in addition to its improvement in quality.
6. Two principles, polluter pays and precautionary principles, are applied for the prevention of pollution.

4.5.2.1.3 Industrial water management

In Chapter VI, sectoral use of water, management of water in industrial sectors is given importance. Industries are encouraged to reduce their water footprint. They are compelled to issue annual reports, including their water utilization per unit produce, details of effluent discharge, and usage of freshwater. New incentives that are not capital-intensive were implemented for the recovery of industrial pollutants.

4.5.3 Institutes for regulation of water resources in India

The regulation of water resources in India is carried out by central institutes as formed under the above-mentioned ministries. These ministries have formed organizations to monitor, regulate, and advise the matters related to pollution and water. The details of these organizations are described below.

4.5.3.1 Central Pollution Control Board and State Pollution Control Board

CPCB is the main organization under MOEFCC, formed under the provision of Water Prevention and Control of Pollution Act of 1974. The main objective of CPCB is to maintain and restore the quality of water. It is the main statutory body at the central level for the prevention and control of pollution in India. Under Section (3) of the water act 1974, CPCB needs to be constituted by the central government, while under Section (4), SPCB should be constituted by the state government. Both the central as well as state boards are the umbrella body for the prevention of water pollution.

The various functions of CPCB and SPCB, as given under Sections 16 and 17 of water act (1974) are as follows:

1. CPCB can advise the central government for the prevention of water pollution.
2. It is responsible to maintain coordination among state boards.
3. It should sponsor research to abate pollution or for its prevention and give technical help to state boards.
4. Provide proper training to persons engaged in schemes for the abatement of water pollution.
5. It is responsible for the publication, compilation, and collection of statistical data related to water pollution.
6. It is accountable for the preparation of manuals, guides, and codes related to discarding of pollutants and effluents.

4.5.3.1.1 Functions of state pollution control boards

1. The main responsibility of state boards is the planning of a broad program to control and prevent the pollution of streams and wells.
2. SPCB should gather and propagate information related to water pollution and also for its prevention and control.
3. SPCB should collaborate with CPCB and should organize various training programs for the personnel involved in the prevention and abatement of water pollution.
4. Inspection of sewage and industrial effluents.
5. Lay down trade effluents standards and also modify it.
6. State boards are involved in an evolving efficient method for sewage disposal and trade effluents and further to lay down standards for the treated sewage to be discharged into streams.

4.5.3.1.2 Powers of Central Pollution Control Board and State Pollution Control Board

Section 18 of the Water (Prevention and Control of Pollution) Act (1974) vested the following powers to CPCB—it can give guidance to state boards. In the case of noncompliance by state boards, central boards have the right to accomplish any function of state boards. Further, it is directed to issue guidelines to any industry and operation for its closure or prohibition. On the other hand, state boards have the right to acquire any information and to take effluent samples for analysis. It can inspect any industries violating standards and further can impose restrictions on new discharges and the establishment of new industries. It can also take emergency measures to eliminate pollution.

4.5.3.2 *Central Water Commission*

CWC is one of India's main technical organizations in the field of water resources. Currently, it is operating as an attached office with MoWR, RD & GR (2016). Its main responsibility is to initiate and coordinate the various schemes for the conservation and protection of water resources. The main functions include the following:

1. appraise water resource projects;
2. plan and manage basins and rivers;
3. safety of dams;
4. resolve interstate disputes;
5. survey water resource projects; and
6. monitor pollution in various water resources.

4.5.3.3 *Central Ground Water Board*

The CGWB is a multidisciplinary and scientific organization that works under the aegis of MoWR, RD & GR. The main role of CGWB is the regulation and management of country's groundwater resources through the development and dissemination of technologies. It is entrusted with the responsibility to carry out different studies on groundwater monitoring, conservation, and pollution of groundwater. Its vision is sustainable development and management of groundwater resources of the country.

4.6 Water quality legislative in India

An act that directly concerns with water pollution is the Water (Prevention and Control of Pollution Act) (1974), the Water Cess Act (1977), and the EPA (1986). The earlier two acts are the main act in the context of water pollution in the country, while EPA has its role in gap filling in the context of industrial pollution. In the act, water pollution is defined as "contamination of water or such alteration of the physical, chemical or biological properties of water or such discharge of any sewage or trade effluent or of any other liquid, gaseous or solid substance into water (whether directly or indirectly) as may, or is likely to, create a nuisance or render such water harmful or injurious to public health or safety, or to domestic, commercial, industrial, agricultural or other legitimate uses, or to the life and health of animals or plants or of aquatic organisms" Water (Prevention and Control of Pollution) Act (1974).

4.6.1 Water (Prevention and Control of Pollution) Act, 1974

This act was formulated in 1974 to combat the problem of water pollution in the country. It is the first legal document that attempts to deal comprehensively with issues related to the environment (Divan and Rosencranz, 2001). This act extends to the whole of India, except Jammu and Kashmir. Some important sections regulating the prevention of water pollution are discussed below.

4.6.1.1 Section 20—Power to obtain information

For the fulfillment of the purpose of Water Act, state boards are conferred with powers to gauge and maintain records of the flow or volume of a stream or any other features such as measurement of rainfall, volume of well, and discharge of streams. State boards should give directions to the person who is in charge of any such establishment.

4.6.1.2 Section 21—Powers to take samples of effluents

State boards have given the power to take samples of effluents, sewage, well, or any stream for analysis purposes. The person taking the sample is an occupier. Various guidelines are given in this section for the occupier such as proper storage of samples, proper marking, and sealing of the container.

4.6.1.3 Section 23—Powers of entry and inspection, and Section 24—Prohibition on use of stream or well for disposal of polluting matter

A person empowered by state boards has the right to enter and inspect any industries, records, and registers or any other object, which they think have committed an offense under this act. No person is allowed to dispose any noxious or poisonous substance in a stream or a well.

4.6.1.4 Section 25—Emergency measures in the case of pollution of stream and well

When it appears that some poisonous and noxious substance has entered into river or streams, state boards should take appropriate measures for its management, such as eliminating that substance from a stream and disposing in an appropriate place and further mitigating the contamination caused by that substance in the stream and the well. They have the right to issue an order on the restraining of the person responsible for discharging that pollutant.

4.6.1.4.1 Limitations of the act

In spite of being the oldest act, it suffers from various drawbacks. For example, drinking water quality was not mentioned under this act. Regulation of water pollution originating from the household and agriculture sectors is not mentioned under this act. There are no provisions for the restoration of polluted water bodies. There are no strict financial or nonfinancial penalties on environment offenders (CAG, 2011). There is no mention of groundwater management policies as well.

4.6.2 The Water (Prevention and Control of Pollution) Bill, 2014

This bill has been passed as an amendment to the Water (Prevention and Control of Pollution) Act 1974. Under this act, there was an addition of a new chapter, Chapter VI A named national and state river conservation authorities. In this, the central government was given responsibility for the formulation of national river conservative authority and state governments would constitute state river conservative authority. Both the national as well as state authorities should bring the necessary assessment of sewage and its treatment facilities. It is the responsibility of these authorities to lay down different parameters for the quality of effluents in sewage plants. This act is amended to minimize the growing pollution of streams and rivers in India. The major reason for polluted streams is municipal sewage. Due to inefficient sewage treatment plants and facilities, sewage has been discharged continuously in rivers and streams. Furthermore, SPCB and CPCB are not able to make much effort in checking water pollution due to municipal sewage. However, considering the alarming level of pollution of stream and rivers and to prohibit any discharging of sewage into river; various national and state river authorities are constituted.

4.6.3 Environment Protection Act, 1986

Despite the presence of several laws that deal with pollution, it was important to have common legislation for the protection of the environment, and hence, EPA was formulated. The EPA of 1986 for the first time laid down standards concerning the quality of water. Science and technology are considered in the identification of pollutants and for the formulation of standards. As a result of this act, the union government has equipped itself with powers for the abatement and prevention of environmental pollution. Industries are not permitted to discharge effluents above standards. Under this act, the Central Government is incorporated with powers to collect samples for analysis purpose. Furthermore, they are directed to establish various environment laboratories and water quality standards including maximum allowable limits.

4.6.4 The Water (Prevention and Control of Pollution) Cess Act, 1977

This act is enacted to furnish the collection of a cess on the amount of water utilized by industries. This collected cess is used to supplement the resources of CPCB and SPCB for the control and prevention of water pollution. This act is more of revenue-producing legislation than an action to abate the utilization of water by the industrial units. Under this act, the state government has the power to collect cess from the person or industries. It extends to the whole of India except Jammu and Kashmir. Industries that are liable to pay tax are the one generating hazardous wastes or the industries having a consumption of water above 10 kiloliters per day (KLD).

4.7 Water pollution monitoring

CPCB and SPCB are responsible for the monitoring of pollution in water resources and performing functions specified under the Water Act, 1974. Water quality monitoring is generally performed by pollution control boards with the objectives such as analyzing and accessing the extent and type of pollution, checking water quality, knowing the environmental fate of different pollutants, and, finally, monitoring the fitness of water. CWC and CGWB also monitor water resources in our country.

4.7.1 Water monitoring for inorganic pollutants

Contamination of groundwater with inorganic pollutants is an emerging problem in India, affecting every state at least. Recently, pollution of water resources with heavy metals like mercury, arsenic, lead, cadmium, iron, zinc, etc. has come into attention in Indian subcontinent. Many parts of India are facing scarcity of potable and clean drinking water like western part of Uttar Pradesh (UP), Jharkhand, West Bengal, Punjab, and Orissa (Bhardwaj and Shukla, 2019). High level of arsenic, more than the upper permissible limit of 10 ppb, is found in 86 districts of 10 states, namely Assam, Bihar, Jharkhand, Chhattisgarh, Haryana, Karnataka, Manipur, Punjab, Uttar Pradesh, and West Bengal (Shukla et al., 2010; Dubey et al., 2012). According to a report, the Ganga—Meghna—Brahmaputra region is widely contaminated with arsenic (Chakraborti et al., 2013). High concentrations of fluorine more than the permissible limit of 1.5 ppm is found in various states in India including Delhi, Bihar, Assam, Jharkhand, Karnataka, Kerala, Madhya Pradesh, Punjab, Uttar Pradesh, and West Bengal (Usham et al., 2018). States like West Bengal, Uttar Pradesh, Rajasthan, Orissa, Assam, Arunachal Pradesh, and Bihar are having high contamination of iron in water. Heavy metals such as cadmium, chromium, and lead are also found to be several times higher than the permissible limit in the groundwater of several states.

As various agencies handle water quality management in the country with no synergy among them, MOEF has constituted water-quality-assessment authority (WQAA) on May 29, 2001. WQAA has performed a different task such as the management and regulation of water resources in the country. It formed an expert group to check the status of the water quality monitoring program. Further, a state-level review community is constituted for the appraisal of monitoring exercise.

The present network of CPCB comprises of 2500 stations in 29 states and 7 union territories, which monitors 445 rivers, 154 lakes, 12 tanks, 78 pounds, 25 canals, 807 wells, and 45 drains. Various water samples are analyzed for 9 core parameters and 19 general parameters. Trace elements (As, Cd, Cu, Pb, Cr, Zn, Hg, and Fe) are also monitored at some locations. CWC also has a role in water monitoring. It monitors the Indian rivers at 371 key locations. CWC maintains a three-tier system of the laboratory for parameter analysis. The level-1 laboratories analyze various physical parameters and is located at 258 monitoring stations. Level-2 laboratories analyze various physiochemical and bacteriological parameters. Heavy metals and toxic compounds and pesticides are being analyzed in four level-3 laboratories.

Recently, CWC published a report on the status of trace and toxic metals in the Indian rivers. It is monitoring and observing river quality at 429 key locations of Indian rivers. A total of eight elements such as arsenic, cadmium, chromium, nickel, copper, iron, zinc, and lead (Table 4.2) are monitored at various locations.

CGWB monitors 15,000 wells in the country for the quality of groundwater once in a year. The main aim of groundwater monitoring is to generate data on different parameters present in groundwater. Soil depth and geological formation of subsurface affect the chemical constituent of groundwater. In deep aquifers, generally, the groundwater is of calcium bicarbonate and mixed type that is suitable for drinking purposes, but at some places, different types of water are also observed that are suitable for consumption.

TABLE 4.2   Presence of heavy metals in Indian rivers from 2014 to 2017 (CWC, 2018).

Element	Permissible limit (BIS) in ppb	Highest observation (river and monitoring station) in ppb
As	10	9.53 (Ganga river, Buxar)
Cd	3	70.51 (Sabarmati river, Vautha)
Cr	50	450.26 (Sharda river, Paliakalan)
Cu	50	314.93 (Dhadher river, Pingalwada)
Fe	300	14,550 (Buridehing River, Chenimari)
Pb	10	374.58 (Sheturni River, Lowara)
Ni	20	184.64 (Sheturni river, Lowara)
Zn	5000	2650 (Narmada River, Manot)

Similarly, different types of water are present in different aquifers, which makes the monitoring of groundwater quite important.

4.8 Reformation of regulatory framework

4.8.1 Strengthening of compliance and monitoring system

There is a need to strengthen the present compliance system in India for the control and abatement of inorganic pollution. There should be self-monitoring and transparency in the compliance system. With rapid urbanization and industrialization, there is a need to monitor industries regularly. Previously, efforts have been made in strengthening the compliance system and encouraging industries for real-time self-monitoring of their effluents. CPCB has taken many steps toward the accomplishment of a real-time self-monitoring system in various industries. For the first time, CPCB installed an effluent quality monitoring system at some outlets in Gujarat and Andhra Pradesh. In 2014 CPCB directed 17 highest polluting industries, such as pulp and paper, sugar, power plants, fertilizers, dye, cement, and pharmaceutical industries, STP, CETP, and incineration of hazardous and biomedical waste to install an online common emission and effluent monitoring quality system to gauge the discharge of contaminants in water resources. The same guidelines were also issued to the Ganga Basin Pollution Control Committee for tracking the discharge of toxic effluents by installing a real-time water monitoring system. Various parameters are measured in such systems such as BOD, pH, total suspended effluents, ammonia, and some heavy metals such as arsenic.

Real-time monitoring systems face various issues, as in many cases, the obtained data are not authentic as a result of nonmaintenance of the system. A manpower should be appointed for such dedicated operations, and industries should provide their support to them. Furthermore, for real-time monitoring of effluents, a precise sensor system should be installed at the discharge locations of ETP. There should be a proper installation of cameras, whose live connectivity should be provided to the CPCB by web portals. For inorganic contaminants, real-time monitoring is not reliable and manual monitoring is practiced in various industries. For some inorganic contaminants like nickel, chromium, and arsenic, well-defined monitoring process is not available. Only six parameters like pH, BOD, COD, ammonia, and TSS can be easily and reliably monitored on a regular basis. There is a need to increase the frequency of monitoring and also to the avail the technologies for real-time monitoring of inorganic pollutants.

4.8.2 Enhancement of sewage treatment capacity

According to the 2001 census, domestic sewage from towns and cities is the most extensive source of contamination and pollution of water resources. Class 1 cities and Class 2 towns together generate 29,129 MLD sewage opposed to 6190 MLD installed sewage capacity. Hence, there is a gap of 22,939 MLD between the installed capacity and the generation of sewage. Very clearly, there is an inadequacy in treatment facilities as a result of which India is facing the problem of river and lake contamination. It is the responsibility of the state government to set up new sewage treatment plants with regard to growing pollution. Various operations of sewage treatment plants and its maintenance are also important for regulating the pollutants. As per CPCB, 39% of sewage treatment plants are discharging their effluents in rivers and streams without complying with standards. Further, there is a need to pay attention to the diversion of sewage. Inorganic pollutants in water bodies cannot be eliminated with conventional treatment methods; hence, untreated or conventionally treated sewage should not be discharged into the rivers or streams at any cost.

Properly treated sewage shall be discharged into the agricultural lands instead of rivers and streams as crops/vegetables can act as a natural filter for some inorganic contaminants. Also the treated sewage has nutrients in it that can further nourish the soil fertility. Industries are also encouraged to use treated sewage water for their operations.

4.8.3 Use of low-cost technologies for metal removal

Many physical and chemical practices such as membrane technologies, chemical oxidation, thermal desorption, chemical precipitation, etc. are being used for the removal of pollutants. However, they are not cost-effective and cannot be fully implemented. As a result, bioadsorbents have enticed significant attention as an effective substitute for the removal of inorganic pollutants from the water bodies and contaminated soil. Many naturally occurring low-cost bioadsorbents are being used for the removal of heavy metals and dyes in water bodies such as rice husk, sawdust, biochar, hydrochar, coconut shells, native algae, peels of native fruits, leaves, roots and bark of many native plants, blast furnace slag, fly ash, bentonite, and recycled paper sludge.

4.8.4 Standards for agricultural wastewater and industrial and mine runoff

Though there are regulations and standards for agricultural irrigation water, there are no regulations on agricultural and industrial runoff water. There are no comprehensive studies on toxic waste storage or abandoned mining activities. These toxic inorganic contaminants from the mining area can pollute both groundwater and surface water; hence, there is a need to regulate them with proper regulating standards. Apart from various reports of pesticides in food and water, there are no national studies or reports implemented for their control. For comprehensive water regulation in the future, there is a need to implement new policies and frameworks for these nonpoint sources of contamination.

In conclusion, the current management and regulation system of India with regard to inorganic pollutants needs to be updated in terms of a strong compliance system, settings of standards, and their enforcement. For every strong regulation, there is a need for a strong monitoring system. Hence, there is a need to strengthen these in terms of frequency of monitoring, parameters, and area of monitoring. Further, there should be proper monitoring of heavy metals in rivers and streams by involving local communities for various pollution abatement activities.

References

Bahadir, T., Bakan, G., Altas, L., Buyukgungor, H., 2007. The investigation of lead removal by biosorption: an application at storage battery industry wastewaters. Enzyme Microb. Technol. 41 (1–2), 98–102.

Bhardwaj, S., Shukla, D.P., 2019. Assessment and Evaluation of Heavy Metal Distribution in Waters of Singrauli Region, India: A Focus on Arsenic and Mercury, Environmental Monitoring and Assessment (under review).

Bhattacharjee, U., Kandpal, T.C., 2002. Potential of fly ash utilisation in India. Energy 27 (2), 151–166.

BIS, 2012. Bureau of Indian Standards drinking water specifications. BIS, 10500, p. 2012.

Blackman, S.A.B., Baumol, W.J., 2008. Natural resources. The Concise Encyclopedia of Economics.

CAG, 2011. Audit report no. 21 of 2011–12, Union Government (Scientific Department) –performance audit of water pollution in India. Report, Comptroller & Auditor General of India, New Delhi.

Chakraborti, D., Rahman, M.M., Das, B., Nayak, B., Pal, A., Sengupta, M.K., et al., 2013. Groundwater arsenic contamination in Ganga-Meghna-Brahmaputra plain, its health effects and an approach for mitigation. Environ. Earth Sci. 70, 1993–2008.

CPCB, 2008. Status of Water Supply, Wastewater Generation and Treatment in Class-I Cities and Class-II Towns of India. Central Pollution Control Board, New Delhi, Control of Urban Pollution Series, CUPS/70/2009-10.

CPCB, 2009. Status of Water Quality in India. Central Pollution Control Board, New Delhi, Monitoring of Indian Aquatic Resources Series, MINARS/2009-10.

Csuros, M., Csuros, C., 2002. Environmental Sampling and Analysis for Metals. CRC Press.

CWC, 2018. Status of Trace and Toxic Metals in Indian River. Government of India ministry of water resources central water commission, pp. 1–225.

Divan, S., Rosencranz, A., 2001. Environmental Law and Policy in India. Oxford University Press, Delhi, India.

Dubey, C.S., Mishra, B.K., Shukla, D.P., Singh, R.P., Tajbakhsh, M., Sakhare, P., 2012. Anthropogenic arsenic menace in Delhi Yamuna Flood Plains. Environ. Earth Sci. 65, 131–139.

EPA, Environment Protection Act, 1986. Government of India, Ministry of Environment and Forest.

Finkelman, R.B., 2007. Health impacts of coal: facts and fallacies. AMBIO 36 (1), 103–107.

MOEF, 2009. State of environment report. Report, Ministry of Environment and Forests, Government of India.

MoWR, RD & GR, 2016. Model Bill for the Conservation, Protection, Regulation and Management of Groundwater, 2016. Draft of 17 May 2016, Ministry of Water Resources, River Development and Ganga Rejuvenation, Govt of India.

Murthy, M.N., Kumar, S., 2011. Water pollution in India: an economic appraisal. India infrastructure report, 19, pp. 285–298.

Nusrath, A., Ahmad, E.M., Rashid, A., 2012. Limnologicall Study of Dal Lake and Bellunder Lake A Brief Comparison.

NWP, 1987. National Water Policy. Government of India, Ministry of Water Resources, New Delhi, India.

NWP, 2002. National Water Policy. Government of India, Ministry of Water Resources, New Delhi, India.

NWP, 2012. National Water Policy. Government of India, Ministry of Water Resources, New Delhi, India.

Rees, W., Wackernagel, M., 2008. Urban ecological footprints: why cities cannot be sustainable—and why they are a key to sustainability. Urban Ecology. Springer, Boston, MA, pp. 537–555.

Shukla, D.P., Dubey, C.S., Singh, N.P., Tajbakhsh, M., Chaudhry, M., 2010. Sources and controls of Arsenic contamination in groundwater of Rajnandgaon and Kanker District, Chattisgarh Central India. J. Hydrol. 395, 49–66.

The Water (Prevention and Control of Pollution) Cess Act, 1977. Government of India, Ministry of Environment and Forest.

The Water (prevention and control of pollution) Bill, 2014. Government of India, Ministry of Environment and Forest.

USEPA, 2009. National Recommended Water Quality Criteria. United States Environmental Protection Agency, Office of Water, Office of Science and Technology.

Usham, A.L., Dubey, C.S., Shukla, D.P., Mishra, B.K., Bhartiya, G.P., 2018. Sources of fluoride contamination in singrauli with special reference to rihand reservoir and its surrounding. J. Geol. Soc. India 91, 441–448.

Water (Prevention and Control of Pollution) Act, 1974. Government of India, Ministry of Environment and Forest.

Water Quality Assessment Authority (WQAA), Ministry of Water Resources, 2005.

WHO, 2011. World Health Organization. Guidelines for drinking water quality, fourth ed. World Health Organization, Geneva, Switzerland.

Further reading

CoI, 2001. Census of India. Population statistics. Ministry of Home Affairs, New Delhi.

CPCB, 2018. Guidelines for Online Continuous Effluent Monitoring Systems (OCEMS), first revised Central Pollution Control Board, Ministry of Environment and Forests, New Delhi, Guidelines for Real-Time Effluent Quality Monitoring System.

Assessment of the negative effects of various inorganic water pollutants on the biosphere—an overview

Priyanshu Verma[1] and Jatinder Kumar Ratan[2]

[1]Department of Chemical and Biochemical Engineering, Indian Institute of Technology Patna, Patna, India [2]Department of Chemical Engineering, Dr. B. R. Ambedkar National Institute of Technology Jalandhar, Jalandhar, India

OUTLINE

Inorganic Pollutants in Water
DOI: https://doi.org/10.1016/B978-0-12-818965-8.00005-6

5.1 Introduction

Over the last few decades, water pollution has become a very alarming concern for anthropoids existing all around the globe. Moreover, this issue is not only limited to anthropoids, it is also affecting all forms of lives, including terrestrial and aquatic that includes plants and unicellular organisms too. In this regard, organic water pollutants have gained a great interest and recognition for their removal from various advanced oxidation processes such as photocatalysis, Fenton process, persulfate oxidation, and adsorption-based removal processes (Verma and Kumar, 2014; Verma and Samanta, 2017, 2018a,b,c,d; Ratan et al., 2018). Apart from the organic pollutants' discharge in water resources via sewage and industrial wastewater streams, a high concentration of heavy metals and other inorganic pollutants may also contaminate the water bodies. In general, these inorganic compounds are persistent and nonbiodegradable in nature. These types of inorganic pollutants may include minerals and their acids, inorganic salts, trace elements, metals, metal compounds, complexes of metals with organic compounds or metal−organic frameworks, cyanides, sulfates, and so on. The concerns of inorganic pollutants are mainly associated with their accumulation in the cells, tissues, and specific body parts of the living organisms. For example, the accumulation of heavy metals may have adverse effect on aquatic flora and fauna and may constitute a public health problem due to their involvement in the food chain, where the contaminated organisms are also being used as food resources for other higher organisms. Besides this, the algal bloom or eutrophication could be considered as the indirect or biological indicator of increasing inorganic pollution in the subjected water resources. Specifically, the algal growth would be significantly increased due to the increasing concentration of nitrogen and phosphorous compounds in the water streams. However, the presence of other metals in water streams, such as Hg, Cd, Cu, As, Pb, and Se (in high concentration), could also be toxic to biological species. For example, the presence of Cu in a concentration greater than 0.1 mg/L is highly toxic for the microorganisms naturally present in water resources.

In other words, inorganic water pollutants (IWPs) are some specific compounds and/or elements, which are mostly found in water supplies, drainage water, and in natural water resources such as groundwater and lakes. The contamination of water supplies and drainage water is very common due to various anthropogenic activities. However, the contamination of natural water resources is caused by intense and commercial level anthropogenic activities such as mining, industrial manufacturing, heavy use of inorganic fertilizers and pesticides in agriculture. In addition, the presence of trace inorganic elements is also very common in natural or man-made water resources. Howbeit, it is very difficult to exact comment about the presence of polluting trace inorganic elements in the pre-1950s, as the modern techniques (such as atomic absorption spectroscopy and inductively coupled plasma mass spectrometry), which are effective to detect the presence and concentrations of trace elements, were not available at that time or under development phase. Nevertheless, their existence could be accessed via study of related disease outbreaks at that time; however, it could be a very different topic of interest (with respect to the scope of this chapter) and deserved to be separately studied by the interested environmental historian (if, it is significantly missing from the existing literature). It has been noticed that any amount above the maximum contaminant levels may cause significant damage to the biosphere. As it may cause a variety of damages to the human body or parts specific such as kidney, liver, central nervous system, circulatory system, blood, gastrointestinal system, bones, and skin depending upon the nature of inorganic contaminants and their level of exposure. It has also been noticed that some specific inorganic contaminants are highly damaging to the health of infants and pregnant women. The major types of inorganic water pollutants, their sources, and impacts on the environment have been enlisted in Table 5.1, whereas the specific human diseases associated to the exposure of inorganic pollutants have been presented in Table 5.2.

TABLE 5.1 Types of inorganic water pollutants, their source, and impacts on the environment (Speight, 2017).

Type	Source	Impacts
Acids and bases	Generated through household, industrial, and laboratory applications.	Less harmful, if neutralized. Have severe effects on the tissues of living organisms including plants.
Metal sulfides	Usually generated from mining, and industrial activities.	Their disposal or presence in the open environment may generate sulfuric acid under the influence of water precipitation, and microbial activity.
Perchlorate (perchloric acids and their salts)	Used in a variety of chemical synthesis applications and explosive production.	It is highly persistent in nature, and may damage thyroid function in the human body.
Ammonia	Generated from various industries such as fertilizer plants, water treatment plants, and food processing industries.	Exposure to an excessive amount of ammonia may cause brain damage and also affect the blood circulation system.
Chemical waste	Chemical processing and laboratory application-based by-products and wastes.	Have mild to severe effect on living organism depending on the characteristics of the chemical waste.
Metals and their salts	Produced from mining and smelting activities followed by waste disposal	May have a variety of adverse effects depending on the characteristic of the metal and their salts.
Heavy metals	Motor vehicles, acid mine drainage, and laboratory applications.	Well known for various associated problems due to cytotoxicity and bioaccumulation.
Inorganic fertilizers (including nitrate and phosphate derivatives)	Mostly used to increase the productivity of agriculture and horticulture.	Excessive presence is harmful to human health. Also responsible for algal blooming or eutrophication.
Silt	Generated from water runoff from construction sites, logging, storm, and land clearing sites.	Worst physical appearance, need for additional filtration and/or sedimentation processes for further use or applications.
Solid trash (plastic, paper, glass, etc.)	Generated for nonrecycled stuff.	May generate leaching pollutants.

TABLE 5.2   Human diseases associated with the exposure of inorganic pollutants (Buzea et al., 2007; Manickam et al., 2017; Sardoiwala et al., 2018).

Administration route	Affected part	Related diseases
Ingestion or oral administration	Gastrointestinal system	• Crohn's disease • Colon cancer • Ulcer
Inhalation	Brain	Neurological diseases:
		• Alzheimer • Parkinson
	Lungs	• Asthma • Bronchitis • Emphysema • Tumor • Cancer
Injection (IV/IM/SC)	Circulatory system	• Artheriosclerosis • Vasoconstriction • Thrombus • High blood pressure
	Lymphatic system	• Podoconiosis • Kaposi's sarcoma
	Heart	• Arrhythmia • Heart failure
	Other organs	• Unknown impairments of kidneys, liver, and spleen.
Topical application	Skin	• Allergy • Autoimmune diseases • Dermatitis
Orthopedic implant residues	Around implant region	• Autoimmune diseases • Dermatitis • Urticaria • Vasculitis
Cellular absorption	Cells and tissues	• Bioaccumulation in cell organelles. • Apoptosis • Necrosis

The major health concerns related to the inorganic pollutants could be as follows:

- *Cancer*: There are enormous types of cancer.
- *Inheritable damage*: Contaminants exposure can alter genetic material like DNA.
- *Birth defects*: Malformation or physiological defects that occur during embryonic development may result in deformed or abnormal offsprings. Certain chemicals may also affect the growth and development of affected person offsprings.
- *Reproductive damage*: Loss or reduction of fertility, possibly due to infected eggs or sperm cells; prenatal exposure may affect the development of reproductive organs and sexual development (that appear during childhood or postadolescence) and may also cause reproductive tract infections. Mainly affected by the presence of hormone-disrupting organochlorines.

- *Developmental and behavioral effects*: Contaminants like lead may also affect the development and function of the central nervous system. Lead adversely affects childhood learning.
- *Damage to the immune system and/or circulatory system*: Contaminants' exposure can reduce the body's ability to protect or fight against the disease.
- *Respiratory problems*: There are direct effects on the lung, heart and blood vessels, lung volume and airways (e.g., asthma).

In this chapter, various inorganic water pollutants such as inorganic ions (such as phosphates, nitrates, chloride, and fluoride), metals and their complexes, solid materials, and impact of the acid or base has been reviewed.

Groundwater is still being considered as the major source of water supply for domestic, industrial, and irrigation section of various countries. Subsequently, the quality of water and its management strategies have gained significant popularity, especially in developing nations. The water quality management mainly includes the identification and analysis of potential contaminants, their emerging sources, and possible remediation processes. In the last few years, there has been a phenomenal extension in the application of fertilizers in almost all kinds of agricultural fields. Due to this, high concentrations of nitrate-, phosphate-, and fluoride-based pollutants in groundwater have been reported by global investigators (Rao and Prasad, 1997). Table 5.3 presents the list of a few selected inorganic water contaminants with their maximum contaminant levels (MCLs).

In general, nitrate (NO_3^-), phosphate (PO_4^{3-}), and ammonia (NH_4^+) are not being considered as the pollutants. All these inorganic ions are essential for the growth and development of plants, trees, and phytoplankton. Hence, they are the major constituents of man-made fertilizers and growth media. Nitrogen is mainly needed for the formation of amino acids, which are further used for the synthesis of proteins and nucleotides that also induce the formation of genetic material such as DNA and RNA. Nitrogen is highly abundant in nature as the atmospheric air contains around 78% of nitrogen gas. However, nitrogen gas could not be directly utilized in its raw form. It is mainly due to the presence of the triple covalent bond between nitrogen atoms in a single nitrogen molecule. These triple covalent bonds are considered quite unbreakable for most of the organisms. Hence, even after being highly abundant in nature, nitrogen gas remains biologically inactive or inert. But still, there are some bacteria or microorganisms, which are capable of converting the inert nitrogen gas into a reactive component like nitrate and ammonia that could be easily utilized by the plants. In addition to that, there are some other forms of nitrogen too, such as nitrite (NO_2^-), organic nitrogen, nitrous oxide (N_2O), and urea. Nevertheless, when these species are found in the high concentrations, they are efficient enough to destroy the normal ecosystem and affect biological compatibility with other living organisms. The major ways of that kind of contaminations are:

- Water runoff from the affected or contaminated land.
- Through precipitation (e.g., in the acid rain).
- Through biological means such as fixation by cyanobacteria.

In the case of biological means, the cyanobacteria or blue-green algae utilize the atmospheric nitrogen and release the fixed nitrogen as ammonia into the water. This is usually

TABLE 5.3   List of inorganic contaminants with maximum contaminant levels (MCLs).

Name	MCLs	Tolerance
Antimony	0.006 mg/L	± 30%
Arsenic	0.01 mg/L	–
Asbestos	7 million fibers/L	–
Barium	2 mg/L	± 15%
Beryllium	0.004 mg/L	–
Cadmium	0.005 mg/L	–
Chromium	0.1 mg/L	–
Cyanide	0.2 mg/L	–
Fluoride	4 mg/L	± 10%
Mercury	0.002 mg/L	–
Nickel	0.1 mg/L	–
Nitrate	10 mg/L	–
Nitrite (as N)	1 mg/L	–
Nitrate & Nitrite (combined)	10 mg/L	–
Selenium	0.05 mg/L	–
Thallium	0.002 mg/L	± 30%

Department of Environment & Natural Resources, South Dakota, US. (https://denr.sd.gov/des/dw/IOCs.aspx).

done in anaerobic or anoxic conditions with the help of some specialized cells known as heterocysts. Heterocyst containing biological species have potential for nitrogen fixation, and are able to fix nitrogen molecule into ammonia, nitrates or nitrites, which can be absorbed by plants and further converted into proteins and nucleic acids as atmospheric nitrogen is not biologically available for plants, except for those having some endosymbiotic interaction with nitrogen-fixing bacteria (For example, *Rhizobium* forms a symbiotic relationship with certain plants such as legumes). The water released ammonia is further oxidized by a number of bacterial species in the nitrification step to form nitrate, which is a preferential source of fixed nitrogen for phytoplankton. Once this nitrate is absorbed inside the cells, it is further reduced to nitrite by an intracellular enzyme known as nitrate reductase, and further to ammonia via nitrite reductase enzyme. Ammonia is a most preferential form of nitrogen for biological assimilation of nitrogen into the cellular metabolic reactions. This ammonia is then further converted into amino acids by the way of a transamination reaction.

Overall, there is no doubt that the excess levels of nitrates, nitrites, and phosphates strongly affect the aquatic ecosystems. More specifically:

- A high concentration of ammonia in water could be toxic for fish and other aquatic animals.

- Eutrophication is another big problem, which is mainly associated with various water bodies like ponds, lake, river, sea, and ocean receiving an excessive amount of nitrate and/or phosphate enriched water.

Eutrophication is basically a condition of water bodies, which induce excessive growth of aquatic plants and algae due to the presence of a high concentration of nutrients, especially phosphates and nitrates. The excessive influx of nitrogen and phosphorus will therefore, act as a stimulus for increased production of algal biomass. This is also referred to an algal bloom. This causes biomass production to such a high extent that the layer of algae becomes so thick on the water surface, and so the lower part water became deprived of sunlight. It results in the inhibition of photosynthetic reactions that ultimately result in the plant or phytoplankton death. Afterward, the heterotrophic bacteria will start to degrade the dead biomass, and start consuming almost all the dissolved oxygen of the subjected water bodies. In this condition, there would be a very little amount of dissolved oxygen available for the aquatic animals such as fish, molluscs, crustaceans, and worms in the infected aquatic ecosystem (which is then start to diminish) (Ansari et al., 2010; Chislock et al., 2013).

Furthermore, if the algal bloom is of red tide type, it may induce some additional issues related to the health of aquatic organisms and humans who used to swim in that water body. In this case, the microalgae also produce some toxic species, such as ichthyotoxins and neurotoxic compounds (e.g., brevitoxin), produced by marine dinoflagellates. These toxins may also be accumulated and magnified in the food chain that induces the danger of the same to the living species on the upper trophic levels (Van Dolah, 2000; Backer et al., 2003; Watkins et al., 2008).

Although, the eutrophication process is not a direct threat to the viability of many ecosystems, it may surely cause a significant drain of various valuable resources, such as:

- Eutrophic water bodies may no longer contribute as the potential source of potable water.
- Algal bloom may negatively impact the water bodies of tourism interests.
- Very expensive and large-scale measures are required to protect the infected cum endangered species.
- Human illness is another emerging problem due to the formation of toxins under the influence of algal bloom.

5.2 Various nonmetallic cum inorganic water pollutants

5.2.1 Phosphates

Phosphorus is a very common part of mineral and manure-based fertilizers. It basically boosts the agricultural crop yields. However, a very large portion of phosphorus, which is being applied as fertilizer is not completely absorbed by the plants. It either builds up in the soil or leach-out into the rivers, groundwater, lakes, and coastal areas. Mekonnen and Hoekstra (2018) recently assessed whether the human activity had surpassed the Earth's ability to dilute and assimilate the excess levels of phosphorus in freshwater bodies or not.

The authors found that phosphorus load has already exceeded the assimilation capacity of freshwater bodies in 38% of Earth's land surface, an area housing 90% of the global human population. Although phosphorus is not a harmful substance for drinking water, its presence in groundwater may be of significant concern and environmental consideration. Phosphorus contamination to the water bodies even in small amounts can cause accelerated growth of algae and aquatic plants, thereby causing eutrophication of the aqueous system (Handa, 1990). This further induces problems related to the odor and taste of water. Eutrophication is a major problem which is associated with the phosphorus pollution that causes algal blooms. This condition further leads to the mortality of fish and aquatic plants due to lack of oxygen and sunlight. In addition to that, it also reduces the availability of suitable water for anthropogenic activities such as drinking and swimming.

5.2.2 Nitrates/nitrites

Nitrates and nitrites are basically nitrogen and oxygen containing compounds. They could be organic or inorganic in nature. In general, when the nitrates are absorbed inside the living organisms, are converted into nitrites. Nitrates basically occur in the animal excreta, human wastes, soil, crop residues, nitrogen fertilizers, and in some industrial wastes too. They are readily soluble in water and is easily transported through the water streams. Excessive exposure of nitrate and nitrite via drinking water may cause severe illness in infants due to their interference with the oxygen-carrying potential of human blood. The infection symptoms may include breath shortness and skin blueness. Nitrates and nitrites also have the potential, in case of very long-time exposure, to induce increased starchy deposits, hemorrhaging of the spleen, and diuresis (increased or excessive production of urine). The major sources of nitrates or nitrites-based contaminations are water run-off from fertilizer used lands, sewage network, leaching from septic tanks, and erosion of natural deposits.

5.2.3 Chloride

Chlorides are very common water pollutants. Most of the time, they are naturally present in water. The presence of excess amounts of chloride salts in ocean water is mainly responsible for its nonapplicability in drinking purpose. However, in the last few decades, the evolution of advanced water desalination techniques has significantly reduced this hurdle. The contamination of chlorides in surface water may emerge due to nearby storage of salts or salty rocks, mixing of freshwater with ocean water, dissolution of salty industrial wastes, and so on. Presence of chloride is a very common cause of well water pollution that occurs due to the leaching of salts from soil to well reservoir of water. Although, chlorides have mild effects on living organisms, their excessive intake may cause some serious damage or poisoning to the living body. The recommended limit of chloride in water is <250 mg/L.

5.2.4 Fluoride

Fluorides are those compounds that contain an ionic form of fluorine element. It occurs naturally in many water resources and also added to the water treatment processes for advanced stage disinfection of drinking water. Fluoride in concentration between 0.5 and 0.9 mg/L is beneficial for the prevention of tooth decay in human. However, the fluoride amount between 2 mg/L and 4 mg/L may cause discolored teeth. Furthermore, the fluoride concentration above 4 mg/L may induce bone damage. To reduce the fluoride-based contamination of small-scale water supply systems, it is better to look for a new water resource or mixing of existing sources with a newer one to reduce the fluoride level. It should also be noted that the removal of fluoride through additional treatment method would be highly expensive for a small-scale water supply system.

5.2.5 Ammonia

Ammonia usually found in trace quantities in nature. It is usually being produced from nitrogenous animal and organic food matter. Ammonia and ammonium salts are also being found in rainwater in very low quantities, whereas ammonium chloride and ammonium sulfate are mostly found in volcanic zones. In addition to that, ammonia is also an integral part of some agricultural fertilizers and their overuse also results in water runoff-based contamination. It may cause a substantial increase in the ammonia concentration of linked water resources. The presence of ammonia and its salts in water resources is quite harmful as the consumption of an excessive amount of ammonia could cause biological toxicity and may also affect the blood circulation system of the body. Hence, to avoid the harmful reactions of ammonia and its salts, it is recommended not to mix the ammonium sources into any liquid containing bleach or other oxidants, as it may cause some chemical reaction and may produce a pungent and poisonous gas.

5.2.6 Hydrogen sulfide

Hydrogen sulfide is basically a gas state chemical that also found in the dissolved form. Due to its good water dissolution nature, it is widely found in several water pollutant resources such as sewage treatment plants, electric power waste, poultry farm waste, oil and gas extraction operations waste, cement plant waste, municipal waste landfills, coal processing plants, sulfur production processes, commercial hydrogen sulfide production, asphalt production, geothermal power plants, and so on. Presence of hydrogen sulfide in water may cause headaches, dizziness, nausea, coma, blurred vision, hemorrhage, and so on. In general, aeration of contaminated water is performed to reduce the level of dissolved hydrogen sulfide in water.

5.2.7 Asbestos (a silicate mineral)

Asbestos is basically a naturally occurring fibrous silicate mineral found in asbestiform with the aspect ratio of about 1:20. These fibrous crystals are resistant to heat and various chemical entities. As per a rough estimation, asbestos has been used in more than 3000 different types of products and construction materials. Asbestos is grouped with various other regulated drinking water contaminants of inorganic nature. It may occur in water streams due to the corrosion of asbestos cement pipe widely being used in water transportation systems. Asbestos has been widely known to be a carcinogen agent if the human body is subjected to long-term exposure of the same (above the maximum contaminant limit). However, a short-term exposure (at levels above the maximum contaminant limit) of asbestos may also cause some serious health problems.

Although, it is very difficult to remove the existing pipes of asbestos from the water supply or transportation system, a corrosion control or asbestos filtration system is highly essential for preventive measures. In addition to that, we could also plan or develop some advanced and inert coating materials for those pipes. But, still, there is a risk of routine damage of pipes that may significantly affect the maintenance workers who are repairing or otherwise coming into contact with those asbestos pipes. Inhalation of those asbestos fibers would also be the most common and serious concern for human health.

5.3 Various metal- and metal complexes-based water pollutants

Another major form of inorganic pollution is metal- and its complex-based contamination. The widely known form of metallic pollution is heavy metal impurity of water. Metals are basically conservative water pollutants, they are not prone to any biodegradation or microbiological attacks, but sometimes they are responsible for the killing of interacted bacterial cells. Metals are widely being recycled from plant to plant and animal to animal without being degraded. Herein, the heavy metals are the most alarming ones. Heavy metals are basically the metal having a density higher than $5\,g/cm^3$. This criterion is very true for five major heavy metals; namely, lead, mercury, chromium, arsenic, and cadmium. However, there are some lighter metals too, which are also substantially harmful to the biosphere. The example of such harmful and lighter metals are zinc and aluminum. It has also been observed that the lighter metals are significantly increasing their abundance in the open aquatic environment in comparison to that of the heavy metals. Hence, they should also be considered for environmental remediation processes. Therefore, the use of metal pollution would be a more generalized and suitable term for referring metal-based pollutions instead of using words like heavy metals only.

Metals are naturally found in the environment either in raw or in oxides form. The substantial increase in metallic pollution is mainly emerged due to the extensive use of metals and alloys in various anthropogenic activity-based applications. This situation is mainly responsible for global metallic water pollution. However, some metals like Zn, Cu, Se, Cr, Al, and Ni are quite necessary for biological growth and reproduction, and so a controlled uptake is also required for a healthy life. They are also termed as the essential metals for living organisms. However, metals like Hg, Cd, and Pb are nonessential kind of metals and their presence inside the living organism act as the additional burden on the organism and have adverse effects on the biological metabolism. If their concentration further increases and crossed the critical limit of the same, the organism may suffer in a more severe manner that ultimately cause the death of the organism. Metal contamination may reach the human in several ways, such as: via inhalation, via absorption through the skin, via blood transportation through the placenta before birth, via infected mother's breast-feeding, via ingestion of contaminated seawater and/or food, and may other potential routes.

The major sources of metal-based inorganic pollution are:

- *Atmospheric metal pollution*: These metals originate from forest fires, volcanic activities, dust particles or anthropogenic emissions such as coal-fired power stations and car exhausts. Metals may enter the atmosphere either in the form of an aerosol or in the form of adsorbed substance on another substrate like dust or particulate matter. Deposition from the atmosphere may occur via precipitation (wet deposition) or just as dry mass (dry deposition).
- *Rivers and other natural waterways:* The metals originate from the erosion of rocks and soils containing metals, and also via water runoff from the contaminated sites.
- *Anthropogenic activities:* The deliberate human actions facilitate the entry of metals into rivers and oceans by various means, such as by the direct discharges of industrial and other wastes via metallic pipelines, and due to the dumping of the contaminated solid wastes of municipal and industrial water treatment plants in natural water bodies.

Fig. 5.1 represents the steps involved in the phytoextraction or phytoaccumulation of metals and metal complexes present in the soil and groundwater. Metals are mainly uptake by plant's roots with water and other nutrients, mostly in metal ion from. The harmful effects of metals and their complexes-based inorganic pollutants have been enlisted in Table 5.4.

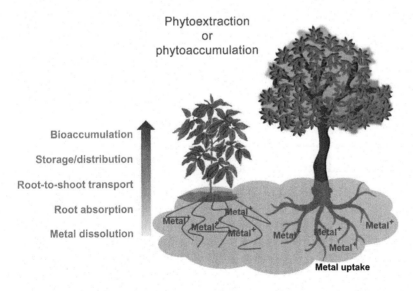

FIGURE 5.1 Mechanism of metal transition in plants and trees.

TABLE 5.4 Harmful effects of metals and their complexes-based inorganic pollutants (Ray et al., 2009; Singh et al., 2011; Sardoiwala et al., 2018; Speight, 2017).

Specific inorganic pollutants	Harmful effects
Arsenic	Dermatitis, bronchitis, poisoning, etc.
Cadmium	Renal dysfunction, lung disease, lung cancer, bone defects (osteomalacia, osteoporosis), increased blood pressure, kidney damage, bronchitis, gastrointestinal disorder, bone marrow infection, cancer, etc.
Lead	Mental retardation in children, developmental delay, fatal infant encephalopathy, congenital paralysis, sensor neural deafness, and acute or chronic damage to the nervous system, epilepticus, liver, kidney, or gastrointestinal damage
Chromium	Damage to the nervous system, fatigue, irritability
Zinc	Zinc fumes have a corrosive effect on skin, may cause damage to the nervous membrane
Manganese	Inhalation or contact causes damage to the central nervous system
Copper	Anemia, liver, and kidney damage, stomach, and intestinal irritation
Mercury	Tremors, gingivitis, minor psychological changes, acrodynia characterized by pink hands and feet, spontaneous abortion, damage to the nervous system, protoplasm poisoning
Carbon nanomaterials, and silica nanoparticles	May induce pulmonary inflammation, granulomas, and fibrosis.
Silver and gold nanoparticles	Widespread biodistribution to different organs and possible passage through the blood–brain barrier.
Iron oxide nanoparticles	Significant distribution in reticuloendothelial system-based organs.
Quantum dots, carbon dots, and titanium dioxide nanoparticles	Skin penetration followed by immunogenic responses.
Manganese dioxide, titanium dioxide, and carbon nanoparticles	May enter inside the brain through the nasal olfactory epithelium.
Titanium dioxide, aluminum oxide, carbon black, cobalt, and nickel nanoparticles	Usually more toxic than micrometer-sized particles.

5.3.1 Impact of pure metal-based nanosubstances

Metallic nanoparticles are submicrometer (1–100 nm) particles of specific metals such as iron (Fe), gold (Au), copper (Cu) and silver (Ag). They are highly reactive due to their nanosized characteristics such as the large surface area to volume ratio, high surface energies, quantum confinement, plasmon excitation and contain a large number of "dangling bonds" that provide exceptional chemical properties and additional electron storage capabilities. In addition, few metallic nanoparticles such as Cu and Ag have antimicrobial properties too. Hence, metallic nanoparticles have been widely used and studied for various pharmaceutical applications such as metallic nanoparticle-based biosensors and nanocarriers for disease diagnosis and therapeutic implementation. However, in recent years, the metallic nanoparticles based environmental and toxicological studies have been reported that suggest their negative aspects. Inevitably, the expeditious development and production of metallic nanoparticle-based pharmaceuticals have equal contribution to the implications arise due to the metallic nanoparticle contamination. However, in the open literature, it has also been reported that most of the environmental or toxicological implications are primarily because of hazardous chemicals (such as reagents, precursors, and capping agents) and the complex fabrication steps used in the production of metallic nanoparticles for various pharmaceutical based applications. Furthermore, a few of metallic nanoparticles have shown toxic effects after the chemical transition. For example, Ag nanoparticles show dissolution behavior by releasing silver ions which reportedly induce a potential toxic effect in cells (Pandiarajan and Krishnan, 2017). However, this concluding remark is still open for intensive research and debate between the concerned researchers and observer field experts. Since a large number of studies are in favor of both silver nanoparticles and silver ions induced cytotoxicity. The combined mechanism of Ag nanoparticle-based cytotoxicity follows a Trojan-horse type mechanism of action (Park et al., 2010). In this mechanism, Ag nanoparticle is supposed to facilitate the release of nontoxic Ag species followed by the entrance of these species into the cell matrices where they get ionized and become toxic in nature and ultimately kill the host cell.

Moreover, exposure to metallic nanoparticles has also been associated with other negative effects such as inflammation, oxidative stress, and genotoxic behavior. In addition, these metallic nanoparticles may accumulate in the living body parts, especially in the liver and/or spleen due to their noncompetitive endothelial barrier. However, metallic nanoparticles may also bioaccumulate in the specific sensitive organ tissues such as the brain, spinal cord, and heart. *In-vitro* and *in-vivo* for both conditions, metallic nanoparticles may lead to the formation of reactive oxygen species (ROS) such as superoxide anions ($O_2^{\cdot-}$) and hydroxyl ($OH\bullet$) free radicals, which are potential health hazard substances due to their rapid protein and cell destruction activity (Brohi et al., 2017; Jahan et al., 2017) (Fig. 5.2). In general, the oxidative stress in the affected tissues or cells may lead to the DNA, protein and membrane damages followed by inflammation that ultimately results in the cell death, that is, apoptosis or necrosis. To minimize the reactive oxygen species mediated side effects, various antioxidants such as ascorbic acid, citric acid, quercetin, α-tocopherol, and lycopene are used in combination with the surface modification of nanocarriers (Khanna et al., 2015; Wang et al., 2016; Brohi et al., 2017).

Furthermore, metallic nanoparticles are also associated with the hypersensitivity of a living organism that may result in the allergic and/or autoimmune response. There are reported studies in the existing literature that ascertained the role of metallic nanoparticles in allergic reactions (Dobrovolskaia and McNeil, 2007; Syed et al., 2013; Yoshioka et al., 2017). In an allergic response, cells of the immune system are activated. Here, the immune cells such as mast cell recognize the foreign substance and trigger the inflammatory response. This response also involves the secretion of cytokines or signaling molecules that attract more cells to destroy the foreign substance. In general, the immune cell recognizes nanoparticles by their surface properties and core composition and accordingly

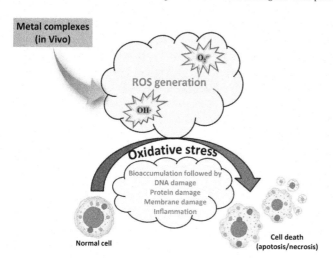

FIGURE 5.2 Reactive oxygen species (ROS) mediated toxicity of metal-based nanosubstances.

produces an inflammatory response. Therefore, Dobrovolskaia and McNeil (2007) recommended a systematic examination of different classes of engineered nanosubstances with their wide range of sizes and surface charges. It may deduce how the change in nanoparticle size and surface charge influence the immune response. In addition to this, the trace impurities present within the nanomaterial-based formulations may also potentially induce the immune response. For example, purified gold and iron oxide nanoparticles do not induce cytokine secretion (Dobrovolskaia and McNeil, 2007). It confirms that the purity of metallic nanoparticles may also affect their toxicity. In general, metallic nanoparticles such as Zn, Au, Al, Ag, carbon-coated silver, and carbon black may lead to inflammation through the activation of tumor necrosis factor alpha (TNF-α). These metallic nanoparticles may also increase the levels of interleukin 6 (IL-6), tumor necrosis factor alpha (TNF-α), and nuclear translocation of nuclear factor kappa-light-chain-enhancer of activated B cells (NF-κB). The different metallic nanoparticles stimulate inflammation through nuclear factor kappa-light-chain enhancer of activated B cells (NF-κB) regulation followed by the release of pro-inflammatory cytokines (Syed et al., 2013). However, the penetration of metallic nanosubstances to the human skin varies from individual to individual. Here, after getting physical contact with the healthy skin, nanosubstances use to penetrate to the stratum corneum or epidermis. In case of damaged skin, nanosubstances may penetrate into the epidermis and dermis layer too (Yoshioka et al., 2017). Therefore, in the case of healthy skin, it is quite difficult to predict whether the topical contact of nanosubstance may result in a negative allergic immune response or not. Since, most of the nanosubstances are quite incompetent to penetrate the healthy skin, which is considerably good for human health.

5.4 Effects of acidic and basic inorganic pollutant species on water bodies

Global researchers have already predicted that the change in pH of the aquatic ecosystem may significantly affect the aquatic life forms, especially those which depend upon a pH sensitive protection shell made up of calcium carbonates such as molluscs and crustaceans. Unfortunately, calcium carbonate dissolves on the influence of increased acidity. In ocean life with rising concentrations of acid, the protective calcium-based shell or skeleton will be damaged or become difficult to be produced. Hence, all the organisms, which form calcium carbonate shells or skeletons may substantially be impacted by this pH imbalance, as acidic water condition may affect the carbonate formation. In addition to that, the phytoplankton that play a vital role in the carbon cycle are also influenced and cause damage to the whole aquatic food chain. For example, the calcareous skeleton of some important phytoplankton species has become comparatively thinner than the preindustrial revolution era (that is now about 30% thinner to the older ones). It could also be considered as concrete evidence of the negative impact of acidification of natural water bodies. In addition to that, there are some other associated considerations of pH imbalance too, such as:

- Loss of Foraminifera species, which are unicellular organisms and very common in the ocean water. These species play a vital role in the sequestration of atmospheric carbon dioxide into the water (almost half of the stored atmospheric carbon dioxide in oceans). In a rough estimation, the oceans alone absorb 30 million tonnes of carbon dioxide per day. Hence, if the population of these species reduced due to any anthropogenic activity, global warming may increase to a more critical level.
- Loss of many species of calcareous coral and shellfish. Generally, these organisms occupy a crucial position in lower levels of the aquatic food chain and any significant decrease in their population may cause an extensive loss of biodiversity.
- Moreover, the decrease in pH could also lead to the decreasing rates of nitrification process which is considered as a bottleneck in the nitrogen cycle.
- Furthermore, if there is a basic or alkaline condition, then it would also cause a significant increase in the pathogenic bacterial growth as bacterial species preferred to be grown in a basic environment.
- Similarly, acidic environment promotes the growth of fungal species, which are most of the time are infectious in nature.

On the other side of the story, not all the environmentalists and researchers have been convinced with the fact that natural water bodies' acidification could cause some severe consequence on the aquatic ecosystems. As per their concerns, those estimations are mainly based on laboratory scale studies, which have been mainly subjected to the sea species to the acute environmental conditions and direct changes in pH, instead of the slow change in the atmospheric conditions. These conditions may not consider the fact of natural evolution and adaptation. Besides this, the level of pH that was used in those experimental studies was also of very high level and too far to the real aquatic conditions.

5.5 Environmental contamination of inorganic nanosubstances

Due to a rapid and sustainable development, production and commercialization of nanopharmaceuticals or nanoparticle-based drug carriers in the current era (Buzea et al., 2007; Wagner et al., 2014), the ecosystem has become a major victim of nanocarrier induced pollution or toxicity, especially the aquatic environment (Klaine et al., 2008; Navarro et al., 2008; Brar et al., 2010; Gottschalk et al., 2011; Miralles et al., 2012; Maurer-Jones et al., 2013; Wang et al., 2016; Brohi et al., 2017; Jahan et al., 2017). The nanosubstances are coming into the environment through different exposure routes starting from the initial manufacturing and production of nanopharmaceuticals to their consumption and use followed by their excretion and repudiation. The soil, groundwater, and surface water are the end sufferer of nanosubstances based contamination. The exposure of the nanosubstances, which are substantially used in nanopharmaceuticals, to the human body through contaminated soil, groundwater, and surface water or through the direct administration as a medicament may induce related diseases depending on the regions of human body get affected (see Table 5.2). Besides this, it should also be noted that the environmental contamination of nanoparticles is also proliferating due to the extended use of various nanoparticles in the wastewater treatment related applications such as photocatalysis, adsorption and advanced oxidation processes (Ghasemzadeh et al., 2014; Ma et al., 2016; Verma and Samanta, 2017, 2018a,b,c). Metal oxides such as TiO_2, ZnO, and carbon nanomaterials including graphene and graphene oxide based nanostructures, and quantum dots are the very established and overused candidates for the water and wastewater treatment based applications (Ghasemzadeh et al., 2014; Lu et al., 2016; Verma and Samanta, 2017, 2018b).

Parthasarathi (2011) and Dev et al. (2018) reviewed the effect of various nanosubstances based toxicity in the plants and food crops which are also widely being used as the nanocarrier for the pharmaceutical compounds. They included latest studies based on phytotoxicity of different nanosubstances such as TiO_2, ZnO, CeO_2, NiO, CuO, Ag, Au, SiO_2, nano zerovalent iron, fullerenes, graphene, graphene oxide, carbon dots and carbon nanotubes, and elaborated individual nanoparticle-based toxic effects observed in the plants. The studies show a clear negative impact on the plant growth, root and shoot lengths, biomass accumulation and seed germination. In addition to this, the oxidative stress, cytotoxic and genotoxic effects of nanoparticles have also been observed in plants (Sardoiwala et al., 2018). Hence, nanosubstance mediated phytotoxicity in plants could also emerge as a major concern for the environment.

Jahan et al. (2017) reviewed the silver, graphene oxide, zinc oxide, titanium dioxide nanoparticles and single-walled or multiwalled carbon nanotubes induced toxicity in aquatic plant, microbial and vertebrate models. They have also enlightened the double-edged sword nature of versatile nanocarriers due to the nanoparticles toxic effects on the aquatic ecosystem. They also summarized that both the physicochemical properties such as shape, size, and surface charge; and environmental factors such as pH, temperature, type of irradiation, dissolved natural organic matter, ionic strength or presence of electrolytes and other contaminants primarily control the transportation, transformation and toxicological behavior of respective nanoparticles. Since, the nanoparticles also have the preeminent potential to cause an oxidative response, cellular toxicity, and inflammatory responses, they have become an impetuous source of damage to the aquatic ecosystem. Therefore, it is always essential to know about the physicochemical properties of nanosubstances followed by their effective concentration prediction in the open environment to calculate the environmental risk quotient (ERQ), nanocarriers transportation and transformational nature. Subsequently, the evaluation of benefits and risks associated with the use of nanosubstances has been always crucial and required to be discussed in a specific manner. Hence, more detailed case-by-case toxicity analyses of versatile nanosubstances are recommended to obtain a more trustworthy predictive model that could estimate and quantify the possible short-term and long-term outcomes of nanosubstances in the open environment.

Furthermore, with the similar opinion of Jahan et al. (2017) it is recommended that the toxicity, fate, and behavior of engineered nanomaterials or nanopharmaceuticals from a large-scale synthesis to industrial application and disposal should be the prime focus of concern, and necessary steps should be taken in the direction of: (1) Nanocarriers' synthesis and modification parameters; (2) Determination of nanocarriers' source, point of entry and end point; and (3) Safety regulations, which are highly essential in case of nanopharmaceuticals. Hence, ecotoxicological tests for nanopharmaceuticals are always required and recommended with the suitable or desired adaptations depending on nanopharmaceuticals usage and applications. The potential risks of nanosubstances, which are widely being used in various commercial applications, could be like oxidative stress, cytotoxicity, phytotoxicity, genotoxicity, and immunogenicity. These are the ultimate toxic response emerging from the biological interaction of various inorganic nanosubstance based contaminants.

In brief, the discarded or runoff nanocontaminants are exposed to the natural aquatic system containing various types of natural organic materials such as humic acid and sulfides in the presence of solar irradiation. These conditions cause an unpredicted change or transformation in the bare nanosubstances that may result in either a positive or negative way depending on the characteristics of transforming or modified nanosubstances (Wang et al., 2016; Jahan et al., 2017). In addition, these nanosubstances may also enter into the food chain due to their bioaccumulating behavior and may cause various diseases in the consuming human as mentioned in Table 5.2. Therefore, it is very essential to study and obtain the data of aquatic life exposure test of versatile inorganic nanosubstances. Here, activated sludge microbes, algae, daphnia, fishes, earthworms, and various sediment organisms should be considered as the test microorganism for the evaluation of respective nanosubstance mediated toxicity in aquatic life. In addition, aquatic plants should also be considered for the estimation of bioaccumulation of nanosubstances using suitable qualitative and quantitative measurement techniques.

There are various categories of tests which are widely being used for the assessment of the negative impacts of various inorganic nanosubstances. The examples of such major categories are cytotoxicity, phytotoxicity, genotoxicity, band gap-based analyses, and quantitative structure activity relationships (QSAR). In contrast, Berkner et al. (2016) enlisted various protocols concerning about the physicochemical properties, fate, and ecotoxicity behavior of nanosubstances that could be further used for the environmental risk assessment. In this regards, Maurer-Jones et al. (2013) enlisted various bacterial monoculture models which are reportedly used for the toxicity assessment of various nanoparticles. In addition, Wang et al. (2016) exclusively reviewed about the alteration of various metallic nanosubstances' toxicity in the presence of natural organic matters with the possible mechanisms. Reviewed studies include bacteria, algae, plant, vertebrates, and invertebrates as the test organism. In addition to that, a few ideal study protocols should also require to be developed to characterize the transformation and alteration in the nanosubstances present inside the human and/or animal body or in other suitable media during their indigestion or infection.

5.6 Qualitative and quantitative measurement of inorganic pollutants

There are various techniques that are widely being used for the estimation of inorganic substances such as metal and their complexes, nano- or microinorganic substances, dissolved inorganic entities, and so on. The most conventional techniques used for the quantification of the same are ICP-MS (inductively coupled plasma mass spectrometry) and ICP-AES (inductively coupled plasma atomic emission spectroscopy) that can quantify the presence of inorganic substances with respect to their elemental composition. These techniques have the advantage of very low concentration-based detection with high precision and accuracy.

Paya-Perez et al. (1993) compared the performances of ICP-AES and ICP-MS for the analysis of trace elements present in soil extracts such as Cr, Ni, Cu, Zn, Cd, and Pb. Overall, Paya-Perez et al. (1993) found that the reproducibility of ICP-AES measurements was relatively better than ICP-MS measurements, possibly due to the less involvement of various reagents. However, for Pb and Ni, the ICP-AES sensitivity was not reported up to the mark. Hence, ICP-MS was recommended for the samples with very less concentration of some elements. Altogether, the ICP-MS provides a fast estimation of the concentration of various trace metals with good precision and higher sensitivity.

Recently, Legat et al (2017) reported capillary electrophoresis combined ICP-MS technique to study the behavior of different gold nanoparticles during the interaction with the serum proteins and their mixtures. This technique reportedly provided a somewhat real-time measurement of bare nanoparticles and different protein conjugates, followed by their conversion into the protein-attached forms with respect to their reaction time. The capillary electrophoresis combined ICP-MS technique looks quite suitable for bioanalysis of inorganic contaminations under more realistic physiological conditions or in bioaccumulated regions.

It should also be noted that the ICP-MS (inductively coupled plasma mass spectrometry) and ICP-AES (inductively coupled plasma atomic emission spectroscopy) are limited to quantify element-based inorganic substances such as pure metal, metal oxide or carbon-based nanostructures only. Moreover, there are other targeted techniques too, such as "mass barcoding" in which specific nanoparticles are tagged with a specific functional group, and their transportation is monitored by LDI-MS (laser desorption/ionization mass spectrometry) (Shi and Deng, 2016). Unfortunately, LDI-MS (laser desorption/ionization mass spectrometry) technique is not widely adopted due to its inherent complexities, uncertainties, and time-intensive behavior.

Furthermore, a new technique was also reported by Lin et al. (2010) that employees cell mass spectrometry (CMS) for the quantitative measurement of micro- and/or nanocarriers uptake in cells. It has a unique ability to rapidly detect the elements from different nanomaterials, simultaneously. This technique exclusively helps in the evaluation of drug targeting efficiency of inorganic nanocarriers, their cellular uptake and associated cytotoxicity emerging due to the differential size and surface properties. Therefore, it is believed that the cell mass spectrometry (CMS) based technology could be efficiently utilized for the rapid and accurate tracking of therapeutic nanocarriers. More interestingly, cell mass spectrometry (CMS) could be used to determine the exact number of inorganic nanocarriers uptake in each cell; whereas ICP-MS (inductively coupled plasma mass spectrometry) can only provide an average uptake of inorganic nanocarriers for all cells. In addition, cell mass spectrometry (CMS) based technique could also be used to measure the cellular uptake of nonmetal based nanotherapeutic agents (Peng et al., 2010).

Moore et al. (2013) reviewed noninvasive measurement based techniques for the assessment of the release of inorganic material based nanosubstances. Mostly, the pharmaceutical inorganic nanoparticles have been studied in laboratory scale for noninvasive measurement of in-situ drug release. However, there are various approaches such as optical upconversion, fluorescence, luminescence, radioluminescence, and magnetic resonance imaging-based techniques that could be utilized for noninvasive measurements of inorganic nanodrug carriers in large scale. These approaches involve the complementation of inorganic nanocarriers with some probes like MRI (magnetic resonance imaging) contrast agents and optically or thermally active species. Besides this, there are some obstacles in the development of noninvasive techniques too, such as the physical limitations of optical techniques, imaging sensitivity, and resolution-based limitations, and toxicity of complemented species. In addition to this, for further details about the available techniques that could be effectively utilized for the physicochemical characterization of nanosubstances, it is suggested to go through the referred literature (Lin et al., 2014; Moore et al., 2013).

In addition, there are various other characterization or analytical techniques which are widely being used to study the qualitative and quantitative properties of versatile inorganic pollutants such as metal and metal complexes-based nanoparticles and nanocarriers. Those suitable and advanced techniques could be categories as follows:

For particle size, size distribution, surface charge, surface area, shape, agglomeration and structure:
- Scanning electron microscopy (SEM)
- Transmission electron microscopy (TEM)
- Atomic force microscopy (AFM)
- Confocal microscopy (CFM)
- Dynamic light scattering (DLS)
- Field flow fractionation (FFF)
- Molecular gas adsorption (BET)
- Electrophoresis particle size

For concentration:
- Inductively coupled plasma and mass spectroscopy (ICP-MS)
- Liquid chromatography and mass spectroscopy (LC-MS)
- Ultraviolet/visible spectroscopy (UV/Vis)
- Fluorescence spectroscopy (FL)

For composition:
- Inductively coupled plasma and mass spectroscopy (ICP-MS)
- Liquid chromatography and mass spectroscopy (LC-MS)
- Ultraviolet/visible spectroscopy (UV/Vis)
- Fluorescence spectroscopy (FL)

Calculation of environmental risk quotient (ERQ)

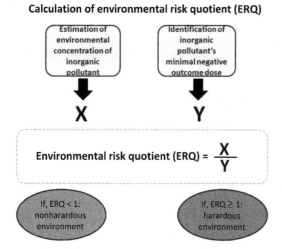

FIGURE 5.3 Environmental risk quotient (ERQ) estimation of inorganic water pollutants.

- Thermogravimetry analysis (TGA)
- Differential scanning calorimetry (DSC)
- Dynamic mechanical analysis (DMA)
- Fourier transform infrared spectroscopy (FT-IR)
- Raman spectroscopy
- Thermogravimetry-gas chromatography/mass spectroscopy (TGA-GC/MS)
- Thermogravimetry and mass spectroscopy (TGA-MS)

The quantification of inorganic pollutants has become quite essential for precise ratification and estimation of the environmental risk factor. This quantification gives information about the environmental availability or presence of the specific inorganic pollutant species and further helps in the estimation of minimum concentration or limit that may induce any toxic or unfavorable effect on the living organism or associated environment. This information could be further used for the estimation of the environmental risk quotient (ERQ). The method of ERQ calculation has been explained in Fig. 5.3.

5.7 Guidelines for inorganic water pollutants

As per the "Department of Environment & Natural Resources, USA" the inorganic contaminant regulations are applicable to all water streams and resources that also include all community and nontransient noncommunity public water supply systems. As per the regulations, water supply systems must be monitored for the presence and level of inorganic contaminants. There are various regulated inorganic contaminants that need to be monitored, however, out of them the fluoride, arsenic, and nitrates are the major ones.

The Department of Environment & Natural Resources suggests the following frequency of sampling:

- Sampling should be done at the entry point, where the source water enters in the distribution system (after treatment or without treatment).
- Samples must be analyzed in an authorized or approved laboratory.
- Regular sampling should be done for regulated inorganic contaminants (yearly or once in 3 years, whichever is applicable).
- For groundwater resources—Once every 3 years.
- For surface water resources—Once in a year.

Things should be done in the alarming state or if analyzed-sample exceeds any of the MCLs:

- Notify to the respective water authority and issuance of drinking water advisory or public notification.
- Work with the water resource agency to determine the best way to reduce the level of contaminants. There are various options that could be considered for the same, such as incorporation of a new treatment process, mixing of contaminated supply stream with another less contaminated supply stream, or use of a new water resource.
- Contact other water resource agencies for help in the planning and finding of other solution or temporary arrangement of direct supply from them.
- Continuous monitoring of the same.

5.8 Conclusive remarks and recommendations

Although the existing regulations are already under enforcement, there is a need of proper inorganic waste management. There are various constraints that need to be addressed, for example, advanced waste sorting facilities are required to be developed and installed for the recovery of recyclable waste having some commercial value. In addition, the ethical support from the industries and manufacturers is also desired to restrict the over usage of hazardous and toxic substances. There should be some provisions for "take-back" culture or system. These systems may enable us to return the unused or waste products to the respective manufacturing agencies, so that it could be properly utilized or recycled. This scheme is highly applicable to solid and electronic goods. Moreover, the government is also requested to conduct routine campaigns and public awareness programs. These programs would be highly beneficial for effective dissemination of information, which aims to highlight the importance of the on-site sorting of hazardous and toxic waste from the remaining types of household wastes. Besides this, it is the government's responsibility to inform and keep updated about the dangers of inappropriate handling and management of hazardous and toxic inorganic wastes. The government should also inform the public about the potential negative outcomes of nearby storage and treatment of inorganic wastes, especially to those who live around the area of treatment plants. The regulatory agencies should clearly disclose the associated risks of inappropriate treatment of inorganic wastes.

References

Ansari, A.A., Singh, G.S., Lanza, G.R., Rast, W., 2010. Eutrophication: Causes, Consequences And Control, Vol. 1. Springer Science & Business Media, Dordrecht, The Netherlands.

Backer, L.C., Fleming, L.E., Rowan, A., Cheng, Y.S., Benson, J., Pierce, R.H., et al., 2003. Recreational exposure to aerosolized brevetoxins during Florida red tide events. Harmful Algae 2 (1), 19–28.

Berkner, S., Schwirn, K., Voelker, D., 2016. Nanopharmaceuticals: tiny challenges for the environmental risk assessment of pharmaceuticals. Environ. Toxicol. Chem. 35 (4), 780–787.

Brar, S.K., Verma, M., Tyagi, R.D., Surampalli, R.Y., 2010. Engineered nanoparticles in wastewater and wastewater sludge–evidence and impacts. Waste Manag. 30 (3), 504–520.

Brohi, R.D., Wang, L., Talpur, H.S., Wu, D., Khan, F.A., Bhattarai, D., et al., 2017. Toxicity of nanoparticles on the reproductive system in animal models: a review. Front. Pharmacol. 8, 606.

Buzea, C., Pacheco, I.I., Robbie, K., 2007. Nanomaterials and nanoparticles: sources and toxicity. Biointerphases 2 (4), MR17–MR71.

Chislock, M.F., Doster, E., Zitomer, R.A., Wilson, A.E., 2013. Eutrophication: causes, consequences, and controls in aquatic ecosystems. Nat. Educ. Knowl. 4 (4), 10.

Dev, A., Srivastava, A.K., Karmakar, S., 2018. Nanomaterial toxicity for plants. Environ. Chem. Lett. 16 (1), 85–100.

Dobrovolskaia, M.A., McNeil, S.E., 2007. Immunological properties of engineered nanomaterials. Nat. Nanotechnol. 2 (8), 469.

Ghasemzadeh, G., Momenpour, M., Omidi, F., Hosseini, M.R., Ahani, M., Barzegari, A., 2014. Applications of nanomaterials in water treatment and environmental remediation. Front. Environ. Science Eng. 8 (4), 471–482.

Gottschalk, F., Ort, C., Scholz, R.W., Nowack, B., 2011. Engineered nanomaterials in rivers–exposure scenarios for Switzerland at high spatial and temporal resolution. Environ. Pollut. 159 (12), 3439–3445.

Handa, B.K., 1990. Contamination of groundwaters by phosphates. Bhu-jal News 5, 24–36.

Jahan, S., Yusoff, I.B., Alias, Y.B., Bakar, A.F.B.A., 2017. Reviews of the toxicity behavior of five potential engineered nanomaterials (ENMs) into the aquatic ecosystem. Toxicol. Rep. 4, 211–220.

Khanna, P., Ong, C., Bay, B., Baeg, G., 2015. Nanotoxicity: an interplay of oxidative stress, inflammation and cell death. Nanomaterials 5 (3), 1163–1180.

Klaine, S.J., Alvarez, P.J., Batley, G.E., Fernandes, T.F., Handy, R.D., Lyon, D.Y., et al., 2008. Nanomaterials in the environment: behavior, fate, bioavailability, and effects. Environ. Toxicol. Chem. 27 (9), 1825–1851.

Legat, J., Matczuk, M., Timerbaev, A., Jarosz, M., 2017. CE separation and ICP-MS detection of gold nanoparticles and their protein conjugates. Chromatographia. 80 (11), 1695–1700.

Lin, H.C., Lin, H.H., Kao, C.Y., Yu, A.L., Peng, W.P., Chen, C.H., 2010. Quantitative measurement of nano-/micro-particle endocytosis by cell mass spectrometry. Angew. Chem. Int. Ed. 49 (20), 3460–3464.

Lin, P.C., Lin, S., Wang, P.C., Sridhar, R., 2014. Techniques for physicochemical characterization of nanomaterials. Biotechnol. Adv. 32 (4), 711–726.

Lu, H., Wang, J., Stoller, M., Wang, T., Bao, Y., Hao, H., 2016. An overview of nanomaterials for water and wastewater treatment. Adv. Mater. Science Eng. 2016.

Ma, Z., Yin, X., Ji, X., Yue, J.Q., Zhang, L., Qin, J.J., et al., 2016. Evaluation and removal of emerging nanoparticle contaminants in water treatment: a review. Desalination Water Treat. 57 (24), 11221–11232.

Manickam, V., Velusamy, R.K., Lochana, R., Rajendran, B., Tamizhselvi, R., 2017. Applications and genotoxicity of nanomaterials in the food industry. Environ. Chem. Lett. 15 (3), 399–412.

Maurer-Jones, M.A., Gunsolus, I.L., Murphy, C.J., Haynes, C.L., 2013. Toxicity of engineered nanoparticles in the environment. Anal. Chem. 85 (6), 3036–3049.

Mekonnen, M.M., Hoekstra, A.Y., 2018. Global anthropogenic phosphorus loads to freshwater and associated grey water footprints and water pollution levels: a high-resolution global study. Water Resour. Res. 54 (1), 345–358.

Miralles, P., Church, T.L., Harris, A.T., 2012. Toxicity, uptake, and translocation of engineered nanomaterials in vascular plants. Environ. Sci. Technol. 46 (17), 9224–9239.

Moore, T., Chen, H., Morrison, R., Wang, F., Anker, J.N., Alexis, F., 2013. Nanotechnologies for noninvasive measurement of drug release. Mol. Pharm. 11 (1), 24–39.

Navarro, E., Baun, A., Behra, R., Hartmann, N.B., Filser, J., Miao, A.J., et al., 2008. Environmental behavior and ecotoxicity of engineered nanoparticles to algae, plants, and fungi. Ecotoxicology 17 (5), 372–386.

Pandiarajan, J., Krishnan, M., 2017. Properties, synthesis and toxicity of silver nanoparticles. Environ. Chem. Lett. 15 (3), 387–397.

Park, E.J., Yi, J., Kim, Y., Choi, K., Park, K., 2010. Silver nanoparticles induce cytotoxicity by a Trojan-horse type mechanism. Toxicol. Vitro 24 (3), 872–878.

Parthasarathi, T., 2011. December. Phytotoxicity of nanoparticles in agricultural crops. In: International Conference on Green Technology and Environmental Conservation (GTEC-2011), IEEE, pp. 51–60.

Paya-Perez, A., Sala, J., Mousty, F., 1993. Comparison of ICP-AES and ICP-MS for the analysis of trace elements in soil extracts. Int. J. Environ. Anal. Chem. 51, 223–230. Available from: https://doi.org/10.1080/03067319308027628.

Peng, W.P., Yu, A.L., Chen, C.H., 2010. Quantification study of drug delivery by nanocarriers: a cell mass spectrometry approach. GIT Lab. J. Europe 14 (9–10).

Rao, N.S., Prasad, P.R., 1997. Phosphate pollution in the groundwater of lower Vamsadhara river basin, India. Environ. Geol. 31 (1–2), 117–122.

Ratan, J.K., Saini, A., Verma, P., 2018. Microsized-titanium dioxide based self-cleaning cement: incorporation of calcined dolomite for enhancement of photocatalytic activity. Mater. Res. Express 5 (11), 115509.

Ray, P.C., Yu, H., Fu, P.P., 2009. Toxicity and environmental risks of nanomaterials: challenges and future needs. J. Environ. Science Health Part C 27 (1), 1–35.

Sardoiwala, M.N., Kaundal, B., Choudhury, S.R., 2018. Toxic impact of nanomaterials on microbes, plants and animals. Environ. Chem. Lett. 16 (1), 147–160.

Shi, C.Y., Deng, C.H., 2016. Recent advances in inorganic materials for LDI-MS analysis of small molecules. Analyst 141 (10), 2816–2826.

Singh, R., Gautam, N., Mishra, A., Gupta, R., 2011. Heavy metals and living systems: an overview. Indian J. Pharmacol. 43 (3), 246–253.

Speight, J.G., 2017. Chapter Five - Sources and types of inorganic pollutants. Environmental Organic Chemistry for Engineers. Butterworth-Heinemann, pp. 231–282. Available from: https://doi.org/10.1016/B978-0-12-849891-0.00005-9.

Syed, S., Zubair, A., Frieri, M., 2013. Immune response to nanomaterials: implications for medicine and literature review. Curr. Allergy. Asthma. Rep. 13 (1), 50–57.

Van Dolah, F.M., 2000. Marine algal toxins: origins, health effects, and their increased occurrence. Environ. Health Perspect. 108, 133–141.

Verma, P., Kumar, J., 2014. Degradation and microbiological validation of meropenem antibiotic in aqueous solution using UV, UV/H_2O_2, UV/TiO_2 and $UV/TiO_2/H_2O_2$ processes. Int. J. Eng. Res. Appl. 4 (7), 58–65.

Verma, P., Samanta, S.K., 2017. Degradation kinetics of pollutants present in a simulated wastewater matrix using UV/TiO_2 photocatalysis and its microbiological toxicity assessment. Res. Chem. Intermed. 43 (11), 6317–6341.

Verma, P., Samanta, S.K., 2018a. Facile synthesis of TiO_2–PC composites for enhanced photocatalytic abatement of multiple pollutant dye mixtures: a comprehensive study on the kinetics, mechanism, and effects of environmental factors. Res. Chem. Intermed. 44 (3), 1963–1988.

Verma, P., Samanta, S.K., 2018b. Microwave-enhanced advanced oxidation processes for the degradation of dyes in water. Environ. Chem. Lett. 16 (3), 969–1007.

Verma, P., Samanta, S.K., 2018c. Continuous ultrasonic stimulation based direct green synthesis of pure anatase-TiO_2 nanoparticles with better separability and reusability for photocatalytic water decontamination. Mater. Res. Express 5 (6), 065049.

Verma, P., Samanta, S.K., 2018d. A novel UV-C/XOH (X = Na or K) based highly alkaline advanced oxidation process (HA-AOP) for degradation of emerging micropollutants. ChemRxiv. Available from: https://doi.org/10.26434/chemrxiv.5777379.v1.

Wagner, S., Gondikas, A., Neubauer, E., Hofmann, T., von der Kammer, F., 2014. Spot the difference: engineered and natural nanoparticles in the environment—release, behavior, and fate. Angew. Chem. Int. Ed. 53 (46), 12398–12419.

Wang, Z., Zhang, L., Zhao, J., Xing, B., 2016. Environmental processes and toxicity of metallic nanoparticles in aquatic systems as affected by natural organic matter. Environ. Sci.: Nano. 3 (2), 240–255.

Watkins, S., Reich, A., Fleming, L., Hammond, R., 2008. Neurotoxic shellfish poisoning. Mar. Drugs. 6 (3), 431–455.

Yoshioka, Y., Kuroda, E., Hirai, T., Tsutsumi, Y., Ishii, K.J., 2017. Allergic responses induced by the immunomodulatory effects of nanomaterials upon skin exposure. Front. Immunol. 8, 169.

Further reading

Aprilia, A., Tezuka, T., Spaargaren, G., 2013. Inorganic and hazardous solid waste management: current status and challenges for Indonesia. Procedia Environ. Sci. 17, 640–647.

Braissant, O., McLin, V.A., Cudalbu, C., 2013. Ammonia toxicity to the brain. J. Inherit. Metab. Dis. 36 (4), 595–612.

Ivanova, E., Staykova, T., Velcheva, I., 2008. Cytotoxicity and genotoxicity of heavy metal and cyanide-contaminated waters in some regions for production and processing of ore in Bulgaria. Bulgarian J. Agric. Science 14 (2), 262–268.

Li, R.Z., Liu, K.F., Qian, J., Yang, J.W., Zhang, P.P., 2014. Nitrogen and phosphate pollution characteristics and eutrophication evaluation for typical urban landscape waters in Hefei City. Huan jing ke xue = Huanjing kexue 35 (5), 1718–1726.

Manuel, J., 2014. Nutrient pollution: a persistent threat to waterways. <https://doi.org/10.1289/ehp.122-A304>.

Martín-González, A., Díaz, S., Borniquel, S., Gallego, A., Gutiérrez, J.C., 2006. Cytotoxicity and bioaccumulation of heavy metals by ciliated protozoa isolated from urban wastewater treatment plants. Res. Microbiol. 157 (2), 108–118.

Analytical methods of water pollutants detection

Majid Peyravi, Mohsen Jahanshahi and Houshmand Tourani

Department of Chemical Engineering, Babol Noshirvani University of Technology, Babol, Iran

6.1 Introduction

Detection of pollutants in drinking water systems is a crucial component in ensuring the safety and security of water. If preventive actions against invasion are defeated, it becomes very important to diagnose the contamination in order to develop an immediate response and a rehabilitation plan. Immediate detection prepares such condition to react immediately which minimizes polluted water transferring to consumer, therefore, decreases outbreak of disease. Detection of the time of water being polluted is only possible using data from public health monitoring and other data. Detection requires monitoring water components qualitatively, pollutants and also systematic process to study collection of qualitative data of water. This essay aims to study on different procedures of water pollutants' detection and usage of water alarm contamination systems.

Inorganic Pollutants in Water
DOI: https://doi.org/10.1016/B978-0-12-818965-8.00006-8

6.2 Detection procedures

Strategies of monitoring quality of running water are developed in accordance with drinking water standards and are not only designed to ensure the safety of water and protecting water against intentional contamination situation. Standard test protocols are not consisting of biological, chemical, and radiological pollutants which may be used in an invasion. Analytical procedures in different contamination detections in order to observe the occurrence of contamination are not fast enough either. Therefore immediate monitoring in order to react immediately is crucial.

There are different technologies of contamination detection emerging these days and this progress is faster in different procedures of testing. Ho et al. (2005) refer to a general survey on sensors that can be used as environmental samples such as technologies which are used to detect rare metals, radioisotope, and other pollutants.

Since pollutants may appear in low density and affect human health, procedures with low detection are needed and as a result, testing compounds of drinking water could be challenging.

6.2.1 Direct analysis of pollutants

Detection of special chemicals which is used in intentional contamination needs to deploy advanced analytical procedures such as gas chromatography/mass spectrometer (GC/MS), accommodation inductive plasma technique, and ion chromatography (IC) (Hall et al., 2007). Some potential chemical pollutants and their related analytical procedures are shown in Table 6.1. These analytical procedures can be useful to detect chemical and biological contamination: GC, GC/MS, liquid chromatography (LC), LC/MS, safety test kit, IC, atomic absorption with graphite furnace, atomic absorption of cool steam, spectroscopy of induced couple plasma (ICP), ICP/MS, ion-selective electrodes, culture-based microbiological experiments,

TABLE 6.1 Analytical procedures for chemical pollutants.

Biotoxins	Risin, botulinum	Immunoassay
Cyanide		Ion-selective electrode or colorimetric
Pesticides		Immunoassay
List 1 of chemical weapons	VX, sarin	Enzyme or colorimetric
Volatile organic chemical	BTEX	GC/MS, LC/MS
Nonorganic chemical	Mercury, lead	Ion chromatography, atomic absorption with graphite furnace, atomic absorption of cool steam, spectroscopy of ICP, and spectroscopy of induced plasma/mass spectrometric (ICP/MS)
Pollutant category	Special samples	Analytical procedures
Pathogens	Tularemia, anthrax, plague	Molecular procedures (based on PCR)

GC, Gas chromatography; *ICP*, induced couple plasma; *LC*, liquid chromatography; *MC*, mass spectrometry; *PCR*, polymerase chain responses.

molecule-based microbiological experiments, implementation of biochemical and serologic tests, immunomagnetic separation, and immunofluorescence.

If any contamination occurs and pollutant is completely unknown, a combination of these analytical procedures is used to detect the contamination factor. However, these procedures are expensive and require a high level of proficiency. These procedures are not used in usual monitoring applications.

6.2.1.1 Biological factors

Among those samples that are polluted by unknown microbiological pollutants, a condensed one must be sent to a qualified laboratory. Both the tests based on culture and molecules have to be done on samples consisting of unknown biological factors. Experiment based on culture consists of an effort on the growth of biological pollutants in selective environments; detection can be based on ability of a culture in growing in special environments and using microscopes. Another tool in screening biological factors is immunomagnetic separation and immunofluorescence microscope. Exploratory procedure consists of those which are molecule based and relied on polymerase chain responses (PCR) and hybridization probe. Molecular test can be used to produce primary results confirmed by sequential analysis. PCR procedures based on molecule turn into results faster than tests based on culture. Considering low density of pathogenic microorganisms in water samples, condensation of samples by ultrafiltration may be needed (U.S. Environmental Protection Agency (U.S. EPA), 2003). Numerous studies have shown that filtration is effective on concentrating samples for a biological test and concentration of samples before analysis can lower detection limits to proper levels of drinking water (States et al., 2006). In Lindquist et al. (2007) the ultrafiltration process is shown using 100-L samples in order to detect biological pollutants.

Screening biotoxins can be achieved using immunoassay confirmed by GC/MS, LC, or LC/MS (U.S. Environmental Protection Agency (U.S. EPA), 2003). Although there are many test kits, all of them do not use standard procedures. These test kits are usually useful for primary screening. In some cases, animals are used to detect biotoxins and other pollutants that are difficult to be detected. In wastewater systems, microbiological immunity is considered as lack of pathogenic microorganisms which is supplied of indicative organisms, including fecal coliforms or *Escherichia coli*. The presence of indicative organisms shows the existence of feces and refers to possible existence of pathogenic microorganisms. On an intentional contamination occurrence, although there may be pathogenic factors used as pollutants, increase in density of indicative organisms is not expected.

6.2.1.2 Chemical factors

In water samples containing unknown chemical compounds, many top-priority pollutants have to be tested while screening. Generally, the following analytical procedures can be used to screen these chemicals (U.S. Environmental Protection Agency (U.S. EPA), 2003): volatile substances (organic): clearing and trap GC/photoionization detector/electrolytic conductivity detector or GC/MS, semivolatile organic compounds (organic including many of pesticides): extraction of solid phase of GC/MS, carbamate pesticide (organic): LC with high function [high-performance LC (HPLC)], fluorescence detection, quaternary compounds of nitrogen (organic): UV detector (UV) HPLC, atomic absorption with graphite furnace (AA), mercury measurement: atomic absorption of cool steam (AA), cyanides: more chemistry, radiation: gross alpha, gross beta, and gross gamma.

Wider screening on organic chemicals needs a set of technics to prepare the sample: microextraction of liquid—liquid, steady extraction on liquid—liquid, solid-phase extraction, microextraction of solid phase, collecting upper space and flow injection (U.S. Environmental Protection Agency (U.S. EPA), 2003). Proper analytical procedures for organic chemicals include multiplex GC detector in screening mode, GC by ionization mass spectroscopy of electron collision, HPLC UV detector, high-performance mass spectroscopy of LC (HPLC/MS), continuous mass spectroscopy (MS/MS), high-resolution MS, and immunoassay (U.S. Environmental Protection Agency (U.S. EPA), 2003). More general screening for nonorganic chemicals consist of the following analytical procedures: ICP AES or ICP MS in semiquantitative phase, IC, more chemistry and selective ion electrodes (U.S. Environmental Protection Agency (U.S. EPA), 2003). Screening cyanides needs distillation. There are developed procedures to detect semivolatile chemicals; therefore chemicals which can be used to invade water system are easily detected (Grimmett and Munch, 2013).

6.2.1.3 *Radiological factors*

Radioactive isotopes can be detected by available procedures. Radioactive detectors are based on high purity germanium, spark crystals, and Geiger counter (Ho et al., 2005). However, referring to Porco (2010) available detector technologies were analyzed to check on radioactive compounds which were disable of monitoring detection on intentional injective pollutants and the result was unavailability of related technologies.

6.2.2 Toxicity test

Toxicity test uses bioorganisms instead of testing components and compounds to detect toxicity in water samples. The main advantage of toxicity test is the fact that it is not necessary to deploy this test on many available pollutants. Base of deploying toxicity test is when special biological species are exposed to toxic compounds in a sample of water; there would be a measurable biological response for each of them. To measure potential effect of polluted water on organisms, there should be a standard to measure biological response. Samples of these standards consist of luminescence of algae, metabolic performance, and respiratory rate. Biosensors based on toxicity may use, algae, mollusks, or fish in water as biological organisms (States et al., 2004, 2006).

Although there are many developed toxicity tests, cellular-based toxicity tests have shown better results (Curtis et al., 2009a,b; Davila et al., 2011; Eltzov and Marks, 2010; Eltzov et al., 2002; Giaever and Keese, 1993; Iuga et al., 2009). The result represents a general study on cellule biosensors that are appropriate for different functions such as safety of water (Eltzov et al., 2002). Cellule toxicity test of bacteria has been compared to other varieties of toxicity test which use higher rate biological indicators, but the results do not show a strong relation between data from human health and animal health (Botsford, 2002). While toxicity test normally takes place in laboratory environment, online biosensors are also available currently. A comparison between available biocommercial sensors such as Eclox, microtox, ToxAlert is done using both toxicity test and *Daphnia magna*

(Dewhurst et al., 2002) and detects that some results are sensitive to changes in temperature and water quality. Microtox test uses *Vibrio fischeri* as biological indicator. Some biological sensors can be used as a quick test in place (U.S. Environmental Protection Agency (U.S. EPA), 2003). The weakness of biological sensors is they may be not appropriate to test drinking water because their used organisms may be sensitive to disinfectants and other drinking water compounds (Ho et al., 2005). A total of 10 toxicity sensors were analyzed (van der Schalie et al., 2006): impedance cellular electrical sensor, severe toxicity sensor Eclox (Severn Trent Services), low-density lipoprotein absorption Hepatocyte, microtox (strategic diagnostics, Inc.), Mitoscan (Harvard Bioscience, Inc.), neurotic micro electrode arrangement, toxicity test of *Sinorhizobium meliloti*, SOS cito sensor, toxic chemo test (Environmental Biodetection Products Inc.), and ToxScreen II (Checklight Ltd.). These tests were done using the following 13 industrial chemicals to analyze the efficiency of detectors: al-dikarb, ammonia, copper sulfate, mercury chloride, metamidophos, nicotine, paraquat dichloride, phenol, sodium arsenite, sodium cyanide, sodium hypochlorite, sodium pentachlorofenate, and toluene (van der Schalie et al., 2006) refer to one of the sensors which do not response to nicotine, one sensor is not able to response no more than six chemicals and combination of sensors is much better than deploying them one by one.

Online sensor toxicity test of Bioscan is a good example which detects contamination by measuring biological response of a bacteria culture (Campbell et al., 2007). Toxicity test can be a useful procedure, but the results may be different based on usage. The results of toxicity test using various biological indicators may be different depended on deployed organisms (Radix et al., 2000). Selection of a proper cellular line to build higher sensitivity sensors is very important (Curtis et al., 2009a). Online toxicity tests have been significantly successful. Referring to de Hoogh et al. (2006), a combination of analytical procedures (HPLC/UV and GC/MS) and online toxicity monitoring is shown which results in detection of available chemical pollutants in a surface water source based in the Netherlands. Deployed online toxicity monitor was bbe Daphnia taximeter which measured behavioral changes in studied organism based on swimming behavior (de Hoogh et al., 2006).

Although online toxicity test is available these days, this technology is yet to be developed. Many efforts have been taken to create technologies with more availability. For instance, Jeon et al. (2008) combined automation and toxicity test using *D. magna*.

6.2.3 New detection technologies

Referring to Foran and Brosnan (2000), new technologies in the field of water safety are represented in order to detect biological pollutants immediately: DNA microchip arrays, immunology technics, micro robots, optical procedures, flow cytometry, and molecular probe.

In another analysis (Ho et al., 2005) on sensors' technology, it was detected that there are many available and developed sensors to detect different kinds of environmental pollutants. To monitor rare metals, there are emerging technologies such as nano electrode arrays, laser breakdown spectroscopy, and miniature chemical flow probe sensors (Ho et al., 2005). To monitor radioisotopes, following technologies are being developed: RadFET (radiation of field-effect transistor), cadmium telluride zinc detectors, beta spectroscopy of low energy positive-intrinsic-negative (PIN) diods, thermoluminescence dosimetry and gamma detectors to detect isotope, and neutron generators to detect nuclear material (Ho et al., 2005). Monitoring volatile organic chemicals (VOCs) which is developing every day consists of following technologies: insecure chemical optical fiber sensor, spectroelectrochemistry of grating light reflection, miniature chemical flow probe sensors, SAW chemical sensor arrays, MictoChemLab (gas phase), resistant to chemical index of hydrogen deficiencies (IHDs) of gold nanoparticles, electrical impedance of double layer lipid in flat electrodes, MicroHound, super spectral imaging, and resistant chemical arrays (Ho et al., 2005).

Chip laboratory consists of chemical pocket equipment along with portable electronic equipment which is crucial for different practical plans, including safety of drinking water (Gardeniers and Van den Berg, 2004; Jang et al., 2011). Machines of chip laboratory can be easily used in screening and announcing primitive results.

There are many advances in detection of biological pollutants, although many of these procedures are concentrated on *E. coli* detection. Quick test for *E. coli* bacteria is analyzed in (Bukhari et al., 2007) which consists of four safety tests and one molecular procedure. Geng et al. (2011) use bioelectrochemical sensors based on molecular structure to detect *E. coli* in which magnetic beads are covered to perform as DNA probe.

In Ercole et al. (2002) a biosensor based on immunoassay is analyzed for *E. coli* bacteria, in which this procedure detection is based on pH changes and because of ammoniac production by an antibody response.

There are other procedures deployed to detect biological pollutants. Ercole et al. (2002) use laser tweezers technic along with Raman spectroscopy to detect bacterial cells in aqueous solution using a quick procedure either. Advanced level Raman spectroscopy using silver nanoparticles suspension, nanocolloid, with 1000 mL/cfu limit detection has been deployed in Sengupta et al. (2006) to detect *E. coli* Bacteria. Referring to Shoji et al. (2000), a biosensor that measures absorption of tagged proteins of fluorescent by human-cultivated cells is analyzed.

There are advances to better detect rare metals and other chemical compounds either. Although many of these tests and machines are currently developed to response to needs rather than safety, for example, problems related to quality of water such as arsenic in wells, these tests can develop the safety of water. For instance, these cases consist of microfluidic sensors which can detect low dense of pollutants as lead (Chang et al., 2005), used biosensor to detect arsenic in drinking water (de Mora et al., 2011), micro chemical enzyme tests to detect mercury in drinking water (Deshpande et al., 2010), self-feeding enzyme biofuel cell on a microchip to detect cyanide (Deng et al., 2010), and cheap developable biosensor to detect arsenic (Joshi et al., 2009). There have been advances on lead test in drinking water either (Wang et al., 2001).

Other machines are being developed to test different pollutants in drinking water systems. Yang et al. (2010) represent nanotube array to detect hydrocarbon in surface water. Surface acoustic wave micro sensors with polymerized layer have been developed as an affordable procedure to detect VOCs in water on-site (Groves et al., 2006). A manual machine to detect gas phase contamination, MicroChemLab, has been built by national laboratory of Sandia (Lewis et al., 2006). MicroChemLab machine consists of a preconcentration (micromachined), a GC channel, and quartz surface acoustic wave (Lewis et al., 2006). Other technologies are being developed to detect.

6.2.3.1 Algae, to detect water contamination

Scientists have found a new procedure to detect water contamination. This procedure is based on analysis of photosynthesis rate and changing sunlight into energy by plants which grow in water. In fact, this procedure studies whether plant uses its photosynthesis potential completely or not and accordingly they can find the disturbing factor in plant growth. When plant uses its photosynthesis potential completely, there would be energy as heat released in water which causes expansion and increases movement of molecules and produces sound waves by change in pressure (Fig. 6.1).

Researchers have simulated required condition for photosynthesis. It starts by shining green laser light on small pieces of floated algae on water and results in making sound waves. Using special microphone under water, severity and wavelength of these waves can be measured, analyzed and they would be able to detect health of algae, its surrounding water, and type and measure of water contamination. It worth mentioning that algae growing in water polluted by wastewater of battery and color production factories have different sound waves comparing to those which grow in water polluted by other toxic chemicals. Considering experiments, it has been proved that this procedure has good accuracy and pace comparing other current procedures and machine used in this experiment needs about one square meter either (http://naturescrusaders.wordpress.com).

6.2.3.2 Carbon nanotube sensors

Researchers try to detect water and food contamination using nanosensor technologies. These sensors are able to easily detect bacteria, virus, and pathogens carried by water. When these pathogens exist in water, the mentioned sensor produces electrical signals and detects type and dense of contamination. These sensors are made of carbon nanotubes which are 10,000 times thinner than human hair, more resistant than steel, more durable than diamond, and with high ability of heat and electric conductivity. This material was

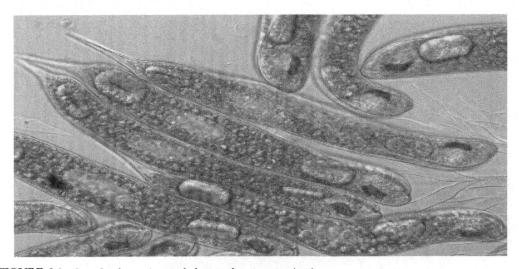

FIGURE 6.1 Sample of experimented algae to detect contamination.

discovered in the early 1990s and was called an amazing material because it was used in different stuff, including reinforced concrete, clothing production, and small and large industries. Nanotubes are mostly made of carbon which means it has the same main element of diamond and pencil graphite (diamond and graphite are two different species of carbon) and the only difference is in its atomic arrangement and link types. This difference between carbon nanotubes, diamonds, and graphite gives them different attributes. In other words, nanotubes are a third form of carbon which have round and spherical model (http://tinyurl.com/22ro2vz).

6.2.3.3 Fluorescent bacteria

Researchers have found a quick innovative procedure that detects whether water is toxic polluted or not. In this procedure by injection of fluorescent gen to bacteria and its genetic changes, bacteria start radiation as soon as it touches toxic chemicals. Common procedures of testing toxic chemicals in water are extinct these days because of being expensive and time consuming. In old procedures, fish were tested in different dense water samples. After some days, those fish were dead and researchers detected the measure of pollutant by counting dead fish. Bacteria have more advantage than fish because it can be massively produced and result of the experiment is detected in some minutes or hours. Instead of counting dead fish, radiated light from bacteria is measured either (Fig. 6.2) (http://www.innovations-report.com; Tecon and van der Meer, 2008).

6.2.3.4 Active electrochemical bacteria (biological sensors)

Arsenic is a pollutant of groundwater and is considered as toxic chemicals which can endanger lives of million people. In the early 1990s a large amount of arsenic was detected in drinking water of Bangladesh and about 20 million of 120 million of its population were suspected to drink arsenic polluted water. By drinking nonarsenic water, this toxicity

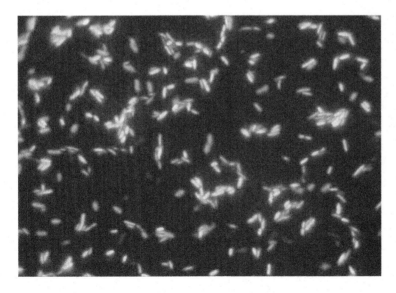

FIGURE 6.2 Sample of fluorescent bacteria.

could be destroyed, but because there was not any healthy water, the toxicities grew quickly and murdered thousands of people. The effects of arsenic are freckles, warts, diarrhea, and ulcers. In very severe toxicities, arsenic can cause kidney disease, cancer, and death. This toxic chemical is considered as a major water pollutant in India, Mexico, Vietnam, and Yugoslavia either. Therefore measuring arsenic in water is very crucial and researchers are looking for procedures to detect and control this pollutant as soon as possible.

Arsenic can be measured by old procedures, but they are too expensive and required machines are not usable in slum areas. Field chemical experiments do not have high accuracy in measuring arsenic and even they can increase mercury and lead in water. Therefore sensors must be used which do not have these limits (Tecon and van der Meer, 2008).

Despite all these problems, researchers designed and created biological sensors after required analysis which are very sensitive and can detect low amounts of arsenic immediately. Researchers are looking for biological sensors that can be used to detect other heavy metals such as cadmium and lead as well. Biological sensors consist of active electrochemical bacteria which produce electricity.

These biological sensors exchange cellular responses to measurable signal by linking bacteria to an electrical transducer. Bacteria are genetically corrected in which has a detector gen (transmitter) inside, gets irritated in the presence of physical or chemical contamination, transmits cellular biochemical responses to transducer and at last produces signals which result in light production. It is clear that existing fluctuations in water quality can cause fluctuation in produced electricity flow (Fig. 6.3) (Tecon and van der Meer, 2008; Singh et al., 2008).

In order to maintain required sensitivity of biological sensors, chelating agent which lowers intended pollutant is transmitted to water and its results are compared with

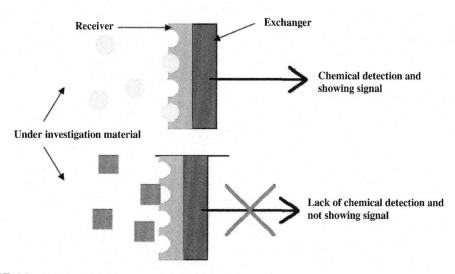

FIGURE 6.3   Pollutant-detection procedure by biological sensor.

samples without chelating agent. This procedure detects whether biological sensors react to pollutant density or not (Tecon and van der Meer, 2008). On the other hand, in order to detect these changes are because of toxic chemicals existence or a result of agent changes such as heat and pH, it is required to test environmental effects in terms of biological and electrochemical on signals carefully.

Advantages of biological sensors compared to other sensors can be mentioned briefly in the following terms (Singh et al., 2008):

1. detection of low amounts of contamination,
2. high accuracy,
3. easy application,
4. using less material compared to traditional procedures,
5. quick detection and prevention of spreading contamination,
6. economic efficiency, and
7. momentary monitoring.

Considering the problems of old procedures in pollutants' detection and their being time consuming, it is required to deploy new procedures in order to accelerate pollutants' detection; therefore process of water contamination detection becomes easier and takes less time. Researches have shown that by creating genetic changes in some aquatics, they can be used as sensors to detect polluted water. It seems that considering the abilities of these biological sensors, they can be used as an appropriate alternative of current procedures. It is also required to analyze their ability and efficiency to recognize costs and other problems which they may have in scales larger than laboratory.

6.2.4 Water quality parameters to detect pollutants

All possible biological, chemical, and radiological threats show a big group of components. Possible contamination is as much as it is not possible to monitor all components at the same time. Contamination is always changing. Although quick monitoring on all or even high percentage of potential pollutants is not possible, practical strategies to eliminate disability in order to measure all contamination only consist of the most important or the most possible pollutants. An alternative procedure of measuring qualitative parameters of water such as pH, heat, and electrical conductivity is to predict the measure of contamination is registered normally or abnormally. Qualitative parameters of standard water and operative condition which potential contamination is increased by its increase are mentioned later: alkalinity, carbon both mineral total inorganic carbon (TIC) and total organic carbon (TOC), remained chlorine both free and total, color, conductivity, dissolved oxygen, fluorescence, nitrogen; ammoniac, nitrate, nitrite, smell, oxygen reduction potential, particle size distribution, pH, pressure changes and problems, taste, temperature, total soluble solid, UV opacity and absorption (such as UV254).

Monitoring qualitative parameters of standard water and operational condition is very advantageous. Monitoring technology of these components is more accessible and cheaper than special contamination monitoring. Many of these components can be measured immediately. There is no need to have professional equipment and staff training. Considering low cost of these technologies, there can be more machines all over water system and collect data immediately. Operational data-management systems are used to trace data which can be used to create required operational condition to detect contamination and calibration of hydraulic models of contribution systems (Helbling and VanBriesen, 2007). A statistical procedure to predict water source in water system is used by Smeti et al. (2009) which shows that water quality data can be used to predict origin on a way that contamination source can be detected. Among tested quality parameters of water, remaining chlorine can be an effective indicator for many pollutants. In laboratory tests, Murray et al. (2008) show that microbial suspension is an attack procedure changing used chlorine which remained chlorine changes can be a reason of contamination.

6.3 Contamination warning systems

An important element in developing a water safety plan is to design and operate a contamination warning system (CWS) to detect contamination occurrence and find an appropriate response. Referring to U.S. Environmental Protection Agency (U.S. EPA) (2007), a structure to design a CWS is suggested which consists of required guidance on detecting numbers of used sensors and system goals. A general CWS has to contain following elements (Skolicki et al., 2006): online monitoring on water and contamination quality, sampling and analysis, advanced safety monitoring, and control on consumer complaint and public health.

Online water quality—monitoring system consists of online monitoring sensors all over the contribution system and collection of water quality data. Information must be transmitted to a centralized network from sensors to be saved and analyzed automatically and electronically. Appropriate software is required to analyze sensor data. When there is possibility of threat sampling protocols must contain common and extra sampling. Advanced safety monitoring consists of equipment and machines which are considered to be physical protection for vital assets. This equipment contains video cameras, motion and light alarms, and motion detectors and intrusion detector alarms. Consumer complaint data can be useful as detector of pollutant if they affect drinking water in aspects of smell, taste, or color. Public health data are also important. The collectable data are sale of over-the-counter medications, reports of hospitals' receptions, and received calls to 110.

Effective CWS has many advantages. A well-designed CWS can help to locate contamination source and also be effective on recovery and rehabilitation of polluted contribution systems. Immediate response is possible by installation of CWS. Other advantages of CWS are better understanding of operational changes, effects on quality of water, and also better dosing of remained disinfectants all over the contribution system. CWS are used to detect other threats either. For instance, instead of contamination analysis, Janke et al. (2006) concentrate on breakdown effects in pipelines using sensor network procedure.

6.3.1 Online pollutant monitoring

Immediate detection is possible by online monitoring of stations. Researchers show advantages of online monitoring on common sampling and analysis (Roberson and Morley, 2005). These needs must be supplied while using CWS: the number of monitoring stations, measuring parameters, analytical procedures, the location of monitoring stations, water contribution modeling software, and data analysis procedures (Foran and Brosnan, 2000). Water contribution modeling software simulates contamination flow and transmission all over the dynamic contribution system. On another hand, water installation usually uses static models for planning aims.

Before installation of CWS, water industry needs to develop standard operational procedures to detect proper responses and trace every alarm and motive (Skolicki et al., 2006). The first step is to collect enough information about installation of online monitoring stations and creation of basic conditions. The data must be collected and analyzed for at least a year to detect seasonal fluctuations. After that, motives that show deviation from baseline and abnormal conditions must be created. Each motive shows a potential threat which needs to be evaluated and detects whether the threat is valid or not. The following items must be considered while evaluating abnormalities: the absolute value of changes based on size and fluctuations about basis and slope of changes. In an ideal form, CWSs should contain monitoring which is "sensitive, devoted, repeatable, and researchable" (Kroll and King, 2010). All data from monitoring stations must be collected and evaluated by a systematic QA/QC program. Although lots of data analyses are done automatically, inputs related to alarms and motives should be entered manually. As positive and negative fake errors are problematic in online pollutant-monitoring technology, an accurate analysis of data is needed before a threat. CWS considerations should contain cost, easy use, automation, and reliability (Hart and Murray, 2010). An online pollutant-monitoring system has to have the following specifications (Foran and Brosnan, 2000): quick response time automatic, ability to save samples, high frequency of sampling, high sensitivity to public health threats, possibility to monitor a wide range of pollutants, high repeatability, low rate of fake positive and negative, resistance, easy use, cheap design, small need to repair and maintenance, and low electricity consumption.

An ideal CWS is very challenging (Skolicki et al., 2006). In a CWS with minimum number of stations, every monitoring station must be able to cover a wide area of contribution system. Considering time sensitive nature of contamination detection and formulation of a response, online monitoring system must detect compounds immediately and response in time. In order to please both employees of water industry and consumers, the numbers of fake positive errors must be minimized in online pollutant-monitoring system.

CWS is being developed every day along with the development of sensors and new strategies of data analysis. There have been many efforts on simulation of contamination and CWS test in water contribution system.

CWSs have more advantages rather than only contamination detection. Some procedures are represented to use sensors and their networks for aims except for intentional contamination detection which are saving remained chlorine all over covered area, taking common needed samples, pressure monitoring and leak detection (Aisopou et al., 2012).

6.3.2 Online sensor technology

Adjustment of water qualitative sensor machines and combining these sensors with contribution system is limited by available technologies and funds. There is not any senor that can directly detect many potential contaminations and still be affordable. Sensors that monitor water qualitative parameters in contribution system can be useful in detection of contamination either (Hall et al., 2007).

Different sensors with various technologies are used for online contamination monitoring stations. Some of them are based on standard parameters of water quality and measure them singularly or in sets of parameters. In addition to available technologies, there are developing sensors based on emerging technologies which are not ready for the market yet. These sensors use tests to detect parameters of water quality, toxicity, and indicators of microbial presence (such as number of particles). Biological sensors that are available use fish, algae, and bacteria. Deploying these sensors requires a significant consideration in users' section that prepares biological species required to show toxicity. Although some microbiological sensors are developed to detect *E. coli* and other common bacteria, it is possible that in future these sensors can be used to detect other microorganisms.

Although online multivariate probes are useful to trace the quality of water in contribution systems, unstable hydraulic condition can affect accuracy and results of the sensor (Murray et al., 2012). To suspend sediment again and mixing conditions in pipes is problematic to detect remaining disinfectants, opacity, and color (Murray et al., 2012). Referring to an analysis of online pollutant-monitoring technology, there are not any detectors or sensors to operate in a complete scale.

The following technologies are analyzed by Ostfeld and Salomons (2004): UV/VIS scan, JMAR Biosentry, Censar, YSI Sonde, Hach Event Monitor (Guardian Blue), Algae Toximeter spectrolyser, TOXcontrol (microloan), Fish Activity Monitoring Scan Water Quality Monitoring Station, ToxProtect (bbe), DaphniaToximeter (bbe), System (FAMS) (BBE), advanced Raman spectroscopy, Raman Laser Tweezers, and Surface acoustic wave machines. Cost is the biggest problem in choosing online sensors.

6.3.3 Computational procedures

There have been significant developments in analytical abilities to design a network of sensors in CWSs in past 10 years (Storey et al., 2011). Previous efforts were on locating computational sensors based on stable models, but recent efforts are mostly concentrating on more complex mechanisms of dynamic contamination transmission (Storey et al., 2011).

There are many procedures to better locate sensors on a network. There are many reports to solve this optimization model (Ostfeld and Salomons, 2005; Propato, 2006; Berry et al., 2009). Over time optimization has changed from single purpose function into multipurpose function. Considerations of delayed response have been considered by Preis and Ostfeld (2008c) either. Referring to Siswana et al. (2008), locating procedure of multipurpose sensor is used in which sensor, redundancy, and response time are considered as well. The possibility of sensor detection is optimized using the following equation:

$$f_1 = \frac{1}{S} \sum_{r=1}^{S} d_r \tag{6.1}$$

S shows all contamination scenarios and f_1 is the maximum amount (Preis and Ostfeld, 2008c). $d_r = 1$ is for a recognized contamination occurrence, r under conditions of flow, density, and specified time and if there will not be recognized contamination $d_r = 0$ (Siswana et al., 2008). Redundancy in a sensor network is shown in the following equation:

$$f_2 = \frac{1}{\sum_{r=1}^{S} d_r} \sum_{r=1}^{S} R_r \qquad (6.2)$$

f_2 is the maximum amount and R_r is the redundancy of the sensor network for r contamination scenario. Here, 1 or 0 is given to at least three sensors until 30 minutes from the first to the third detection of a nonzero pollutant (Siswana et al., 2008). Time to detect an occurrence, t_d, is computed by the following equation:

$$t_d = \min_{i} t_i \qquad (6.3)$$

t_i is the time of i sensor detection (Siswana et al., 2008). The time of sensor detection is optimized by the following equation:

$$f_3 = E(t_d) \qquad (6.4)$$

f_3 is minimized in it and $E(t_d)$ is math hope for t_d (Siswana et al., 2008).

6.4 Conclusion

The detection of pollutants in drinking water systems is an important component to defeat the safety of water. This detection needs common monitoring of qualitative water and contamination components and also systematic process to detect abnormalities. Available water quality–monitoring strategies are not only designed to protect against contamination but also to adjust with standards of drinking water. Nowadays, detection of special chemicals and biological pollutants which can be used in an intentional contamination occurrence generally needs advanced analytical procedures. These analyses need to be sent to out of laboratory most of the times, which results in the increase of time detection and later response. Advances in immediate technology include biological and online sensors which have facilitated contamination detection, but there are still limits by detectable contaminations. Despite these limits, immediate response is crucial because it can eliminate total contamination by preparation of a quick response and renovation program.

Despite ongoing researches, pollutant detection in drinking water systems is still limited in water industry by available technology and its costs. Costs can be decreased by deploying dual detection technology. However, there should be much more researches to widen the ability to detect contamination, decrease the costs, locating sensors optimized in order to create quick response sensors more achievable and general in our country's water systems.

References

Aisopou, A., Stoianov, I., Graham, N.J.D., 2012. In-pipe water quality monitoring in water supply systems under steady and unsteady state flow conditions: a quantitative assessment. Water Res. 46 (1), 235–246.

Berry, J., Carr, R.D., Hart, W.E., Leung, V.J., Phillips, C.A., Watson, J.P., 2009. Designing contamination warning systems for municipal water networks using imperfect sensors. J. Water Resour. Plann. Manage. ASCE 135 (4), 253–263.

Botsford, J.L., 2002. A comparison of ecotoxicological tests. Altern. Lab. Anim. 30 (5), 539–550.

Bukhari, Z., Weihe, J.R., Lechevallier, M.W., 2007. Rapid detection of *Escherichia coli* O157:H7 in water. J. Am. Water Works Assoc. 99 (9), 157–167.

Campbell, C.G., Mascetti, M.M., Hoppes, W., Stringfellow, W.T., 2007. Measurement reproducibility of the BioscanTM flow-through respirometer applied as a toxicity-based early warning system for water contamination. Environ. Pract. 9 (1), 42–53.

Chang, I.H., Tulock, J.J., Liu, J.W., Kim, W.S., Cannon, D.M., Lu, Y., et al., 2005. Miniaturized lead sensor based on lead-specific DNAzyme in a Nano capillary interconnected microfluidic device. Environ. Sci. Technol. 39 (10), 3756–3761.

Curtis, T.M., Tabb, J., Romeo, L., Schwager, S.J., Widder, M.W., van der Schalie, W.H., 2009a. Improved cell sensitivity and longevity in a rapid impedance-based toxicity sensor. J. Appl. Toxicol. 29 (5), 374–380.

Curtis, T.M., Widder, M.W., Brennan, L.M., Schwager, S.J., van der Schalie, W.H., Fey, J., et al., 2009b. A portable cell-based impedance sensor for toxicity testing of drinking water. Lab on a Chip 9 (15), 2176–2183.

Davila, D., Esquivel, J.P., Sabate, N., Mas, J., 2011. Silicon-based microfabricated microbial fuel cell toxicity sensor. Biosens. Bioelectron. 26 (5), 2426–2430.

de Hoogh, C.J., Wagenvoort, A.J., Jonker, F., van Leerdam, J.A., Hogenboom, A.C., 2006. HPLC- DAD and Q-TOF MS techniques identify cause of Daphnia biomonitor alarms in the River Meuse. Environ. Sci. Technol. 40 (8), 2678–2685.

de Mora, K., Joshi, N., Balint, B.L., Ward, F.B., Elfick, A., French, C.E., 2011. A pH-based biosensor for detection of arsenic in drinking water. Anal. Bioanal. Chem. 400 (4), 1031–1039.

Deng, L., Chen, C.G., Zhou, M., Guo, S.J., Wang, E.K., Dong, S.J., 2010. Integrated self-powered microchip biosensor for endogenous biological cyanide. Anal. Chem. 82 (10), 4283–4287.

Deshpande, K., Mishra, R.K., Bhand, S., 2010. A high sensitivity micro format chemiluminescence enzyme inhibition assay for determination of Hg(II). Sensors 10 (7), 6377–6394.

Dewhurst, R.E., Wheeler, J.R., Chummun, K.S., Mather, J.D., Callaghan, A., Crane, M., 2002. The comparison of rapid bioassays for the assessment of urban groundwater quality. Chemosphere 47 (5), 547–554.

Eltzov, E., Marks, R.S., 2010. Fiber-optic based cell sensors. Adv. Biochem. Eng. Biotechnol. 117, 131–154.

Eltzov, E., Marks, R.S., 2011. Whole-cell aquatic biosensors. Anal. Bioanal. Chem. 400 (4), 895–913.

Ercole, C., Del Gallo, M., Pantalone, M., Santucci, S., Mosiello, L., Laconi, C., et al., 2011. A biosensor for *Escherichia coli* based on a potentiometric alternating biosensing (PAB) transducer. Sens. Actuators B—Chem. 83 (1–3), 48–52.

Ercole, C., Del Gallo, M., Pantalone, M., Santucci, S., Mosiello, L., Laconi, C., et al., 2002. A biosensor for *Escherichia coli* based on a potentiometric alternating bio sensing (PAB) transducer. Sens. Actuators B—Chem. 83 (1–3), 48–52.

Foran, J.A., Brosnan, T.M., 2000. Early warning systems for hazardous biological agents in potable water. Environ. Health Perspect. 108 (10), 993–996.

Gardeniers, H., Van den Berg, A., 2004. Micro- and nanofluidic devices for environmental and biomedical applications. Int. J. Environ. Chem. 84 (11), 809–819.

Geng, P., Zhang, X.A., Teng, Y.Q., Fu, Y., Xu, L.L., Xu, M., et al., 2011. A DNA sequence-specific electrochemical biosensor based on alginic acid-coated cobalt magnetic beads for the detection of *E. coli*. Biosens. Bioelectron. 26 (7), 3325–3330.

Giaever, I., Keese, C.R., 1993. A morphological biosensor for mammalian cells. Nature 366 (6455), 591–592.

Grimmett, P.E., Munch, J.W., 2013. Development of EPA Method 525.3 for the analysis of semi volatiles in drinking water. Anal. Methods 5 (1), 151–163.

Groves, W.A., Grey, A.B., O'Shaughnessy, P.T., 2006. Surface acoustic wave (SAW) micro sensor array for measuring VOCs in drinking water. J. Environ. Monit. 8 (9), 932–941.

Hall, J., Zaffiro, A.D., Marx, R.B., Kefauver, P.C., Krishnan, E.R., Herrmann, J.G., 2007. On-line water quality parameters as indicators of distribution system contamination. J. Am. Water Works Assoc. 99 (1), 66−77.

Hart, W.E., Murray, R., 2010. Review of sensor placement strategies for contamination warning systems in drinking water distribution systems. J. Water Resour. Plann. Manage. ASCE 136 (6), 611−619.

Helbling, D.E., VanBriesen, J.M., 2007. Free chlorine demand and cell survival of microbial suspensions. Water Res. 41 (19), 4424−4434.

Ho, C.K., Robinson, A., Miller, D.R., Davis, M.J., 2005. Overview of sensors and needs for environmental monitoring. Sensors 5 (1−2), 4−37.

<http://naturescrusaders.wordpress.com>.

<http://tinyurl.com/22ro2vz>.

<http://www.innovations-report.com>.

Iuga, A., Lerner, E., Shedd, T., Van der Schalie, W.H., 2009. Rapid responses of melanphore cell line to chemical contaminants in water. J. Appl. Toxicol. 29, 346−349.

Jang, A., Zou, Z.W., Lee, K.K., Ahn, C.H., Bishop, P.L., 2011. State-of-the-art lab chip sensors for environmental water monitoring. Meas. Sci. Technol. 22 (3).

Janke, R., Murray, R., Uber, J., Taxon, T., 2006. Comparison of physical sampling and real-time monitoring strategies for designing a contamination warning system in a drinking water distribution system. J. Water Resour. Plann. Manage. ASCE 132 (4), 310−313.

Jeon, J., Kim, J.H., Lee, B.C., Kim, S.D., 2008. Development of a new biomonitoring method to detect the abnormal activity of *Daphnia magna* using automated Grid Counter device. Sci. Total Environ. 389 (2−3), 545−556.

Joshi, N., Wang, X., Montgomery, L., Elfick, A., French, C.E., 2009. Novel approaches to biosensors for detection of arsenic in drinking water. Desalination 248 (1−3), 517−523.

Kroll, D., King, K., 2010. Methods for evaluating water distribution network early warning systems. J. Am. Water Works Assoc. 102 (1), 79−89.

Lewis, P.R., Manginell, R.P., Adkins, D.R., Kottenstette, R.J., Wheeler, D.R., Sokolowski, S.S., et al., 2006. Recent advancements in the gas-phase MicroChemLab. IEEE Sensors J. 6 (3), 784−795.

Lindquist, H.D.A., Harris, S., Lucas, S., Hartzel, M., Riner, D., Rochele, P., et al., 2007. Using ultrafiltration to concentrate and detect *Bacillus anthraces*, *Bacillus atrophies* subspecies globigii, and *Cryptosporidium parvum* in 100-liter water samples. J. Microbial. Methods 70 (3), 484−492.

Murray, R., Janke, R., Hart, W.E., Berry, J.W., Taxon, T., Uber, J., 2008. Sensor network design of contamination warning systems: a decision framework. J. Am. Water Works Assoc. 100 (1), 97−109.

Murray, S., Ghazali, M., McBean, E.A., 2012. Real-time water quality monitoring: assessment of multisensory data using Bayesian belief networks. J. Water Resour. Plann. Manage. ASCE 138 (1), 63−70.

Ostfeld, A., Salomons, E., 2004. Optimal layout of early warning detection stations for water distribution systems security. J. Water Resour. Plann. Manage. ASCE 130 (5), 377−385.

Ostfeld, A., Salomons, E., 2005. Securing water distribution systems using online contamination monitoring. J. Water Resour. Plann. Manage. ASCE 131 (5), 402−405.

Porco, J.W., 2010. Municipal water distribution system security study: recommendations for science and technology investments. J. Am. Water Works Assoc. 102 (4), 30−32.

Preis, A., Ostfeld, A., 2008c. Multiobjective contaminant sensor network design for water distribution systems. J. Water Resour. Plann. Manage. ASCE 134 (4), 366−377.

Propato, M., 2006. Contamination warning in water networks: General mixed-integer linear models for sensor location design. J. Water Resour. Plann. Manage. ASCE 132 (4), 225−233.

Radix, P., Leonard, M., Papantoniou, C., Roman, G., Saouter, E., Gallotti-Schmitt, S., et al., 2000. Comparison of four chronic toxicity tests using algae, bacteria, and invertebrates assessed with sixteen chemicals. Ecotoxicol. Environ. Saf. 47 (2), 186−194.

Roberson, J.A., Morley, K.M., 2005. Contamination Warning Systems for Water: An Approach for Providing Actionable Information to Decision-Makers. AWWA, Denver, CO.

Sengupta, A., Mujacic, M., Davis, E.J., 2006. Detection of bacteria by surface-enhanced Raman spectroscopy. Anal. Bioanal. Chem. 386 (5), 1379−1386.

Shoji, R., Sakai, Y., Sakoda, A., Suzuki, M., 2000. Development of a rapid and sensitive bioassay device using human cells immobilized in macro porous micro carriers for the on-site evaluation of environmental waters. Appl. Microbial. Biotechnol. 54 (3), 432−438.

Singh, M., Verma, N., Garg, A.K., Redhu, N., 2008. Urea biosensors, urea biosensors. Sens. Actuators B: Chem. 134 (1), 345–351.

Siswana, M., Ozoemena, K.I., Nyokong, T., 2008. Electro catalytic detection of amitrole on the multi-walled carbon nanotube – iron (II) tetra-aminophthalocyanine platform. Sensors 8, 5096–5105.

Skolicki, Z., Wadda, M.M., Houck, M.H., Arciszewski, T., 2006. Reduction of physical threats to water distribution systems. J. Water Resour. Plann. Manage. ASCE 132 (4), 211–217.

Smeti, E.M., Thanasoulias, N.C., Lytras, E.S., Tzoumerkas, P.C., Golfinopoulos, S.K., 2009. Treated water quality assurance and description of distribution networks by multivariate chemo metrics. Water Res. 43 (18), 4676–4684.

States, S., Newberry, J., Wichterman, J., Kuchta, J., Scheuring, M., Casson, L., 2004. Rapid analytical techniques for drinking water security investigations. J. Am. Water Works Assoc. 96 (1), 52–64.

States, S., Wichterman, J., Cyprych, G., Kuchta, J., Casson, L., 2006. A field sample concentration method for rapid response to security incidents. J. Am. Water Works Assoc. 98 (4), 115–121.

Storey, M.V., van der Gaag, B., Burns, B.P., 2011. Advances in on-line drinking water quality monitoring and early warning systems. Water Res. 45 (2), 741–747.

Tecon, R., van der Meer, J.R., 2008. Bacterial biosensors for measuring availability of environmental pollutants. Sensors 8 (7), 4062–4080.

U.S. Environmental Protection Agency (U.S. EPA), 2003. Module 4: Analytical guide, EPA- 817-D-03–004. Response Protocol Toolbox (RPTB) Interim Final: Planning for and Responding to Contamination Threats to Drinking Water Systems. EPA, Washington, DC.

U.S. Environmental Protection Agency (U.S. EPA), 2007. Water Security Initiative: Interim Guidance on Planning for Contamination Warning System Deployment, EPA 817-R-07-002. EPA, Washington, DC.

van der Schalie, W.H., James, R.R., Gargan, T.P., 2006. Selection of a battery of rapid toxicity sensors for drinking water evaluation. Biosens. Bioelectron. 22 (1), 18–27.

Wang, J., Lu, J.M., Hocevar, S.B., Ogorevc, B., 2001. Bismuth-coated screen-printed electrodes for stripping voltammetric measurements of trace lead. Electroanalysis 13 (1), 13–16.

Yang, L.X., Chen, B.B., Luo, S.L., Li, J.X., Liu, R.H., Cai, Q.Y., 2010. Sensitive detection of poly- cyclic aromatic hydrocarbons using CdTe quantum dot-modified TiO_2 nanotube array through fluorescence resonance energy transfer. Environ. Sci. Technol. 44 (20), 7884–7889.

Further reading

Hillaker, T.L., Botsford, J.L., 2004. Toxicity of herbicides determined with a microbial test. Bull. Environ. Contam. Toxicol. 73 (3), 599–606.

Serjeantson, B., McKenny, S., van Buskirk, R., 2011. Leverage operations data and improve utility performance. AWWA Opflow 37 (2), 10–15.

van der Schalie, W.H., Shedd, T.R., Knechtges, P.L., Widder, M.W., 2001. Using higher organisms in biological early warning systems for real-time toxicity detection. Biosensor. Bioelectron. 16 (7–8), 457–465.

van der Schalie, W.H., Shedd, T.R., Widder, M.W., Brennan, L.M., 2004. Response characteristics of an aquatic bio monitor used for rapid toxicity detection. J. Appl. Toxicol. 24 (5), 387–394.

Xie, C., Mace, J., Dinno, M.A., Li, Y.Q., Tang, W., Newton, R.J., et al., 2005. Identification of single bacterial cells in aqueous solution using confocal laser tweezers Raman spectroscopy. Anal. Chem. 77 (14), 4390–4397.

7

Methods of inorganic pollutants detection in water

Chhavi Sharma and Yuvraj Singh Negi

Department of Polymer and Process Engineering, Indian Institute of Technology (IIT Roorkee), Roorkee, India

7.1 Introduction

Water is essential to sustain life on this blue planet Earth. As we all know, water is considered to be the best universal solvent that supports various physiochemical reactions to occur (Henry, 2005). Based on various chemical compositions of human bodies, study reveals that our body is made up of 70% of water (Mitchell et al., 1945). On the other hand, as per the WHO guidelines, for proper functioning of cells of a normal person, daily intake of water must be equivalent to the daily loss. Hence, on an average a person weighing 70 kg must consume water about 3 L/day. The amount varies depending on body weight, age, gender etc. (Noronha et al., 2019).

But, nowadays, gradual decrease of clean water resources is highly alarming and is gaining everyone's attention. Although if we look back to the history of water contaminants, Spiegel et al. in the year 1900 (Noronha et al., 2019) had published a research article indicating the contamination of water. Their main focus was on the detection of nitrite

Inorganic Pollutants in Water
DOI: https://doi.org/10.1016/B978-0-12-818965-8.00007-X

contamination in drinking water. Later, Savage (1902) highlighted the importance of bacteria (*Bacillus coli*) detection in drinking water. Later, snowballing economic development and World War II had added chemicals, for example, TNT (2,4,6-trinitrotoluene), RDX (hexahydro1,3,5-trinitro-1,3,5-triazine), and HMX (octahydro-1,3,5,7-tetranitro-1,3,5,7-tetrazocine), which have increased the level of ground and surface water contamination. Moreover, after this incidence, all the pollutants got slowly accumulated into the ecosystem (Noronha et al., 2019). Later in 1974 and onward, various research articles have mentioned various surface and ground water pollutants and entitled them as "emergent contaminants," and US EPA had given the definition of water contaminant as "any substance or matter present in physical, chemical, biological or radiological form." In this context, we can characterize water pollutants as (1) organic, (2) inorganic, (3) biological, and (4) radiological.

Organic contaminants may be defined as chemicals that are carbon based, for example, organic solvents, pesticides, petroleum-based wastage, timber, and gas or liquid phase volatile compounds. All these substances produce toxic by-products during decomposition. Pathogens are also included into it, and together, they cause various life-threatening diseases and produce toxins during disinfecting (Ram, 1990; Sharma et al., 2003).

Inorganic contaminants are a group of contaminants that consist fluorine, heavy metals, nitrides, etc. The existence or addition of these metals and metalloids into the main stream is because of natural (volcanic eruption, rock weathering, etc.) and man-made resources (industrial utilization, untreated effluent discharge, mining, etc.) (World Health Organization, 2003a).

The term "heavy metals" has no specific definition, but these metals are classified as the elements that have high atomic weight as well as higher relative density (at least five times with respect to water). They impart a higher significance to living creatures because some of the metals such as Cu and Se are essential for human development, but on the other hand, metals such as As and Pb impart high health related risks and issues. Toxicity imparted by all these elements depends upon multiple factors such as up taken amount, route of exposure, chemical species, age, hereditary, and nutritive status. In relevance with environmental exposure, heavy metals and metalloids [As(III, V), Cd(II), Cr(V), Pb(II), and Hg(II)] are kept at high risk priority by various public health agencies (WHO, U.S. Environmental Protection Agency, etc.) (Aragay et al., 2011; Gumpu et al., 2015; Lagarde and Jaffrezic-Renault, 2011) (Table 7.1).

Fluoride exists as mineral due to its high reactivity (World Health Organization, 2003a). It is one of the widely used compounds in various industrial sectors such as aluminum smelting, glass fiber drawing and rowing, pharmaceutical manufacturing, and fertilizers. All the discharges when get added to round and surface water bodies will result in up shooting of fluorides. Although fluorides concentration is decisive not only based on human activities but also based on geography, water origin, and type of rock bed, it is flowing through, and people taking up fluoride (through water or medication) will irritates GI tract, brain functioning (memory related issues), etc. (Susheela, 1999; Fawell et al., 2006; Water Standards and World Health Organization, 2004; Gopal and Ghosh, 1985). Unfortunately, among various countries, India has more than 19 states which are suffering from acute fluorosis (Eswar et al., 2011).

Cyanide is a compound which is viable and essentially used in the metal extraction (gold and silver), coal gasification, decontamination of ships, buildings, grain silos, seeds in

TABLE 7.1 Heavy metals with their toxicity and MCL limits based on the MCL standards (established by US EPA) (Barakat, 2011).

Metal ions	Described adversarial effects	Conc. (mg/L)
Arsenic (III)	Skin appearances, cancer, vascular syndrome	0.050
Cadmium (II)	Kidney destruction, renal ailment, human cancer-causing agent	0.01
Chromium (IV)	Headache, diarrhea, motion sickness, vomit, malignancy causing	0.05
Copper (II)	Liver mutilation, Wilson ailment, sleeplessness	0.25
Nickel (II)	Dermatitis, unsettled stomach, chronic asthma, carcinogenic	0.20
Zinc (II)	Despair, laziness, nervous breakdown, and greater than before thirst	0.80
Lead (II)	Impairment to the fetus, kidneys impairment, circulatory and nervous system	0.006
Mercury (II)	Rheumatoid arthritis, ailments related to kidney, neurological disorders	0.00003

vacuum cavities, electroplating, etc. During water disinfection process through chlorination, by-products such as cyanogen chloride and in situ chloramines are produced to uphold the sterile conditions of water supply provisions (World Health Organization, 2003b).

Nitride (MCL 10 mg/L) is found in sewage from humans and/or farmhouse and commonly overflows into water sources. Extreme intake of nitrate can cause serious ailments such as methemoglobinemia or blue-baby syndrome in infants because of the interference made in the transportation of oxygen in blood (McClellan and Halden, 2010; Gupta et al., 2000).

Biological contamination is a type of contamination that occurs due to algae, bacteria, pathogens, viruses, etc. These parasites are well known for affecting human health, for example, *E. coli* is a well-known coliform that causes bloody diarrhea, stomach aches, etc. Similarly, bacteria, viruses, and protozoa are also the pathogens that cause water contamination resulting in water-borne diseases such as dysentery, cholera, and gastroenteritis (Sharma et al., 2003; https://www.safewater.org/fact-sheets-1/2017/1/23/pesticides).

Radiological contaminants are undesirable radionuclides that include naturally occurring radiological isotopes of radium, uranium, and radon. Apart from that, granite rocks and some distinct minerals are responsible for such contamination. Moreover, manual contribution is termed as "technologically enhanced radioactive material (TENORM)," which collectively addresses industries dealing with coal (mining and combustion), oil and gas industry, and fertilizer (phosphate) production sectors (Fawell et al., 2006; https://www.berkeywater.com/news/a-look-at-radiological-water-contaminants/).

In this chapter, our prime concern is to focus on inorganic class of pollutants due to its abovementioned potential effects on living creatures.

7.2 Various detection methodologies

7.2.1 Conventional methodologies

Various conventional techniques such as inductively coupled plasma—mass spectrometry (ICP—MS), mass spectroscopy, X-ray fluorescence spectroscopy, and atomic absorption spectroscopy (AAS) are commonly used for heavy metal ion detection. Although they have a high degree of selectivity, sensitivity, and multiple analyte handling, a certain set of limitations such as expensive instrumentation and infrastructures, trained manpower, and time consuming sample preparation procedures emphasizes on finding alternative cost-effective approaches (Aragay et al., 2011; Gumpu et al., 2015).

7.2.1.1 Mass spectroscopy

Mass spectroscopy in a destructive characterization technique in which the structural properties and chemical composition of unknown sample is done by converting sample into fragments of ions, and these fragments will be studied by segregating them on the basis of its mass-to-charge (m/z) ratio with respect to relative abundance of each ion type. With the help of this technique, we are basically studying the effect of ionizing energy on the sample molecule or compound (http://premierbiosoft.com/tech_notes/mass-spectrometry.html).

7.2.1.1.1 Ionization sources

Various methods such as electron impact (EI), fast atomic bombardment (FAB), atmospheric pressure chemical (APCI), electrospray (ESI), matrix-assisted laser desorption (MALDI), and also ICP are used for various analyte assessments. Among all, EI and FAB are the techniques which are used only with gas chromatography (GC−MS). APCI works with liquid chromatography (LC−MS) because it produces protonated fragments. ESI and MALDI are well thought out as "soft"-ionizing source, that is, reasonably slight energy is imparted to the analyte, and hence, tiny fragmentation come about (Pitt, 2009). MALDI is very a sensitive and expensive technique that is used by biological specimen (usually) (http://chemistry.emory.edu/msc/tutorial/mass-spectrometry-ionization.html;   https://www.labcompare.com/18-Mass-Spectrometry-Ionization-Sources/). ICP are the methods mostly reported for the analysis of inorganic analytes for environmental applications (Tsednee et al., 2016).

7.2.1.2 Atomic absorption spectroscopy

AAS is a qualitative methodical approach that can measure the concentration of components present in the analyte. Then light of specific wavelength penetrates through the analyte, and then ground state electrons will be kicked up to excited state. Hence, the change in the incident and absorbed light is being recorded and evaluated by the spectrometer (http://faculty.sdmiramar.edu/fgarces/labmatters/instruments/aa/AAS_Theory/AASTheory.htm). It is one of the conventional techniques invented in 1950 (http://faculty.sdmiramar.edu/fgarces/labmatters/instruments/aa/AAS_Theory/AASTheory.htm;   Koirtyohann,   1991) but it is still popular because of its rapid, easy, precise, and extremely sensitive determination of inorganic, metallic, metalloid, and others to a very low level of detection. It has diversified applications in the field of tablets, agricultural science, metallurgy, biochemical technology, and ecological observation (http://faculty.sdmiramar.edu/fgarces/labmatters/instruments/aa/AAS_Theory/AASTheory.htm).

Toxicologists have persistently spotted a high heavy metal ion concentration in a number of water resources, plants, vegetables, etc. In the context of technique, a number of literature are available for various different analytes, and as heavy metals are having high

TABLE 7.2 Comparison between all conventional methods (Tyler and Jobin Yvon, 1995).

Description	ICP–MS	ICP–OES	Flame AAS	GF–AAS
LOD	Brilliant for utmost elements	Great for utmost elements	Great for more or less elements	Brilliant for some elements
Sample throughput	All elements: 6 min	5–30 elements/min	15 s/element	4 min/element
Accuracy (%)	1–3	0.3–1	0.1–1	1–5
Isotopes examination	Available	Not available	Not available	Not available
Max. dissolved solids conc.	0.4–0.1	30–1	3–0.5	Greater than 30
Elements	Greater than 75	Greater than 75	Greater than 68	Greater than 50
Material requirement	Small	Small	Huge	Very small
Semiquant study	Yes	Yes	No	No
Routine procedure	Simple	Simple	Simple	Simple
Technique development	Skill compulsory	Skill compulsory	Easy	Skill compulsory
Unattended process	Yes	Yes	No	Yes
Ignitable airs	No	No	Yes	No
Operational cost	High	High	Low	Medium
Investment cost	Very expensive	Expensive	Cost effective	Reasonable

AAS, Atomic absorption spectroscopy; *GF*, gas flame; *ICP*, inductively coupled plasma; *LOD*, limit of detection; *MS*, mass spectrometry; *OES*, optical emission specroscopy.

adversities, many sources of food and water are constantly being monitored by the toxicologists and environmentalists. Many researchers had conducted studies on various fish samples collected from various distant sites such as lakes and water bodies of Indonesia (Koirtyohann, 1991), Lake of Victoria waters (Koirtyohann, 1991), River Galma (https://csiropedia.csiro.au/atomic-absorption-spectroscopy/), Mediterranean Sea (Widianarko et al., 2000), Ogba River (Tole and Shitsama, 2003), Poompuhar coast (Nnaji et al., 2007), Parangipettai coast (Tyler and Jobin Yvon, 1995), and many more. All have used AAS and/or FAAS (Flame AAS) to estimate the concentration of elements, namely, Zn, Pb, Cu, Mn, V, Mg, As, Cd, and others, present in various body parts of fishes (Table 7.2).

7.2.2 Absorption spectrometry

Absorption spectroscopy is a technique in which light intensity change interacted with the sample is observed. The recorded data is then plotted in the coordinate axes in terms of intensity and wavelength or frequency (https://physics.tutorvista.com/matter/absorption-spectroscopy.html).

7.2.2.1 Colorimetric/plasmonic sensing

Colorimetric sensing is popularly gaining attention because of its hassle-free visual sensing. These sensors follow optical detection methodology in which color reflected (or visible to eye) after the binding of the analyte of interest with the sensor molecule depends upon the wavelength or frequency of light absorbed by the analyte. Its working principle is surface plasmon resonance (SPR) and, hence, mentioned as plasmonic devices too. SPR is an optical effect that arises if plane-polarized wave strikes the surface of noble metallic film at a specific angle, will result in total internal reflection, and hence at this frequency of light (referred as resonating frequency), the free electrons present in the outer shell of atom will start oscillating in resonance. It happens only at resonance frequency, and at the moment here is slight variation occurs at frequency state, electrons comes back to normal state, and this results in measurable variation to be utilized in sensor designing. It is used to study the molecular binding interactions in real-time monitoring using any label (https://www.sprpages.nl/spr-overview/spr-theory; Piriya et al., 2017). For example, Stolarczyk et al. (2016) had developed a colorimetric sensing toward Pb(II), Cd(II), and Hg(II) using gold particles capped with 11-mercaptoundecanoic acid as chromophore, and reversibility of the binding was checked using EDTA solution. During binding of Pb(II), red solution of Au-functionalized nanoparticles (NPs) turned to blue and changed back to red after the addition of EDTA. In this context, various polymers in combination with Au NPs were also used by the researchers to design SPR-based sensors not only in solution but also in paper (Kim et al., 2001). Terra et al. (2017) presented a distinctive review based on reported electrospun nanofibers. This review is unique in its own way because of its comparative study on mat-type woven membranes that did not contaminate even the sample used and it allows easy cleaning and treatment of membrane after use. In addition, the porosity of the mesh, functionalization as well as morphological properties are also liable to be customized (Terra et al., 2017). Various review articles are present in the literature addressing the appreciable work by scientists in sensing of heavy metals (Kim et al., 2001; Storhoff et al., 1998; Lee et al., 2007; Li et al., 2010b; Liu and Lu, 2005; Chai et al., 2010; He et al., 2005; Kalluri et al., 2009; Xue et al., 2008; Yu and Tseng, 2008; Pitarke et al., 2006), nitrites (Gong et al., 2009; Farag et al., 2005; Gapper et al., 2004; Fang et al., 2002; Chen et al., 2008; Zhang et al., 2009), and others (Terra et al., 2017; Li et al., 2013c) using nanomaterials clubbed with various polymeric, biological, organic, and inorganic entities.

7.2.2.2 Surface-enhanced Raman scattering–based sensing

The term surface-enhanced Raman scattering "SERS" is combined together to describe the effect associated with the surface property of the matter. It may be defined as a spectroscopic technique in which Raman signals (produced due to inelastic collisions between light and target atom) produced by atoms lying in close vicinity of (or onto) the surface will be enhanced due to plasmon resonance. It is a nondestructive technique and provides molecular imprints of the analyte (Li et al., 2013c; Uskoković-Marković et al., 2017; Sur, 2010; Eric Le Ru, 2008). Most of the SERS sensors reported for inorganic ion detection were designed by clubbing the noble metal NPs (gold or silver) with polymers and/or chemical structures (Li et al., 2010a, 2011a, 2011c; Mulvihill et al., 2008; Han et al., 2010; Ding et al., 2013; Kim et al., 2012; Ianoul et al., 2002; Nie and Emory, 1997).

7.2.2.3 Fluorescence spectroscopy

Fluorescence spectrometry is an investigative system that identifies species concentration (prepared in water, ethanol, and hexane as solvent), which is dependent on that one fluorescent property. It is an inexpensive, reliable technique with quick response to estimate about solution-based analyte (known as fluorophore) concentration. The spectra obtained are dependent upon various parameters such as excitation/emission wavelength, concentration of analyte in experimental sample, path length, and self-absorption of analyte (Albrecht, 2008). The interactions of fluorophores were explained and/or supported by the mechanisms such as Förster resonance energy transfer (FRET), nanometal surface energy transfer (NSET), and chemiluminescent resonant energy transfer (CRET), which will be discussed next.

1. FRET observed with dipole–dipole contact takes place among an energy donor and acceptor pair of atoms. It follows one-sixth distance rule to occur. It is a frequently used phenomenon for illumination of heavy metal sensing (usually but not always) (Baslak, 2019; Zhou et al., 2005; Lu et al., 2003; Clapp et al., 2004; Ono and Togashi, 2004).
2. NSET happens because of dipole-induced interband transitions in thin film of metal instead of resonating electron overlap as that of FRET. It is observed iff (if and only if) the interaction distance is one-fourth of the diameter of the atom. It is similar to that of FRET but the only difference lies within use of fluorophore used, that is, in the case of FRET fluorophore used is an organic dye, while in the case of NSET, organic dye is replaced by noble metal (Au) NPs which are very persistent to show SPR (Li et al., 2011b; Persson and Lang, 1982; Yun et al., 2005; Jennings et al., 2006).
3. CRET produces illumination without using any external source (of excitation) but occurs due to chemical reaction involved in binding. The electron transfer mechanism is explained and understood by Dexter recombination, intramolecular, and/or interfacial photo-induced electron transfer. The electron transfer rate constant obtained usually varies exponentially with respect to its distance from nuclei (Freeman et al., 2011a,b; Li et al., 2013b).

In order to design robust sensor, various nanomaterials were constantly monitored, while using them in combination with oligonucleotide in order to make various SERS, colorimetric, fluorescence-based sensors. For designing optical sensors, different quenching materials such as organic dye, for example, Rhodamine B (Ono and Togashi, 2004; Wang et al., 2009), gold NPs (Au NPs) (Li et al., 2011b; Jennings et al., 2006; Ray et al., 2006; Pons et al., 2007), graphene (Li et al., 2013a; Swathi and Sebastian, 2008; Zhang et al., 2011; Freeman et al., 2011a; Li et al., 2013b,c), quantum dots (Li et al., 2011b, 2013a; Pons et al., 2007; Freeman et al., 2011a), luminescent molecules (Freeman et al., 2011a; Li et al., 2013b), and others were reported in literature in ample amount toward Pb(II), Hg(II), As(III), Ni(II), and others. There are various nicely written reviews that give a very clear picture about these sensing methodologies (Li et al., 2013b,c; Verma and Singh, 2005; Zhou et al., 2016).

7.2.3 Electrochemical detection

It is an impactful analytical technique that can sense electrical signals (voltage or current) produced due to redox reaction involved in the experiment. This detection methodology comprises three electrodes: (1) working electrode (WE) at which desired set of reactions occurs, (2) auxiliary electrode (AE) that balance WE potential and current of the cell, and (3) reference electrode (RE) acts as a reference source across which all measurements take place (Long et al., 2013; Figeys and Pinto, 2000).

Among various electrochemical techniques, most commonly used are amperometric, voltammetry, and electrochemical impedance spectroscopy (EIS).

7.2.3.1 Amperometric

Amperometric is a mode of electrochemical study in which constant potential is applied between WE and AE dipped into electrolyte containing electroactive species and then time varying current signals are monitored and recorded. As electron transmissions happen in very close by vicinity to the electrode, the current at the WE is linearly proportionate to analyte concentration deposited. First, analyte adjacent to double-layer is worn-out, causing high current. But with time, concentration gradient drops and current declines. As WE have a fixed potential, hence only one selected component of the analyte can be identified or studied. The analyte shows faradic response at set potential and applied polarity (Figeys and Pinto, 2000; Rackus et al., 2015).

7.2.3.2 Voltammetry

Voltammetry is the most far and widely followed procedure in electrochemical recognition of analyte in several mediums with high precision and sensitivity. Fig. 7.1 is representing the schematic diagram for working of an electrochemical system. In brief, analyte first deposited or adsorbed on the WE at specific applied potential, allowing for assembly of analyte as a time function. Afterward, at applied potential, species gets off resulting in

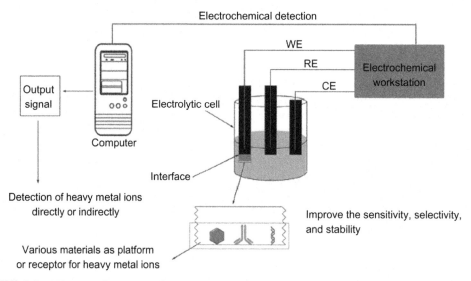

FIGURE 7.1 Schematic illustration of general principle of electrochemical sensing of heavy metal ions (Cui et al., 2015).

a peak current that correlates the analyte concentration. Unlike amperometric, voltammetry is a study of potential as a time varying quantity at fixed current application between the WE and AE. This set of techniques improves limit of detection by suppressing the background currents. Its added an advantage of checking reversibility of the reaction is helpful in qualitative analysis of the analyte (Yilong et al., 2015; Kissinger and Heineman, 1996; Pujol et al., 2014; Krolicka et al., 2003; Timur and Anik, 2007).

These techniques can be classified as AC voltammetry, cyclic voltammetry, DC polarography, differential pulse, linear sweep, square-wave, and stripping [anodic stripping voltammetric (ASV) and cathodic stripping voltammetry (CSV)] voltammetries, respectively.

Heavy metal determination carried out using ASV is a preferably used method for water quality determination. This technique has various advantages such as (1) multiple metal analyses present in the sample in one run cycle; (2) fast, accurate, and sensitive detection; and (3) compatibility with on-site monitoring system by connecting potentiostat with your mobile or laptop (https://www.palmsens.com/anodic-stripping-voltammetry/). During the electrode preparation for ASV analysis, in certain volume of sample solution, electrodes are dipped and stirred for certain amount of time (normally 60–300 seconds) at low applied potential; during this step, multiple metal ions (present in sample) get accumulated on the electrode surface. All accumulated metal particles have a fixed oxidizing potential, which when achieved potentiostat will measure the current related to the oxidized metal and hence a peak determining the concentration will be obtained (https://www.palmsens.com/anodic-stripping-voltammetry/). Various materials such as carbon derivatives (Zuo et al., 2019; Pandey et al., 2016), nanocomposites (Pandey et al., 2016), polymer/NP (Zuo et al., 2019), and various others were reported toward heavy metals (Zuo et al., 2019; Pandey et al., 2016), respectively.

7.2.3.3 *Electrochemical impedance spectroscopy*

EIS is an analytical technique in which study related to electrode–electrolyte interfacial properties are carried out. The device used is termed as EIS analyzer that measures not only difference between current (amplitude of the wave) but also potential and current time relation. These obtained values are helpful in calculating impedance of the electrolyte, and hence, resultant RLC circuits can be simulated (Jovanovic et al., 2013). In other words, EIS is nothing, but the AC analysis of circuit response in which a DC potential and small sinusoidal AC pulses are applied among WE and RE, resulting in magnitude and phase angle measured, and represent analysis data in the form of Bode plot and Nyquist plot. These obtained values will be used to draw equivalent RC circuits for the cell known as electrical equivalent cell (Christensen et al., 1992; Orazem and Tribollet, 2017). This technique is popular due to its inexpensive setup to provide sensitive analysis for distinctive available matrices as related to other analytical techniques.

7.2.4 Potentiometric detection

Potentiometry can be defined as a branch of electroanalytical tool in which potential change at WE is measured with respect to RE under the constant current flow condition. In other words, we can say that the concentration of inorganic ionic species can be identified with the change in potential at WE dipped into electrolyte containing the ions, potential varies with increase or decrease of ion concentration (Orazem and Tribollet, 2017). Potentiometric techniques can be subdivided as follows:

1. *Ion-selective electrode (ISE)* consists of polymeric membranes that were made ion selective by incorporating customized supramolecule. These ISEs result in better selectivity toward ions because they bind the analyte and reduce matrix interference. It is interestingly gaining attention because of it high accuracy, low cost, quick response, and low cost analysis (Aragay et al., 2011; Radu and Diamond, 2007).
2. *Field effect transistor (FET)* is an electronic device that controls the current flow from drain terminal to source terminal under the influence of applied electric field. Therefore sensing of organic and/or inorganic species is performed by adjusting gate material or by placing a selective membrane or by using (bio/chemical)-recognition component onto it. The resulting sensor is known as chemical field effect transistor (CHEMFET). When ion selective material is used, then it is known as ion selective field effect transistor (ISFET). When organic sensing platform is used as sensing material then it is known as OFET (Ono and Togashi, 2004). ISFET-based sensors have numerous possible applications in diversified fields of chemistry, sensor designing and development, microbiology, flexible and wearable electronic devices, and many more (Jimenez-Jorquera et al., 2010).

Kaisti (2017) had presented a review based on FET-based sensors. In Fig. 7.2, the author had summarized the entire sensing process through one diagram itself. This figure had shown various types of sensor binding sites (e.g., oligonucleotides and antigen-antibody interaction represents bio-FETs, and ions make ion selective FET or ISFET) As we have discussed about working of FET sensors, hence as of now we can make a note to understand a fundamental prospect, that is, working and precision (and/or selectivity) of the sensor is fully dependent upon the selection of gating material because gate terminal provides a bias voltage to control the current flow between drain and source terminal This effect is termed as gating effect (https://www.electronics-tutorials.ws/transistor/tran_1.html). In the context, various materials such as supramolecule (Cobben et al., 1992; Gupta et al., 2002), carbon derivatives [graphene, reduced graphene, electrochemically reduced graphene, and carbon nanotube (CNTs)] (Forzani et al., 2006), polymeric membranes functionalized with NPs, supramolecule or with both (Borraccino et al., 1992; Taillades et al., 1999; Forzani et al., 2006) were extensively used by the researchers to develop a wide range of FETs, that is, ISFETs, CHEMFETs, and OFETs, toward heavy metal ions, nitrides, and other with appreciable limit of detection and linearity of the device in acceptable domains. Direct incorporation of sensing material onto the gate terminal might be sometimes difficult during FET fabrication; hence, extended gate terminal devices are also grasping the attention because it had simplified the utilization of bulky supramolecular structure to enhance selectivity of the sensor without hindering the rest of the performance of the sensor (Table 7.3).

7.2.4.1 Biosensors

Biosensors may be defined as the synergistic combination of biotechnology and microelectronics, in which a transducer unit converts the biochemical signals coming from binding site of the device into measurable electrical signals (Verma and Singh, 2005).

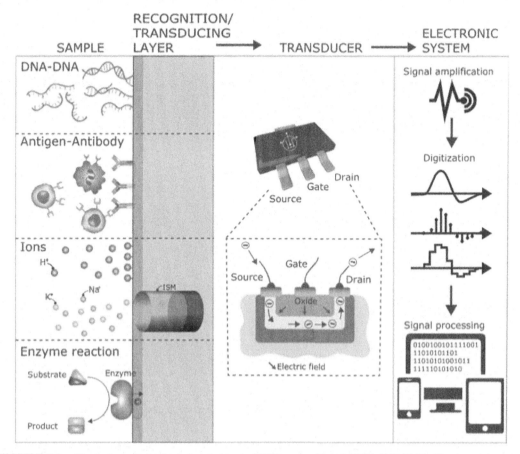

FIGURE 7.2 Illustration of a biological and chemical FET sensor (Kaisti, 2017). *FET*, Field effect transistor.

7.2.4.2 Description about biosensors

Biosensors are the specified class of sensors which uses bio-molecules at sensor surface which will provide electrical signals in form of varied current or voltages. These varied signals were processed and converted to readable form with the help of transducers. Finally, signals are displayed on the screen (Fig. 7.3).

7.2.4.3 Block diagram description

1. *Stage I:* It represents the exposed surface of the sensing unit, that is, the binding site of the sensor which was functionalized by the immobilized biomolecules (e.g., antibodies, enzymes, cell, protein, polymeric, and supramolecule) which, when interacts with the analyte of interest, results in the change of physiochemical properties and hence utilized by the second stage.
2. *Stage II:* Transducer unit may be defined as an electronic device that converts one form of signals (such as mechanical, electrical, light, chemical, thermal, acoustic, and

TABLE 7.3 Various types of field effect transistors (FETs) along with its sensing material and analyte.

Analyte	Type of FET	Sensing material(s)	LOD/Nernstian selectivity	Reference
Ag, Pb, Cd, Cu	CHEMFET	*For Ag,* calix[4]arene with two diametrically substituted thioether functionalities	60 mV/dec	(Cobben et al., 1992)
		For Cu, The calix[4]arene, with four dithiocarbamoyl groups	30 mV/dec	
		For Cd, The calix[4]arene, with four dimethylthiocarbamoylmethoxyethoxy substituents	30 mV/dec	
		For Pb(II), oxamide, and thioamide ionophores		
Pb(II), Cd (II)	ISFET	Membrane: PVC-PVA-PVAc with dipropyloctadecyl dioxadiamide *for Pb(II)* dipropyloctadecyl ditioamide *for Cd(II)*		(Borraccino et al., 1992)
Cu(II)	ISFET	Membrane: Cu–Ge–Sb–Se	LOD: 10^{-6} mol/L	(Taillades et al., 1999)
Pb(II)	ISFET	Membrane: 4-*tert*-butylcalix[4]arene in PVC	30.0 ± 1.0 mV/dec	(Gupta et al., 2002)
Ni(II) Cu (II)	CHEMFET	Peptide modified PVC incorporated with SWCNTs		(Forzani et al., 2006)
Hg(II)	CHEMFET	Octadecyltrichlorosilane monolayer with SWCNT	LOD: 10 nM	(Kim et al., 2009)
Hg(II), Cd (II)	SiNW-FETs	Si nanowires with MPTES	LOD: 10^{-7} M (Hg) 10^{-4} M (Cd)	(Luo et al., 2009)
Hg(II)	OFET	Membrane: PII2T-Si		(Knopfmacher et al., 2014)
Hg(II)	ISFET	rGO–PF nanohybrids	LOD: 10 pM	(Park et al., 2014)
Heavy metals	FET	Review		(Li et al., 2013c)
Hg(II)	ISFET	ERGO using *N*-[(1-pyrenyl-sulfonamido)-heptyl]-gluconamide	LOD: 0.1 nM	(Yu et al., 2013)
Hg(II)	ISFET	Graphene functionalized with aptamer	LOD: 10 pM	(An et al., 2013)
Hg(II)	OFET	Extended-gate gold electrode functionalized with L-cysteine	LOD: 31 ppb	(Minami et al., 2016)
Hg(II)	HEMT	Thioglycolic acid functionalized Au-gated AlGaN/GaN HEMTs	LOD: 1.5×10^{-8} M	(Minami et al., 2016)
Hg(II)	FET	rGO/TGA functionalized Au NPs	LOD: 2.5×10^{-8} M.	(Chen et al., 2012)
Hg(II)	CHEMFET	Thermally rGO sheets with thioglycolic acid functionalised Au NPs	LOD: 1 nM	(Zhou et al., 2014)

Analyte	Device	Material	Performance	Reference
Hg(II)	CNT-FET	Review		(Pokhrel et al., 2017)
Pb(II)	ISFET	Graphene film		(Wen et al., 2013)
Zn(II)	HEMT	Schiff base functionalized EG–AlGaN/GaN	LOD: 1 fM	(Gu et al., 2019)
As(III)	ISFET	CPPy-coated flower-like MoS_2 nanospheres	LOD: 1 pM	(An and Jang, 2017)
As(III)	FET	DTT-functionalized Au NPs	LOD: 1 nM	(Zhou et al., 2018)
Pb(II)	ISFET	Black phosphorous	LOD: 1 ppb	(Li et al., 2015)
As(III)	ISFET	MoS_2	LOD: 0.1 ppb	(Li et al., 2016)
Pb(II)	ISFET	Au NP–graphene and DNAzyme–graphene	LOD: 20 pM	(Wen et al., 2013)
Pb(II)	HEMT	Semiempirical model	LOD: 10^{-10} M	(Chen et al., 2018)
Hg(II)	ISFET	DNA–MoS_2 nanosheet/Au NPs	LOD: 0.1 nM	(Schulthess et al., 1984)
Pb(II)	ISFET	rGO/GSH	LOD: 10 nM	(Malinowska and Meyerhoff, 1995)
Hg(II)	OFET	Thiol group functionalized pyrene	LOD: 1 mM to 0.01 µM	(Prasad et al., 2004)
NO_2^- (nitrite)	ISFET	Aquocyanocobalt(III)-hepta(2-phenylethyl)cobyrinate	LOD: $10^{-4.6}$ M	(Stepánek et al., 1986)
NO_2^-	ISFET	Co(II)-corrins (**1** and **2**) *incorporated into* poly(vinyl chloride)/bis(1-butylpentyl) adipate liquid membranes	Nernst: 10^{-6} to 10^{-1} M	(Badr, 2006)
NO_2^-	ISFET	Co(III)TPPX (I) and (II), with PVC film	Nernst:$10^{-1}-10^{-5}$ M	(Ganjali et al., 2003)
NO_2^-	ISFET	Iron(III) and cobalt(III) complexes (OBTAP) in PVC matrix		(Gao et al., 1995)
NO_2^- SCN^-, F^-	ISFET	Co(II) (Co-Sal), Cr(III) (Cr-Sal), and Al(III) (Al-Sal) with PVC		(Badr et al., 1995)
NO_2^-	ISFET	Co(III)–Schiff-based complex with PVC membrane	LOD:5×10^{-7} M	(Stepánek et al., 1986)
NO_2^-	ISFET	Some metalloporphyrin derivatives	LOD:5.0×10^{-6} mol/L	(Verma and Singh, 2005)
NO_2^-	ISFET	Palladium organophosphine complex with PVC membrane	5.0 µM	(Han et al., 2010)

CNT, Carbon nanotube; *CPPy*, carboxylic polypyrrole; *DTT*, dithiothreitol; *EG*, extended gate; *ERGO*, electrochemically reduced graphene; *GSH*, L-glutathione; *HEMTs*, high electron mobility transistors; *LOD*, limit of detection; *MPTES*, mercaptopropyltriethoxysilane; *NP*, nanoparticle; *PF*, polyfuran; *rGO*, reduced graphene oxide; *SWCNT*, single wall carbon nanotube; *TGA*, thioglycolic acid.

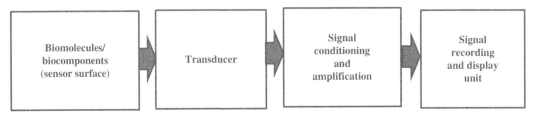

FIGURE 7.3   Block diagram representation of biosensors.

TABLE 7.4   Classification based on types of transducers.

Transducer	Benefits	Shortcomings	Use
ISE	Simple, dependable, easy handling	1. Larger response time 2. Prone to electronic noise	1. Amino acids 2. Carbohydrates and alcohols 3. Inorganic ions
Amperometric	1. Simple, wide variation redox reaction are supported for designing of biosensors 2. Device miniaturization is possible	1. Low sensitivity 2. Multiple membranes or enzyme are needed to achieve desirable selectivity	1. Biological species (hisamine, glucose, etc.) 2. Inorganic species (heavy metals etc.)
FET	1. Low cost, mass production, stable output 2. Small amount of sensing material needed 3. Multiple sample analysis and handling possible	1. Temperature sensitive 2. Difficulty in fabrication of gate material	1. Biological species 2. Organic and inorganic species 3. Environmental aspects
Optical	1. Remote sensing, economical 2. Supported manifold types: absorbance, reflectance, fluorescent, and EM spectrum	1. Obtained signals are interference prone with stray light from unwanted source 2. Sophisticated light/energy sources required 3. Miniaturization might hinder the signal magnitude range	1. Organic and inorganic pollutants 2. Biological species. 3. Chemical species
Thermal	1. Simple, robust designing 2. No optical intrusions	Discrimination issues	Biological samples (carbohydrates, sucrose etc.)
Piezoelectric	1. Simple, robust and economical device designing 2. Easy sample preparation and handling 3. Compatible with gas sensing array designing	1. Low sensitivity toward liquid samples 2. Nonspecific binding results in interference in signals	1. Biological samples 2. Organic and inorganic species 3. Bacterial contamination and toxins analysis

EM, electromagnetic spectrum; FET, Field effect transistor; *ISE,* ion-selective electrode.

electromagnetic) into electrical signals; and type of transducer usually symbolizes the type of biosensor, for example, if stage I is producing optical signals, then it can be named as optical biosensor.

3. *Stage III:* The obtained signals were processed by using multiple or dual channel ratio or subtraction followed by comparison with respect to calibrated signals and then these conditioned signals were amplified in order to make them recognizable electrical signals.

4. *Stage IV:* The processed signal will be recorded and displayed on the display unit such as computer monitor (Table 7.4).

Biosensors are nothing but the incorporation of all the abovementioned techniques (e.g., electrochemical, and optical) were entitled with bio-inspired species that will form a biosensor. In this context, carbon nanostructures such as reduced grapheme (Han et al., 2010; Ding et al., 2013; Kim et al., 2012; Ianoul et al., 2002; Li et al., 2010a, 2011a), CNT (single wall and multiwall) structures (Nie and Emory, 1997; Albrecht, 2008) were modified or anchored with biomolecules such aptamers (Baslak, 2019), peptides (Han et al., 2010), DNAzyme (Zhou et al., 2005; Lu et al., 2003), NP (Kalluri et al., 2009; Gapper et al., 2004), and with various polymers (Liu and Lu, 2005) to make sensor more sensitive and selective toward Hg (II) (Han et al., 2010; Ding et al., 2013; Nie and Emory, 1997; Albrecht, 2008; Baslak, 2019), Cd (II) (Kim et al., 2012), Cu (II) (Han et al., 2010), Ni (II) (Han et al., 2010), ad Pb (II) (Ianoul et al., 2002; Baslak, 2019).

7.3 Conclusion and future prospect

During the complete comprehensive study, various techniques starting from conventional techniques to latest trendy techniques such as FETs and biosensors were discussed. In the conclusion, we can get a picture that there is still a greater room for evolution in miniaturizations of device development. As on the landing truth, we all are aware that even today, we have a huge source of literature library available, but on grounds not more than 5% of the reported materials were shaped into a commercialized product. Hence, it's our responsibility to nurture our ability and innovation toward practical and commercially viable device development.

Acknowledgment

The authors appreciatively acknowledge the Ministry of Human Resource Development (MHRD), New Delhi, for providing financial aid to conduct the study.

References

Albrecht, C., 2008. Joseph R. Lakowicz: principles of fluorescence spectroscopy. Anal. Bioanal. Chem. 390 (5), 1223–1224.

An, J.H., Jang, J., 2017. A highly sensitive FET-type aptasensor using flower-like MoS_2 nanospheres for real-time detection of arsenic(iii). Nanoscale 9 (22), 7483–7492.

An, J.H., Park, S.J., Kwon, O.S., Bae, J., Jang, J., 2013. High-performance flexible graphene aptasensor for mercury detection in mussels. ACS Nano 7 (12), 10563–10571.

Aragay, G., Pons, J., Merkoçi, A., 2011. Recent trends in macro-, micro-, and nanomaterial-based tools and strategies for heavy-metal detection. Chem. Rev. 111 (5), 3433–3458.

Badr, I.H., 2006. Potentiometric anion selectivity of polymer-membrane electrodes based on cobalt, chromium, and aluminum salens. Anal. Chim. Acta 570 (2), 176–185.

Badr, I.H., Meyerhoff, M.E., Hassan, S.S., 1995. Potentiometric anion selectivity of polymer membranes doped with palladium organophosphine complex. Anal. Chem. 67 (15), 2613–2618.

Barakat, M.A., 2011. New trends in removing heavy metals from industrial wastewater. Arab. J. Chem. 4 (4), 361–377.

Baslak, C., 2019. Development of fluorescence-based optical sensors for detection of Cr (III) ions in water by using quantum nanocrystals. Res. Chem. Intermed. 45 (7), 3633–3640.

Borraccino, A., Campanella, L., Sammartino, M.P., Tomassetti, M., Battilotti, M., 1992. Suitable ion-selective sensors for lead and cadmium analysis. Sens. Actuators, B: Chem. 7 (1–3), 535–539.

Chai, F., Wang, C., Wang, T., Li, L., Su, Z., 2010. Colorimetric detection of Pb2 + using glutathione functionalized gold nanoparticles. ACS Appl. Mater. Interfaces 2 (5), 1466–1470.

Chen, G., Yuan, D., Huang, Y., Zhang, M., Bergman, M., 2008. In-field determination of nanomolar nitrite in seawater using a sequential injection technique combined with solid phase enrichment and colorimetric detection. Anal. Chim. Acta 620 (1–2), 82–88.

Chen, K., Lu, G., Chang, J., Mao, S., Yu, K., Cui, S., et al., 2012. Hg(II) ion detection using thermally reduced graphene oxide decorated with functionalized gold nanoparticles. Anal. Chem. 84 (9), 4057–4062.

Chen, Y.T., Sarangadharan, I., Sukesan, R., Hseih, C.Y., Lee, G.Y., Chyi, J.I., et al., 2018. High-field modulated ion-selective field-effect-transistor (FET) sensors with sensitivity higher than the ideal Nernst sensitivity. Sci. Rep. 8 (1), 8300.

Christensen, B.J., Mason, T.O., Jennings, H.M., 1992. Influence of silica fume on the early hydration of Portland cements using impedance spectroscopy. J. Am. Ceram. Soc. 75 (4), 939–945.

Clapp, A.R., Medintz, I.L., Mauro, J.M., Fisher, B.R., Bawendi, M.G., Mattoussi, H., 2004. Fluorescence resonance energy transfer between quantum dot donors and dye-labeled protein acceptors. J. Am. Chem. Soc. 126, 301–310.

Cobben, P.L., Egberink, R.J., Bomer, J.G., Bergveld, P., Verboom, W., Reinhoudt, D.N., 1992. Transduction of selective recognition of heavy metal ions by chemically modified field effect transistors (CHEMFETs). J. Am. Chem. Soc. 114 (26), 10573–10582.

Cui, L., Wu, J., Ju, H., 2015. Electrochemical sensing of heavy metal ions with inorganic, organic and bio-materials. Biosens. Bioelectron. 63, 276–286.

Ding, X., Kong, L., Wang, J., Fang, F., Li, D., Liu, J., 2013. Highly sensitive SERS detection of Hg2 + ions in aqueous media using gold nanoparticles/graphene heterojunctions. ACS Appl. Mater. Interfaces 5 (15), 7072–7078.

Le Ru, E. and Etchegoin, P., 2008. Principles of Surface-Enhanced Raman Spectroscopy: and related plasmonic effects. Elsevier.

Eswar, P., Nagesh, L., Devaraj, C.G., 2011. Intelligence quotients of 12-14 year old school children in a high and a low fluoride village in India. Fluoride 44 (3), 168.

Fang, Y., Chen, H., Gao, Z., Jing, X., 2002. Flow injection determination of nitrite in food samples by dialysis membrane separation and photometric detection. Int. J. Environ. Anal. Chem. 82 (1), 1–6.

Farag, A.B., Moawed, E.A., El-Shahat, M.F., 2005. Sensitive detection, selective determination, and removal of nitrite from water using the reactive function group of polyurethane foam. Anal. Lett. 38 (5), 841–856.

Fawell, J., Bailey, K., Chilton, J., Dahi, E., Magara, Y., 2006. Fluoride in Drinking-Water. IWA Publishing.

Figeys, D., Pinto, D., 2000. Lab-on-a-Chip: A Revolution in Biological and Medical Sciences.

Forzani, E.S., Li, X., Zhang, P., Tao, N., Zhang, R., Amlani, I., et al., 2006. Tuning the chemical selectivity of SWNT-FETs for detection of heavy-metal ions. Small 2 (11), 1283–1291.

Freeman, R., Liu, X., Willner, I., 2011a. Chemiluminescent and chemiluminescence resonance energy transfer (CRET) detection of DNA, metal ions, and aptamer–substrate complexes using hemin/G-quadruplexes and CdSe/ZnS quantum dots. J. Am. Chem. Soc. 133 (30), 11597–11604.

Freeman, R., Willner, B., Willner, I., 2011b. Integrated biomolecule–quantum dot hybrid systems for bioanalytical applications. J. Phys. Chem. Lett. 2 (20), 2667–2677.

Ganjali, M.R., Rezapour, M., Pourjavid, M.R., Salavati-Niasari, M., 2003. Highly selective PVC-membrane electrodes based on Co (II)-Salen for determination of nitrite ion. Anal. Sci. 19 (8), 1127–1131.

Gao, D., Gu, J., Yu, R.Q., Zheng, G.D., 1995. Nitrite-sensitive liquid membrane electrodes based on metalloporphyrin derivatives. Analyst 120 (2), 499–502.

Gapper, L.W., Fong, B.Y., Otter, D.E., Indyk, H.E., Woollard, D.C., 2004. Determination of nitrite and nitrate in dairy products by ion exchange LC with spectrophotometric detection. Int. Dairy J. 14 (10), 881–887.

Gong, W., Mowlem, M., Kraft, M., Morgan, H., 2009. A simple, low-cost double beam spectrophotometer for colorimetric detection of nitrite in seawater. IEEE Sens. J. 9 (7), 862–869.

Gopal, R., Ghosh, P.K., 1985. Fluoride in drinking water-its effects and removal. Def. Sci. J. 35 (1), 71–88.

Gu, L., Yang, S., Miao, B., Gu, Z., Wang, J., Sun, W., et al., 2019. Electrical detection of trace zinc ions with an extended gate-AlGaN/GaN high electron mobility sensor. Analyst 144 (2), 663–668.

Gumpu, M.B., Sethuraman, S., Krishnan, U.M., Rayappan, J.B.B., 2015. A review on detection of heavy metal ions in water—an electrochemical approach. Sens. Actuators, B: Chem. 213, 515–533.

Gupta, A., Maranas, C.D., McDonald, C.M., 2000. Mid-term supply chain planning under demand uncertainty: customer demand satisfaction and inventory management. Comput. Chem. Eng. 24 (12), 2613–2621.

Gupta, V.K., Mangla, R., Agarwal, S., 2002. Pb (II) selective potentiometric sensor based on 4-*tert*-butylcalix[4] arene in PVC matrix. Electroanalysis 14 (15–16), 1127–1132.

Han, D., Lim, S.Y., Kim, B.J., Piao, L., Chung, T.D., 2010. Mercury (II) detection by SERS based on a single gold microshell. Chem. Commun. 46 (30), 5587–5589.

He, X., Liu, H.U.I.B.I.A.O., Li, Y., Wang, S., Li, Y., Wang, N., et al., 2005. Gold nanoparticle-based fluorometric and colorimetric sensing of copper (II) ions. Adv. Mater. 17 (23), 2811–2815.

Henry, M., 2005. The state of water in living systems: from the liquid to the jellyfish. Cell. Mol. Biol. (Noisy-le-Grand, Fr.) 51 (7), 677–702.

Ianoul, A., Coleman, T., Asher, S.A., 2002. UV resonance Raman spectroscopic detection of nitrate and nitrite in wastewater treatment processes. Anal. Chem. 74 (6), 1458–1461.

Jennings, T.L., Singh, M.P., Strouse, G.F., 2006. Fluorescent lifetime quenching near d = 1.5 nm gold nanoparticles: probing NSET validity. J. Am. Chem. Soc. 128 (16), 5462–5467.

Jimenez-Jorquera, C., Orozco, J., Baldi, A., 2010. ISFET based microsensors for environmental monitoring. Sensors 10 (1), 61–83.

Jovanovic, Z., Buica, G.O., Miskovic-Stankovic, V., Ungureanu, E.M., Amarandei, C.A., 2013. Electrochemical impedance spectroscopy investigations on glassy carbon electrodes modified with poly (4-azulen-1-yl-2, 6-bis (2-thienyl) pyridine). Univ. Politehnica Buchar. Sci. Bull. Ser. B: Chem. Mater. Sci. 75 (1), 125–134.

Kaisti, M., 2017. Detection principles of biological and chemical FET sensors. Biosens. Bioelectron. 98, 437–448.

Kalluri, J.R., Arbneshi, T., Afrin Khan, S., Neely, A., Candice, P., Varisli, B., et al., 2009. Use of gold nanoparticles in a simple colorimetric and ultrasensitive dynamic light scattering assay: selective detection of arsenic in groundwater. Angew. Chem. Int. Ed. 48 (51), 9668–9671.

Kim, Y., Johnson, R.C., Hupp, J.T., 2001. Gold nanoparticle-based sensing of "spectroscopically silent" heavy metal ions. Nano Lett. 1 (4), 165–167.

Kim, T.H., Lee, J., Hong, S., 2009. Highly selective environmental nanosensors based on anomalous response of carbon nanotube conductance to mercury ions. J. Phys. Chem. C 113 (45), 19393–19396.

Kim, K., Kim, K.L., Shin, K.S., 2012. Selective detection of aqueous nitrite ions by surface-enhanced Raman scattering of 4-aminobenzenethiol on Au. Analyst 137 (16), 3836–3840.

Kissinger, P., Heineman, W.R., 1996. Laboratory Techniques in Electroanalytical Chemistry, Revised and Expanded. CRC Press.

Knopfmacher, O., Hammock, M.L., Appleton, A.L., Schwartz, G., Mei, J., Lei, T., et al., 2014. Highly stable organic polymer field-effect transistor sensor for selective detection in the marine environment. Nat. Commun. 5, 2954.

Koirtyohann, S.R., 1991. A history of atomic absorption spectroscopy from an academic perspective. Anal. Chem. 63 (21), 1024A–1031A.

Krolicka, A., Bobrowski, A., Kalcher, K., Mocak, J., Svancara, I., Vytras, K., 2003. Study on catalytic adsorptive stripping voltammetry of trace cobalt at bismuth film electrodes. Electroanalysis 15 (23–24), 1859–1863.

Lagarde, F., Jaffrezic-Renault, N., 2011. Cell-based electrochemical biosensors for water quality assessment. Anal. Bioanal. Chem. 400 (4), 947.

Lee, J.S., Han, M.S., Mirkin, C.A., 2007. Colorimetric detection of mercuric ion (Hg2 +) in aqueous media using DNA-functionalized gold nanoparticles. Angew. Chem. Int. Ed. 46 (22), 4093–4096.

Li, J.F., Huang, Y.F., Ding, Y., Yang, Z.L., Li, S.B., Zhou, X.S., et al., 2010a. Shell-isolated nanoparticle-enhanced Raman spectroscopy. Nature 464 (7287), 392.

Li, T., Wang, E., Dong, S., 2010b. Lead (II)-induced allosteric G-quadruplex DNAzyme as a colorimetric and chemiluminescence sensor for highly sensitive and selective Pb2 + detection. Anal. Chem. 82 (4), 1515–1520.

Li, J.F., Ding, S.Y., Yang, Z.L., Bai, M.L., Anema, J.R., Wang, X., et al., 2011a. Extraordinary enhancement of Raman scattering from pyridine on single crystal Au and Pt electrodes by shell-isolated Au nanoparticles. J. Am. Chem. Soc. 133 (40), 15922–15925.

Li, M., Cushing, S.K., Wang, Q., Shi, X., Hornak, L.A., Hong, Z., et al., 2011b. Size-dependent energy transfer between CdSe/ZnS quantum dots and gold nanoparticles. J. Phys. Chem. Lett. 2 (17), 2125–2129.

Li, J., Chen, L., Lou, T., Wang, Y., 2011c. Highly sensitive SERS detection of As3 + ions in aqueous media using glutathione functionalized silver nanoparticles. ACS Appl. Mater. Interfaces 3 (10), 3936–3941.

Li, M., Zhou, X., Guo, S., Wu, N., 2013a. Detection of lead (II) with a "turn-on" fluorescent biosensor based on energy transfer from CdSe/ZnS quantum dots to graphene oxide. Biosens. Bioelectron. 43, 69–74.

Li, M., Zhou, X., Ding, W., Guo, S., Wu, N., 2013b. Fluorescent aptamer-functionalized graphene oxide biosensor for label-free detection of mercury (II). Biosens. Bioelectron. 41, 889–893.

Li, M., Gou, H., Al-Ogaidi, I., Wu, N., 2013c. Nanostructured Sensors for Detection of Heavy Metals: A Review.

Li, P., Zhang, D., Liu, J., Chang, H., Sun, Y.E., Yin, N., 2015. Air-stable black phosphorus devices for ion sensing. ACS Appl. Mater. Interfaces 7 (44), 24396–24402.

Li, P., Zhang, D., Sun, Y.E., Chang, H., Liu, J., Yin, N., 2016. Towards intrinsic MoS2 devices for high performance arsenite sensing. Appl. Phys. Lett. 109 (6), 063110.

Liu, J., Lu, Y., 2005. Stimuli-responsive disassembly of nanoparticle aggregates for light-up colorimetric sensing. J. Am. Chem. Soc. 127 (36), 12677–12683.

Long, F., Zhu, A., Shi, H., 2013. Recent advances in optical biosensors for environmental monitoring and early warning. Sensors 13 (10), 13928–13948.

Lu, Y., Liu, J., Li, J., Bruesehoff, P.J., Pavot, C.M.B., Brown, A.K., 2003. New highly sensitive and selective catalytic DNA biosensors for metal ions. Biosens. Bioelectron. 18 (5–6), 529–540.

Luo, L., Jie, J., Zhang, W., He, Z., Wang, J., Yuan, G., et al., 2009. Silicon nanowire sensors for Hg 2 + and Cd 2 + ions. Appl. Phys. Lett. 94 (19), 193101.

Malinowska, E., Meyerhoff, M.E., 1995. Role of axial ligation on potentiometric response of Co (III) tetraphenylporphyrin-doped polymeric membranes to nitrite ions. Anal. Chim. Acta 300 (1–3), 33–43.

McClellan, K., Halden, R.U., 2010. Pharmaceuticals and personal care products in archived US biosolids from the 2001 EPA national sewage sludge survey. Water Res. 44 (2), 658–668.

Minami, T., Minamiki, T., Tokito, S., 2016. Detection of mercury (II) ion in water using an organic field-effect transistor with a cysteine-immobilized gold electrode. Jpn. J. Appl. Phys. 55 (4S), 04EL02.

Mitchell, H.H., Hamilton, T.S., Steggerda, F.R., Bean, H.W., 1945. The chemical composition of the adult human body and its bearing on the biochemistry of growth. J. Biol. Chem. 158 (3), 625–637.

Mulvihill, M., Tao, A., Benjauthrit, K., Arnold, J., Yang, P., 2008. Surface-enhanced Raman spectroscopy for trace arsenic detection in contaminated water. Angew. Chem. Int. Ed. 47 (34), 6456–6460.

Nie, S., Emory, S.R., 1997. Probing single molecules and single nanoparticles by surface-enhanced Raman scattering. Science 275 (5303), 1102–1106.

Nnaji, J.C., Uzairu, A., Harrison, G.F.S., Balarabe, M.L., 2007. Evaluation of cadmium, chromium, copper, lead and zinc concentrations in the fish head/viscera of Oreochromis niloticus and Synodontis schall of River Galma, Zaria, Nigeria. Ejeafche 6 (10), 2420–2426.

Noronha, V.T., Aquino, Y.M., Maia, M.T. and Freire, R.M., 2019. Sensing of water contaminants: from traditional to modern strategies based on nanotechnology. Nanomaterials Applications for Environmental Matrices (pp. 109–150). Elsevier.

Ono, A., Togashi, H., 2004. Highly selective oligonucleotide-based sensor for mercury (II) in aqueous solutions. Angew. Chem. Int. Ed. 43 (33), 4300–4302.

Orazem, M.E., Tribollet, B., 2017. Electrochemical Impedance Spectroscopy. John Wiley & Sons.

Pandey, S.K., Singh, P., Singh, J., Sachan, S., Srivastava, S., Singh, S.K., 2016. Nanocarbon-based electrochemical detection of heavy metals. Electroanalysis 28 (10), 2472–2488.

Park, J.W., Park, S.J., Kwon, O.S., Lee, C., Jang, J., 2014. High-performance Hg2+ FET-type sensors based on reduced graphene oxide–polyfuran nanohybrids. Analyst 139 (16), 3852–3855.

Persson, B.N.J., Lang, N.D., 1982. Electron-hole-pair quenching of excited states near a metal. Phys. Rev. B 26 (10), 5409.

Piriya, V.S.A., Joseph, P., Daniel, SCG, Kiruba, Lakshmanan, S., Kinoshita, T., Muthusamy, S., 2017. Colorimetric sensors for rapid detection of various analytes. Mater. Sci. Eng. C 78, 1231–1245.

Pitarke, J.M., Silkin, V.M., Chulkov, E.V., Echenique, P.M., 2006. Theory of surface plasmons and surface-plasmon polaritons. Rep. Prog. Phys. 70 (1), 1.

Pitt, J.J., 2009. Principles and applications of liquid chromatography-mass spectrometry in clinical biochemistry. Clin. Biochemist Rev. 30 (1), 19.

Pokhrel, L.R., Ettore, N., Jacobs, Z.L., Zarr, A., Weir, M.H., Scheuerman, P.R., et al., 2017. Novel carbon nanotube (CNT)-based ultrasensitive sensors for trace mercury (II) detection in water: a review. Sci. Total Environ. 574, 1379–1388.

Pons, T., Medintz, I.L., Sapsford, K.E., Higashiya, S., Grimes, A.F., English, D.S., et al., 2007. On the quenching of semiconductor quantum dot photoluminescence by proximal gold nanoparticles. Nano Lett. 7 (10), 3157–3164.

Prasad, R., Gupta, V.K., Kumar, A., 2004. Metallo-tetraazaporphyrin based anion sensors: regulation of sensor characteristics through central metal ion coordination. Anal. Chim. Acta 508 (1), 61–70.

Pujol, L., Evrard, D., Groenen-Serrano, K., Freyssinier, M., Ruffien-Cizsak, A., Gros, P., 2014. Electrochemical sensors and devices for heavy metals assay in water: the French groups' contribution. Front. Chem. 2, 19.

Rackus, D.G., Shamsi, M.H., Wheeler, A.R., 2015. Electrochemistry, biosensors and microfluidics: a convergence of fields. Chem. Soc. Rev. 44 (15), 5320–5340.

Radu, A., Diamond, D., 2007. Ion-selective electrodes in trace level analysis of heavy metals: potentiometry for the XXI century. Compr. Anal. Chem. 49, 25–52.

Ram, R., 1990. Educational expansion and schooling inequality: International evidence and some implications. Rev. Econ. Stat. 72 (2), 266–274.

Ray, P.C., Fortner, A., Darbha, G.K., 2006. Gold nanoparticle based FRET assay for the detection of DNA cleavage. J. Phys. Chem. B 110 (42), 20745–20748.

Savage, W.G., 1902. The significance of *Bacillus coli* in drinking water. Epidemiol. Infect. 2 (3), 320–357.

Schulthess, P., Ammann, D., Simon, W., Caderas, C., Stepánek, R., Kräutler, B., 1984. A lipophilic derivative of vitamin B12 as selective carrier for anions. Helv. Chim. Acta 67 (4), 1026–1032.

Sharma, S., Sachdeva, P., Virdi, J.S., 2003. Emerging water-borne pathogens. Appl. Microbiol. Biotechnol. 61 (5–6), 424–428.

Stepánek, R., Kräutler, B., Schulthess, P., Lindemann, B., Ammann, D., Simon, W., 1986. Aquocyanocobalt (III)-hepta (2-phenylethyl)-cobyrinate as a cationic carrier for nitrite-selective liquid-membrane electrodes. Anal. Chim. Acta 182, 83–90.

Stolarczyk, J.K., Deak, A., Brougham, D.F., 2016. Nanoparticle clusters: assembly and control over internal order, current capabilities, and future potential. Adv. Mater. 28 (27), 5400–5424.

Storhoff, J.J., Elghanian, R., Mucic, R.C., Mirkin, C.A., Letsinger, R.L., 1998. One-pot colorimetric differentiation of polynucleotides with single base imperfections using gold nanoparticle probes. J. Am. Chem. Soc. 120 (9), 1959–1964.

Sur, U.K., 2010. Surface-enhanced Raman spectroscopy. Resonance 15 (2), 154–164.

Susheela, A.K., 1999. Fluorosis management programme in India. Curr. Sci. 77 (10), 1250–1256.

Swathi, R.S., Sebastian, K.L., 2008. Resonance energy transfer from a dye molecule to graphene. J. Chem. Phys. 129 (5), 054703.

Taillades, G., Valls, O., Bratov, A., Dominguez, C., Pradel, A., Ribes, M., 1999. ISE and ISFET microsensors based on a sensitive chalcogenide glass for copper ion detection in solution. Sens. Actuators, B: Chem. 59 (2–3), 123–127.

Terra, I., Mercante, L., Andre, R., Correa, D., 2017. Fluorescent and colorimetric electrospun nanofibers for heavy-metal sensing. Biosensors 7 (4), 61.

Timur, S., Anik, Ü., 2007. α-Glucosidase based bismuth film electrode for inhibitor detection. Anal. Chim. Acta 598 (1), 143–146.

Tole, M.P., Shitsama, J.M., 2003. Concentrations of Heavy Metals in Water, Fish and Sediments of the Winam Gulf, Lake Victoria, Kenya. Lake Victoria Fisheries: Status, Biodiversity and Management. Aquatic Ecosystem Health and Management Society, pp. 1–9.

Tsednee, M., Huang, Y.C., Chen, Y.R., Yeh, K.C., 2016. Identification of metal species by ESI-MS/MS through release of free metals from the corresponding metal-ligand complexes. Sci. Rep. 6, 26785.

Tyler, G., Jobin Yvon, S., 1995. ICP-OES, ICP-MS and AAS techniques compared. In: ICP Optical Emission Spectroscopy Technical Note, 5.

Uskoković-Marković, S., Kuntić, V., Bajuk-Bogdanović, D., Holclajtner-Antunović, I., 2017. Surface-Enhanced Raman Scattering (SERS) Biochemical Applications.

Verma, N., Singh, M., 2005. Biosensors for heavy metals. Biometals 18 (2), 121–129.

Wang, G., Lim, C., Chen, L., Chon, H., Choo, J., Hong, J., et al., 2009. Surface-enhanced Raman scattering in nanoliter droplets: towards high-sensitivity detection of mercury (II) ions. Anal. Bioanal. Chem. 394 (7), 1827–1832.

Water Standards and World Health Organization, 2004. Guidelines for Drinking-Water Quality. Vol. 1, Recommendations.

Wen, Y., Li, F.Y., Dong, X., Zhang, J., Xiong, Q., Chen, P., 2013. The electrical detection of lead ions using gold-nanoparticle- and DNAzyme-functionalized graphene device. Adv. Healthc. Mater. 2 (2), 271–274.

Widianarko, B., Van Gestel, C.A.M., Verweij, R.A., Van Straalen, N.M., 2000. Associations between trace metals in sediment, water, and guppy, *Poecilia reticulata* (Peters), from urban streams of Semarang, Indonesia. Ecotoxicol. Environ. Saf. 46 (1), 101–107.

World Health Organization, 2003a. *Nitrate and Nitrite in Drinking-Water: Background Document for Development of WHO Guidelines for Drinking-water Quality* (No. WHO/SDE/WSH/04.03/56). World Health Organization.

World Health Organization, 2003b. *Atrazine in Drinking-Water: Background Document for Development of WHO Guidelines for Drinking-Water Quality* (No. WHO/SDE/WSH/03.04/32). World Health Organization.

Xue, X., Wang, F., Liu, X., 2008. One-step, room temperature, colorimetric detection of mercury (Hg2 +) using DNA/nanoparticle conjugates. J. Am. Chem. Soc. 130 (11), 3244–3245.

Yilong, Z., Dean, Z., Daoliang, L., 2015. Electrochemical and other methods for detection and determination of dissolved nitrite: a review. Int. J. Electrochem. Sci. 10, 1144–1168.

Yu, C.J., Tseng, W.L., 2008. Colorimetric detection of mercury (II) in a high-salinity solution using gold nanoparticles capped with 3-mercaptopropionate acid and adenosine monophosphate. Langmuir 24 (21), 12717–12722.

Yu, C., Guo, Y., Liu, H., Yan, N., Xu, Z., Yu, G., et al., 2013. Ultrasensitive and selective sensing of heavy metal ions with modified graphene. Chem. Commun. 49 (58), 6492–6494.

Yun, C.S., Javier, A., Jennings, T., Fisher, M., Hira, S., Peterson, S., et al., 2005. Nanometal surface energy transfer in optical rulers, breaking the FRET barrier. J. Am. Chem. Soc. 127 (9), 3115–3119.

Zhang, M., Yuan, D., Chen, G., Li, Q., Zhang, Z., Liang, Y., 2009. Simultaneous determination of nitrite and nitrate at nanomolar level in seawater using on-line solid phase extraction hyphenated with liquid waveguide capillary cell for spectrophotometric detection. Microchim. Acta 165 (3–4), 427–435.

Zhang, M., Yin, B.C., Tan, W., Ye, B.C., 2011. A versatile graphene-based fluorescence "on/off" switch for multiplex detection of various targets. Biosens. Bioelectron. 26 (7), 3260–3265.

Zhou, D., Piper, J.D., Abell, C., Klenerman, D., Kang, D.J., Ying, L., 2005. Fluorescence resonance energy transfer between a quantum dot donor and a dye acceptor attached to DNA. Chem. Commun. 38, 4807–4809.

Zhou, G., Chang, J., Cui, S., Pu, H., Wen, Z., Chen, J., 2014. Real-time, selective detection of Pb2 + in water using a reduced graphene oxide/gold nanoparticle field-effect transistor device. ACS Appl. Mater. Interfaces 6 (21), 19235–19241.

Zhou, Y., Tang, L., Zeng, G., Zhang, C., Zhang, Y., Xie, X., 2016. Current progress in biosensors for heavy metal ions based on DNAzymes/DNA molecules functionalized nanostructures: a review. Sens. Actuators, B: Chem. 223, 280–294.

Zhou, G., Pu, H., Chang, J., Sui, X., Mao, S., Chen, J., 2018. Real-time electronic sensor based on black phosphorus/Au NPs/DTT hybrid structure: application in arsenic detection. Sens. Actuators, B: Chem. 257, 214–219.

Zuo, Y., Xu, J., Zhu, X., Duan, X., Lu, L., Yu, Y., 2019. Graphene-derived nanomaterials as recognition elements for electrochemical determination of heavy metal ions: a review. Microchim. Acta 186 (3), 171.

Further reading

Anık, Ü., Timur, S., Cubukcu, M., Merkoci, A., 2008. The usage of a bismuth film electrode as transducer in glucose biosensing. Microchim. Acta 160 (1–2), 269–273.

Devi, P., Sharma, C., Kumar, P., Kumar, M., Bansod, B.K., Nayak, M.K., et al., 2017. Selective electrochemical sensing for arsenite using rGO/Fe$_3$O$_4$ nanocomposites. J. Hazard. Mater. 322, 85–94.

Organic linkers for colorimetric detection of inorganic water pollutants

Nancy Sidana[1,2], Harminder Kaur[2] and Pooja Devi[1]

[1]CSIR-Central Scientific Instruments Organisation, Chandigarh, India [2]Punjab Engineering College (Deemed to be University), Chandigarh, India

8.1 Introduction

The escalated contamination of water resources by various inorganic pollutants has emerged out as a global nuisance, posing thereby a critical impact onto human life/environmental risk (Walcarius and Mercier, 2010; Serife et al., 2012; Periyasamy et al., 2013; Zhang et al., 2017; Gupta et al., 2006). These pollutants have paved their ways into water bodies due to increased industrialization, urbanization, and population boom. Therefore there is a critical need for efficient detection and removal of these pollutants. To overcome these pollutants, several materials based upon activated carbon adsorption method technique are employed for the effective trapping of traditional pollutants such as heavy-metal ions (Kaur et al., 2018a). However, it is important to measure the performance of remedial methods and materials for the field adaptation. Several techniques based upon conventional methods such as inductive coupled plasma mass spectroscopy (Lin and Huang, 2001; Pourreza and Hoveizavi, 2005; Lin et al., 2005; Becker et al., 2005), atomic adsorption spectroscopy (Shi et al., 2013; Shtoyko et al., 2004; Ensafi et al., 2006), plasmon resonance Rayleigh scattering spectroscopy, and electrochemical techniques are also employed for their detection (Sharma et al., 2016; Cui et al., 2018; Patil et al., 2014). However, these methods are more complicated owing to requirement of pretreatment procedures, costly equipment, and time consumption, which makes them difficult to handle for onsite analysis (Li et al., 2018; Wan et al., 2019). On the other hand, colorimetric techniques offer advantages of naked eye quantification of water pollutants for the qualitative as well as quantitative analysis of pollutants.

Schiff base colorimetric chemosensors (Kaur and Kumar, 2012) have been widely used to provide naked eye detection of toxic heavy-metal ions (Kaur et al., 2018b). This is due to their advantages such as high selectivity, cost-effectiveness, fast and simple operation, and user-friendliness (Gupta et al., 2014a,b, 2015). In addition, many sensors have been developed for multiion sensing, which have added more advantages toward analysis strategies.

The selection mode of Schiff base is indispensable on the basis of the functional group available in the ligand that can act as a binding site for metal ion, whereas in dye-based ligands such as rhodamine derivatives, metal ion−triggered spirolactam ring opening of rhodamine takes place (Xiang et al., 2006; Yu et al., 2009). Rhodamine derivatives are colorless and on binding with metal ring opening of spirolactam ring leads to strong emission and appearance of a pink color.

With the increasing demand of the techniques for the detection of heavy-metal ions, significant research has been done to meet the development of a field kit. This chapter gives an overview on recent advancements in synthesis of organic linkers based on Schiff base and dye derivatives, and their implication on paper strips for onsite analysis. This chapter also discusses the significant limitations of the techniques and provides suggestions for future development in the water quality monitoring. Therefore this chapter also summarizes the recent developments in the organic linker−based sensors for sensing inorganic pollutants.

8.2 Inorganic pollutants

Major inorganic pollutants affecting water and soil are released into the environment due to rapid industrialization and population boom. These inorganic pollutants have a greater environmental risk and health issues, which vary on interaction at both extracellular and intracellular levels. Metals along with their biological importance also have harmful effects associated with their degree of consumption. Alkali and alkaline earth metal salts degrade both physical and chemical environment of soil leading to difficulty in water and nutrients uptake. Toxic heavy metals compared to alkali metal interact strongly with soil constituent depending on the element and their physiochemical characteristics. Despite lower mobility of these elements in soils, they could disrupt biochemical phenomenon in organism when consumed at a lower concentration affecting both physiological and biological activities (Sharma et al., 2016). Major inorganic pollutants includes alkali metals and heavy-metal ions discussed in this chapter with special emphasis on their colorimetric detection (Walcarius and Mercier, 2010; Scozzafava, 1989; Dzombak and Morel, 1987) (Fig. 8.1).

Inorganic pollutants comprise mainly of heavy metals, which are toxic or poisonous even at low concentrations. Example of heavy metals includes arsenic, mercury, lead, and chromium. They can enter body system through water, food, and air and cause health issues. For instance, arsenic metal (Chauhan et al., 2017) is a pervasive element in Earth's crust, the presence of which higher than permissible limit in ground water as well as surface water is becoming hazardous to human health (Dey et al., 2017). Maximum permissible limit of arsenic in water is 10 ppb as recommended by the World Health Organization (WHO) (Lohar et al., 2014). In water system, arsenic exists as arsenite As(III) and arsenate As(V). Similarly, as trace amount, many elements such as copper and cobalt are even though essential for physiological and biological processes, but at higher concentration they can lead to poisoning (Cui et al., 2018; Gonzales et al., 2009; Wu et al., 2007). Copper is an essential metal in micronutrients in all living forms and is used in a variety of enzymes as a cofactor. But its long-term exposure to high concentration can cause Alzheimer's disease, gastrointestinal disturbance, Wilson diseases, injury to kidney, and liver damage (Madsen and Gitlin, 2007; Hung et al., 2010; Lee et al., 2008; Vulpe et al., 1993). While deficiency of the same can cause anemia-like symptoms, neutropenia, and other health problems (Becker et al., 2007; Song et al., 2012; Ghanei-Motlagh et al., 2016).

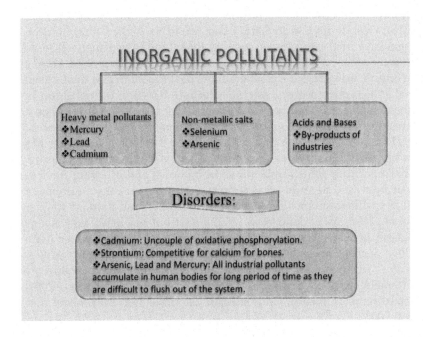

FIGURE 8.1 Main components of inorganic pollutants present in environmental matrixes and their harmful health effects.

Likewise, iron is the major element for all the essential metabolic functioning (Shellaiah et al., 2013; Theil and Goss, 2009) of living organisms, but its excess amount can cause detrimental effects onto human health. Iron homeostasis may lead to the progression of Alzheimer's disease (Mohamed et al., 2019; Chen et al., 2009b) and neuro-inflammation. As the detection of iron involves interference from other metal ions, there is a need to develop specific sensory material for iron. Cobalt is also an essential element, which plays an important role in living species as it is a major constituent of vitamin B_{12} (Tavallali et al., 2013). However, its deficiency in humans can cause anemia, on the other hand excess amount of this can cause cardiomyopathy, vasodilatation, and cobalturia (Maity and Govindaraju, 2011; Little et al., 1982; Leonard and Lauwerys, 1990). Detection of cobalt ion in drinking water is of immense concern as mainly incorporation of cobalt ion in human body happens through indigestion (Basketter et al., 2003; Zeng and Jewsbury, 1998; Fanny et al., 2002). Similarly, mercury being a major element in scientific research is used in thermometers, manometers, mercury switches, etc. (Akshay Krishna et al., 2018), although concerned element's toxicity has largely phased out its use in many clinical environments (Park et al., 2015; Fang et al., 2015; Yıldız et al., 2017). Taking in view the toxicity of mercury in terms of adverse impacts on human health and environment, EPA has settled a permissible limit of 0.002 ppm in drinking water. Chromium is considered as a most common ion present in environment. However, its higher demand and usage in industrial processes such as oxidative dying, pigments, and textile manufacturing has resulted in hazardous environmental contamination problems (Serife et al., 2012; Periyasamy et al., 2013; Zhang et al., 2017). Due to its excessive deposition in populated areas, it has long-lasting effects such as epigastric pain (Yang et al., 2018), carcinogenic effects, hemorrhage, and acute diarrhea. At higher concentrations, it can cause DNA damage (Gonzales et al., 2009; Gupta et al., 2006; Anamika et al., 2015). Nickel is also a vital trace element, which plays an important role in various physiological and biological phenomena (Aruna et al., 2016). Despite its less toxicity, exposure to its higher concentration can lead to many disorders and injurious effects such as Prion and Alzheimer's disease (Dhaka et al., 2015; Kaur and Sareen, 2011), central nervous system disorders, and kidney problems. Based on these effects, nowadays researchers have developed an azo dye–based (Kirk, 1991; Goswami et al., 2013) chemosensor for the selective detection of nickel ion in the presence of many other metal ions. Another heavy-metal ion, that is, cadmium has a great application in electronics like in making of electroplated steel, electrical batteries, and pigments in plastics (Kim et al., 2012). Along with its great number of usage, it has also many adverse effects on environment. It is a toxic metal and is carcinogenic in nature. It is often found in environment due to its applications in batteries and all other devices (Lauwerys et al., 1994). An increased amount of cadmium exposure can cause number of health related issues such as cardiovascular diseases, cancer mortality, and damage to liver and kidneys (Nordberg et al., 1992; McFarland et al., 2002). Last but not least, lead is also a major pollutants of concern in water resources, mostly in soil and ground water (Kim et al., 2012). In a very low concentration, it can cause severe health problems such as reproductive, neurological, cardiovascular disorders, which mainly affect children's growth and cause hypertension and decreased IQs (World Health Organization, 2004; Claudio et al., 2003; Department of Health and Human Services and Prevention and Center for Disease Control, 2003).

8.3 Organic linkers (chemosensor)

Among various applications of organic linkers (Kaur and Kumar, 2011b), molecular analysis is one of the most studied area, which is related to the detection of heavy-metal ions through host–guest noncovalent interaction (Anslyn, 2007; Beer, 1998; Bargossi et al., 2000; Ellis and Walt, 2000; Prodi, 2000). Formally, a chemosensor is a species, which binds with the analyte and shows measurable signal. The binding mode leads to the electronic change in the host species, which induces signal in the chromophoric unit. There are three entities constituting chemosensor, a receptor, and a photoactive unit.

Chemosensors as colorimetric sensors are concerned with the change in electronic properties, which include ICT (intra/intermolecular charge transfer), MLCT (metal-to-ligand charge transfer), and LMCT (ligand-to-metal charge transfer) transitions. Colorimetric chemosensor is a progressive technique, which allows qualitative and quantitative detection of toxic as well as medically and environmentally important metal ions without use of any sophisticated instrumentation. It can be utilized as a test kit for onsite analysis along with its other advantages such as short detection time, reversibility, simplicity along with high selectivity and sensitivity, which does not require any pretreatment preparations and manual expertise.

For colorimetric analysis of inorganic pollutants mainly metal ions, there are two main categories of chemosensors. First, organic ligands that contain either an electron-donating (ED) group or an electron withdrawing (EW) group at appropriate positions and second, rhodamine-containing chemosensors, which are based on spirolactam ring opening. The concept of binding of a metal ion with EW or ED group depends on the hard and soft acids and bases (HSAB), that is, hardness and softness of the binding sites and the analyte. In binding mode if metal ion binds with ED group then it results in decrease in its donating ability thus reducing the conjugation and results in blueshift in the absorption spectra. While if metal binds with EW group there will be redshift. Therefore, along with IC, other modes, that is, MLCT and LMCT also contribute in the color change.

8.3.1 Characteristic of chemosensors

An efficient chemosensor should exhibit following characteristics:

- *Sensitivity*: Receptor should be sensitive to a particular metal ion and a minute change in concentration should produce a huge change in optical response. Sensor should not be affected by inference of environment.
- *Selectivity*: Sensor should be highly selective for a particular metal ion even in the presence of competing metal ions. Binding interaction of target metal and receptor should not be affected by competing metals.
- *Response time and binding constant*: Receptor should have high binding constant and response time of short duration.
- *Water solubility*: Since most of the inorganic pollutants are detected in aqueous medium, solubility of a chemosensor in water is an important point. A sensor must have a large number of electronegative atoms for hydrogen bonding to avoid precipitation and aggregation in aqueous medium.

8.3.2 Schiff base sensors

Schiff base ligands are considered as an indispensable ligand because they can be easily synthesized by the condensation reaction between aldehydes and imines. In the synthetic design of these ligands, stereogenic centers of chirality can be introduced. Schiff bases are capable of forming coordinate bond with the metal ion through their azomethine groups and phenolic groups, because this way they can develop a sensor as an optical sensing material. They can detect the metal ions through their ED binding sites, which produce a photophysical change in the form of color and shift in absorbance value.

Designing a Schiff base is an important step for the development of a chemosensor. They are synthesized according to the kind of analyte such as anion or cation. In the case of anionic sensor, binding sites of receptors are enhanced with N—H or O—H groups for performing hydrogen bonding. For this, moieties such as amides, urea, thiourea, and pyrrole can be introduced in receptor, which can involve in hydrogen bonding. While, in the case of cationic sensor, electrostatic interaction takes place between analyte and receptor. Therefore in the design of metal detection Schiff bases, ED groups such as N, S, and O are introduced at the binding sites.

Metals have empty orbitals and are capable of accepting electron density from ED sites of Schiff base. This interlinkage alters the electronic design of the receptor, which causes a change in the wavelength or intensity of the absorption spectra as well as an optical change. This mode of binding shows a naked eye color change and formation of Schiff sensor (Fig. 8.2).

8.3.3 Schiff base sensors for colorimetric detection of heavy-metal ions

The common ligands for the detection of metal ions are mainly based upon isatin, benzothiazole-and carbazide-based derivatives. The following sections discuss various Schiff bases that have been reported for colorimetric detection of heavy-metal ions.

8.3.3.1 Colorimetric chemosensor for Hg^{2+} ions

Akshay Krishna et al. (2018) reported three isatin-based chemosensors for the detection of Hg^{2+} and AsO_2^- through colorimetric/UV–visible (UV–Vis) techniques and achieved detection up to ppm levels. Among the synthesized chemosensors, two chemosensors showed dual ion selectivity for Hg^{2+} and AsO_2^-. Further they used these chemosensors for paper strip–based detection of these metal ions using 10^{-3} M concentration of chemosensor in DMSO solvent. Another simple sensor based upon anthracene (cost-effective and highly selective) is reported for colorimetric detection of Hg^{2+} ions (Kaur et al., 2015) even in the presence of several interfering ions. Sensing study of metal ion by fluorescence and ^{1}H NMR titration suggested that metal ion (Hg^{2+}) induced deprotonation of imidazole NH group. Moreover, sensor is successfully demonstrated for the analysis of real water samples, which supported its field application for Hg^{2+} ions monitoring.

8.3.3.2 Colorimetric chemosensor for Cu^{2+} ions

Rajaswathi et al. (2019) reported the synthesis of easily accessible chemo probe by simple admixture of 4-nitrobenzaldehyde and 2,4-dimethylpyrrole for the detection of

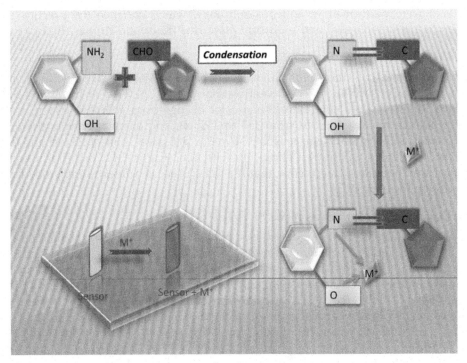

FIGURE 8.2 Mechanism of heavy-metal ions sensing with Schiff base chemosensors.

Cu^{2+} ion in double-distilled deionized water. They analyzed an appearance of red color even in the absence of other metal ions. The lowest detection of limit for Cu^{2+} ion is found to be 2.51 μM, which is significantly lower than that of prescribed WHO level (<30 μM of Cu^{2+}) in potable water. Also new cation receptor using oligoether linkage is reported, which entraps metal ion in cavity formed inside sensor. They investigated their ability to sense transition metal colorimetrically using this cavity. Addition of metal ion in methanolic solution of sensor result in a color change from colorless to yellow brown for Cu^{2+} and yellow for Co^{2+}, Ni^{2+}, Cd^{2+}, and Zn^{2+}. They also observed UV—Vis change in short interval of time. The binding ability toward Cu^{2+} is more compared to other d-metal ions, with the observed detection limit of 1.0×10^{-5} and 1.0×10^{-6} M level of Cu^{2+}, 1.0×10^{-4} and 1.0×10^{-5} M level of Co^{2+}, Ni^{2+}, Cd^{2+}, Zn^{2+} according to visible color change and UV—Vis spectral change, respectively. This indicate that this receptor is more selective toward these metal ions over other such as Ca^{2+}, Mg^{2+}, Mn^{2+}, Hg^{2+}, Cr^{3+}, Fe^{2+}, and Pb^{2+} (Gupta et al., 2013).

Similarly, carboxylic group—based chromogenic chemosensor is constructed for selective detection of copper ions even in the presence of other metal ions, including heavy-metal ions. Synthesized carboxylic acid—diamine—based chemosensors showed both naked eye as well as optical response for Cu^{2+} ions. The selective deprotonation of NH proton by Cu^{2+} is responsible for the drastic color switching from red to blue. UV-spectral responses were used to check the sensitivity of the sensor for Cu^{2+} ions. All the sensing experiments were performed in aqueous solutions, which enhance the practical use of chemosensor in any biological and industrial platforms (Kaur and Kumar, 2011a).

8.3.3.3 Colorimetric chemosensor for Co^{2+} ions

A chemosensor based upon phthalazine is reported for highly selective and sensitive response toward Co^{2+} in the mixed solvent condition {CH_3CN/H_2O (1:1) v/v} (Patil et al., 2015). On addition of Co^{2+} ion, there was a color change from yellow to green, a shift in absorption maximum was redshift from 383 to 435 nm, and enhancement in fluorescence of chemosensor at 550 nm. Chemosensors showed a detection limit of 23 nM and sensing mechanism is assigned to complex formation with Co^{2+} ion in 1:1. Further it was proved by confocal laser scanning microscope that it can be used as a fluorescent probe for Co^{2+} ion monitoring in living cells.

8.3.3.4 Colorimetric chemosensor for Fe^{2+}/Fe^{3+} ions

Kaur et al. (2017) has studied a new tripod Schiff base chemosensor for the detection of Zn^{2+}, Cd^{2+}, Fe^{2+}, and Fe^{3+} metal ions, which resulted in a dual probe for the visual detection of these ions. Both iron species showed chromogenic response in both aqueous as well as methanol solvent. Reimann et al. (2019) reported four Schiff base ligands based on sulfone aldehyde soluble in water capable of sensing selective metal ions in water. These synthesized linkers have practical application in the detection of metal ions Cu^{2+}, Ni^{2+}, Cr^{3+}, Co^{2+}, and Pb^{2+} in water. In addition, pH sensing was performed for application in biological studies.

In another work, Shellaiah et al. (2013) synthesized two novel Schiff bases derived from pyrene and anthracene. These two Schiff bases were utilized as fluorescence turn-on chemosensor for Cu^{2+} and Fe^{3+} ions, respectively. UV–Vis titrations were used to check the binding mode of Schiff base sensors and the analytes, which showed the 2:1 stoichiometry. Sensing behavior of the synthesized Schiff bases was also supported by ^1H NMR titrations and fluorescence reversibility with the addition of metal ions and PMDTA (tridentate organic ligand). Schiff bases showed sensing behavior in a wide range of pH (1–14), which enhances the importance of the particular sensors. Time-resolved photoluminescence spectra were used to calculate the fluorescence life time values.

Another julolidine derivative has been reported and utilized as a chemosensor for the sensing of Fe^{2+} ions. The mechanism of interaction between sensor and Fe^{2+} ions was predicted on the basis of UV–Vis spectral studies. The complete bonding of Fe^{2+} ions with sensor took place in two steps. Initially Fe^{2+} ions showed slight interaction with the chemosensor, whereas in the second step complete chelation was happened. The detection limit of chemosensor is reported as 0.14×10^{-14} M (Kim et al., 2014).

8.3.3.5 Colorimetric chemosensor for Cr^{3+} ions

Lee et al. (2016) prepared a chemosensor for detection of Cr^{3+} in DMSO solution. Sensing response was studied using UV–Vis spectroscopy and naked eye changes in color from violet to blue on addition of Cr^{3+} ions. On the other hand, negligible response was noted for other metal ions. At 1:1 binding between sensor and metal ion was confirmed from absorption spectroscopy. The designed chemosensor is deployed for real-time sensing of Cr^{3+} ions in contaminated water sources.

8.3.3.6 Colorimetric chemosensor for Mn^{2+} ions

A novel ethylenediamine-based water-soluble sensor is reported for the detection of Mn^{2+} ions (Gou et al., 2011). The presence of available binding sites, that is, two C—N bonds and two OH group in the chemosensor made an interesting chelate ring for the binding of target analyte. On addition of Mn^{2+} ions, the sensor showed a color change from colorless to red. However, further studies could be explored on similar type of organic linkers as a promising sensor material for inorganic pollutants.

Similarly, a new sensor containing two julolidine moieties is designed for selective response toward Mn^{2+} ions (Kim et al., 2014). A color change from pale yellow to orange, visible to naked eye, was observed selectively only for Mn^{2+} ions. UV—Vis absorption study further supported the colorimetric change. The sensing mechanism is proposed through MLCT for the binding between guest and host at 1:1.

8.3.3.7 Colorimetric chemosensor for Ni^{2+} ions

Wang et al. (2012) prepared and reported a novel Schiff base containing coumarin derivatives for the sensing of Ni^{2+} ions in acetonitrile solution. Schiff base was structurally interesting due to presence of two coumarin units. In addition to visible color change, this Schiff base also showed a bathochromic shift of 51 nm in UV—Vis absorption spectroscopy. Schiff base was tested for different metal ions, including rare earth metal ions, transition metal ions, and alkali metal ions. But the selective response is reported only for Ni^{2+} ions with an excellent detection limit of $2.9 \times 10^{-4} M^{-1}$.

8.3.3.8 Colorimetric chemosensor for Zn^{2+} ions

For sensing of Zn^{2+} ions, Zhu et al. (2010) synthesized a multifunctional spirobenzopyran derivative. The designed chemosensor exhibited a chromogenic response from colorless to red upon addition of Zn^{2+} ions. Also, in UV—Vis absorption spectroscopy study, a new peak appeared at 650 nm. NMR titrations in addition to UV—Vis and fluorescence spectroscopy were used to depict the binding mode of sensor with Zn^{2+} ions. FRET mechanism is proposed for the system for charge transfer from quinoline (donor) → merocyanine (acceptor), which is observed as visible color change. The change in intensity is also reported w.r.t I_{650}/I_{398}. To further interpret data of this sensor, logic gates OR and INHIBIT of truth tables were created using the color response of Schiff base upon addition of zinc ions.

In another study, Xie et al. (2011) designed a new chemosensor containing 8-aminoquinoline moiety for predicting Zn^{2+} ions in living cells and organisms. The interaction between chemosensor and Zn^{2+} ions was very rapid and sensitive. Thus this chemosensor could be used for real sample analysis for selective and sensing detection of Zn^{2+} ions. However, it is yet to be studied for real sample analysis.

8.3.3.9 Colorimetric chemosensor for Ag^{2+} ions

Taking into advantage of plasmon resonance properties of metal ions, Alizadeh et al. (2014) designed novel series of gold nanoparticle—based chemosensors for the detection of Ag^+ ions. The addition of Ag^+ ions to these sensor systems is observed as change is color from brown to purple, which is also further mapped with TEM and UV—Vis spectroscopy.

The presence of two 2-aminopyridine moieties provided two coordination sites for the coordination of Ag^+ ions of $AgClO_4$. Interestingly, the coordinating behavior of Ag^+ ions changed with change in counter ion, that is, in the case of $PhCOO^- \, Ag^+$ ions showed three coordination behaviors with two 2-aminopyridine ligands; this major difference between three coordination and two coordination behaviors was due to the poor coordination nature of ClO_4^- anions. Thus in the case of benzoate zwitterion Ag^+ ion coordinated more strongly with the pyridine derivative so aggregation took place at a much faster rate in the case of PhCOOAg salt.

8.4 Dye-based colorimetric sensor

Rhodamine is most commonly used as a tracer dye in water to determine its rate and flow. These dyes show fluoresce and thus can be easily detected with fluoremeters. Fluorescent sensors for metal ions have gained a great potential in various fields such as environmental sensors, gas sensors, biosensors, and chemosensors. Rhodamine being a chromophore and a fluorophore probe has been extensively explored by researchers on account of its excellent photophysical properties such as high fluorescence quantum yield, large extinction coefficient, long absorption and emission wavelength, and high photostability. Derivatives (Sun et al., 2017, 2018; Bao et al., 2015; Zayed and Terry, 2003) of rhodamine are favorably nonfluorescent and colorless, so ring opening of these ligands gives rise to a strong fluorescence emission and an appearance of pink color (Dujols et al., 1997) in acidic solution due to activation of carbonyl part in a spirolactam moiety. Based on this process, fluorescent increment of rhodamine probes has been extensively used in the detection of many metal ions. Many probes have been designed by researchers by incorporation of ligands on spirolactam ring, which on addition of metal ion induce a change in color as well as fluorescent change (Kim et al., 2008). This application of ring opening of rhodamine B derivatives has attained a great attention by organic researchers. However, compared to organic ligands, these derivatives have some disadvantages in synthesis (Yu et al., 2009; Chen et al., 2009a; Kim et al., 2006; Kaur and Kumar, 2002; Xu et al., 2005; Royzen et al., 2005), long response time, and interference from other metal ions. Few case studies on their usage as colorimetric sensors for inorganic pollutants are discussed below.

8.4.1 Copper ion sensors

A rhodamine derivative designed by formation of rhodamine B hydrazide and its condensation with pyrrole-2-carboxaldehyde is demonstrated for heavy-metal ions sensing (Puangploy et al., 2014). The sensor showed a linear response toward copper ions in a concentration range from 0.4 to 10 μM with a detection limit of 280 nM. This probe is successfully applied for real water samples and serums as well as living cell imaging of Cu^{2+}. The same system is successfully transferred onto filter paper and shown for Cu^{2+} sensing in spiked solutions.

Similarly, other RhB derivates such as rhodamine hydrazone derivative from rhodamine hydrazine and rhodamine diacetic derivative are shown for Cu^{2+} sensing in water

(Xiang et al., 2006; Zhang et al., 2007). The earlier has shown change in fluorescence, while later showed chromogenic change as well as green fluorescence on addition of Cu^{2+} ions.

8.4.2 Chromium ion sensors

Yang et al. (2018) designed a new chemosensor based on rhodamine derivatives for the detection of Cr^{3+} ions. It was based on the "off−on" effect of rhodamine and its derivatives. The sensor gave the rapid and selective color response (colorless to dark pink) for Cr^{3+} ions in water: ethanol (4:1) solution. It was concluded from Benesi−Hildebrand equation that rhodamine derivation and Cr^{3+} ions interact in 1:1 stoichiometry. The presence of other metal ions did not affect the binding between Cr^{3+} ions and the synthesized derivative.

8.4.3 Lead ion sensors

Kwon et al. (2005) synthesized rhodamine B derivative sensor for the sensing of Pb^{2+}. Spirolactam ring formation was confirmed with help of single crystal X-ray analysis. Neglecting the small response of the synthesized chemosensor toward Cu^{2+} and Zn^{2+}, rhodamine B derivative showed significantly large response for Pb^{2+} ions in acetonitrile solution. Complex formation and its mechanism were elaborated using mass, FTIR, and ^{13}C NMR spectroscopies. Reversible ring opening was the purposed mechanism for color change observed.

8.4.4 Mercury ion sensors

Yang et al. (2005) developed a rhodamine 6G derivative, a chemodosimeter for the sensing of Hg^{2+}. Designed derivation showed very selective and sensitive response for mercury ions in aqueous medium. Sensing was monitored using colorimetric and fluorescence response of rhodamine 6G derivative at room temperature and it was concluded that mercury promoted oxadiazole formation was the key step in the detecting procedure. 1:1 binding mode was confirmed for the interaction between sensor and the Hg^{2+} ions. The sensitivity of synthesized rhodamine derivative is less than 2 ppb. Zheng et al. (2006) prepared and utilized rhodamine B thiohydrazide derivative as fluorescent sensor for mercury ions. This sensor worked as mercury detector at pH 3.4 and showed the selective and sensitive results for mercury ions in the presence of other metal ions. Results obtained from spectroscopy determined the 1:2 stoichiometry between sensor and mercury ions due to presence of two donor sites, that is, N and S in rhodamine B thiohydrazide derivative.

Lee et al. (2007) synthesized a new chemosensor containing tren(triethylenetetramine) and diethylenetriamine with tosyl group−based rhodamine derivative and utilized as a sensor for mercury ions in acetonitrile derivative. The sensor exhibits visual color change and fluorescence intensity changes for Hg^{2+}. Similar structures without tosyl groups were also synthesized and sensing mechanism was repeated for these derivatives. But no spectral responses were recorded for Hg^{2+} ions. This proved the role of tosyl groups in ring opening of spirolactam in the chemosensor. Shi and Ma (2008) and Zhan et al. (2008)

independently studied a highly sensitive and selective rhodamine derivative, that is, rhodamine B thiolactone for the sensing of Hg^{2+} ion in 20 mM phosphate buffer (pH 7) and 10 mM acetate buffer (pH 4), respectively. Various metal ions were examined with sensor under these conditions, while only Hg^{2+} responded selectively showing fluorescent enhancement as well as colorimetric change. Xu et al. reported this ring opening as a reversible process under this condition, whereas Ma and Shi reported that on introducing KI in the system can change/reverse the color in the presence of less than half equivalent of Hg^{2+}.

8.5 Conclusion

Controlling of inorganic pollutant is becoming more and more challenging and thus it is important to synthesize a highly selective and sensitive chemosensor, which not only detects the pollutant but also quantifies the amount of pollutant. The sensor should be selective, sensitive, quick, cost-effective, less time-consuming, and does not demand laborious process. Moreover, it should not require pretreatment of the sample for onsite monitoring. Though selective and real-time monitoring of water samples is hard, remarkable achievements have been made in this area. The introduction of ligands in rhodamine dyes have been explored for selective sensing in monitoring metal ions, but they have their own shortcomings and are in developing stage for inorganic pollutants. Also, paper strip−based technologies are more appreciated as they are more advantageous in sensing of heavy-metal ions as compared to organic linker in liquid phase. Recent trend is to integrate the organic linker to cellulose-based paper and the same to smartphones/other portable electronic devices and convert the signals to useful information using embedded system app. Plenty of literature is available for techniques based on metal ion detection, but commercialization of the techniques is still a challenge. For a successful design of sensor, binding mode of different metals with ligands and their acceptability on cellulose-based paper is of more concern, which needs to be addressed.

8.6 Future outlook

Increasing amount of inorganic pollutants in drinking water is a severe problem that urgently needs concepts for monitoring their detection as well as their quantification. Deriving solutions for this plan involves the need for design of highly selective and sensitive chemosensor. Developing naked eye chemosensors will administer sustainable monitoring of water resources as well as remedial systems being deployed for the recovery of toxic metals that lack any involvement in physiological process or beneficial effect.

References

Akshay Krishna, T.G., Tekuri, V., Mohan, M., Trivedi, D.R., 2018. Selective colorimetric chemosensor for the detection of Hg^{2+} and arsenite ions using Isatin based Schiff's bases; DFT Studies and Applications in test strips. Sens. Actuators B: Chem. Available from: https://doi.org/10.1016/j.snb.2018.12.003.

Alizadeh, A., Khodaei, M.M., Hamidi, Z., Shamsuddin, M.B., 2014. Naked-eye colorimetric detection of Cu^{2+} and Ag^+ ions based on close-packed aggregation of pyridines-functionalized gold nanoparticles. Sens. Actuators B: Chem. 190, 782−791. Available from: https://doi.org/10.1016/j.snb.2013.09.02.

Anamika, D., Nikhil, G., Kar, S.K., 2015. A novel Cr^{3+} fluorescence turn-on probe based on rhodamine and isatin framework. J. Fluoresc. 25 (6), 1921−1929.

Anslyn, E.V., 2007. Supramolecular analytical chemistry. J. Org. Chem. 72 (3), 687−699. Available from: https://doi.org/10.1021/jo0617971.

Aruna, J.W., Agozie, N.O., Fasi, A.A., Andre, R.V., Ekkehard, S., 2016. Rhodamine based turn-on sensors for Ni^{2+} and Cr^{3+} in organic media: detecting CN − via the metal displacement approach. J. Fluoresc. 28 (3), 891−898.

Bao, X.F., Cao, Q.S., Nie, X.M., et al., 2015. Design and synthesis of a novel chromium(III) selective fluorescent chemosensor bearing a thiodiacetamide moiety and two rhodamine B fluorophores. Sens. Actuators B: Chem. 221, 930−939.

Bargossi, C., Fiorini, M.C., Montalti, M., Prodi, L., Zaccheroni, N., 2000. Recent developments in transition metal ion detection by luminescent chemosensors. Coord. Chem. Rev. 208 (1), 17−32. Available from: https://doi.org/10.1016/s0010-8545(00)00252-6.

(a) Basketter, D.A., Angelini, G., Ingber, A., Kern, P.S., Menne, T., 2003. Contact Dermatitis 49, 1.

(b) Barceloux, D.G., Barceloux, D., 2003. Clin. Toxicol. 37, 201.

(c) El-Safty, S.A., 2003. Adsorption 15, 227.

Becker, J.S., Zoriy, M.V., Pickhardt, C., Palomer, G.N., Zilles, K., 2005. Imaging of copper, zinc, and other elements in thin section of human brain samples (hippocampus) by laser ablation inductively coupled plasma mass spectrometry. Anal. Chem. 77, 3208−3216.

Becker, J.S., Matusch, A., Depboylu, C., Dobrowolska, J., Zoriy, M.V., 2007. Quantitative imaging of selenium, copper, and zinc in thin sections of biological tissues (slugs-genus arion) measured by laser ablation inductively coupled plasma mass spectrometry. Anal. Chem. 79, 6074−6080.

Beer, P.D., 1998. Transition-metal receptor systems for the selective recognition and sensing of anionic guest species. Acc. Chem. Res. 31 (2), 71−80. Available from: https://doi.org/10.1021/ar9601555.

Chauhan, K., Singh, P., Kumari, B., Singhal, R.K., 2017. Synthesis of new benzothiazole Schiff base as selective and sensitive colorimetric sensor for arsenic on-site detection at ppb level. Analyt. Methods 9 (11), 1779−1785. Available from: https://doi.org/10.1039/c6ay03302d.

Chen, X., Jia, J., Ma, H., Wang, S., Wang, X., 2009a. Characterization of rhodamine B hydroxylamide as a highly selective and sensitive fluorescence probe for copper (II). Anal. Chim. Acta 632, 9−14.

Chen, Y., Pu, K.Y., Fan, Q.L., Qi, X.Y., Huang, Y.Q., Lu, X.M., et al., 2009b. Water-soluble anionic conjugated polymers for metal ion sensing: effect of interchain aggregation. J. Polym. Sci. A: Polym. Chem. 47 (19), 5057−5067. Available from: https://doi.org/10.1002/pola.23558.

Claudio, E.S., Godwin, H.A., Magyar, J.S., 2003. Prog. Inorg. Chem. 51, 1. and references therein.

Cui, Y., Wang, X., Zhang, Q., Zhang, H., Li, H., Meyerhoff, M., 2018. Colorimetric copper ion sensing in solution phase and on paper substrate based on catalytic decomposition of S-nitrosothiol. Anal. Chim. Acta . Available from: https://doi.org/10.1016/j.aca.2018.11.050.

Department of Health and Human Services and Prevention, Center for Disease Control, 2003. Surveillance for elevated blood lead levels among children: United States, 1997−2001. Morb. Mortal. Wkly. Rep. 52, 1.

Dey, S., Sarkar, S., Maity, D., Roy, P., 2017. Rhodamine based chemosensor for trivalent cations: synthesis, spectral properties, secondary complex as sensor for arsenate and molecular logic gates. Sens. Actuators B: Chem. 246, 518−534. Available from: https://doi.org/10.1016/j.snb.2017.02.094.

Dhaka, G., Kaur, N., Singh, J., 2015. A facile ratiometric and colorimetric azo-dye possessing chemosensor for Ni^{2+} and AcO − detection. Supramolecular Chem. 27 (10), 654−660. Available from: https://doi.org/10.1080/10610278.2015.1068314.

Dujols, V., Ford, F., Czarnik, A.W., 1997. A long-wavelength fluorescent chemodosimeter selective for Cu(II) ion in water. J. Am. Chem. Soc. 119, 7386−7387.

Dzombak, D.A., Morel, F.M.M., 1987. Adsorption of inorganic pollutants in aquatic systems. J. Hydraul. Eng. 113 (4), 430−475. Available from: https://doi.org/10.1061/(asce)0733-9429(1987)113:4(430).

Ellis, A.B., Walt, D.R., 2000. Guest editorial. Chem. Rev. 100 (7), 2477−2478. Available from: https://doi.org/10.1021/cr990025k.

Ensafi, A.A., Khayamiam, T., Benvidi, A., Mirmomtaz, E., 2006. Simultaneous determination of copper, lead and cadmium by cathodic adsorptive stripping voltammetry using artificial neutral network. Anal. Chim. Acta 251, 225−232.

Fang, Y., Zhou, Y., Li, J.Y., Rui, Q.Q., Yao, C., 2015. Naphthalimide-rhodamine based chemosensors for colorimetric and fluorescent sensing Hg^{2+} through different signaling mechanisms in corresponding solvent systems. Sens. Actuators B: Chem. 215, 350−359.

(a) Fanny, M., Rivera, J.D., 2002. Anal. Bioanal. Chem. 374, 1105.

(b) Lin, W.Y., Yuan, L., Long, L.L., Guo, C.C., Feng, J.B., 2002. Adv. Funct. Mater. 18, 2366.

(c) Yao, Y., Tian, D., Haibing, L., 2002. ACS Appl. Mater. Interfaces 2, 684.

(d) JunZhen, S., Guo, F.L., QiangChen, L., Li, Y.F., Zhanga, Q., Huang, C.Z., 2002. Chem. Commun. 47, 2562.

Ghanei-Motlagh, M., Karami, Ch, Taher, M.A., Hosseini-Nasab, S.J., 2016. Stripping voltammetric detection of copper ions using carbon paste electrode modified with aza-crown ether capped gold nanoparticles and reduced graphene oxide. RSC Adv. 6, 89167−89175.

Gonzales, A.P.S., Firmino, M.A., Nomura, C.S., Rocha, F.R.P., Oliveira, P.V., Gaubeur, I., 2009. Peat as a natural solid-phase for copper preconcentration and determination in a multicommuted flow system coupled to flame atomic absorption spectrometry. Anal. Chim. Acta 636, 198−204.

(a) Goswami, S., Das, A.K., Aich, K., Manna, R., 2013. Tetrahedron Lett. 54, 4215−4220.

(b) Martínez-Máñez, R., Sancenón, F., 2013. Chem. Rev. 103, 4419−4476.

(c) Gunnlaugsson, T., Glynn, M., Tocci, G.M., Kruger, P.E., Pfeffer, F.M., 2013. Coord. Chem. Rev. 250, 3094−3117.

(d) Suksai, C., Tuntulani, T., 2013. Chem. Soc. Rev. 32, 192−202.

(e) Gunnlaugsson, T., Davis, A.P., O'Brien, J.E., Glynn, M., 2013. Org. Lett. 4, 2449−2452.

Gou, C., Wu, H., Jiang, S., Yi, C., Luo, J., Liu, X., 2011. A highly selective colorimetric chemosensor for Mn^{2+} based on Bis(N-salicylidene)ethylenediamine in pure aqueous solution. Chem. Lett. 40 (10), 1082−1084. Available from: https://doi.org/10.1246/cl.2011.1082.

Gupta, V.K., Jain, A.K., Kumar, P.K., et al., 2006. Chromium(III)-selective sensor based on tri-o-thymotide in PVC matrix. Sens. Actuators B: Chem. 113 (1), 182−186.

Gupta, V.K., Singh, A.K., Ganjali, M.R., Norouzi, P., Faridbod, F., Mergu, N., 2013. Comparative study of colorimetric sensors based on newly synthesized Schiff bases. Sens. Actuators B: Chem. 182, 642−651. Available from: https://doi.org/10.1016/j.snb.2013.03.062.

Gupta, V.K., Mergu, N., Singh, A.K., 2014a. Fluorescent chemosensors for Zn^{2+} ions based on flavonol derivatives. Sens. Actuators B 202, 674−682.

Gupta, V.K., Singh, A.K., Kumawat, L.K., 2014b. A turn-on fluorescent chemosensor for Zn^{2+} ions based on anti-pyrineschiff base. Sens. Actuators B 204, 507−514.

Gupta, V.K., Mergu, N., Kumawat, L.K., Singh, A.K., 2015. Selective naked-eye detection of magnesium(II) ions using a coumarin-derived fluorescent probe. Sens. Actuators B 207, 216−223.

Hung, Y.H., Bush, A.I., Cherny, R.A., 2010. Copper in the brain and Alzheimer's disease. J. Biol. Inorg. Chem. 15, 61−76.

Kaur, N., Kumar, S., 2011a. Insights into the photophysics, protonation and Cu^{2+} ion coordination behaviour of anthracene-9,10-dione-based chemosensors. Supramolecular Chem. 23 (11), 768−776. Available from: https://doi.org/10.1080/10610278.2011.622386.

Kaur, N., Kumar, S., 2011b. Colorimetric metal ion sensors. Tetrahedron 67 (48), 9233−9264. Available from: https://doi.org/10.1016/j.tet.2011.09.003.

Kaur, N., Kumar, S., 2012. Aminoanthraquinone-based chemosensors: colorimetric molecular logic mimicking molecular trafficking and a set−reset memorized device. Dalton Trans. 41 (17), 5217. Available from: https://doi.org/10.1039/c2dt12201d.

Kaur, P., Sareen, D., 2011. Dyes Pigm. 88, 296−300.

(a) Mulrooney, S.B., Hausinger, R.P., 2011. Microbiol. Rev. 27, 239−261.

(b) Jaouen, G., 2011. Bioorganometallics: Biomolecules, Labeling, Medicine. Wiley-VCH, Weinheim, Germany.

(c) Wang, H.X., Wang, D.L., Wang, Q., Li, X.Y., Schalley, C.A., 2011. Org. Biomol. Chem. 8, 1017−1026.

(d) Feng, L., Zhang, Y., Wen, L.Y., Chen, L., Shen, Z., Guan, Y.F., 2011. Analyst 136, 4197−4203.

(e) Frausto da Silva, J.J.R., Williams, R.J.P., 2011. The Biological Chemistry of Elements: The Inorganic Chemistry of Life. Clarendon Press, Oxford, pp. 388−397.

(f) Li, Z., Li, Q., Peng, M., Li, N., Qin, J., 2011. Sens. Actuators B: Chem. 173, 580–584.

(g) Maity, D., Kumar, V., Govindaraju, T., 2011. Org. Lett. 14, 6008–6011.

Kaur, S., Kumar, S., 2002. Photoactive chemosensors 3: a unique case of fluorescence enhancement with Cu(II). Chem. Commun. 2840–2841.

Kaur, N., Dhaka, G., Singh, J., 2015. Hg^{2+}-induced deprotonation of an anthracene-based chemosensor: set–reset flip-flop at the molecular level using Hg^{2+} and I^- ions. New J. Chem. 39 (8), 6125–6129. Available from: https://doi.org/10.1039/c5nj00683j.

Kaur, P., Singh, J., Singh, R., Kaur, V., Talwar, D., 2017. Extending photophysical behavior of Schiff base tripod for the speciation of iron and fabrication of INHIBIT type molecular logic gate for fluorogenic recognition of Zn(II) and Cd(II) ions. Polyhedron 125, 230–237. Available from: https://doi.org/10.1016/j.poly.2016.12.044.

Kaur, B., Kaur, N., Kumar, S., 2018a. Colorimetric metal ion sensors – a comprehensive review of the years 2011–2016. Coord. Chem. Rev. 358, 13–69. Available from: https://doi.org/10.1016/j.ccr.2017.12.002.

Kaur, P., Singh, R., Kaur, V., Talwar, D., 2018b. Reusable Schiff base functionalized silica as a multi-purpose nanoprobe for fluorogenic recognition, quantification and extraction of Zn^{2+} ions. Sens. Actuators B: Chem. 254, 533–541. Available from: https://doi.org/10.1016/j.snb.2017.07.077.

Kim, S.H., Kim, J.S., Park, S.M., Chang, S.K., 2006. Hg^{2+}-selective off-on and Cu^{2+}-selective on-off type fluoroionophore based upon cyclam. Org. Lett. 8, 371–374.

Kim, H.N., Lee, M.H., Kim, H.J., Kim, J.S., Yoon, J., 2008. A new trend in rhodamine-based chemosensors: application of spirolactam ring-opening to sensing ions. Chem. Soc. Rev. 37 (8), 1465–1472. Available from: https://doi.org/10.1039/b802497a.

Kim, H.N., Ren, W.X., Kim, J.S., Yoon, J., 2012. Fluorescent and colorimetric sensors for detection of lead, cadmium, and mercury ions. Chem. Soc. Rev. 41 (8), 3210–3244. Available from: https://doi.org/10.1039/c1cs15245a.

Kim, K.B., Park, G.J., Kim, H., Song, E.J., Bae, J.M., Kim, C., 2014. A novel colorimetric chemosensor for multiple target ions in aqueous solution: simultaneous detection of Mn(II) and Fe(II). Inorg. Chem. Commun. 46, 237–240. Available from: https://doi.org/10.1016/j.inoche.2014.06.009.

(a) Kirk, K.L., 1991. Biochemistry of the Halogens and Inorganic Halides. Plenum Press, New York, p. 58.

(b) Kleerekoper, M., 1991. Endocrinol. Metab. Clin. North Am. 27, 441–452.

(c) Kim, S.Y., Hong, J.I., 1991. Org. Lett. 9, 3109–3112.

(d) Enderby, J.E., Collins, K.D., Neilson, G.W., 1991. Biophys. Chem. 128, 95–104.

(e) Muller, R.N., Laurent, S., Forge, D., Port, M., Roch, A., Robic, C., et al., 1991. Chem. Rev. 108, 2064–2110.

(f) Jianlong, W., 1991. Process Biochem 37, 847–850.

(g) Changa, M.C., Koa, C.-C., Douglasa, W.H., 1991. Biomaterials 24, 2853–2862.

Kwon, J.Y., Jang, Y.J., Lee, Y.J., Kim, K.M., Seo, M.S., Nam, W., et al., 2005. A highly selective fluorescent chemosensor for Pb^{2+}. J. Am. Chem. Soc. 127 (28), 10107–10111. Available from: https://doi.org/10.1021/ja051075b.

Lauwerys, R.R., Bernard, A.M., Reels, H.A., Buchet, J.-P., 1994. Clin. Chem. (Washington, DC) 40, 1391.

Lee, M.H., Wu, J.-S., Lee, J.W., Jung, J.H., Kim, J.S., 2007. Highly sensitive and selective chemosensor for Hg^{2+} based on the rhodamine fluorophore. Org. Lett. 9 (13), 2501–2504. Available from: https://doi.org/10.1021/ol0708931.

Lee, J.C., Gray, H.B., Winkler, J.R., 2008. Copper(II) binding to α-synuclein, the Parkinson's protein. J. Am. Chem. Soc. 130, 6898–6899.

Lee, S.Y., Bok, K.H., Kim, J.A., Kim, S.Y., Kim, C., 2016. Simultaneous detection of Cu^{2+} and Cr^{3+} by a simple Schiff-base colorimetric chemosensor bearing NBD (7-nitrobenzo-2-oxa-1,3-diazolyl) and julolidine moieties. Tetrahedron 72 (35), 5563–5570. Available from: https://doi.org/10.1016/j.tet.2016.07.051.

(a) Leonard, A., Lauwerys, R., 1990. Mutat. Res. 239, 17.

(b) Selden, A.I., Norberg, C., Karlson-Stiber, C., Hellstrom-Lindberg, E., 1990. Environ. Toxicol. Pharmacol. 23, 129.

Li, Y., Chen, Y., Yu, H., Tian, L., Wang, Z., 2018. Portable and smart devices for monitoring heavy metal ions integrated with nanomaterials. TrAC Trends Anal. Chem. 98, 190–200. Available from: https://doi.org/10.1016/j.trac.2017.11.011.

Lin, T.W., Huang, S.D., 2001. Direct and simultaneous determination of copper, chromium, aluminium, and manganese in urine with a multielement graphite furnace atomic absorption spectrometer. Anal. Chem. 73, 4319–4325.

Lin, Y., Liang, P., Guo, L., 2005. Nanometer titanium dioxide immobilized on silica gel as sorbent for preconcentration of metal ions prior to their determination by inductively coupled plasma emission spectrometry. Talanta 68, 25−30.

(a) Little, C., Aakre, S.E., Rumsby, M.G., Gwarsha, K., 1982. Biochem. J. 207, 117.

(b) Dennis, M., Kolattukudy, P.E., 1982. Proc. Natl. Acad. Sci. U.S.A. 89, 5306.

(c) Maret, W., Vallee, B.L., 1982. Methods Enzymol. 226, 52.

(d) Walker, K.W., Bradshaw, R.A., 1982. Protein Sci. 7, 2684.

Lohar, S., Pal, S., Sen, B., Mukherjee, M., Banerjee, S., Chattopadhyay, P., 2014. Selective and sensitive turn-on chemosensor for arsenite ion at the ppb level in aqueous media applicable in cell staining. Anal. Chem. 86 (22), 11357−11361. Available from: https://doi.org/10.1021/ac503255f.

Madsen, E., Gitlin, J.D., 2007. Copper and iron disorders of the brain. Annu. Rev. Neurosci. 30, 317−337.

Maity, D., Govindaraju, T., 2011. Highly selective colorimetric chemosensor for Co^{2+}. Inorg. Chem. 50 (22), 11282−11284. Available from: https://doi.org/10.1021/ic2015447.

McFarland, C.N., Bendell-Young, L.I., Guglielmo, C., Williams, T.D., 2002. J. Environ. Monit. 4, 791.

Mohamed, T.M., Nasef, S.M., Mahmoud, G.A., 2019. Preparation of high sensitive colorimetric sensing film for detection of iron ions using gamma irradiation. Polym.-Plastics Technol. Mater. 1−10. Available from: https://doi.org/10.1080/25740881.2019.1599940.

Nordberg, G.F., Herber, R.F.M., Alessio, L., 1992. Cadmium in the Human Environment. Oxford University Press, Oxford.

Park, J., In, B., Lee, K.-H., 2015. Highly selective colorimetric and fluorescent detection for Hg^{2+} in aqueous solutions using a dipeptide-based chemosensor. RSC Adv. 5, 56356−56361.

Patil, S., Fegade, U., Sahoo, S.K., Singh, A., Marek, J., Singh, N., et al., 2014. Highly sensitive ratiometric chemosensor for selective 'Naked-Eye' nanomolar detection of Co^{2+} in semi-aqueous media. Chem. Phys. Chem. 15 (11), 2230−2235. Available from: https://doi.org/10.1002/cphc.201402076.

Patil, S., Patil, R., Fegade, U., Bondhopadhyay, B., Pete, U., Sahoo, S.K., et al., 2015. A novel phthalazine based highly selective chromogenic and fluorogenic chemosensor for Co^{2+} in semi-aqueous medium: application in cancer cell imaging. Photochem. Photobiol. Sci. 14 (2), 439−443. Available from: https://doi.org/10.1039/c4pp00358f.

Periyasamy, P., Girish, C., Anitha, K., et al., 2013. Potential of novel bacterial consortium for the remediation of chromium contamination. Water Air Soil Pollut. 224, 1716.

Pourreza, N., Hoveizavi, R., 2005. Simultaneous preconcentration of Cu, Fe and Pb as methylthymol blue complexes on naphthalene adsorption and flame atomic absorption determination. Anal. Chim. Acta 549, 124−128.

Prodi, L., 2000. Luminescent chemosensors for transition metal ions. Coord. Chem. Rev. 205 (1), 59−83. Available from: https://doi.org/10.1016/s0010-8545(00)00242-3.

Puangploy, P., Smanmoo, S., Surareungchai, W., 2014. A new rhodamine derivative-based chemosensor for highly selective and sensitive determination of Cu^{2+}. Sens. Actuators B: Chem. 193, 679−686. Available from: https://doi.org/10.1016/j.snb.2013.12.037.

Rajaswathi, K., Jayanthi, M., Rajmohan, R., Anbazhagan, V., Vairaprakash, P., 2019. Simple admixture of 4-nitrobenzaldehyde and 2,4-dimethylpyrrole for efficient colorimetric sensing of copper(II) ions. Spectrochim. Acta, A: Mol. Biomol. Spectrosc. Available from: https://doi.org/10.1016/j.saa.2019.01.014.

Reimann, M.J., Salmon, D.R., Horton, J.T., Gier, E.C., Jefferies, L.R., 2019. Water-soluble sulfonate Schiff-base ligands as fluorescent detectors for metal ions in drinking water and biological systems. ACS Omega 4 (2), 2874−2882. Available from: https://doi.org/10.1021/acsomega.8b02750.

Royzen, M., Dai, Z., Canary, J.W., 2005. Ratiometric displacement approach to Cu(II) sensing by fluorescence. J. Am. Chem. Soc. 127, 1612−1613.

Scozzafava, A., 1989. Handbook on toxicity of inorganic compounds. Inorg. Chim. Acta 165 (1), 139. Available from: https://doi.org/10.1016/s0020-1693(00)83413-1.

Serife, S., Senol, K., Yakup, Y., et al., 2012. A new chelating resin: synthesis, characterization and application for speiation of chromium (III)/(VI) speies. Chem. Eng. J. 181, 746−753.

Sharma, H., Kaur, N., Singh, A., Kuwar, A., Singh, N., 2016. Optical chemosensors for water sample analysis. J. Mater. Chem. C 4 (23), 5154−5194. Available from: https://doi.org/10.1039/c6tc00605a.

Shellaiah, M., Wu, Y.-H., Singh, A., Ramakrishnam Raju, M.V., Lin, H.-C., 2013. Novel pyrene- and anthracene-based Schiff base derivatives as Cu^{2+} and Fe^{3+} fluorescence turn-on sensors and for aggregation induced emissions. J. Mater. Chem. A 1 (4), 1310−1318. Available from: https://doi.org/10.1039/c2ta00574c.

Shi, W., Ma, H., 2008. Rhodamine B thiolactone: a simple chemosensor for Hg^{2+} in aqueous media. Chem. Commun. (16), 1856. Available from: https://doi.org/10.1039/b717718f.

Shi, L., Jing, C., Ma, W., Li, D.W., Halls, J.E., Marken, F., et al., 2013. Plasmon resonance scattering spectroscopy at the single-nanoparticle level: real-time monitoring of a click reaction. Angew. Chem. Int. Ed. 125, 6127–6130.

Shtoyko, T., Conklin, S., Maghasi, A.T., Richardson, J.N., Piruska, A., Seliskar, C.J., et al., 2004. Spectroelectrochemical sensing based on attenuated total internal reflectance stripping voltammetry. 3. Determination of cadmium and copper. Anal. Chem. 76, 1466–1473.

Song, W.J., Wang, X.W., Ding, J.W., Zhang, J., Zhang, R.M., Qin, W., 2012. Electrochemical sensing system for determination of heavy metals in seawater. Chin. J. Anal. Chem. 40, 670–674.

Sun, D.Q., Lu, T., Xiao, F.P., Zhu, X.Y., Sun, G.Q., 2017. Formulation and aging resistance of modified bio-asphalt containing high percentage of waste cooking oil residues. J. Clean. Prod. 161, 1203–1214.

Sun, D.Q., Sun, G.Q., Zhu, X.Y., Ye, F.Y., Xu, J.Y., 2018. Intrinsic temperature sensitive self-healing character of asphalt binders based on molecular dynamics simulations. Fuel 211 (1), 609–620.

Tavallali, H., Deilamy-Rad, G., Parhami, A., Mousavi, S.Z., 2013. A novel development of dithizone as a dual-analyte colorimetric chemosensor: detection and determination of cyanide and cobalt(II) ions in dimethyl sulf-oxide/water media with biological applications. J. Photochem. Photobiol. B: Biol. 125, 121–130. Available from: https://doi.org/10.1016/j.jphotobiol.2013.05.013.

Theil, E.C., Goss, D.J., 2009. Chem. Rev. 109, 4568.

Vulpe, C., Levinson, B., Whitney, S., Packman, S., Gitschier, J., 1993. Isolation of a candidate gene for Menkes disease and evidence that it encodes a copper-transporting ATPase. Nat. Genet. 3, 7–13.

Walcarius, A., Mercier, L., 2010. Mesoporous organosilica adsorbents: nanoengineered materials for removal of organic and inorganic pollutants. J. Mater. Chem. 20 (22), 4478. Available from: https://doi.org/10.1039/b924316j.

Wan, X., Ke, H., Tang, J., Yang, G., 2019. Acid environment-improved fluorescence sensing performance: a quino-line Schiff base-containing sensor for Cd^{2+} with high sensitivity and selectivity. Talanta . Available from: https://doi.org/10.1016/j.talanta.2019.01.101.

Wang, L., Ye, D., Cao, D., 2012. A novel coumarin Schiff-base as a Ni(II) ion colorimetric sensor. Spectrochim. Acta, A: Mol. Biomol. Spectrosc. 90, 40–44. Available from: https://doi.org/10.1016/j.saa.2012.01.017.

World Health Organization, Guidelines for Drinking-Water Quality, third ed., vol. 1, Geneva, 2004, 188. 4.

Flegal, A.R., Smith, D.R., 2004. Environ. Res. 58, 125.

Wu, J., Yu, J., Li, J., Wang, J., Ying, Y., 2007. Detection of metal ions by atomic emission spectroscopy from liquid-electrode discharge plasma. Spectrochim. Acta B 62, 1269–1272.

Xiang, Y., Tong, A., Jin, P., Ju, Y., 2006. New fluorescent rhodamine hydrazone chemosensor for Cu(II) with high selectivity and sensitivity. Org. Lett. 8 (13), 2863–2866. Available from: https://doi.org/10.1021/ol0610340.

Xie, G., Shi, Y., Hou, F., Liu, H., Huang, L., Xi, P., et al., 2011. A highly selective fluorescent and colorimetric chemosensor for ZnII and its application in cell imaging. Eur. J. Inorg. Chem. 2012 (2), 327–332. Available from: https://doi.org/10.1002/ejic.201100804.

Xu, Z., Xiao, Y., Quain, X., Cui, J., Cui, D., 2005. Ratiometric and selective fluorescent sensor for Cu(II) based on internal charge transfer (ICT). Org. Lett. 7, 889–892.

Yang, Y.-K., Yook, K.-J., Tae, J., 2005. A rhodamine-based fluorescent and colorimetric chemodosimeter for the rapid detection of Hg^{2+} ions in aqueous media. J. Am. Chem. Soc. 127 (48), 16760–16761. Available from: https://doi.org/10.1021/ja054855t.

Yang, Z., Chen, S., Li, F., Bu, Y., Du, Y., Zhou, P., et al., 2018. A rhodamine derivative based chemosensor with high selectivity and quick respond to Cr^{3+} in aqueous solution. J. Fluoresc. 28 (3), 809–814. Available from: https://doi.org/10.1007/s10895-018-2243-2.

Yıldız, M., Demir, N., Ünver, H., Sahiner, N., 2017. Synthesis, characterization, and application of a novel water-soluble polyethyleneimine-based Schiff base colorimetric chemosensor for metal cations and biological activity. Sens. Actuators B: Chem. 252, 55–61.

Yu, F., Zhang, W., Li, P., Xing, Y., Tong, L., Ma, J., et al., 2009. Cu^{2+}-selective naked-eye and fluorescent probe: its crystal structure and application in bioimaging. Analyst 134, 1826–1833.

Zayed, A.M., Terry, N., 2003. Chromium in the environment: factors affecting biological remediation. Plant Soil 249, 139–156.

Zeng, Z., Jewsbury, R.A., 1998. Analyst 123, 2845.

Zhan, X.-Q., Qian, Z.-H., Zheng, H., Su, B.-Y., Lan, Z., Xu, J.-G., 2008. Rhodamine thiospirolactone. Highly selective and sensitive reversible sensing of Hg(II). Chem. Commun. (16), 1859. Available from: https://doi.org/10.1039/b719473k.

Zhang, X., Shiraishi, Y., Hirai, T., 2007. Cu(II)-selective green fluorescence of a rhodamine—diacetic acid conjugate. Org. Lett. 9 (24), 5039—5042. Available from: https://doi.org/10.1021/ol7022714.

Zhang, P., Gong, J.L., Zeng, G.M., Deng, C.H., Yang, H.C., Liu, H.Y., et al., 2017. Cross-linking to prepare composite graphene oxdeframework membranes with high-flux for dyes and heavy metal ions removal. Chem. Eng. J. 322, 657—666.

Zheng, H., Qian, Z.-H., Xu, L., Yuan, F.-F., Lan, L.-D., Xu, J.-G., 2006. Switching the recognition preference of rhodamine B spirolactam by replacing one atom: design of rhodamine B thiohydrazide for recognition of Hg(II) in aqueous solution. Org. Lett. 8 (5), 859—861. Available from: https://doi.org/10.1021/ol0529086.

Zhu, J.-F., Yuan, H., Chan, W.-H., Lee, A.W.M., 2010. A colorimetric and fluorescent turn-on chemosensor operative in aqueous media for Zn^{2+} based on a multifunctionalized spirobenzopyran derivative. Org. Biomol. Chem. 8 (17), 3957. Available from: https://doi.org/10.1039/c004871b.

Further reading

Eastmond, D.A., MacGregor, J.T., Slesinski, R.S., 2008. Trivalent chromium: assessing the genotoxic risk of an essential trace element and widely used human and animal nutritional supplement. Crit. Rev. Toxicol. 38, 173—190.

Mahapatra, A.K., Roy, J., Manna, S.K., Kundu, S., Sahoo, P., Mukhopadhyay, S.K., et al., 2012. Hg2 + -selective "turn-on" fluorescent chemodosimeter derived from glycine and living cell imaging. J. Photochem. Photobiol. A: Chem. 240, 26—32.

Materials in surface-enhanced Raman spectroscopy-based detection of inorganic water pollutants

Prachi Rajput[1,2], R.K. Sinha[1,2] and Pooja Devi[2]

[1]Academy of Scientific and Innovative Research, New Delhi, India [2]CSIR-Central Scientific Instruments Organisation, Chandigarh, India

9.1 Introduction: water pollutants

Safe drinking water is the basic human right that is internationally accepted. Therefore an extensive and stringent monitoring and implementation of the adequate remedial strategies are needed to avoid negative ecological effects associated with contaminated water. As much as 2 million tons of wastewater is produced everyday by industrial and agricultural sector. This produced sewage constitutes of a variety of water pollutants, which are mainly categorized on the basis of their properties and sources as shown in Fig. 9.1.

The origin of certain pollutants and ecological impact (environmental and human health) is not yet known. Ultimately the pollutants land up in to the surface and even

Inorganic Pollutants in Water

DOI: https://doi.org/10.1016/B978-0-12-818965-8.00009-3

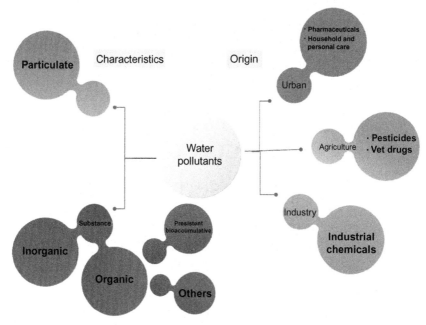

FIGURE 9.1   Various types of water pollutants.

ground water by various courses of actions such as soil erosion, run offs, or even leaching from the industrial, domestic, or agricultural wastewater (Geissen et al., 2015). The contamination from the microorganisms is rapidly considered, as their health effects are short term in comparison to certain inorganic pollutants. But certainly, for the strategic development for safe drinking water and pollution management, all the aspects must be kept under consideration. Inorganic water pollutants to some extent are vital elements for metabolic activities, while a few of them are not beneficial at all. For instance, Hg is highly toxic, even when consumed in trace amounts. These inorganic pollutants exist in natural ores through adsorption on sediment, soil, or rocks, rich in organic carbon content. Various physiochemical parameters such as pH and redox potential plays a key role in their bioavailability. Several of these pollutants are of geogenic origin such as arsenic and selenium and make their way into water bodies due to disturbance in redox conditions due to microbial as well as human originated activities. Because of the natural origin, these inorganic chemicals have widespread distribution in ground water and in turn surface water as well. They are found responsible for various severe health hazards attributable to contaminated drinking water ingestion. Water cycle is mainly regulated by the rainfall, which plays a role as the carrier of the particulate matter present in the atmosphere. The inorganic ions present in the air hold a notable portion of the particulate matter and are water soluble. They then get solubilized in the rain water and reach the surface of earth and in turn into ground water. Therefore to prevent the contamination through environmental precipitation, the source and chemical mechanisms responsible for producing these ions have to be monitored (Shen et al., 2012). Inorganic pollutants pose a high risk on health and environment by contaminating water, soil, and air.

Many a times, the inorganic pollutants are induced directly into the water sources through anthropogenic contribution. For instance, the currently found water pollutants such as persistent organic compounds, pharmaceutical and personal care products are as a result of their uncontrolled and unregulated disposal in water bodies. For instance, sunscreen waste dumped in water bodies comprises inorganic chemicals such as TiO_2 and ZnO and other chemical preservatives, colorants, surfactants, film forming agents, chelators, fragrances, and viscosity controllers that make their ways into ground and surface water systems. Most of these chemicals possess a characteristic of persistence and bioaccumulation (Tovar-Sánchez et al., 2013). Heavy metals and metalloids on one hand are entering into the ecosystems and, on the other hand, being nonbiodegradable, they are also getting accumulated in the environment and posing threat to it (Li et al., 2013).

Another category of inorganic pollutants, that is, nitrogen and its associates also adds up to water pollution due to its unregulated level in the environment. The presence of excess nitrogen in environment due to anthropogenic activities causes the alterations in the forest N cycle, eutrophication in the water, tropospheric ozone formation, which in turn damages plants and causes acidification of surface water (Driscoll et al., 2003; Dumont et al., 2005). However, its detection methodologies are not covered in this chapter, and major focus is done on heavy metals and metalloids. In order to ensure safe delivery of drinking water for human consumption, a set of limits has been settled by environmental bodies, namely, World Health Organization and US Environment Protection Agency. For the sustenance of quality of water, these guidelines are to be implemented stringently. The guidelines are applicable to and mostly adopted by all nations worldwide. This is because criteria and standards formulation hold same numerical values for pollutants due to the following facts (Chang et al., 2002):

- The mode of exposure is predominantly drinking water.
- Population that is exposed to the pollutant is defined clearly, that is, the entire community.
- Same database is utilized for deriving the chemical's dose–response relationships.
- The execution of the water treatment process is reasonably unswerving worldwide.

The standards and limits of the maximum contaminant levels for the various inorganic pollutants in water are being presented in Table 9.1.

To prevent the hazards caused by the inorganic pollutants, there is a need of their monitoring which can be fulfilled by the sensitive detection techniques. From last few decades, SERS has emerged as an effective option for the detection of heavy metals and metalloids as it provides the enhancements in the Raman signals it is capable of identifying the compounds specifically. This is resultant of the interaction of light with the vibrational modes or states of metallic structures. Although there are various analytical techniques existing currently for the detection of heavy metals in different environmental matrixes besides, SERS is gaining interest, the reason being that the other techniques are tedious, need preconcentration of the sample, relatively expensive, and need expert handling, whereas SERS enables the scope of portable, real time, and easy handling detection to be carried out (Eshkeiti et al., 2012a). It is a very sensitive technique, capable of detecting the trace

TABLE 9.1   Maximum permissible limits of inorganic pollutants in water.

Pollutant	Maximum permissible limit
Antimony	0.006 mg/L
Arsenic	0.01 mg/L
Asbestos	7 million fibers/L
Cadmium	0.005 mg/L
Chromium	0.1 mg/L
Cyanide	0.2 mg/L
Fluoride	4 mg/L
Mercury	0.002 mg/L
Nickel	0.1 mg/L
Nitrate	10 mg/L
Nitrite (as N)	1 mg/L
Nitrate and nitrite (combined)	10 mg/L
Selenium	0.05 mg/L
Thallium	0.002 mg/L

From Department of Environment and Natural Resources, South Dakota <https://denr.sd.gov/des/dw/IOCs.aspx>.

inorganic pollutants present in water (Sackmann and Materny, 2006). Implication of nanostructures in the SERS techniques is further giving rise to solutions for limitations associated with this technique in terms of reproducibility (Ullah et al., 2018).

This chapter discusses in detail the inorganic pollutants in water, their associated health effects, and detection approach through SERS. An insight on detection of various inorganic pollutants and the different substrates as well as materials utilized in the SERS has been provide in this chapter.

9.2 Health effects of inorganic pollutants

As, mentioned earlier in this chapter that the inorganic pollutants accountable for the water pollution pose severe health risks to the population consuming it. The health effects associated with them are usually visible after a prolonged time of the consumption of the contaminated water. However, certain inorganic pollutants also have health effects observable in a shorter time period. These pollutants find their way into the drinking water through agricultural, domestic, and industrial activities. Natural occurrence of these pollutants depends on several factors such as geological settings, source of water, and climate (Thompson et al., 2007). The clinical health effects of certain inorganic pollutants are summarized in Table 9.2.

TABLE 9.2 Health effects of various inorganic water pollutants along with their sources.

Inorganic water pollutant	Source	Health hazards	References
Antimony	Flame retardant in textile, rubber, plastics, etc.	• Weight loss • Head ache • Dizziness • Nausea • Vomiting • Diarrhea	Fu et al. (2011)
Arsenic	Natural (rocks and ores) anthropogenic (mining, wood preservatives, pesticides, and smelting wastes)	• Acute paralytic syndrome • Acute gastro intestinal syndrome • Diabetes • Central nervous disruption • Death	Abernathy et al. (2003)
Asbestos	Mining, electrical generation and distribution industry, chemical industry, metal products manufacturing industry, shipping industry, construction industry, and railroad industry	• Respiratory diseases • Immunological disorder • Lung cancer • Malignant mesothelioma • Reduction of tumor immunity • Auto immune disorders	Otsuki et al. (2007) and Van den Borre and Deboosere (2015)
Barium	Natural and anthropogenic activities (petroleum industries, semiconductors production, and steel industry)	• Blood pressure • Renal failure • Peribronchial sclerosis • Anemia • Hearing loss • Endocrine disruption • Hepatic failure • Effect on fetus	Kravchenko et al. (2014)
Beryllium	Natural (rocks, volcanic dust, oil, and coal) and industry (biomedical, fire prevention, manufacturing, automotive, scrap recovery, nuclear energy sites, and aerospace)	• Lung disorder • Cardiovascular system disorder • Cardiac enlargement • Cancer • Death	Cooper and Harrison (2009)
Cadmium	Industry, cigarette smoke, food from contaminated water or soil	• Vertigo • Choking or vomiting • Head ache • Abdominal pain • Pneumonitis • Leg pains • Asthma	Johri et al. (2010)

(Continued)

TABLE 9.2 (Continued)

Inorganic water pollutant	Source	Health hazards	References
Chromium	Nuclear facilities, metal finishing, industry (leather, textile, and dyes), mining	• Cancer (head and neck) • Mutagenic	Gao and Xia (2011)
Cyanide	Natural and anthropogenic (extraction of gold and silver from ores, fumigants in ship, flour mills, and railroad cares), tobacco smoke	• Damage to respiratory system • Cardiovascular damage • Central nervous system damage	World Health Organization (2004)
Fluoride	Natural (volcanic magma, rocks, and minerals) and anthropogenic (coal ash, phosphate pesticides)	• Impaired thyroid function • Decreased birth rate • Dental fluorosis • Skeletal fluorosis • Kidney stones • Lower intelligence in children • Bladder cancer	Ozsvath (2009)
Mercury	Natural (Hg deposits), anthropogenic (pulp and paper, ceramic, chlor-alkali industries)	• Coronary heart disease • Heart attack • Stroke • Death • Cardiovascular disease	Monteiro et al. (2010) and Driscoll et al. (2013)
Nickel	Natural (in earth's crust, soils, volcanoes, meteorites), anthropogenic (metallurgical processes, alloy production, electroplating, and nickel–cadmium batteries)	• Cancer • Mutagenic • Developmental abnormality • Anemia • Immunotoxicity • Neurological effects • Reproductive abnormalities	Das et al. (2008)
Nitrate	Natural and agricultural (nitrogen based fertilizers)	• Methemoglobinemia	Wongsanit et al. (2015)
Nitrite (as N)	Nitrate gets reduced to nitrite	• Methemoglobinemia (more potent as compared to nitrate)	Yousefi et al. (2016)
Selenium		• Prostate cancer • Skin cancer • Type 2 diabetes • Proteomic changes	Jablonska and Vinceti (2015)
Thallium	Natural (soil, fresh water, marine water and air) and industry (glass, pharmaceutics, cement, and electronic)	• Cytotoxicity • Genotoxicity • Developmental problems in children	Campanella et al. (2016) and Rodríguez-Mercado et al. (2017)

CNS, Central nervous system.

9.3 Surface-enhanced Raman scattering

Raman spectroscopy that was named after its discoverer himself Sir C.V. Raman is a milestone in physics as he was awarded the Nobel Prize in 1930 for this discovery. It is an analytical tool that gives away the information of chemical composition and molecular structure of a molecule. The signals of Raman spectroscopy are very weak as this phenomenon occurs for one time in a million cases.

Surface-enhanced Raman spectroscopy (SERS) overcomes this hurdle by enhancing the Raman signals and making the process more sensitive by the virtue of which even a single molecule can be detected at a time. Even the larger enhancements can be gained if SERS is coupled along with resonance Raman. SERS for the first time was observed by Martin Fleischmann, Patrick J. Hendra, and A. James McQuillan in the adsorption of pyridine on electrochemically roughened silver. SERS works on the principle that the Surface Plasmon enhances the Raman scattering in order to increase the signals to improve the sensitivity of the technique. The mechanism that leads to the surface enhancement is not yet understood completely, whereas the two main contributions are proposed from electromagnetic and charge transfer (CT) mechanism. The significance of Raman spectroscopy elevated after the discovery of LASER. Laser light is made to be incident on the metal surface, which causes the plasmon excitation that in turn leads to the enhancement of the electromagnetic field, whereas the mechanism of CT involves the transfer of charge in between the metal surface and the adsorbed analyte (Hering et al., 2007). The increased sensitivity of SERS enables the detection of various water pollutants. It has been already proven as a dynamic analytical tool for biological and chemical sensing. These newly found techniques include the capabilities of detecting bacteria and viruses even at very low concentrations. They also exhibit the distinct quality of discriminating even between the types and strains of the microorganisms/pathogens together with the pathogens with the gene deletions. SERS has a potential for biosensing as it has the ability of providing the unique fingerprints of the analytes in its spectrum (Tripp et al., 2008).

Mostly the SERS substrates are designed by nanofabrication techniques and oblique angle deposition to ensure uniformity and reproducibility in the substrates. Despite the limitations in SERS such as substrate stability, reproducibility, and homogeneity, it is a favored technique in the chemical sensing as well. The reason behind its robustness is that SERS can detect a wide range of chemicals, including chlorinated solvents, explosives, drugs and pharmaceuticals, ionic nutrients, and pesticides (Mosier-Boss, 2017). Fig. 9.2 delineates the various applications of this technique, supporting its versatility as an analytical tool.

In the last decade SERS has emerged as a powerful tool in the field of analysis and detection. Its versatile applications in the areas such as chemical sensing, biological sensing, real world applications, and even the detection of a single molecule prove it to be a worthy technique. SERS can be employed in the detection of small molecules as well as large biomolecules. It can be implemented in the detection of the biomolecules such as proteins, DNA, biomarkers, drugs, microorganism, nucleobases, and viruses. SERS is also competent in the detection of low concentration melamine, where SERS can detect it at a single molecule level. Despite the different target analytes, there are certain factors that hugely influence the substrate quality and those are linear detection range, selectivity, and

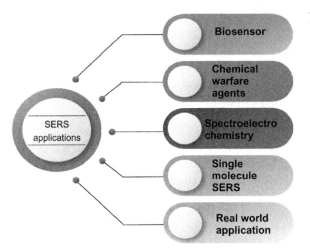

FIGURE 9.2   Multiple applications of SERS.

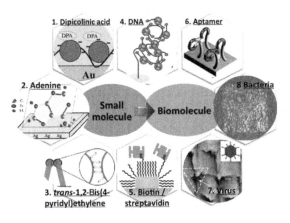

FIGURE 9.3 SERS facilitated detection of numerous small molecules and biomolecules (Luo et al., 2014).

sensitivity. Fig. 9.3 portrays the different small molecules and biomolecules detection by the application of SERS. Development of smart or intelligent material for SERS is of foremost interest in biomedical field. In near future, SERS could be implemented in the differentiation of the specific and nonspecific binding. It could also be helpful in not only the analysis of cell signals but also the behavior of whole tissues and interaction of intracellular behavior (Luo et al., 2014).

9.4 Surface plasmon in surface-enhanced Raman spectroscopy

Surface plasmon is a phenomenon in which the delocalized electrons behave as dielectric—metal interface on the surface of the metal, where electromagnetic wave interaction takes place. If these plasmons are in nanometer range, the electron density can couple

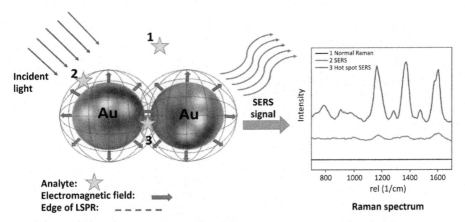

FIGURE 9.4 Schematic view of SERS phenomenon for an organic analyte onto Au NPs (Wei et al., 2015).

with the electromagnetic radiations, which are way larger as compared to the particle size. These properties of plasmons administer them with the excellent application in the SERS.

As depicted in Fig. 9.4, gold nanoparticles (Au NPs) exhibited SERS enhancement by virtue of surface plasmon resonance (SPR) of Au NPs. The creation of "hot spots" leads to stronger enhancement in the signals as compared to the monomer surface. These "hot spots" are the localized areas, where the particle distance is less than 10 nm although the exact distant is still a topic of debate. Different shapes of nanoparticles, namely, nanorods, nano stars, and nano prisms, produce different types of hot spots due to the difference of gap in them w.r.t. to their shapes and edges (Wei et al., 2015).

9.4.1 Materials in surface-enhanced Raman spectroscopy-based sensing

As discussed above, the phenomenon of surface plasmon is the one providing SERS with the signal enhancement or it will not be false to say that SERS is a nanoscale effect. The major role is played by the material, which is used in the signal enhancement. The prerequisite property the material or metal should possess is plasmonic behavior. There are many metal nanosurfaces that are utilized as the substrate for the SERS signal enhancement. The metals showing the plasmonic behavior such as Au, Ag, Cu, and Al have also made their place in the list in this category. Several fabrication methods have been used for the design of variety of SERS active nanomaterials. As with variation in size and shape, SERS enhancement factor changes. For example, polystyrene latex microspheres adhered on a glass slide is coated with the silver metal and their performance is found to be dependent upon the size of spheres, silver thickness, and excitation wavelength, which in turn affects the signal magnitude (Moody et al., 1987). In addition, nanorods are also utilized for the detection of environmental pollutants by SERS technique (Huang et al., 2013). It is observed that the amplitude of the enhancement through CTs is hugely specific to the molecules. The major parameters with utmost significance in the SERS application are SERS substrate and the laser source. SERS substrate is the basic dictating factor of the signal enhancement. The plasmonic behavior of the metal substrate causes the electromagnetic enhancements and by the virtues of which the SERS signal gets enhanced. After the signal enhancement the further interpretation of the signals is similar to that of Raman spectroscopy. In addition, the laser source (the wavelengths of the incident laser) also has a significant effect on the signal enhancements (Sharma et al., 2012).

Different plasmonic nanomaterials are used in the detection of heavy metals. An interesting category is metal containing nanocomposites for metal ion detection in water samples by employing SERS. To introduce selectivity to the SERS substrate, various chemical and biological linkers such as cysteine and 3,5-dimethoxy-4-(6-azobenzotriazolyl) phenol, EDTA, DNA, citrate, 2-mercaptoisonicotinic acid, 2-mercaptoethanesulfonate, and dendrimer, have been reported for analyte selective sensing (Pinheiro et al., 2018). Fig. 9.5 broadly categorizes the class of materials for SERS-based detection of environmental pollutants.

Activity of SERS is highly morphology dependent and hence it is influenced by the substrate's morphology. Various experiments have been done and are still going on with the substrate development for SERS. In order to enhance the SERS activity, there has been a recent development in the substrate where it exhibits a sea urchin such as morphology. Here, MoO_3 is structured with a sea urchin like morphology where it consists of hundreds of long spikes originating from a core forming a globular structure of nearly 20−40 μm. These spikes taper to constitute 20 nm sharp tips and are reported to enhance the SERS activity by the order of 10^5 and achieve a detection limit of 100 nM (Prabhu et al., 2019).

As discussed in previous section, plasmons are the basics in SERS technique and are behind the signal enhancement. But the precise modeling of Plasmon physics is challenging. A need has emerged for the correction of standard electromagnetic approaches. In gold nanoparticles the applications of Plasmon absorption and scattering are enormous, extended from sensing photothermal effects to cell imaging. Plasmon resonance is very much affected by the dielectric of the matrix. Hence, the fabrication of Au NPs with absorbing materials, namely, metal oxides, graphene, or semiconductor quantum dots is frequently pursued for the hybrid multifunctional materials (Amendola et al., 2017). Metal oxides doped with charged impurities or crystal defects can be utilized as the substitutes for the traditional metals. Metals are inherently conducting and exhibit a carrier density of

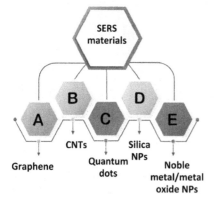

FIGURE 9.5 Various materials utilized in the formulation of SERS substrate.

the order of 10^{23} cm^{-3} and the Fermi level within the conduction band. In contrast, the metal oxides have a bandgap, which provides an opportunity of tuning the conductivity by generating the free charge carriers. These metal oxides can be either n-type or p-type doped although most of the conductive metal oxides derive the conductivity from free electrons hence, n-type. Doping gives the privilege of controlling carrier concentration. Certain metal oxides with local Plasmon resonance properties find their applications in the SERS, to mention a few; commonly known ITO (Sn:In$_2$O$_3$), Al:ZnO, etc., are the metal oxides that can find application as the base for the substrate (Agrawal et al., 2017).

9.4.2 Surface-enhanced Raman spectroscopy for detection of inorganic pollutants

SERS is basically grounded on the signal enhancement if molecules adsorbed on the nanostructures metallic surface in Raman spectroscopy. SERS is capable of detecting very low concentration up to single molecule of any analyte and have proven itself as a sensitive technique for the detection purposes. This section will discuss the various detection applications of SERS for varied inorganic pollutants utilizing a number of different substrates. The substrates in SERS hold an important place in the detection process as they incorporate the enhancement factor which increases the signal strength and sensitivity. Arsenic being a very toxic metal can pose higher health risks at very lower concentrations, and polyhedral Ag nanoparticles were developed for the detection of As in aqueous medium. These polyhedral Ag nanoparticles were obtained with varying shapes and the sizes. The shape of the particles was varied as cubes, cuboctahedral, and octahedral, using a polymer capping agent. The obtained structures are shown to detect both form of arsenic, that is, arsenite and arsenate in water. The authors report a detection limit as low as 1 ppb while ensuring high reproducibility and robustness of the method, ensuring its field applicability (Mulvihill et al., 2008). Similarly, Ag nanoparticles are studied for the detection of As(V) by using coffee ring effect technique. It is a dynamic tool for the nanomaterials self-structuring, which involves the evaporation of the aqueous droplet containing the nonvolatile compounds. A three phase contact line exists between the atmosphere, droplet, and the solid substrate. In this study the droplet of Ag nanoparticles and the analyte [As(V)] together were studied for the coffee ring effect. The levels of detection were dependent on the concentration of Ag nanoparticles in the aqueous solution. This work demonstrated the high enhancement factor for the signal and high enrichment factor for the target analyte. This platform proves to be simple, effective and sensitive (Wang et al., 2014). Furthermore Ag nanoparticles capped with polyvinylpyrrolidone exhibiting the octahedral shape manifested the largest enhancement factor of 10^7–10^8 depending on wavelength by monolayer for the detection of As(V). The substrate having Ag nanowires fabricated by applying two phase interfacial self-assembly technique and were proved to be selective, stable, reproducible, sensitive, and uniform. Another substrate was fabricated by multilayer Ag films developed by one step electroless plating process and have a characteristic of depressing the background peaks of the substrates. There is a linear relationship between the As(V) SERS signals and the concentration of multilayer Ag nanofilms substrates. Same as in As(V), the two characteristics SERS bands are seen in As(III) as well.

The positions of these peaks vary on certain factors such as excitation laser wavelength, SERS substrate, and experimental conditions. It has also been understood that in the spiked samples of As(III) species formulated in pure water, it is subjected to oxidation during SERS measurements, specifically the lower ppb level concentrations, whereas no oxidation was seen in the case of raw groundwater even for a high power laser and long laser irradiation time. There is an indirect detection strategy developed for the analysis of As(III) selectively at trace levels using glutathione/4-mercaptopyridine (4-MPY) modified Ag nanoparticles. In this approach, As(III) binds with 4-MPY and the accumulation of Ag nanoparticles happens by the virtue of which the Raman signal enhances and in turn is proportional to the As(III) concentration in the sample which was 0−700 ppb for this study. The limit of detection (LOD) was found to be 0.76 ppb which is quite low concentration. The substrate is also selective and holds good reproducibility (Hao et al., 2015).

In addition, a turn on surface enhances Raman scattering sensor of a new class was reported for the selective and sensitive detection of cadmium [Cd(II)]. This sensor is based upon the interparticle plasmonic coupling created in the Cd(II) selective nanoparticles self-aggregation. The sensor consisted of gold nanoparticles of size 41 nm and also encrypted with a Raman active dye adhered through a disulfide group. The selectivity of the nanoparticles enables them to self-aggregate followed by the signal enhancement in SERS up to 90 folds. The SERS sensor was capable of detection even in the colored samples. The interference studies revealed that there is no interference from any other ion in the detection of Cd(III) except for Zn^{2+} due to the reason that there is an electronic structure similarity among them. This sensor validates its unique benefits as its combination with portable Raman spectroscopy will provide a scope for onsite sampling and quantitative detection of Cd (Yin et al., 2011). Another sensor for detection of Cd as well as Hg was developed by ink jet printing of Ag nanoparticles on to the surface of silicon wafer. The average particle size of Ag nanoparticles is reported to be 150 nm and the thickness of the Ag nanoparticles film deposited on the surface of silicon wafer was evaluated to be 400 nm by the means of vertical scanning interferometry. Due to the formation of hot spots in between the nanoparticles aggregates, the signal enhancement of 3−5 times could be observed for 25 mM of CdS, 45 mM of HgS, and 75 mM of ZnO in comparison to the CdS, HgS, and ZnO when immobilized on a bare silicon wafer (with no film of Ag nanoparticles printed on it). This sensor can be tuned for the selectivity of a wide range of biochemical detection (Eshkeiti et al., 2012b). Further, Alizarin which is a highly Raman active dye was utilized to functionalize Au NPs to form a strong SERS substrate. Once the dye Alizarin is functionalized on the Au NPs, 3-mercaptopropionic acid, 2,6-pyridinedicarboxylic acid were also immobilized on the NPs for the selective coordination of Cd(II). This sensor is sensitive toward Cd(II) against alkali and other heavy metals. The LOD of this sensor for Cd(II) detection was reported to be as low as 10 ppt and designed probe is really simple and rapid and also holds prospective of scale up for field applications (Dasary et al., 2016).

On moving forward for the other surface-enhanced Raman spectroscopy (SERS) substrates a label free Au NPs−based system is developed for the detection of highly toxic lethal poison known that is cyanide. This system is competent for the detection of cyanide in parts per trillion (ppt) levels with high discrimination against supplementary cations and anions even in the environmental samples. The detection is based on the specificity that CN^- and ascorbic acid—coated Au NPs form aggregates as a result of Au—CN complex formation. This leads to the formation of several hot spots and hence causes the enhancement in the signal by the order of the magnitude nearly 5.8×10^7 (Senapati et al., 2011).

A novel method for the detection of Hg was developed by utilizing SERS—active probe which is DASS (dimethyldithiocarbamic acid sodium salt) modified Ag nanoparticles. This provides a highly sensitive sensor for Hg as mercury species have a tendency to specifically bind with sulfur atoms and this promotes a shift in frequency. As compared to the intensity—dependent quantitative determination by SERS the frequency—shift-based method of SERS is more advantageous due the fact that it has higher accuracy and sensitivity, perfect linear relationship, and smaller standard deviation. The LOD of this probe for mercury is as low to 10^{-8} M. This probe possesses a potential of on-site mercury detection in environmental samples (Chen et al., 2016). SERS active probe by utilizing oligonucleotide functionalized magnetic silica sphere at Au NPs were developed for the selective detection of Hg^{2+} and Ag^+ ions. This assembly utilizes the mismatched T—Hg—T and C—Ag—C bridges to capture Hg^{2+} and Ag^+. The responses are in the range of $0.1-1000$ nM for Hg^{2+} ions and in the range of $10-1000$ nM for Ag^+ ions. This assay is highly selective for mercury(II) and silver(I), and there is no interference of other ions. The probe also exhibits the recyclable property and can be recycled up to 80% by the help of cysteine (Liu et al., 2014). Similar to the previous substrate, this substrate also exploits the mechanism of T—Hg—T formation. Here DNA technology is utilized with the silicon nanowire array fabricated with gold nanoparticles. The basic mechanism behind the working of the sensor is that the enhancement of the SERS signals takes place when DNA structure converts to hairpin structure in the presence of mercury ion. Resultant to this method is capable of detecting very low concentrations of mercury, in the range of ppt. The concentrations of Hg detected here were 1 pM (0.2 ppt). The salient features of this sensor include selectivity, recyclability, and also real time applicability for detection in real water samples such as river waters (Sun et al., 2015).

In addition, there was development of an extraordinary enhancement factor providing substrate. This substrate is capable of detecting mercury ions up to a detection limit of 0.8 pg/mL. The substrate is a self-assembled dimer constituting of gold nano star and ssDNA. Structure of the substrate includes several "hot spots" and the multiple sharp tips counterfeits lighting rods and enhances the EM field around the particle (Ma et al., 2013). Another substrate for the detection of Hg^{2+} composed of gold nanorods—polycaprolactone nanocomposites fibers fabricated with ion-binding ligands. Bridging molecules such as 2,5-dimercapto-1,3,4-thiadiazole dimer and trimercaptotriazine were also incorporated for capturing the metal ions. This substrate achieves the detections of vibrational spectroscopically silent heavy metals (Tang et al., 2015). A novel SERS-based sensor was reported for the chemical speciation of metal ions in water. This is the first example reported for such detection strategy. The sensor was studied for the detection of inorganic mercury (Hg^{2+}) and methyl mercury (CH_3Hg^+). The segments of the sensing platform for SERS are 4-MPY on gold nanoparticles anchored onto polystyrene microbeads. Under the same detection conditions the LOD obtained for Hg^{2+} and CH_3Hg^+ were 0.1 and 1.5 ppb, respectively (Guerrini et al., 2014). Another substrate for the mercuric ion detection based on T—Hg^{2+}—T interaction was investigated. The basis of the strategy for developing this sensor was the fact that Hg^{2+} mediates the ssDNA transformation to double helical DNA. Substrate was formulated by modifying Au NPs by DNA and assembled into chains. These chains displayed considerable SERS signal. The LOD reported was 0.45 pg/mL where the length of the Au nanochains was directly proportional to the Hg^{2+} concentrations (Xu et al., 2015). Again the $T-Hg^{2+}-T$ interaction was utilized in developing a sandwich structured substrate for the detection of Hg^{2+} ions in water as well as soil samples. The substrate was constructed in a sandwich structure by Au triangular nano arrays (Au TNAs)/n-layer grapheme/Au NPs. Graphene layer employed here constructed the uniform subnanometer gaps in between Au TNAs and Au NPs which in turn created more hot spots and smaller interparticle distance. As we know the smaller interparticle distance mediates the "hot spots" formation by the virtue of which signal strength enhances. The reported substrate was successful in detecting the Hg^{2+} ions at the low levels of 8.3×10^{-9} M (LOD). This SERS substrate also possesses the versatility of detecting the Hg^{2+} ions in contaminated sandy soil in addition to the water (Zhang et al., 2017).

A multimodal nanosensor was developed for the colorimetric detection of Ni along with SERS. This type of sensor fosters the sample detection by employing different techniques thus for diverse analytical method to be quick and sensitive the surface chemistry design holds its own importance. As usual, the nanoparticle size exhibits an important role in the functional nanoparticle designing. Such an assembly has been reported for the trace detection of Ni(II) in an aqueous matrix. Combination of the nitrilotriacetic acid moieties and L-carnosine dipeptide covalently bonded with each other and then combined with nanoparticle surface provides a highly sensitive, selective, and rapid platform for Ni (II) detection at trace levels (Krpetić et al., 2012). Also, a very sensitive SERS substrate was formulated for developing a quantitative analysis method to detect Ni^{2+}. The nanogold reaction in trisodium citrate and chloroauric acid ($HAuCl_4$) is very slow. Whereas, graphene oxide nano ribbon (GONR) manifests a strong catalysis during the reaction in the presence of Victoria blue 4R (VB4r) molecular probe. GONR concentration increase caused the increase in the SERS peak due the elevated formation of Au NPs. On adding the dimethylglyoxime (DMG) ligand the SERS peak got diminished as a result of the inhibited GONR catalysis. Ni^{2+} forms stable complexes with DMG, that is, $[Ni(DMG)_2]^{2+}$ attributable to the increase in the SERS peak as free GONR is released. The change in peak is linear and the LOD reported is 0.036 μM (Liang et al., 2018) (Table 9.3).

An interesting SERS platform has been reported for the detection of multiple heavy metal ions at same time. The substrate consists of gold surface functionalized by diethylenetriaminepentaacetic acid (DTPA). The grafted DTPA molecules have a high SERS response and they show an efficient chelation of metal ions as well. Depending on the metal ions the SERS response changes, where the peak shift is found to be depending on the atomic number of the heavy metal ion. These DTPA are covalently attached to the gold surface by following a two-step protocol: (1) using diazonium chemistry the grafting of 4-aminophenylene groups and (2) their acylation by DPTA anhydride. The LOD for the developed substrate is reported to be 10^{-14} M. Below presented Fig. 9.6 is a graphical representation of the SERS platform and its detection process (Guselnikova et al., 2017).

Moreover, for the detection of nitrate, a cationic coated silver substrate for SERS was developed for the detection of nitrate and sulfate in the aqueous matrix or environments. The reported method was executed at near infrared excitation wavelength; it showed a

TABLE 9.3 Materials reported for SERS-based detection of heavy metals in water.

Heavy metal	SERS material	LOD	References
Arsenic (As)	Polyhedral Ag nanoparticles	1 ppb	Mulvihill et al. (2008)
	Ag nanoparticles by using coffee ring effect		Wang et al. (2014)
	Ag nanoparticles capped with polyvinylpyrrolidone	0.76 ppb	Hao et al. (2015)
Cadmium (Cd)	Raman active dye encrypted gold nanoparticles	–	Yin et al. (2011)
	Ink jet–printed Ag nanoparticles on Si wafer	25 mM	Eshkeiti et al. (2012b)
	Alizarin functionalized gold nanoparticles	10 ppt	Dasary et al. (2016)
Mercury (Hg)	Ink jet–printed Ag nanoparticles on Si wafer	45 mM	Eshkeiti et al. (2012b)
	Dimethyldithiocarbamic acid sodium salt–modified Ag nanoparticles	10^{-8} M	Chen et al. (2016)
	Oligonucleotide-functionalized magnetic silica sphere at Au nanoparticles	–	Liu et al. (2014)
	Silicon nanowire array fabricated with gold nanoparticles	1 pM (0.2 ppt)	Sun et al. (2015)
	GNS–DNA dimers	0.8 pg/mL	Ma et al. (2013)
	Au NR–PCL nanocomposite fibers	–	Tang et al. (2015)
	Monolayer of 4-MPY on Au nanoparticles anchored on polystyrene microbeads	0.1 ppb	Guerrini et al. (2014)
	DNA modified Au nanochains	0.45 pg/mL	Xu et al. (2015)
	Au TNAs/n-layer grapheme/Au NPs in sandwich structure	8.3×10^{-9} M	Zhang et al. (2017)
Nickel (Ni)	Nitrilotriacetic acid moieties and L-carnosine dipeptide covalently bonded and then combined with nanoparticle		Krpetić et al. (2012)
	Graphene oxide nanoribbon catalytic gold nanoreaction regulated by dimethylglyoxime reaction	0.036 μM	Liang et al. (2018)

Au NPs, Au nanoparticles; *Au NR*, gold nanorods; *Au TNAs*, Au triangular nano arrays; *DASS*, dimethyldithiocarbamic acid sodium salt; *GNS*, gold nano star; *LOD*, limit of detection; *MPY*, mercaptopyridine; *PCL*, polycaprolactone.

linear concentration response and the detection limits (LOD) were obtained to be 260 ppm for nitrate and 440 ppm for sulfate, respectively. The nitrate and sulfate ions get adsorbed on to the cationic coatings and this adsorption is described by Frumkin isotherm (Mosier-Boss and Lieberman, 2000). Here a commercially available gold nanosubstrate that was a gold-coated silicon material was examined for its applicability of nitrate detection in SERS.

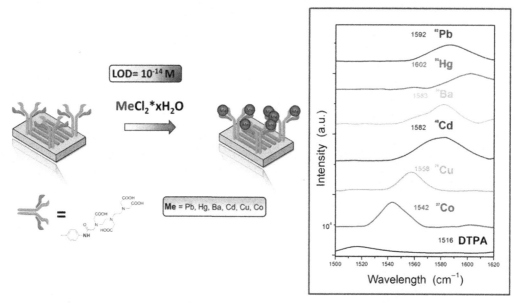

FIGURE 9.6   Surface plasmon-polariton−based functionalized SERS substrate for multiple heavy metal ions detection (Guselnikova et al., 2017).

The detection of nitrate was carried in water and wastewater. The enhancement in the signal was found to be by a factor of $\sim 10^4$ as compared to Raman spectroscopy. It was investigated that phosphate ion emerged as the major interfering ion in the nitrate detection. The LOD of nitrite in water and wastewater was achieved about 0.5 mg/L (0.5 ppm). This study depicts that the nanosubstrates coupled with SERS can be a promising way of detection of nitrate concentrations in water (Gajaraj et al., 2013). Nitrate in the water sources usually gets reduced to nitrite, and it is known that the nitrite ion is more toxic than nitrate ion. The concentrations of nitrite ion required for causing the blue baby syndrome are far lower than that of nitrate ion concentrations. For the selective, rapid, and sensitive detection of nitrite a SERS sensor is developed from citrate-coated silver nanoparticles. The detection method is based on the diazo reaction of nitrite with p-nitroaniline and 1-naphthylamine in acidic medium. The SERS substrate basically detects the product from this azo reaction that is the azo dye (4-(4-nitrophenyldiazenyl) naphthalene-1-aminium), which directly gives the nitrite concentrations present in the samples. The detection limits for nitrides were obtained as 0.01 mg/L (0.01 ppm) at 720 cm^{-1}, 0.03 mg/L (0.03 ppm) at 1459 cm^{-1} and 0.05 mg/L (0.05 ppm) at 1609 cm^{-1}, respectively. This method in combination with the portable Raman spectrometer can be utilized and applied in on-site natural samples detections (Ma et al., 2014).

Certain explosives also contribute to the class of inorganic water pollutants, there are various explosives listed as in organic water pollutants. Perchlorate is one of the inorganic explosive that contaminates the water. The detection of perchlorate was carried out on gold nanoparticles as SERS substrate functionalized by 2-dimethylaminoethanethiol, the LOD achieved was 10^{-9} M. Another substrate for the detection of perchlorate employed was Ag nanoparticles SAM on rough Cu foil which was fabricated by HS$-$CH$_2$$-CH_2$$-NH_2$$-$HCl. This substrate proved to be more sensitive with a LOD of ʻ1 μM (Gillibert et al., 2018). Another approach for the detection of inorganic explosives such as perchlorates (ClO$_4^-$), nitrates (NO$_3^-$), chlorates (ClO$_3^-$), picric acid, and 2, 4-dinitrophenol was made by employing Ag nanowires as SERS substrate. These Ag nanowires were synthesized by the hydrothermal synthesis process using polyvinylpyrrolidone as a negatively charged stabilizer. The SERS substrate that is Ag nanowires was modified by positively charged diethyldithiocarbamate. This membrane serves as an efficient SERS substrate for the detection of both typical oxidizers in inorganic explosives as well as organic nitro explosives, through electrostatic interaction. The limits of detection of the employed substrate for different explosives were obtained: for perchlorates (ClO$_4^-$) it was 2.0 ng, for nitrates (NO$_3^-$) it was found 0.1 ng, for chlorates (ClO$_3^-$) it was detected 1.7 ng, for picric acid it was 45.8 ng and for 2,4-dinitrophenol it was found 36.6 ng which can be considered as a satisfactorily low levels. This substrate is also capable of detecting the typical explosives in the waters samples likely black powders, match heads, and fire crackers, which represent the versatility of the substrate. To top it all, it can also be perceived as an efficient alternative for implication in the onsite or field detection of the inorganic explosives in natural water samples (Shi et al., 2016).

9.5 Conclusion

A million tons of sewage is produced in combination by industry and domestic wastewater per day that can stand equal to the weight of the whole world's population. These anthropogenic wastes as well as natural geogenic disturbances are adding up to the inorganic pollutants concentrations in water. SERS is a highly sensitive and dynamic techniques, which can enable monitoring of trace amount of pollutants in water. It is therefore important to design SERS active substrates, which have biological and chemical compatibility with the analyte to be detected, temporal, and chemical stability and should be economical and reproducible. Such substrates should also be robust, possess high-enhancement factors, cheap, reproducible, and can be commercialized. Thus this technique holds significant potential for field usability for inorganic pollutants detection in water.

Acknowledgment

PR acknowledges DST (Department of Science and Technology), New Delhi, India for the financial support through WOS-A fellowship for this study.

References

Abernathy, C.O., Thomas, D.J., Calderon, R.L., 2003. Health effects and risk assessment of arsenic. J. Nutr. 133 (5), 1536S$-$1538S.

Agrawal, A., Johns, R.W., Milliron, D.J., 2017. Control of localized surface plasmon resonances in metal oxide nanocrystals. Annu. Rev. Mater. Res. 47, 1$-$31.

Amendola, V., Pilot, R., Frasconi, M., Marago, O.M., Iati, M.A., 2017. Surface plasmon resonance in gold nanoparticles: a review. J. Physics: Condens. Matter 29 (20), 203002.

Campanella, B., Onor, M., D'Ulivo, A., Giannecchini, R., D'Orazio, M., Petrini, R., et al., 2016. Human exposure to thallium through tap water: a study from Valdicastello Carducci and Pietrasanta (northern Tuscany, Italy). Sci. Total Environ. 548, 33–42.

Chang, A.C., Pan, G., Page, A.L., Asano, T., 2002. Developing Human Health-Related Chemical Guidelines for Reclaimed Water and Sewage Sludge Applications in Agriculture. World Health Organization.

Chen, L., Zhao, Y., Wang, Y., Zhang, Y., Liu, Y., Han, X.X., et al., 2016. Mercury species induced frequency-shift of molecular orientational transformation based on SERS. Analyst 141 (15), 4782–4788.

Cooper, R.G., Harrison, A.P., 2009. The uses and adverse effects of beryllium on health. Indian J. Occup. Environ. Med. 13 (2), 65.

Das, K.K., Das, S.N., Dhundasi, S.A., 2008. Nickel, Its Adverse Health Effects & Oxidative Stress. Indian J. Med. Res. 128 (4), 412.

Dasary, S.S.R., Jones, Y.K., Barnes, S.L., Ray, P.C., Singh, A.K., 2016. Alizarin dye based ultrasensitive plasmonic SERS probe for trace level cadmium detection in drinking water. Sens. Actuators B: Chem. 224, 65–72.

Driscoll, C.T., Whitall, D., Aber, J., Boyer, E., Castro, M., Cronan, C., et al., 2003. Nitrogen pollution in the northeastern United States: sources, effects, and management options. BioScience 53 (4), 357–374.

Driscoll, C.T., Mason, R.P., Chan, H.M., Jacob, D.J., Pirrone, N., 2013. Mercury as a global pollutant: sources, pathways, and effects. Environ. Sci. Technol. 47 (10), 4967–4983.

Dumont, E., Harrison, J.A., Kroeze, C., Bakker, E.J., Seitzinger, S.P., 2005. Global distribution and sources of dissolved inorganic nitrogen export to the coastal zone: results from a spatially explicit, global model. Global Biogeochem. Cycles 19 (4), 14.

Eshkeiti, A., Reddy, A.S.G., Narakathu, B.B., Joyce, M.K., Bazuin, B.J., Atashbar, M.Z., 2012a. Gravure printed surface enhanced Raman spectroscopy (SERS) substrates for detection of toxic heavy metal compounds. SENSORS, 2012 IEEE. IEEE, pp. 1–4.

Eshkeiti, A., Narakathu, B.B., Reddy, A.S.G., Moorthi, A., Atashbar, M.Z., Rebrosova, E., et al., 2012b. Detection of heavy metal compounds using a novel inkjet printed surface enhanced Raman spectroscopy (SERS) substrate. Sens. Actuators B: Chem. 171, 705–711.

Fu, Z., Wu, F., Mo, C., Liu, B., Zhu, J., Deng, Q., et al., 2011. Bioaccumulation of antimony, arsenic, and mercury in the vicinities of a large antimony mine, China. Microchem. J. 97 (1), 12–19.

Gajaraj, S., Fan, C., Lin, M., Hu, Z., 2013. Quantitative detection of nitrate in water and wastewater by surface-enhanced Raman spectroscopy. Environ. Monit. Assess. 185 (7), 5673–5681.

Gao, Y., Xia, J., 2011. Chromium contamination accident in China: viewing environment policy of China. Environ. Sci. Technol. 45, 8605–8606.

Geissen, V., Mol, H., Klumpp, E., Umlauf, G., Nadal, M., van der Ploeg, M., et al., 2015. Emerging pollutants in the environment: a challenge for water resource management. Int. Soil Water Conserv. Res. 3 (1), 57–65.

Gillibert, R., Huang, J.Q., Zhang, Y., Fu, W.L., de La Chapelle, M.L., 2018. Explosive detection by surface enhanced Raman scattering. TrAC Trends Anal. Chem. 105, 166–172.

Guerrini, L., Rodriguez-Loureiro, I., Correa-Duarte, M.A., Lee, Y.H., Ling, X.Y., Garcia de Abajo, F.J., et al., 2014. Chemical speciation of heavy metals by surface-enhanced Raman scattering spectroscopy: identification and quantification of inorganic-and methyl-mercury in water. Nanoscale 6 (14), 8368–8375.

Guselnikova, O., Postnikov, P., Erzina, M., Kalachyova, Y., Švorčík, V., Lyutakov, O., 2017. Pretreatment-free selective and reproducible SERS-based detection of heavy metal ions on DTPA functionalized plasmonic platform. Sens. Actuators B: Chem. 253, 830–838.

Hao, J., Han, M.-J., Han, S., Meng, X., Su, T.-L., Wang, Q.K., 2015. SERS detection of arsenic in water: a review. J. Environ. Sci. 36, 152–162.

Hering, K., et al., 2007. SERS: a versatile tool in chemical and biochemical diagnostics. Anal. Bioanal. Chem. 390, 113–124.

Huang, Z., et al., 2013. Large-area Ag nano rod array substrates for SERS: AAO template-assisted fabrication, functionalization, and application in detection PCBs. J. Raman Spectrosc. 44, 240–246.

Jablonska, E., Vinceti, M., 2015. Selenium and human health: witnessing a Copernican revolution. J. Environ. Sci. Health, C 33 (3), 328–368.

Johri, N., Jacquillet, G., Unwin, R., 2010. Heavy metal poisoning: the effects of cadmium on the kidney. Biometals 23 (5), 783–792.

Kravchenko, J., Darrah, T.H., Miller, R.K., Lyerly, H.K., Vengosh, A., 2014. A review of the health impacts of barium from natural and anthropogenic exposure. Environ. Geochem. Health 36 (4), 797–814.

Krpetić, Ž., Guerrini, L., Larmour, I.A., Reglinski, J., Faulds, K., Graham, D., 2012. Importance of nano particle size in colorimetric and SERS. Based multimodal trace detection of Ni(II) ions with functional gold nano particles. Small 8 (5), 707–714.

Li, M., Gou, H., Al-Ogaidi, I., Wu, N., 2013. Nanostructured sensors for detection of heavy metals: a review. ACS Sustain. Chem. Eng. 1, 713–723.

Liang, A., Li, X., Zhang, X., Wen, G., Jiang, Z., 2018. A sensitive SERS quantitative analysis method for Ni^{2+} by the dimethylglyoxime reaction regulating a graphene oxide nanoribbon catalytic gold nanoreaction. Luminescence 33 (6), 1033–1039.

Liu, M., Wang, Z., Zong, S., Chen, H., Zhu, D., Wu, L., et al., 2014. SERS detection and removal of mercury(II)/silver(I) using oligonucleotide-functionalized core/shell magnetic silica sphere@ Au nanoparticles. ACS Appl. Mater. Interfaces 6 (10), 7371–7379.

Luo, S.-C., Sivashanmugan, K., Liao, J.-D., Yao, C.-K., Peng, H.-C., 2014. Nanofabricated SERS-active substrates for single-molecule to virus detection in vitro: a review. Biosens. Bioelectron. 61, 232–240.

Ma, W., Sun, M., Xu, L., Wang, L., Kuang, H., Xu, C., 2013. A SERS active gold nanostar dimer for mercury ion detection. Chem. Commun. 49 (44), 4989–4991.

Ma, P., Liang, F., Li, X., Yang, Q., Wang, D., Song, D., et al., 2014. Development and optimization of a SERS method for on-site determination of nitrite in foods and water. Food Anal. Methods 7 (9), 1866–1873.

Monteiro, D.A., Rantin, F.T., Kalinin, A.L., 2010. Inorganic mercury exposure: toxicological effects, oxidative stress biomarkers and bioaccumulation in the tropical freshwater fish matrinxã, *Brycon amazonicus* (Spix and Agassiz, 1829). Ecotoxicology 19 (1), 105.

Moody, R.L., et al., 1987. Investigation of experimental parameters for surface-enhanced Raman scattering (SERS) using silver-coated microsphere substrates. Appl. Spectrosc 41, 966–970.

Mosier-Boss, P., 2017. Review of SERS substrates for chemical sensing. Nanomaterials 7 (6), 142.

Mosier-Boss, P.A., Lieberman, S.H., 2000. Detection of nitrate and sulfate anions by normal Raman spectroscopy and SERS of cationic-coated, silver substrates. Appl. Spectrosc. 54 (8), 1126–1135.

Mulvihill, M., Tao, A., Benjauthrit, K., Arnold, J., Yang, P., 2008. Surface-enhanced Raman spectroscopy for trace arsenic detection in contaminated water. Angew. Chem. Int. Ed. 47 (34), 6456–6460.

Otsuki, T., Maeda, M., Murakami, S., Hayashi, H., Miura, Y., Kusaka, M., et al., 2007. Immunological effects of silica and asbestos. Cell Mol. Immunol. 4 (4), 261–268.

Ozsvath, D.L., 2009. Fluoride and environmental health: a review. Rev. Environ. Sci. Bio/Technol. 8 (1), 59–79.

Pinheiro, P.C., Daniel-da-Silva, A.L., Nogueira, H.I.S., Trindade, T., 2018. Functionalized inorganic nanoparticles for magnetic separation and SERS detection of water pollutants. Eur. J. Inorg. Chem. 2018 (30), 3443–3461.

Prabhu, R., Bramhaiah, K., Singh, K.K., John, N.S., 2019. Single sea urchin–MoO_3 nanostructure for surface enhanced Raman spectroscopy of dyes. Nanoscale Adv. 1, 2426–2434.

Rodríguez-Mercado, J.J., Mosqueda-Tapia, G., Altamirano-Lozano, M.A., 2017. Geno toxicity assessment of human peripheral Lymphocytes induced by thallium(I) and thallium(III). Toxicol. Environ. Chem. 99 (5–6), 987–998.

Sackmann, M., Materny, A., 2006. Surface enhanced Raman scattering (SERS)—a quantitative analytical tool. J. Raman Spectrosc. 37 (13), 305–310.

Senapati, D., Dasary, S., Singh, A.K., Senapati, T., Yu, H., Ray, P.C., 2011. A label free gold nano particle based SERS assay for direct cyanide detection at the parts per trillion level. Chemistry–A Eur. J. 17 (30), 8445–8451.

Sharma, B., Frontiera, R.R., Henry, A.-I., Ringe, E., Van Duyne, R.P., 2012. SERS: materials, applications, and the future. Mater. Today 15 (1–2), 16–25.

Shen, Z., Zhang, L., Cao, J., Tian, J., Liu, L., Wang, G., et al., 2012. Chemical composition, sources, and deposition fluxes of water-soluble inorganic ions obtained from precipitation chemistry measurements collected at an urban site in northwest China. J. Environ. Monit. 14 (11), 3000–3008.

Shi, Y.-E., Wang, W., Zhan, J., 2016. A positively charged silver nano wire membrane for rapid on-site swabbing extraction and detection of trace inorganic explosives using a portable Raman spectrometer. Nano Res. 9 (8), 2487–2497.

Sun, B., Jiang, X., Wang, H., Song, B., Zhu, Y., Wang, H., et al., 2015. Surface-enhancement Raman scattering sensing strategy for discriminating trace mercuric ion(II) from real water samples in sensitive, specific, recyclable, and reproducible manners. Anal. Chem. 87 (2), 1250–1256.

Tang, W., Chase, D.B., Sparks, D.L., Rabolt, J.F., 2015. Selective and quantitative detection of trace amounts of mercury(II) ion (Hg^{2+}) and copper(II) ion (Cu^{2+}) using surface-enhanced Raman scattering (SERS). Appl. Spectrosc. 69 (7), 843–849.

Thompson, T., Fawell, J., Kunikane, S., Jackson, D., Appleyard, S., Callan, P., et al., 2007. Chemical Safety of Drinking Water: Assessing Priorities for Risk Management. Sanitation Water, and World Health Organization.

Tovar-Sánchez, A., Sánchez-Quiles, D., Basterretxea, G., Benedé, J.L., Chisvert, A., Salvador, A., et al., 2013. Sunscreen products as emerging pollutants to coastal waters. PLoS One 8 (6), e65451.

Tripp, R.A., Dluhy, R.A., Zhao, Y., 2008. Novel nanostructures for SERS biosensing. Nano Today 3 (3–4), 31–37.

Ullah, N., Mansha, M., Khan, I., Qurashi, A., 2018. Nanomaterial-based optical chemical sensors for the detection of heavy metals in water: recent advances and challenges. TrAC Trends Anal. Chem. 100, 155–166.

Van den Borre, L., Deboosere, P., 2015. Enduring health effects of asbestos use in Belgian industries: a record-linked cohort study of cause-specific mortality (2001–2009). BMJ Open 5 (6), e007384.

Wang, W., Yin, Y., Tan, Z., Liu, J., 2014. Coffee-ring effect-based simultaneous SERS substrate fabrication and analyte enrichment for trace analysis. Nano Scale 6 (16), 9588–9593.

Wei, H., et al., 2015. Plasmonic colorimetric and SERS sensors for environmental analysis. Environ. Sci.: Nano 2, 120–135.

Wongsanit, J., Teartisup, P., Kerdsueb, P., Tharnpoophasiam, P., Worakhunpiset, S., 2015. Contamination of nitrate in groundwater and its potential human health: a case study of lower Mae Klong river basin, Thailand. Environ. Sci. Pollut. Res. 22 (15), 11504–11512.

World Health Organization, 2004. Concise International Chemical Assessment Document 61, Hydrogen Cyanide and Cyanides: Human Health Aspects, Geneva, pp. 4–5.

Xu, L., Yin, H., Ma, W., Kuang, H., Wang, L., Xu, C., 2015. Ultrasensitive SERS detection of mercury based on the assembled gold nanochains. Biosens. Bioelectron. 67, 472–476.

Yin, J., Wu, T., Song, J., Zhang, Q., Liu, S., Xu, R., et al., 2011. SERS active nano particles for sensitive and selective detection of cadmium ion (Cd^{2+}). Chem. Mater. 23 (21), 4756–4764.

Yousefi, N., Fatehizedeh, A., Ghadiri, K., Mirzaei, N., Ashrafi, S.D., Mahvi, A.H., 2016. Application of nano filter in removal of phosphate, fluoride and nitrite from groundwater. Desalin. Water Treat. 57 (25), 11782–11788.

Zhang, X., Dai, Z., Si, S., Zhang, X., Wu, W., Deng, H., et al., 2017. Ultrasensitive SERS substrate integrated with uniform subnanometer scale "hot spots" created by a graphene spacer for the detection of mercury ions. Small 13 (9), 1603347.

Further reading

de Perio, M.A., Durgam, S., Caldwell, K.L., Eisenberg, J., 2010. A health hazard evaluation of antimony exposure in fire fighters. J. Occup. Environ. Med. 52 (1), 81–84.

Habermeyer, M., Roth, A., Guth, S., Diel, P., Engel, K.-H., Epe, B., et al., 2015. Nitrate and nitrite in the diet: how to assess their benefit and risk for human health. Mol. Nutr. Food Res. 59 (1), 106–128.

Pradhan, J.K., Kumar, S., 2014. Informal e-waste recycling: environmental risk assessment of heavy metal contamination in Mandoli industrial area, Delhi, India. Environ. Sci. Pollut. Res. 21, 7913–7928.

World Health Organization, 2003. No. WHO/SDE/WSH/04.03/56 Nitrate and Nitrite in Drinking-Water: Background Document for Development of WHO Guidelines for Drinking-Water Quality. World Health Organization.

Low-cost adsorbents for removal of inorganic impurities from wastewater

Surinder Singh[1], Kailas L. Wasewar[2] and
Sushil Kumar Kansal[1]

[1]Dr. S. S. Bhatnagar, University Institute of Chemical Engineering and Technology, Panjab
University, Chandigarh, India [2]Advance Separation and Analytical Laboratory (ASAL),
Department of Chemical Engineering, Visvesvaraya National Institute of Technology (VNIT),
Maharashtra, India

OUTLINE

Inorganic Pollutants in Water
DOI: https://doi.org/10.1016/B978-0-12-818965-8.00010-X

10.1 Introduction

Water qualifies to be the most significant global resource required for life on earth. Clean drinking and sanitation water is a great socially, economically, environmentally, and politically important issue of the current times. Due to the advent of industrialization, agricultural, and other municipal activities, lot of vivid chemical species, that is, dyes, heavy metals, chlorinated, phenolic, and phosphatic compounds and inorganic chemicals are continuously discharged into water bodies, namely, lakes, ponds, canals, and rivers apart from their entry in municipal and industrial sewage treatment influents. Clean water for drinking and other purposes is the need of the hour and hence there is a great need of removal of these pollutants from the water and wastewater. Treatment of polluted water is a crucial issue and a global problem for the environment and the water ecosystem (Reddy and Lee, 2013; Mo et al., 2018). The researchers worldwide have utilized different techniques for the removal of contaminants from wastewater which include photocatalysis, precipitation, Fenton process, incineration, sedimentation, chlorination, coagulation, reverse osmosis, membrane filtration, electrochemical process, activated carbon technologies, ion exchange, and adsorption (Anastopoulos et al., 2017; Mo et al., 2018). Most of these technologies except adsorption have inherent limitations such as high operational and equipment cost, high energy requirement, high maintenance cost, partial metal ions removal, and the generation of waste after treatment (Bahadur et al., 2019).

Huge amounts of inorganic, biological, and organic contaminants have infiltrated into our water bodies due to natural or anthropogenic activities. Some of these, having high rates of seepage, have moved into the soil and have got biomagnified in plants, sea foods, and animals and are ultimately becoming a part of our food chain. It is well established that most of these pollutants are highly carcinogenic and have toxic effects. For example, cobalt poisoning is the reason for thyroid, liver, and gastrointestinal ailments owing to its high-toxicity and increased accumulation. Zinc is normally required as micronutrient for human growth, but high concentrations of zinc larger than 3.0 mg/L cause mental fever, poor skeletal, and body growth.

The disastrous impacts on ecology and environment caused by distinct inorganic pollutants such as heavy metal ions, chlorides and fluorides, nitrates, bromates, and phosphates have posed serious questions and threats to human and organisms' lives. It is estimated that most counties of the world will suffer from acute potable water scarcities in 2020 at the time when world's population will cross the 8 billion mark. There will be huge pressure on the natural resources in general and water in specific. Hence, the mitigation of these inorganic pollutants from potable and wastewater is necessary for providing clean drinking water and healthy environment to the society.

Many techniques and methods for wastewater treatment have been developed worldwide but still adsorption remains the most efficient and commonly utilized technique due to its inherent merits such as simplicity, ease of operation, and large range of adsorbents availability and high efficiency. It can be utilized for the treatment of soluble and insoluble organic, biological, as well as inorganic contaminants in wastewater. The adsorption method is useful for physical, chemical, and biological process systems and is generally utilized for industrial wastewater and potable water decontamination.

The common inorganic pollutants of grave concern regarding wastewater treatment include nitrates, nitrites, selenium, cyanides, heavy metals arsenic, lead, copper, cadmium, mercury, iron, phosphates, fluorides, calcium, aluminum, sodium and titanium compounds, perchlorates, nitric acid, chromium oxides, sulfates, and chlorides. Even trace amounts of these heavy metals can pose serious threats to aquatic organisms and health problems to humans. The use of nitrate or nitrogen fertilizers has given birth to nitrate pollution in water bodies. Great amounts of nitrates can cause diseases such as "blue baby syndrome" and phosphates and nitrates collectively cause eutrophication (Bombuwala et al., 2018). Excess nitrate concentrations above permissible levels of 10 mg/L adopted by Environment Protection Agency (United States) have been found in public water systems and drinking water wells in United States. Selenium is another pollutant that seeps into water bodies and water wells through geogenic causes and weathering of rocks and sediments. Similarly, fluorides contamination is caused due to above reasons and due to fluoride rich soils (Banerjee, 2015). The fluoride contamination in water gives rise to dental ailments and fluorosis disease which affects millions of citizens globally.

Cadmium is another toxic heavy metal impurity that is nonbiodegradable in nature. Cadmium assembles in the organisms and its levels enhance as the trophic levels increases (Godt et al., 2006). Toxic effects of cadmium are more visible in the organisms higher in the food web. Cadmium gives rise to Itai–Itai ailment, tumors, weight loss, pneumonitis, hypertension, bronchitis, bone lesions, pulmonary edema, and cancer.

A large number of inorganic (anionic) species, such as, nitrite, phosphate, nitrate, cyanide, perchlorate, and fluorides enter the water bodies via discharge of various industrial effluents from fertilizer, mining, tanneries, refining ores, and paper and batteries industries. These species are considered as actual contaminants if their concentration exceeds the permissible limits in water bodies, for example, phosphate and nitrite exceeding 0.5–1.0 mg/L will be considered as active pollutants. At these concentrations, phosphates and nitrates stimulate algae growths in most water ecosystems, thereby polluting them and causing harm to aquatic life such as fish and other aquatic organisms. The progressively building up amounts of these contaminants in undersurface water, lakes, rivers, wells, and ponds, causes extensive degradation of quality of our water resources.

The inorganic anions present in different concentrations compete for their selective removal that causes typical problem controlling and removal of these anions from potable water supplies. The major inorganic species of grave importance are (PO_4^{3-}), (F^-), (BrO_3^-), (NO_3^-), and (ClO_4^-), posing severe health issues and environmental degradation. Heavy metals such as arsenic is a carcinogenic metalloid that causes diseases such as melanosis, tumors, hypertension, lung diseases, and various types of cancers of kidney, liver, skin, lung, and bladder. It enters the human body and other organisms through different routes, and then large amounts of arsenic prevalent in the body can cause severe poisoning, which deteriorates the health and well-being of humans, plants, and animals.

High amounts of nitrate impurities in potable water have generated a grave concern throughout the world. The pollution due to nitrate from the nonpoint sources includes chemicals and fertilizers from agricultural lands and urban runoff and point sources consists of spills of nitrogen-rich compounds, disposal of untreated sewage, animal manure, sinkholes, leakages in septic systems, improperly constructed wells, and open boreholes and landfill leachates. Because of the high aqueous solubility of nitrates and relatively lean adsorption attraction toward the soils, the nitrate pollution has become the most serious groundwater impurity throughout the world, which poses a great threat to potable water and its normal supplies. The inorganic phosphate impurity in potable water and wastewater can be treated by using common techniques such as settling/sedimentation, crystallization, and chemical precipitation using suitable chemicals and coagulants, for example, lime, alum, and ferric chloride. However, to achieve very low phosphate concentrations and effective removal, advanced water-treatment techniques, for example, adsorption and ion exchange need to be utilized.

Some of the most commonly used adsorbent materials include activated carbon both chemically synthesized and made from biomaterials, chitosan, zeolites, molecular sieves, metal organic frameworks (MOFs), and nano- and biocomposites. But due to increased cost of these materials and certain complexities involved in their synthesis, there is now a rising trend toward the use of low-cost agricultural and bio-based waste materials that can be suitably activated or used as such for removal of inorganic impurities from wastewater.

Alternative cheap adsorbents are needed to be developed for the treatment of inorganic impurities and metal-ion containing wastewater. There is now a reported utilization of adsorbents made from natural materials, for example, natural clays, natural zeolites, and polysaccharides such as chitosan and its derivatives. Other adsorbents include biomass and industry waste materials, for example, wheat straw and rice straw—derived activated carbon, polymeric, and boron nitride—based materials and their composites, calcined waste egg shells, aloe vera, and sugar industry waste. Nowadays, a series of new agricultural and forest waste materials, such as grouts, crumbs, husks, corn cobs, dross and straws, barks, fruit peels, nuts, dates, fruit seeds, and biowaste adsorbents, for example, biochars and biorefinery wastes, have initiated a great interest in the use of such materials as alternate adsorbents (Köse and Kıvanc, 2011; Zhu et al., 2016; Anastopoulos et al., 2017; Sizmur et al., 2017; Bombuwala et al., 2018; Giannakoudakis et al., 2018; Mo et al., 2018; Yu et al., 2018; Bahadur et al., 2019; Filote et al., 2019; Nabbou et al., 2019). The inexpensive (low-cost) adsorbents utilized for removal of inorganic contaminants have been shown in Fig. 10.1.

10.2 Significance of low-cost adsorbents

There has been a great emphasis on developing inexpensive adsorbents for treatment of inorganic impurities from wastewater and water bodies due to the inability of developing countries to use available expensive treatment technologies on mass scales. Although photocatalysis and some other treatment methods have been utilized, but by and large adsorption techniques have been the choice for most of the researchers in the area of wastewater treatment on large scales.

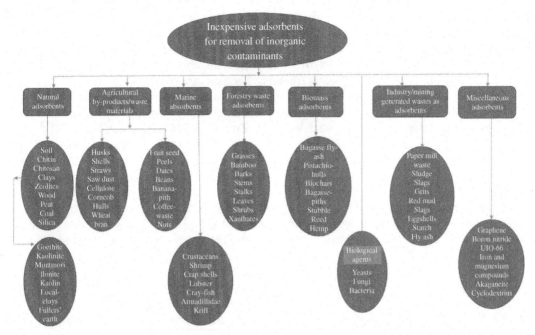

FIGURE 10.1 Inexpensive adsorbents utilized for the removal of inorganic impurities from water/wastewater.

A lot of attention has been devoted to develop adsorbents that can bind and remove unwanted heavy metals from polluted water at least cost. These low-cost materials are easily and locally available, can be activated using low-cost and accessible reagents or used as such, and have high adsorption efficiencies. Natural materials, for example, chitosan, biomass waste, zeolites, clays, and other waste products originating from different industries, such as bagasse, fly ash, slags, and grits, and agricultural and biowastes, such as straws, barks, grouts, seeds, peels, biochar, and biorefinery wastes, have been classified as cheap adsorbents.

A large amount of readily available agricultural waste materials have distinct advantages during the treatment of wastewater such as large availability, eco-friendly, easily regenerable, chemical stability, structural properties, and low cost. The structural properties available in these materials include large porosity, presence of diverse function groups, and high specific surface area which are required for adsorption of inorganic metals (Anastopoulos et al., 2017). The agricultural by-products contain some active materials such as cellulose, lignin, pectin, and hemicellulose that can react with metal ions and thus help in their removal. Currently, there is a lot of emphasis on research related to optimal use of agricultural wastes materials and biomass materials for treating water and wastewater contaminants. A large number of researchers have done their research studies on the application of agricultural waste for adsorbing the heavy metals.

Industrial waste materials are also potentially useful adsorbents for heavy metal removal. They can be activated or modified, and some processing is needed to increase the adsorption capacity. Normally industrial waste materials are obtained as by-products. Being locally available in huge amounts, they are obtained at extremely low cost. In India, different categories and types of industrial wastes, for example, waste slurry, biochar, lignin, biorefinery waste, ferrous(III) hydroxide, slag, grits, and red mud have been utilized to treat wastewater containing heavy metals. Zeolites are another naturally occurring promising materials which are basically crystalline aluminosilicates tetrahedrons interlinked by Al, Si, and oxygen atoms. These are extremely useful for the treatment of different heavy metals from wastewater such as Cu^{2+}, Zn^{2+}, Hg^{2+}, Pb^{2+}, Cd^{2+}, Cu^{2+}, Ni^{2+}, and Co^{2+}.

Therefore the need of the hour is to develop more such low-cost adsorbents to treat contaminated water and wastewater containing inorganic impurities. Research done in the water-treatment area suggests that low-cost adsorbents can be utilized to replace expensive synthetic adsorbents owing to their physical, chemical, and structural characteristics, high efficiencies of impurities removal and available usage for repeated cycles. Some of these agricultural and biowastes, natural and industry waste–based adsorbents are discussed below.

10.3 Inexpensive adsorbents utilized for treating inorganic contaminants

10.3.1 Chitosan

Chitosan as a biomaterial is of great interest pertaining to adsorbing inorganic metals from wastewater owing to its superior metal-binding ability, being available in large amounts and its low cost when compared with other materials, for example, activated carbon, and biochars (Babel and Kurniawan, 2003). In countries such as China, Bangladesh, Japan, Bhutan, Thailand, and Myanmar, large amounts of fishery wastes such as shrimp, crab shells and lobster have been utilized to produce low-cost chitosan. These fisheries waste is available at free of cost from local fishery units in these countries. Chitosan is similar to cellulose in molecular structure, hence chitosan showcase properties similar to that of cellulose. It contains large number of hydroxyl bonds that enables it to be a good ion exchanger. Initially it was applied for adsorption of Cd^{2+} ions. Though it showed a decent level of adsorption, it was later found out that the presence of EDTA in water can reduce its adsorption capacity toward Cd^{2+} to a great extent. Similar researches also showed affinity of chitosan toward metal ions, for example, Hg^{2+}, Cu^{2+}, Ni^{2+}, Cr^{6+}, Cd^{2+}, Pt^{6+}, Zn^{2+}, and nitrate ions (Juang and Shao, 2002; Ng et al., 2002; Bhatnagar and Sillanpää, 2011; Al-Sherbini et al., 2019). The maximum obtained adsorption capacities utilizing chitosan came out to be 815, 222, 164, 273, 280, and 75 mg/g for the ions Hg^{2+}, Cu^{2+}, Ni^{2+}, Cr^{6+}, Pt^{6+}, and Zn^{2+}, respectively (Mckay, Blair and Findon, 1989; Babel and Kurniawan, 2003). It was observed that reducing the particle size increases the adsorption capacity. Many chemical modifications have been done to chitosan to increase its performance as well as make it selective toward a particular metal ion (Kyzas and Bikiaris, 2015). Hence, chitosan has emerged as a potential low-cost adsorbent for the treatment of wastewater in present times.

10.3.1.1 *Magnetic chitosan*

Magnetic chitosan—based adsorbents are being utilized as novel adsorbents to treat toxic impurities being present in aqueous solutions. Chitosan, a low-cost and efficient adsorbent, has excellent structural properties due to which it is utilized in removing various metal ions from wastewater, thus making the selectivity of chitosan a highly desirable property. Also after adsorption, it is hard to separate the spent adsorbent which can lead to blockage of the filters and discarding the adsorbent along with sludge induces secondary pollution. These practical problems have promoted the use of magnetic chitosan—based adsorbents (Reddy and Lee, 2013). The distinct merit of the magnetic separation technology is the fact that huge amounts of wastewater can be treated in relatively short intervals of time utilizing low energy inputs and without production of any contaminants. Magnetic chitosan composites show high adsorption capacity and significant biological, physical, and chemical properties (Huang et al., 2009). They are eco-friendly as well as reusable. Analysis showed that amine and hydroxyl functional groups were responsible for binding organic and inorganic impurities. The optimum pH value for removal depends on the type of magnetic chitosan composites and target ion. Yu et al. (2013) obtained increased adsorption capacities for Cr(VI) of 144.9 mg/g which increased to 169.5 mg/g with elevation in temperature from 303K to 323K utilizing magnetic chitosan (Yu et al., 2013). Another study reported 29.6 mg/g adsorption capacity for the removal of copper ions at temperature of 288.15K which was further increased to 35.5 mg/g by increasing the temperature up to 308.15K (Yuwei and Jianlong, 2011).

10.3.2 Zeolites

Zeolites are aluminosilicates formed within a structural cage such as framework, consisting of molecules placed in tetrahedrons, interlinked to each other by the shared aluminum, oxygen, and silicon atoms. There are about 40 naturally occurring zeolites originating from volcanic and sedimentary rocks. Zeolites are highly useful for removal of heavy metals that have generated keen interest among scientists, owing to their significant properties such as ion exchange-capacity and definite pore sizes. Main varieties of zeolites adsorbents are clinoptilolite and chabazite. Clinoptilolite is represented by more than 40 naturally occurring zeolite species. Clinoptilolite has exhibited high selectivity for heavy metals such as Zn^{2+}, Pb^{2+}, Cu^{2+}, and Cd^{2+}. It was reported in the literature that zeolites are very effective for treating wastewater contaminated with heavy metals, for example, Zn^{2+}, Pb^{2+}, Cu^{2+}, Cd^{2+}, Co^{2+}, and Ni^{2+} (Jiménez-Castañeda and Medina, 2017; Obaid et al., 2018; Hong et al., 2019; Joseph et al., 2019; Li et al., 2019). Both clinoptilolite and chabazite zeolites showed almost 100% adsorption for metal ions having concentrations of 10 ppm. The authors also reported that the zeolites chabazite and clinoptilolite showed varying selectivities for almost all the heavy metals treated excluding lead (Pb^{2+}) (Kesraoui and Kavannagh, 1997). Large reserves of natural zeolites are found in countries such as, Jordan, Italy, United Kingdom, Mexico, Greece, and Iran. These reserves give local industries valuable benefits such as readily availability of natural zeolites, low cost, and high efficiency of removal. The maximum adsorption capacity for removal of Pb^{2+} by zeolite ZIF-8 obtained was 1119.80 mg/g and zeolite ZIF-67 was 1348.42 mg/g, respectively. However, adsorption capacities of 454.72 and 617.51 mg/g for Cu^{2+} removal was obtained for ZIF-8 and ZIF-67 zeolites, which was found to be higher as compared to other zeolitic adsorbents (Mo et al., 2018).

10.3.3 Natural clays

Clays are another potential replacement of expensive activated carbon and other synthetic adsorbents. Similar to zeolites, natural clay minerals are constituted from inorganic components in the soil. The high adsorption capacities of clays are due to their large surface area and high ion-exchange capabilities. There exists a net negative charge on the clay minerals that is responsible for their ability to attract heavy metal ions (Alshameri et al., 2019). United States, Lithuania, Russia, Kazakhstan, and Georgia are the countries having large reserves of natural clay minerals.

The clays have gained interest due to their high cation exchange properties and large surface area. Smectite, kaolinite, and micas are three main species of natural clay minerals used for adsorption. These are very low-cost materials and have market price about 20 times cheaper than synthetic activated carbon. Smectite clays such as montmorillonite show highest adsorption and cation exchange capacities (CECs) and thus can be used for the removal of Zn^{2+}, Cd^{2+}, Cu^{2+}, Pb^{2+}, Al^{3+}, etc., heavy metals, dyes, and phosphate (Wilson et al., 2006; Bhattacharyya and Gupta, 2011; Ma et al., 2015, 2017; Adeyemo et al., 2017; Cherif et al., 2017; Park et al., 2019). Some heavy metals such as lead (Pb^{2+}) are temperature sensitive and exhibit higher adsorption at low temperatures only. The affinity of natural clays toward different heavy metal ions and their adsorption is dependent on process conditions such as process temperature, competing ions present, pH, and initial metal ion strength (Bahadur et al., 2019).

The scientists have utilized the natural clay minerals for adsorption by activating the natural clays using acid or alkali pretreatment or additional chemicals treatment to get modified sorption characteristics such as surface functional group, CEC, pore size, high surface area per unit mass, and volume of pores (Bahadur et al., 2019). The natural clays utilized for this purpose are kaolinite, bentonite, and montmorillonite, and after their chemical treatment or modification, these are used to adsorb inorganic impurities (Bhattacharyya and Gupta, 2011).

The reported adsorption removal capacities of Cr^{6+} on natural kaolinite clay was 11.6 mg/g and modified kaolinite clay by treatment with acid, TBA, and ZrO were 13.9, 10.6, and 10.9 mg/g, respectively (Bahadur et al., 2019). The adsorption of Pb^{2+} on natural montmorillonite obtained was 28 mg/g (de Pablo et al., 2011) and modified montmorillonite was 131.579 mg/g, respectively (Sdiri et al., 2011). Such high adsorption capacities were due to the effect of availability of active porous sites at the surface of clay after acetic acid treatment. The utilization of bentonite natural clay mineral and the modified bentonite clay as better adsorbent for treating copper, nickel, lead, cobalt, chromium, nickel, mercury, cadmium, and zinc metals from water and aqueous solutions have been considered widely in the last decade (Arfaoui et al., 2008; Anirudhan et al., 2012; Liu et al., 2015; Heydari et al., 2017). A high adsorption capacity of 59.7 mg/g for Pb^{2+} ions by virgin bentonite clay and 123.3 mg/g by modified bentonite clay was achieved using aqueous solution (Anirudhan et al., 2012).

Nabbou et al. (2019) applied kaolinite clay, for the treatment of fluoride ion potable water in Tindouf, Algeria as high concentrations of fluoride were detected in that region. The natural clay from Béchar province of Algeria was used without any further treatment. The clay particle size used was less than 2 μm. During the adsorption fluoride ion removal efficiency was 0.442 and 0.448 mg/g at solution pH of 4.5 and 6, respectively. Pseudo-second-order kinetics for the adsorption of fluoride was reported, and adsorption data was fitted well by Freundlich adsorption isotherm. The adsorption process was reported to be endothermic in nature. The effect of presence of nitrate, sulfate, carbonate, and chloride ions was also studied and was found that nitrate and chloride ions interference was negligible, although sulfate and carbonate ions tend to lower the adsorption capacity (Nabbou et al., 2019).

10.3.4 Biochar

Biochar is similar to charcoal that is basically a carbon-enriched substance made from pyrolyzing the agricultural/forest waste biomass in closed container with no air and possesses properties such as fully developed pore structure, specific functional groups, high specific surface area, and high adsorption efficiency (Ahmad et al., 2014; Yang et al., 2019). It is obtained using thermochemical reduction of biomass at temperature range of 200°C–900°C in the low availability or absence of oxygen. It is sometimes used as soil amendment and has great environmental significance also. It helps in the reduction of greenhouse gases, enhances soil nutrients, and improves soil fertility, thereby increasing crop yields and water-holding capacity of the soil (Woolf et al., 2010). As biochar is obtained directly from biomass, it is very inexpensive and can be applied to adsorb heavy metals as well as nitrate-based and phosphatic contaminants. Zhang and Gao (2013) have utilized nano-sized AlOOH particles for modification of biochar and got the increased surface area and other structural properties. They have been able to efficiently remove methylene blue, phosphate, and arsenic impurities using the combination of biochar and nano-sized AlOOH. The procedure of treatment of the heavy metals ions using biochar can be understood by two methods: first, adsorption takes place in porous structure of biochar as the second step exchange of ions occurs. Redox reactions can also take place in between the active components of the modified biochar and the metal ions, thereby stabilizing the metal ions precipitates and detoxification of heavy metals by converting them to the low-valency states (Yang et al., 2019). The adsorption capacity of biochar can be increased by modifying the surface properties of biochars using nanomaterials and other chemical compounds, for example, magnesium and iron oxides (Zhang and Gao, 2013; Usman and Ahmad, 2015).

Biochar possess various functional groups such as hydroxyl, phenolic, and carboxyl which can bind soil impurities, for example, heavy metals and nitrates (Ahmad et al., 2014). These characteristics of biochar depend upon various factors such as pyrolysis temperature, base material, and process time required for making biochar. There are different mechanisms of immobilization or attachment of inorganic compounds with biochar which are schematically represented in Fig. 10.2.

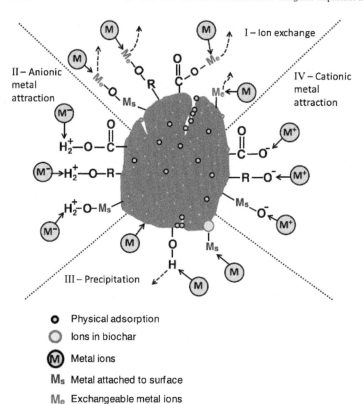

FIGURE   10.2   Postulated mechanisms of biochar interactions with inorganic contaminants. Circles on biochar particle show physical adsorption. I—ion exchange between target metal and exchangeable metal in biochar, II—electrostatic attraction of anionic metal, III—precipitation of target metal, and IV—electrostatic attraction of cationic metal. Source: *Reproduced with permission from reference Ahmad, M., et al., 2014. Chemosphere Biochar as a sorbent for contaminant management in soil and water: a review. Chemosphere 99, 19–33. Available from: https://doi.org/ 10.1016/j.chemosphere.2013.10.071.*

10.3.4.1 Dairy manure–derived biochar

Biochar obtained from waste dairy manure through slow pyrolysis can be used directly as an adsorbent attributing to PO_4^{3-} and CO_3^{2-} groups present on it which cause heavy metal ions such as Zn^{2+}, Cu^{2+}, and Cd^{2+} to precipitate on it (Xu et al., 2013a). This secondary biochar can also efficiently adsorb lead, having maximum adsorption rate of 141 mg/g of adsorbent (Cao et al., 2009). Adsorption of lead using dairy manure biochar is due to the precipitation of lead phosphate. The dairy manure derive biochar produced at temperature of 200°C was capable of adsorbing Cu^{2+}, Cd^{2+}, and Zn^{2+} with the maximum adsorption capacity of 54.4, 51.4, and 32.8 mg/g, respectively (Xu et al., 2013a). Even though it did not possess a large surface area for adsorption, the formation of metal phosphates and carbonates through precipitation has compensated for its capacity.

Dairy manure–derived biochar was prepared at two different temperatures of 200°C and 350°C for comparison (Xu et al., 2013a). The one prepared at higher temperature showed better results, that is, more than 93% of metals was precipitated on it due to higher concentration of CO_3^{2-} present on it as compared to only 78% of metal adsorption on low temperature base biochar. The adsorption capacity also varies for different metal ions. It also has a high pH buffer capacity, that is, pH does not need to be controlled as it remains almost constant throughout the process. Apart from the metals mentioned previously, it is also efficient in removing Pb^{2+} from aqueous solutions and thus can give a promising solution for the removal of heavy metals from wastewater.

Douglas fir biochar is another low-cost adsorbent employed for fluoride and nitrate impurities removal. This biochar impregnated with ferrous chloride, having α-iron(II) oxide and iron(III) oxide on the char surface was fabricated at 1173K. The fir biochar possessed 663 m^2/g surface area and was utilized for the adsorption of fluoride and nitrate ions in the pH range of 2−10 (Bombuwala et al., 2018). The biochar was converted into magnetic biochar by pyrolysis. α-Iron(II) oxide and iron(III) oxide were responsible for the magnetic characteristics of the biochar and allowed numerous active sites for adsorption to take place in the temperature range of 25°C−45°C. A maximum of 9 mg/g fluoride and 15 mg/g nitrate adsorption was achieved using the magnetic Douglas fir biochar which was reported to be higher than other iron oxide−based adsorbents and biochars (Bombuwala et al., 2018).

10.3.4.2 Sugarcane bagasse−based biochar

Sugarcane bagasse is another suitable and inexpensive adsorbent for the removal of inorganic impurities, for example, nitrates from wastewater. It is available in abundance everywhere as sugarcane is a major cash crop around the globe and bagasse is a useful by-product that can be utilized for adsorption purpose. It can be used as such, modified chemically, or converted into biochar for adsorption of organic and inorganic impurities.

Hafshejani et al. (2016) have utilized chemically modified sugarcane bagasse−based biochar using epichorohydrin and *N,N*-dimethylformamide, for removing nitrate from wastewater. They have characterized the resultant biochar using different analytical techniques, for example, scanning electron microscopy (SEM) and fourier transform infrared spectroscopy (FTIR). The process of converting the raw bagasse to biochar, its chemical modification, characterization, and nitrate removal capacity using various adsorption isotherms have been depicted in Fig. 10.3.

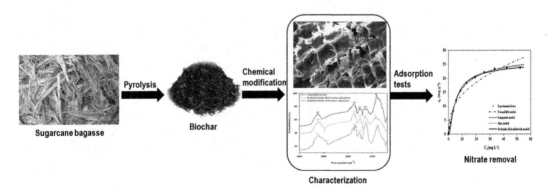

FIGURE 10.3 Removal of nitrate from aqueous solution using modified sugarcane bagasse−based biochar with maximum adsorption capacity of 28.21 mg/g (Langmuir model). Source: *Reproduced with permission from reference Hafshejani et al. (2016).*

10.3.4.3 Biomass-based biochar

Uchimiya et al. (2010) have utilized broiler litter manure biochar having biomass origin that underwent different degrees of carbonization at 350°C and 700°C for heavy metals (Cd^{2+}, Cu^{2+}, Ni^{2+}, and Pb^{2+}) immobilization in soil and in water (Uchimiya et al., 2010). Biomass-based biochars are resistant to biological and/or chemical degradation, and they possess large mean residence times (thousands of years) when obtained in soils. These biochars possess specific functional groups, for example, hydroxyl, phenolic, carboxylic, quinones, and carbonyl along with porous structures. They also possess high CEC, retention of water and plant nutrients, and pH resistance due to their distinct structures. The physical, structural, and chemical characteristics of these biochars differ according to the source of biomass utilized, pre- and post-treatment methods, and pyrolysis conditions (Uchimiya et al., 2010).

10.3.5 Wheat straw

Wheat straw (WS) is an agricultural by-product obtained after harvesting of wheat. Wheat straw, an unrequired by-product, is utilized as ingredient for paper-making process and also used as food for livestock. It is a waste material and is mostly burnt in the fields itself which is harmful for the environment as it causes immense smog and air pollution in favorable conditions. Due to the presence of cellulose in wheat straw, it possesses hydroxyl groups on it which can be suitably replaced by inorganic impurities such as phosphate from water and hence removal of phosphate can be achieved. The wheat straw possess good potential for treatment of inorganic impurities such as zinc, nickel, lead, chromium copper, and cadmium (Chojnacka, 2006; Aydin et al., 2008; Doan et al., 2008; Farooq et al., 2010, 2011; Najiah et al., 2014). Farooq et al. investigated the treatment of Cd^{2+} using simple wheat straw and also urea-modified wheat straw as novel and alternative adsorbent material from aqueous solutions. The adsorption capacity achieved to treat Cd^{2+} in the study reported was 39.22 mg/g using modified wheat straw that was better than simple WS (Farooq et al., 2011).

Xue et al. (2016) utilized modified (WS) utilizing twin layered and double hydroxide particles for the mitigation of nitrate species from water solutions. Hydroxides utilized for the purpose were of iron and magnesium. The layered double hydroxide (LDH) is a potential anion-exchange medium utilized for treating inorganic impurities from contaminated wastewater. LDH-wheat straw biochar incorporated composite was prepared using liquid-phase deposition. The prepared wheat straw—based biochar-LDH composite was utilized for adsorption of nitrate from aqueous solutions already containing sulfates and phosphates (Xue et al., 2016). The biochar—LDH composite exhibited 24.8 mg/g adsorption capacity for nitrate removal from aqueous solutions.

10.3.5.1 UiO-66 immobilized on wheat straw

Wheat straw (WS) can also be converted to an anion or cation exchanger to give rise to biomass-based adsorbent and can further be fabricated with MOFs such as UiO-66. Water contains many ions other than phosphate as impurities such as sulfates and nitrates which can compete with phosphorus for occupying the active sites and lowering the efficiency of phosphate adsorption. The research reports have indicated that when fabricated with UiO-66 the biomass or wheat straw (WS) becomes more selective toward phosphate due to the occurrence of specific complex reactions between UiO-66 and phosphate. Also it can operate within a wide pH range with slight reduction in its adsorption capacity and can be regenerated easily.

Qiu et al. (2019) utilized biomass (WS) waste supported UiO-66 nanoparticles for treating phosphate impurities from wastewater. WS-derived UiO-66 hybrids displayed better nitrogen adsorption potential as compared to simple WS. It was reported that incorporating UiO-66 nanoparticle on WS enhances total amount of available micropores within the obtained hybrids. In presence of competitive anions the adsorption capacity of biomass composite remained at 60% of original capacity. The biomass hybrid-composites exhibited stable reduction capability for phosphate impurities at near neutral and acidic, pH while 2.0–5.5 was obtained as the optimum pH range. The kinetic study showcased that the equilibrium could be attained in 3 hours' time interval (Qiu et al., 2019).

10.3.6 Walnut shell

An alternative bioadsorbent obtained from agricultural waste, that is, walnut shell (WS) was reported to adsorb heavy metals from water (Ding et al., 2013; Pathirana et al., 2019; Vardhan et al., 2019). Many synthetic and natural substances have been utilized to remove cesium (Cs) from water owing to its hazardous nature. The other ions such as nitrates and sulfates present in water also compete for active sites which results in low efficiency of removal of cesium. Nickel hexacyanoferrate (NiHCF) incorporated into walnut shell has been reported to be selective for cesium. NiCHF possesses small particle size but walnut shell is an abundant bioresidue, so it was utilized as host on which NiCHF was incorporated. Because of its distinct structure, only the cesium ions can permeate through it, thereby resulting in their effective removal. The stability of walnut shells and its composite with NiCHF solved the issue of difficulty in the removal of NiCHF nanoparticles from water, thereby providing a cost-effective method to remove Cs selectively (Ding et al., 2013).

10.3.7 Magnetized-activated carbon

Activated carbon is being utilized as an adsorbent widely. One major problem encountered with its usage is the difficulty in separation of the used adsorbent and waste generated which results in the loss of carbon. Nethaji et al. (2013) utilized activated carbon–derived from corn cob, magnetized using the magnetite nanoparticles, for the removal of chromium(VI). The magnetic particles employed were super paramagnetic in nature and could be recovered and used again without any damage of the active sites utilizing external magnetic field. The prepared adsorbent exhibited heterogeneous surface. The obtained monolayer adsorption capacity of the chromium(VI) adsorption amounted to be 57.37 mg/g (Nethaji et al., 2013). The adsorption was dependent on process parameters such as pH, initial chromium strength, dose, and particle size. Percentage adsorption declined with enhancement in chromium concentration. Increased dose and small particle size both increased the chromium removal percentage (Nethaji et al., 2013). The adsorption was attributed to film diffusion, intra-particle diffusion and external mass transfer; which was observed to be the rate-limiting step. A quantity of 3.5 g of prepared adsorbent was sufficient for the treatment of 1.38 L of effluent having Cr(VI) concentration of 25 mg/L.

10.3.7.1 Rice straw

Rice straw is lingo-cellulosic biowaste from agriculture. Similar to wheat straw, rice straw possesses large number of hydroxyl groups which provides it the required ion-exchange capacity. A typical composition (lignocellulosic) of rice straw consists of mainly cellulose (30%−45%), then hemicellulose (20%−32%), and significant amount of lignin (7%−20%) (Jin and Chen, 2007). To increase its adsorption capacity, it can be treated with NaOH, quaternary amino groups, and/or other chemical agents which can increase its reactivity. The modified rice straw can be used to adsorb inorganic impurities such as nitrates, phosphates and sulfates from water, and wastewater; however, its adsorption efficiency decreased significantly in the presence of other competing ions.

Cao et al. (2011) investigated the use of chemically modified rice straw for treatment of sulfate ions from water. The rice straw was modified using chemical treatment with sodium hydroxide, then with epichlorohydrin to introduce the epoxy group and further with trimethylamine to substitute the amine group to make it a strong and effective anion exchanger. An exchange capacity of 1.32 mEq/g was enhanced by chemical treatment of modified rice straw; also the required amino functional sites at the fibrous surface of rice straw were obtained after chemical treatment. During the removal of sulfate using chemically treated rice straw in batch mode, highest adsorption strength of 74.76 mg/g was obtained as compared to 11.68 mg/g for the simple rice straw (Cao et al., 2011). Modified rice straw exhibited good regeneration capacity after a number of cycles while possessing simultaneously good selectivity for sulfate adsorption from water solutions.

10.3.7.2 Rice straw–based activated carbon

Similar to other heavy metals such as Na^+, K^+, Li^+, Ca^{2+}, Mg^{2+}, Fe^{3+}, uranium is also harmful to humans and livestock. Uranium can cause severe damage to the kidney through its biological pathways of entering the human body. Being a radioactive material, uranium poses high threats and toxicity to the environment and biolife.

Different methods such as ion exchange, precipitation, solvent extraction, and adsorption have been employed to concentrate on nuclear wastewater solutions. Adsorption technique has been utilized for the removal of uranium using low-cost adsorbents such as rice husk–based activated carbon which is highly important from nuclear and environmental point of view. Wastewater generated from nuclear industries and radioactive effluents have high amounts of uranium which is required to be treated before disposal. Activated carbon–derived using steam pyrolysis of rice straw was utilized for adsorption of uranium (Yakout and Rizk, 2015). But the activated carbon prepared by this procedure possessed low surface area as per commercial requirements and hence was modified utilizing oxidation with HNO_3 and KOH. When competing ions such as Ca^{2+}, Mg^{2+}, and Fe^{3+} are present in wastewater, adsorption capacity of activated carbon decreases while its retention capacity increases. The adsorption experiments produced at maximum 100 mg/g adsorption capacity for uranium and 71.1 mg/g for thorium using rice husk—modified activated carbon (Yakout and Rizk, 2015).

Sharma and Singh (2013) utilized modified rice straw to adsorb nickel in their study using a fixed bed column. Nickel pollution is being produced by many industries, for example, smelting, electroplating, mining, ore processing, and refining, and the higher concentrations are carcinogenic and can cause threats to human health and related issues. Modification of rice straw by treatment with 0.1 M NaOH was useful in increasing its adsorption capacity as NaOH destroys lignin and makes the straw more reactive. This modified rice straw is further utilized in fixed bed columns to adsorb nickel from wastewater. The authors have studied effects of important bed conditions affecting the adsorption process such as depth of the bed column, initial Ni(II) ions strength by experiments in a fixed-bed adsorption column. It was reported that high influent concentration and more contact time enhanced the rate of adsorption. The highest adsorption capacity achieved with modified rice straw was 43 mg/g at influent nickel strength of 75 mg/L and bed column having 2 cm depth (Sharma and Singh, 2013).

10.3.7.3 Maize tassel—based activated carbon

Activated carbon is being utilized for adsorption of various inorganic impurities. Maize is a crop that is grown in abundance. Activated carbon generated from maize tassel powder is used for treating Pb^{2+} from water (Moyo et al., 2013). It was documented by the authors that percentage adsorption was time dependent up to a certain point but then became constant. The adsorption capacity also varied with initial concentration of Pb^{2+}, that is, it declined with increase in its concentration. The adsorption capacity tends to decrease with higher pH (9—14) values. It was observed that a high adsorption capacity (37.31 mg/g) for lead ions was achieved at optimum conditions, that is, adsorbent dose-1.2 g, pH-5.4, and initial strength of Pb^{2+}-10 mg/L (Moyo et al., 2013). The obtained results showcased the adsorption potential of maize tassel—derived activated carbon for Pb^{2+} ions removal from wastewater and also to solve the problem of disposal of agricultural waste.

10.3.8 Monomer-grafted cellulose-based adsorbents

Cellulose is another agricultural residue—based adsorbent that is freely available and can be utilized for the removal of heavy metal impurities. A large number of hydroxyl groups are present on the cellulose surface which can be chemically modified to use it as adsorbents for removal of inorganic pollutants from wastewater (Hokkanen et al., 2016). Chemical modification can be done by either grafting the monomers to cellulose or direct modification of hydroxyl groups of cellulose structures by various modifying agents. Grafting using various argents such as sodium hydroxide, Imidazole, and amines onto cellulose can chemically modify the structure of cellulose thereby imparting specific structural and physicochemical properties to the modified adsorbent. Fig. 10.4 depicts the process of chemically modified cellulosic adsorbents and their utility in treating inorganic impurities from wastewater.

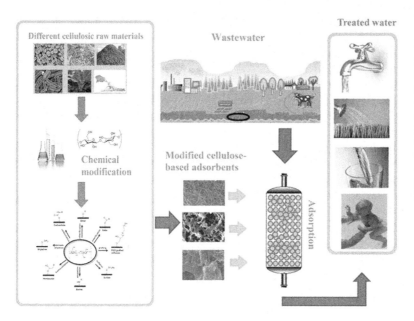

FIGURE 10.4 Use of modified cellulosic materials for adsorption of inorganic impurities. Source: *Reproduced with permission from Hokkanen, S., Bhatnagar, A., Sillanpää, M., 2016. A review on modification methods to cellulose-based adsorbents to improve adsorption capacity. Water Res. 91, 156–173. Available from: https://doi.org/10.1016/j.watres.2016.01.008.*

10.3.9 Aloe vera waste biomass—based adsorbent

Aloe vera is well known for its health benefits and medicinal properties. Recent researches have shown that aloe vera also acts as a bioadsorbent and helps in treatment of toxic metals from water. Its roots showed accumulation properties for metals present in soil and water. Giannakoudakis et al. (2018) have discussed in detail about the use of aloe vera waste—based biosorbents. For its use as an adsorbent to treat wastewater, five different materials were derived from aloe vera. The biosorbents had been categorized in five distinct groups depending upon the treatment given to the aloe vera, that is, without any treatment or no treatment, air dried, chemically treated, thermally treated, and carbonized and functionalized. Activated carbon was made from the abovementioned aloe vera—derived materials. Adsorbent produced from modified aloe vera leaves showed the maximum adsorption capacity toward heavy metals such as Ag^{1+}, Pb^{2+}, Cu^{2+}, Cr^{3+}, and Zn^{2+} (Giannakoudakis et al., 2018). The optimum pH observed for adsorption varied for different materials. The monolayer adsorption capacities reported by the authors in their discussion range from 2.27 to 500 mg/g. Besides metal ions, aloe vera also proved to be successful in adsorbing dyes such as methylene blue from water.

10.3.10 Biorefinery waste *Fucus spiralis*

Biomass-derived bioadsorbents such as macroalgae or seaweeds are being utilized as low-cost alternative adsorbents due to their distinct properties such as chemical binding, ion-exchange ability, and electrostatic affinity toward heavy metals (Davis et al., 2003). Different algae species uptake heavy metal ions amounts differently depending upon the adsorbate, but the general range of seaweeds, including algal species, have been reported to be 0.1−2 mmol/g (Santos et al., 2018). Amongst different divisions of algal species such as red, green, and brown, the brown macroalgae adsorb heavy metals to a great extent (Romera et al., 2008). Filote et al. have utilized biorefinery macroalgae waste *Fucus spiralis* to adsorb lead ions from wastewater collected from Viana do Castelo region near Northern coast of Portugal. Marine algae have been drawing attention because of its effective use as a renewable material. *F. spiralis* seaweed for treating lead from water was studied and tested experimentally which proved to be very successful as the adsorption was very fast, that is, equilibrium was achieved within few hours (Filote et al., 2019). Overgrowth of sea weeds lead to eutrophication and thus cause an ecological imbalance. This makes usage of seaweed as an adsorbent not only cost-effective but also environment friendly. The presence of other ions in water such as calcium, magnesium, and sulfate reduced to some extent the adsorption capacity of seaweed wherein calcium had the maximum effect (19%). The adsorption experiments at optimum conditions, that is, 20 mg/L initial metal concentration, 0.50 g/L dose, and 4.5 pH were conducted. Under these conditions, Langmuir adsorption capacity of 132 ± 14 mg/g for lead adsorption from aqueous medium was obtained (Filote et al., 2019). Desorption studies performed utilizing EDTA 0.1 mol/L gave best results, thereby generating $95 \pm 4\%$ desorption.

10.3.11 Sugar industry waste

Bagasse, beet pulp, and filter cake are the major by-products generated from sugar industry. Huge amount of waste production leads to difficulty in disposing the waste-generated environmental issues. Three types of materials derived from sugar waste are considered as effective bioadsorbent for the treatment of inorganic metal impurities ions from wastewater (Anastopoulos et al., 2017).

Raw sugar: Beet pulp and bagasse obtained as sugar waste were found to possess methoxy, carboxy, and phenolic groups which participated in adsorption. These were used to adsorb metals ions such as Mn^{2+}, Cu^{2+}, Pb^{2+}, Cd^{2+}, and Hg^{1+}. The adsorption was observed to be spontaneous.

Chemically modified sugar waste: For the process of chemical modification, raw sugar is treated with various chemicals such as NaOH, citric acid, acetonitrile, hydroxylamine, and sulfuric acid, depending upon the metal ion which is to be adsorbed. Compared to untreated sugar waste, modified sugar waste showed higher CEC and showed better results for adsorption of the metals mentioned above.

Sugar waste−based adsorbents: Different sugar-based adsorbents that include iron(3) impregnated sorbent, activated carbon obtained from SCB, activated carbon obtained from modified SCB using phosphoric acid, and biochar fabricated from SCB were used to remove Cr^{6+}, Cr^{3+}, Cd^{2+}, and Pb^{2+}, respectively.

10.3.12 Medicinal herb waste (chicory)

Jokar et al. (2019) utilized medicinal herb waste to adsorb Cd^{2+} and Pb^{2+} ions from wastewater. Huge amounts of medicinal herbs, for example, chicory waste is generated in distillation and extraction units. Chicory waste acts as an ion exchanger, and research studies show that it can be used in removing Pb^{2+} and Cd^{2+}. The use of chicory waste as an ion exchanger is beneficial both from environmental as well as economic perspective. It was observed that surface modification utilizing chemical compounds could increase its ion-exchange capacity by making it chemically more reactive. Chicory waste was subjected to surface modification by treatment with $CaCl_2$ (Jokar et al., 2019).

10.3.13 Red mud waste

Red mud waste has been used for treating water which contains high phosphorus impurities such as phosphates. Since the goal was to make use of waste materials as much as possible, red mud generated from Bayer's process in alumina refining industries was utilized. Modified granules of red mud have shown a good adsorption capacity for phosphate. Recently, it has been modified to a nanocrystalline material called red mud akaganeite (RMA). RMA showed much higher rate of adsorption (12.9 mg P/g) than granular ferric hydroxide and a wider range of optimum pH values (Pepper et al., 2018). In the presence of carbonate species, granules lost half the percentage of the phosphate adsorbed, whereas RMA remained phosphate specific. Also, certain experiments showed that both the materials also remove sulfate from water efficiently without affecting the phosphate uptake.

10.3.14 Waste ginkgo shells as adsorbent

Biomaterials such as waste ginkgo shells serve as low-cost adsorbents and have unique spherical structure. However, to increase their adsorption capacity and make their surfaces more reactive, they better need to be treated with specific chemicals and undergo modifications. By incorporating carbonyl group and adding the amine groups the biosorbent was modified to have multiple functional groups on its surface to promote better adsorption. Compared to carbonized shells, these modified shells have much higher activity, chemical stability, reusability, and capability to adsorb Cu^{2+} ions. Various other modifications using different functional groups can be done according to the metal ion which needs to be adsorbed on the biosorbent. The biosorbent can be easily regenerated as it shows no significant loss in its capacity even after four cycles of reusing. Four different ginkgo shells−based adsorbents, namely, raw, carbonized, oxidized, and both oxidized and ammonized ginkgo shells were studied for the adsorption of Cu^{2+} by Qiu et al. (2019). The obtained adsorption capacities of the four ginkgo shells−based adsorbents reported for were 8.62 (raw), 15.5 (carbonized), 31.8 (oxidized), and 41.6 (oxidized and ammonized) mg/g, respectively.

10.3.15 Fly ash

Fly ash obtained from a paper mill was incorporated as a base material for making effective bioadsorbent (Shadbahr and Husain, 2019). The ash used was mainly wood ash that was else dumped in landfills. The composition of ash showed that it has very high carbon content. After carbonization and activation with CO_2 alone, and both CO_2 and steam, metal impregnated activated carbon filter was produced. This was used for treatment of arsenic from wastewater. The adsorbent was able to reduce not only arsenic but also to remove it to a concentration which was acceptable as per the permissible arsenic concentration. Also iron impregnation was observed to increase the removal of arsenic to a great extent. Its capacity was observed to be very close to commercial activated carbon. Net adsorption strength of the modified fly ash for local well water obtained was $35.6\,\mu g/g$ and for laboratory synthesized water was $1428.6\,\mu g/g$ (Shadbahr and Husain, 2019).

10.3.16 Sawdust

Sawdust is another inexpensive and useful biosorbent for treating of inorganic metal impurities from wastewater. It is available in large amounts at all places, environment friendly, and is a very efficient adsorbent for treating the dyes and inorganic contaminants from water. It can be modified by different chemical treatments to increase its adsorption characteristics. Saw dust modified using acid hydrolysis can be utilized to treat pollutants, for example, dyes, heavy metal ions, and inorganic impurities from wastewater (Shukla et al., 2002). The wastewater generated from mining sites contains a lot of metal-sulfate impurities that are toxic in nature. These need to be converted into less toxic form or sulfides and can be precipitated. Sawdust cell walls are composed of cellulose, lignin, and some hydroxyl functional groups such as tannins. These functional groups and compounds act as active ion-exchanger agents. Ca^{2+}, Mg^{2+}, K^+, and Na^+ are the competing cations present in wastewater. Sawdust modified with 0.1 M HNO_3 was utilized to adsorb Cu^{2+} ions from heavy metal solutions resulting in 97% copper and lead adsorption. The adsorbed heavy metal ions got fully removed from the adsorbent in a period of 6 hours (Shukla et al., 2002).

10.3.17 Graphene-based nanoadsorbents

Nanoparticles possess large surface area and wide applications as catalysts as well as adsorbents. Graphene has been thoroughly studied for its distinct physicochemical and structural characteristics such as chemical strength, structural mutability, porosity, and low density. Graphenes are thinnest materials produced and simplest form of carbon. They possess high surface area to volume ratios along with excellent, chemical, mechanical, thermal, and electrical properties. Recently fabricated graphene oxides have been employed in adsorbing inorganic impurities from water since they contain numerous hydroxyl and epoxy groups. Graphenes and graphene oxides cannot be used directly as the aggregate and are difficult to separate from water. To overcome this, graphene-based nanocomposites (GN) have been used. Many metal ions such as Sb, Pb, Co, Cd, and Hg can be removed using these GN. Different metal ions were observed to have different pH values for which they showed maximum adsorption. Also, one of the major qualities that make using GN cost efficient is its high reusability (Kim et al., 2018).

10.3.18 Boron nitride—based materials

Boron nitride (BN) is environment friendly chemical compound derived from same numbers of boron and nitrogen atoms. Since the synthesis of boron nitride in the year 1842, extensive research has been undertaken to prepare different boron nitride structures, for example, nanosheets and nanoparticles. These BN structures possess distinct physical, structural, and chemical characteristics, for example, large surface area per unit mass, thermal and chemical stability, UV photoluminescence, insulation properties, large band gap, high conductivity, and exceptional oxidation resistance. These characteristics make boron nitrides potential substance for their utility in various domains such as sensor applications, pollutants removal and treatment, hydrogen energy storage, and heterogeneous catalysis (Yu et al., 2018). A lot of literature has been published on BN structures in the last few years, thereby stressing the need of utilization of BN materials as effective adsorbents in wastewater treatment. For a typical single-layered boron nitride material the adsorption sites are available on both sides of the BN. But large-scale manufacturing of single-layered BN sheets is a complex job due to factors such as oxidation resistance and structural stability. At the same time, multilayer boron nitride sheets possess great potential for the treatment of heavy metal pollutants persistent in the aqueous environment. Normally, adsorption of contaminants utilizing boron nitride materials involves a physicochemical interaction present at BN—water interface owing to its high surface area per unit mass and surface-active characteristics.

Chen et al. (2011) utilized Fe_3O_4 nanoparticles for modifying boron nitride for adsorbing As(V) from wastewater and reported 32.2 mg/g maximum adsorption capacity at pH 6.9. By incorporating structural properties of BN and Fe_3O_4's magnetic characteristics, the resulting adsorbent derived was potential candidate for wastewater treatment. Porous BN structure was utilized to mitigate Cu^{2+} ions from wastewater, and a very high removal capacity of 373 mg/g was achieved (Li et al., 2013). The high removal efficiency of BN was attributed to its large specific surface area of 1687 m^2/g, pore volume of 0.45 cm^3/g, and the resulting structural defects. BN materials were also utilized to adsorb Cr^{3+} ions utilizing modified fluorinated BN. A percentage of 82.6 of Cr^{3+} adsorption was achieved using fluorinated BN in 3 hours and at temperature of 30°C (Li et al., 2014).

10.3.19 Silica

In the category of inorganic minerals, silica possesses hydrophilic characteristics due to the presence of silanol (Si—O—H) functional groups. It exhibits high surface area per unit mass, fully porous structure, high mechanical stability, and high tendency to adsorb inorganic impurities. It possesses small resistance in alkaline mediums; therefore, it must be utilized for adsorption in acidic mediums, that is, pH less than 8. The acidic silanol groups on silica favor nonspecific adsorption; hence, to improve the adsorption capacity and reduce the irreversible adsorption effects, the silica adsorbent was modified by addition of amino group on its surface which enhanced its treatment potential for certain inorganic impurities and dyes. Another silica-containing mineral adsorbent is alunite containing 50% silica. Treated or modified alunite was successful in removing heavy metal impurities from wastewater. Dolomite and perlite minerals have also been utilized in the area of water treatment. Silica nanotubes can also be employed for heavy metal impurities.

10.3.20 Acinetobacter johnsonii

Acinetobacter johnsonii is amongst the popular microorganisms for treatment of inorganic impurities from spent oil (Jiang et al., 2012). The bacterium *A. johnsonii* separated from spent oil utilizing oxygen rich conditions was used to treat sulfide impurities. The experiments were designed having starting inoculum volume within a range from 2% to 15% and heating up to 323K till aeration (8−16 hours) was terminated, followed by sedimentation process for a period of 8 hours. The removal rates achieved ranged from 16.5% to 48.7% in oil phase at these conditions and the natural material reduced the environmental impurities. The highest removal rate achieved was 79.2% with aeration period extended to 16 hours. Further, the authors observed that by increasing aeration period from 12 to 14 hours, high removal efficiency (75%) was attained for both oil and water treatment (Jiang et al., 2012).

10.3.21 Calcined waste eggshell

Kose and Kivanc in 2011 reported the treatment of inorganic impurity, that is, phosphate ions (PO_4^{3-}) from waste solutions by utilizing biosorbent calcined waste eggshell (CWE) which was calcined prior to its use. The egg shells were obtained from domestic and commercial sources such as bakery. Since the shells were having low porosity so calcination at 800°C for about 2 hours of obtained egg shells were performed. For the batch adsorption experiments utilizing the CWEs the total reduction of phosphate attained increased from 99.4% to 99.6% with increase in strength of phosphate ions from 50 to 200 mg (Köse and Kıvanc, 2011). The adsorbent dose utilized was (0.1−0.5) g/50 mL. It was observed that adsorption obtained was physical adsorption as the adsorption energy value obtained (0.4 kJ/mol) corresponded to physical adsorption. The removal percentage for phosphate on CWEs was 99% for the pH limit of 2−10. The desorption isotherm highlighted that adsorption was not fully reversible due to strong bonding between the CWE and phosphate impurity (Köse and Kıvanc, 2011).

10.3.22 Miscellaneous agricultural and industrial wastes

Peat is a basically a soil matter originally formed due to partial decomposition of vegetable materials. It is also called turf, fens, bogs, peatlands, mires, etc. Four different varieties of peat are known: wood peat, soft (herbaceous) peat, moss peat, and sedimentary (marine) peat. Peat is having large availability, is cheap and has a strong affinity to adsorb organic and heavy metals impurities (Mo et al., 2018).

Xanthate is another cost-effective adsorbent for the treatment of inorganic impurities from wastewater. It consists of sulfur compounds, containing hydroxyl substrate, formed by reaction with carbon disulfide. Xanthate straw having 4.1% sulfur content was applied to adsorb Cr^{3+} ions from wastewater with more than 80% removal efficiency.

The underlying mechanism for chromium removal was monosulfur chelation and ion exchange. Xanthate has been also utilized for the removal of different metal ions (Ni^{2+}, Pb^{2+}, Cd^{2+}, Zn^{2+}, and Cu^{2+}) present in wastewater (Liang et al., 2010; Li et al., 2015; Yang et al., 2018).

Chitin $(C_8H_{13}O_5N)_n$ and shell polyfluorene are adsorbents in the biopolymer category which are utilized for water treatment. Chitin is a polysaccharide having long chain and is available abundantly in krill, crustaceans, lobster, fungi, crabs, insects, annelids, shrimps, and mollusks. Presently, chitosan and chitin are produced from crustaceans, that is, from sea food industry waste. At an estimate 1.2 million tons, crustacean was extracted to make chitin and protein annually, which mitigated the problem of solid waste disposal, saved the environment, and also generated large economic gains.

Biosorption involves the use of some biological agents such as yeasts, fungi, for example, white rot fungi and bacteria to treat heavy metal impurities and radio-nuclides in solution by the mechanism of chelation and complexation. Yeasts, for example, *S. cerevisiae*, and fungus, for example, *A. niger* and *P. chrysogenum*, have been utilized as biosorption materials. The presence of various functional groups on these biomass species could adsorb inorganic impurities as well as dyes, etc.

Starch and cyclodextrins relate to family of cyclic oligosaccharides and happen to be the cheapest adsorbents utilized for treating inorganic metal ions. Modified cyclodextrin, that is, carboxymethyl b-cyclodextrin or γ-cyclodextrin metal—organic framework is derived from natural materials and used for the treatment of heavy metal impurities such as cadmium and copper ions and also used as chelating agent for further use. The biggest advantage of cyclodextrin is the ability of its molecules to get attached with aromatic molecules to give rise to inclusion compounds.

Cyclodextrins can act as coupling agents by reacting with OH^- groups to form water-insoluble cross-linked structures which can easily capture heavy metal impurities. The modified γ-cyclodextrin-MOF-nanoporous carbon adsorbent was effectively utilized for cadmium ions removal in short period of time (1 minute), exhibiting very high adsorption potential under varied process conditions, for example, pH, dose, temperature, and competing ions. 140.85 mg/g cadmium ions treatment efficiency was obtained by the modified adsorbent owing to the oxygen rich functional groups on the modified biosorbent and ion-exchange mechanism (Liu et al., 2018).

Iron and iron-derived low-cost adsorbents had been utilized in treatment of heavy metal impurities. Iron oxides are abundant in nature and are the cause of many geological and physiological activities in nature. Hematite, magnetite, akaganeite, ferrihydrite, etc., iron compounds are potent adsorbents for removal of oxyanions from potable/wastewater. Akaganeite (β-FeOOH) having a structure like hollandite with general formula $FeO_{0.80}(OH)_{1.20}Cl_{0.20} \cdot 0.13H_2O$ is a nano-sized iron oxide. Many researchers have reported the potential of Akaganeite for phosphate and heavy metals removal (Lazaridis et al., 2005; Chitrakar et al., 2006; Deliyanni et al., 2007a; Zhao et al., 2012). Phosphorous removal up to 451.20 mg P/g and bromate removal capacity up to 41.8 mg/g has also been reported in the literature (Chitrakar et al., 2006, 2009; Deliyanni et al., 2007b).

Like akaganeite, hematite (α-Fe_2O_3) is also widely found iron oxide in natural systems such as sedimentary rocks and soils. It is thermodynamically the most stable form of iron oxides and greatly affects the water chemistry. Magnetite the first ever mineral used for

X-ray diffraction is a ferrimagnetic iron oxide having formula Fe_3O_4, contains both Fe^{2+} and Fe^{3+} ions in its structure consisting of tetrahedral and octahedral layers. The divalent iron in magnetite can be substituted by other divalent ions such as Cu^{2+}, Zn^{2+}, and Mn^{2+}. The phosphate removal capacity of magnetite reported is around 5.2 P mg/g (Daou et al., 2007).

Nano-sized magnesium oxide (MgO) adsorbents are capable of removing inorganic impurities, that is, arsenates and phosphates from wastewater. Li et al. (2016) documented exceptionally high adsorption of arsenic from wastewater utilizing 0.5 g/L mesoporous MgO with a removal efficiency of 813 and 912 mg/g for and for As^{3+} and As^{5+}, respectively (Li et al., 2016). Recently, MgO nanowires were utilized for treating phosphate and arsenate ions from wastewater and 962 ± 8.6 and 620 ± 6.2 mg/g maximum adsorption capacities at pH 11 for phosphate and arsenate ions, respectively, were reported (Ma et al., 2018). Adsorption column or batch process incorporating magnesium oxide nanowires can effectively be utilized for the removal of phosphates and arsenate from potable water.

Lead-based adsorbent materials such as lead oxides and mimetite having chemical formula ($Pb_5(AsO_4)_3Cl$) have been utilized to remove arsenic(V) impurities from wastewater. Mimetite is a very stable material at high pH conditions and can be used for adsorbing low strength (ppm and ppb)−based arsenate impurities from natural water decreasing arsenic concentration as low as 0.2 μg/L. Long et al. (2019) have utilized lead oxide (mimetite) for adsorption of As^{5+} out of aqueous chloride solution and have compared it with other lead-containing materials such as $Pb(NO_3)_2$, lead powder, and $Pb(OH)_2$. Three distinct pathways for lead addition, for 250 ml As^{5+} impurity stock solution, have been reported, namely, one time addition (0.35 g lead), intermittent addition (0.25 g initially and 0.10 g after half an hour), and addition followed filtration and then final addition (0.25 g initially, filtration and 0.10 g after half an hour) to optimize the mimetite ability to adsorb arsenic along with coprecipitation of Pb^{2+} ions by hydrolysis. 0.2 mg/L maximum As(V) removal was reported in the pH range of 1.9−12.3 in maximum time of 48 hours (Long et al., 2019).

Other low-cost adsorbents utilized for inorganic impurities removal include pulp and paper waste slags, grits, sugarcane bagasse, plant stems, fruit peels, stalks, bagasse fly ash, peat, corn cob, red mud, sludges, cellulose, coffee waste, straws, brans, coconut waste, lignin, husks, clino-pyrrhotite, barks, leaves, dates, fruit peels and seeds, grasses, and aragonite shells (Liang et al., 2010; Thirumavalavan et al., 2010; Ahmaruzzaman, 2011; Kumar et al., 2013; De Gisi et al., 2016; Jain and Yadav, 2017; Li et al., 2017; Regkouzas and Diamadopoulos, 2019).

Some of the low-cost adsorbents utilized for heavy metals removal and other inorganic impurities such as nitrates, fluorides, and phosphates removal with their adsorption capacities have been given in Table 10.1.

Future recommendations:

1. Currently due to the widespread inorganic pollutants prevalent in water bodies and being discharged from industrial effluents having complex configurations, no single adsorbent whether raw or modified or technique is efficient to treat pollutants from water and wastewater. Hence, effective utilization of different techniques, treatment methods, and modified adsorbents can solve the potable and wastewater treatment problem.

TABLE 10.1 Low-cost adsorbents utilized for inorganic pollutants removal.

Name of the adsorbent/biosorbents	Inorganic pollutant removed	Modification/treatment applied	Adsorption capacity (mg/g)	Reference
Cotton stalk	perchlorate	Amine cross-linking	42.6 at pH 6	Xu et al. (2013b)
Pistachio hull waste	Cr^{6+}	—	116.3 at pH 2	Moussavi and Barikbin (2010)
Walnut shell	Zn^{2+}	Acetic acid	27.86 at pH 3–6	Segovia-Sandoval et al. (2018)
Boron nitride	Cr^{3+}	Fluorinated with NH_4BF_4 solution	387 at pH 5.5	Li et al. (2014)
Carbon microspheres	Cr^{3+}, Pb^{2+}, Cd^{2+}, Cu^{2+}	Under nitrogen flow and from walnut shells	792, 638, 574, and 345 for respective ions at pH 5	Zbair et al. (2019)
Walnut shell	Cr^{6+}	—	40.99 at pH 2	Banerjee et al. (2018)
Rice husk	Cd^{2+}	Functionalized with xanthate	138.85	Qu et al. (2017)
Orange peel xanthate	Pb^{2+}	—	218.34	Liang et al. (2010)
Lignin xanthate resin	Pb^{2+}	—	64.9 at pH 5	Li et al. (2015)
Lemon peel	Cu^{2+}	Cellulose acid carboxylation after saponification	70.92 at pH 4–6	Thirumavalavan et al. (2010)
Granular ferric hydroxide	Phosphorous	Fixed bed adsorption	23.3 at pH 5.5	Genz et al. (2004)
Wheat bran	Cd^{2+}	—	15.71 at pH 5	Nouri et al. (2007)
Kraft lignin	Cd^{2+}	—	137.14 at pH 4.5	Mohan et al. (2006)
Sugar beet pulp	Pb^{2+}	—	0.356 mmol/g at pH 2–5.5	Reddad et al. (2002)
Pristine biochar from municipal solid waste	As^{5+}	Activated with 2 M KOH	24.49 mg/g at pH 8.5	Jin et al. (2014)
Ground burnt patties	Phosphorous	—	0.41 mg/g at pH 5–7	Rout et al. (2014)
Biochar-chitosan composite	Phosphorous and nitrate	Activated chitosan with soya bean husk and embedded with Zr^{4+} ions	131.29 mg/g and 90.09 mg/g respectively at pH 3–7	Banu et al. (2019)
Conocarpus green waste–based biochar	Nitrate	Modified with MgO and FeO	39.7 (for MgO) and 44.2 (for FeO) mmol/kg at pH 2	Usman et al. (2016)

(Continued)

TABLE 10.1 (Continued)

Name of the adsorbent/ biosorbents	Inorganic pollutant removed	Modification/treatment applied	Adsorption capacity (mg/g)	Reference
Sugarcane bagasse–based biochar	Pb^{2+}	–	86.96 mg/g at pH 5	Abdelhafez and Li (2016)
Graphene-biochar composite	Hg^{2+}	–	853 μg/g at pH 6.8–7.0	Tang et al. (2015)
Sugarcane bagasse biochar	Nitrate	Chemical modification	28.21 mg/g at pH 4.64	Hafshejani et al., (2016)
Cellulose aerogels/ MOF/ZIF-8	Cr^{6+}	Zeolitic imidazolate framework (ZIF-8)	41.8 mg/g	Bo et al. (2018)

2. Due to limitations of various other wastewater-treatment techniques such as high cost, need of equipment, and waste generation and inherent benefits of adsorption, it is a very useful, easy, and inexpensive technique to be applied for treating inorganic impurities from wastewater.

3. Broad range of low-cost, renewable adsorbents, and biosorbents are available from biomass, agricultural crops and residues, fruit and vegetable wastes, seaweeds, marine adsorbents, industrial wastes, and bioorganisms such as fungi and yeasts from different natural and industrial sources.

4. The agricultural residues and other organic wastes possess excellent physical, chemical, and structural properties, contain active functional groups which can effectively treat inorganic impurities such as heavy metal ions and phosphates, nitrates, and chlorides from waste effluents and water.

5. Adsorption technique banks upon on process conditions such as temperature, pH, initial strength of influents, and other factors such as competing ions. Sometimes the raw or naive adsorbents are not much effective in removal of the inorganic impurities, thus treatment of these adsorbents with certain chemicals, calcification, carbonization, immobilization on bio- or nano-structures/supports, etc., modification can be done to drastically enhance the adsorption efficiency in addition to enhancing the removal efficiencies of the modified adsorbents.

6. The underlying mechanisms of contaminant adsorption are ion exchange, chelation effects, electrostatic interaction, structural modification, complex formation, and use of nano- and biosupports. The development of efficient green adsorbent technology is the need of hour, and it will encompass all sources of natural, agricultural, biomass, and industrial waste adsorbents.

7. Utilization of agricultural residues/biowaste/forest waste as low-cost raw/modified adsorbents can reduce the environmental degradation, help in treating complex inorganic impurities from wastewater, and to arrive at the task of "treatment of waste by waste."

References

Abdelhafez, A.A., Li, J., 2016. Removal of Pb(II) from aqueous solution by using biochars derived from sugar cane bagasse and orange peel. J. Taiwan Inst. Chem. Eng. 61, 367–375. Available from: https://doi.org/10.1016/j.jtice.2016.01.005.

Adeyemo, A.A., Adeoye, I.O., Bello, O.S., 2017. Adsorption of dyes using different types of clay: a review. Appl. Water Sci. 7 (2), 543–568. Available from: https://doi.org/10.1007/s13201-015-0322-y.

Ahmad, M., et al., 2014. Chemosphere biochar as a sorbent for contaminant management in soil and water: a review. Chemosphere 99, 19–33. Available from: https://doi.org/10.1016/j.chemosphere.2013.10.071.

Ahmaruzzaman, M., 2011. Industrial wastes as low-cost potential adsorbents for the treatment of wastewater laden with heavy metals. Adv. Colloid Interface Sci. 166 (1–2), 36–59. Available from: https://doi.org/10.1016/j.cis.2011.04.005.

Alshameri, A., et al., 2019. Understanding the role of natural clay minerals as effective adsorbents and alternative source of rare earth elements: adsorption operative parameters. Hydrometallurgy 185, 149–161. Available from: https://doi.org/10.1016/j.hydromet.2019.02.016.

Al-Sherbini, A.S.A., et al., 2019. Utilization of chitosan/Ag bionanocomposites as eco-friendly photocatalytic reactor for Bactericidal effect and heavy metals removal. Heliyon 5 (6), e01980. Available from: https://doi.org/10.1016/j.heliyon.2019.e01980.

Anastopoulos, I., et al., 2017. A review on waste-derived adsorbents from sugar industry for pollutant removal in water and wastewater. J. Mol. Liq. 240, 179–188. Available from: https://doi.org/10.1016/j.molliq.2017.05.063.

Anirudhan, T.S., Jalajamony, S., Sreekumari, S.S., 2012. Adsorption of heavy metal ions from aqueous solutions by amine and carboxylate functionalised bentonites. Appl. Clay Sci. 65–66, 67–71. Available from: https://doi.org/10.1016/j.clay.2012.06.005.

Arfaoui, S., Frini-Srasra, N., Srasra, E., 2008. Modelling of the adsorption of the chromium ion by modified clays. Desalination 222 (1–3), 474–481. Available from: https://doi.org/10.1016/j.desal.2007.03.014.

Aydin, H., Bulut, Y., Yerlikaya, Ç., 2008. Removal of copper(II) from aqueous solution by adsorption onto low-cost adsorbents. J. Environ. Manage. 87 (1), 37–45. Available from: https://doi.org/10.1016/j.jenvman.2007.01.005.

Babel, S., Kurniawan, T.A., 2003. Low-cost adsorbents for heavy metals uptake from contaminated water: a review. J. Hazard. Mater. 97, 219–243.

Bahadur, V., Gadi, R., Kalra, S., 2019. Clay based nanocomposites for removal of heavy metals from water: a review. J. Environ. Manage. 232, 803–817. Available from: https://doi.org/10.1016/j.jenvman.2018.11.120.

Banerjee, A., 2015. Groundwater fluoride contamination: a reappraisal. Geosci. Front. 6 (2), 277–284. Available from: https://doi.org/10.1016/j.gsf.2014.03.003.

Banerjee, M., Basu, R.K., Das, S.K., 2018. Cr(VI) adsorption by a green adsorbent walnut shell: Adsorption studies, regeneration studies, scale-up design and economic feasibility. Process. Saf. Environ. Protection. Inst. Chem. Eng. 116, 693–702. Available from: https://doi.org/10.1016/j.psep.2018.03.037.

Banu, H.T., Karthikeyan, P., Meenakshi, S., 2019. International Journal of Biological Macromolecules Zr^{4+} ions embedded chitosan-soya bean husk activated bio-char composite beads for the recovery of nitrate and phosphate ions from aqueous solution. Int. J. Biol. Macromol. 130, 573–583. Available from: https://doi.org/10.1016/j.ijbiomac.2019.02.100.

Bhatnagar, A., Sillanpää, M., 2011. A review of emerging adsorbents for nitrate removal from water. Chem. Eng. J. 168 (2), 493–504. Available from: https://doi.org/10.1016/j.cej.2011.01.103.

Bhattacharyya, K.G., Gupta, S.S., 2011. Removal of Cu(II) by natural and acid-activated clays: an insight of adsorption isotherm, kinetic and thermodynamics. Desalination 272 (1–3), 66–75. Available from: https://doi.org/10.1016/j.desal.2011.01.001.

Bo, S., et al., 2018. Flexible and porous cellulose aerogels/zeolitic imidazolate framework (ZIF-8) hybrids for adsorption removal of Cr (IV) from water. J. Solid State Chem. 262, 135–141. Available from: https://doi.org/10.1016/j.jssc.2018.02.022.

Bombuwala, N., et al., 2018. Bioresource technology fast nitrate and fluoride adsorption and magnetic separation from water on $\alpha\text{-}Fe_2O_3$ and Fe_3O_4 dispersed on Douglas fir biochar. Bioresour. Technol. 263, 258–265. Available from: https://doi.org/10.1016/j.biortech.2018.05.001.

Cao, X., et al., 2009. Dairy-manure derived biochar effectively sorbs lead and atrazine. Environ. Sci. Technol. 43 (9), 3285–3291. Available from: https://doi.org/10.1021/es803092k.

Cao, W., et al., 2011. Removal of sulphate from aqueous solution using modified rice straw: preparation, characterization and adsorption performance. Carbohydr. Polym. 85 (3), 571−577. Available from: https://doi.org/10.1016/j.carbpol.2011.03.016.

Chen, R., et al., 2011. Arsenic(V) adsorption on Fe_3O_4 nanoparticle-coated boron nitride nanotubes. J. Colloid Interface Sci. 359 (1), 261−268. Available from: https://doi.org/10.1016/j.jcis.2011.02.071.

Cherif, M.A., et al., 2017. A robust and parsimonious model for caesium sorption on clay minerals and natural clay materials. Appl. Geochem. 87, 22−37. Available from: https://doi.org/10.1016/j.apgeochem.2017.10.017.

Chitrakar, R., et al., 2006. Phosphate adsorption on synthetic goethite and akaganeite. J. Colloid Interface Sci. 298 (2), 602−608. Available from: https://doi.org/10.1016/j.jcis.2005.12.054.

Chitrakar, R., et al., 2009. Bromate ion-exchange properties of crystalline akaganéite. Ind. Eng. Chem. Res. 48, 2107−2112.

Chojnacka, K., 2006. Biosorption of Cr(III) ions by wheat straw and grass: a systematic characteristics of new biosorbents. Pol. J. Environ. Stud. 15 (6), 845−852.

Daou, T.J., et al., 2007. Phosphate adsorption properties of magnetite-based nanoparticles. Chem. Mater. 19 (18), 4494−4505. Available from: https://doi.org/10.1021/cm071046v.

Davis, T.A., Volesky, B., Mucci, A., 2003. A review of the biochemistry of heavy metal biosorption by brown algae. Water Res. 37 (18), 4311−4330. Available from: https://doi.org/10.1016/S0043-1354(03)00293-8.

De Gisi, S., et al., 2016. Characteristics and adsorption capacities of low-cost sorbents for wastewater treatment: a review. Sustain. Mater. Technol. 9, 10−40. Available from: https://doi.org/10.1016/j.susmat.2016.06.002.

Deliyanni, E.A., Peleka, E.N., Matis, K.A., 2007a. Removal of zinc ion from water by sorption onto iron-based nanoadsorbent. J. Hazard. Mater. 141 (1), 176−184. Available from: https://doi.org/10.1016/j.jhazmat.2006.06.105.

Deliyanni, E.A., Peleka, E.N., Lazaridis, N.K., 2007b. Comparative study of phosphates removal from aqueous solutions by nanocrystalline akaganéite and hybrid surfactant-akaganéite. Sep. Purif. Technol. 52 (3), 478−486. Available from: https://doi.org/10.1016/j.seppur.2006.05.028.

de Pablo, L., Chávez, M.L., Abatal, M., 2011. Adsorption of heavy metals in acid to alkaline environments by montmorillonite and Ca-montmorillonite. Chem. Eng. J. 171 (3), 1276−1286. Available from: https://doi.org/10.1016/j.cej.2011.05.055.

Ding, D., et al., 2013. Adsorption of cesium from aqueous solution using agricultural residue e Walnut shell: equilibrium, kinetic and thermodynamic modeling studies. Water Res. 47 (7), 2563−2571. Available from: https://doi.org/10.1016/j.watres.2013.02.014.

Doan, H.D., et al., 2008. Removal of Zn^{2+} and Ni^{2+} by adsorption in a fixed bed of wheat straw. Process. Saf. Environ. Prot. 86 (4), 259−267. Available from: https://doi.org/10.1016/j.psep.2008.04.004.

Farooq, U., et al., 2010. Biosorption of heavy metal ions using wheat based biosorbents − a review of the recent literature. Bioresour. Technol. 101 (14), 5043−5053. Available from: https://doi.org/10.1016/j.biortech.2010.02.030.

Farooq, U., et al., 2011. Effect of modification of environmentally friendly biosorbent wheat (*Triticum aestivum*) on the biosorptive removal of cadmium(II) ions from aqueous solution. Chem. Eng. J. 171 (2), 400−410. Available from: https://doi.org/10.1016/j.cej.2011.03.094.

Filote, C., et al., 2019. Bioadsorptive removal of Pb(II) from aqueous solution by the biorefinery waste of *Fucus spiralis*. Sci. Total. Environ. 648, 1201−1209. Available from: https://doi.org/10.1016/j.scitotenv.2018.08.210.

Genz, A., Kornmüller, A., Jekel, M., 2004. Advanced phosphorus removal from membrane filtrates by adsorption on activated aluminium oxide and granulated ferric hydroxide. Water Res. 38 (16), 3523−3530. Available from: https://doi.org/10.1016/j.watres.2004.06.006.

Giannakoudakis, D.A., et al., 2018. Aloe vera waste biomass-based adsorbents for the removal of aquatic pollutants: a review. J. Environ. Manage. 227, 354−364. Available from: https://doi.org/10.1016/j.jenvman.2018.08.064.

Godt, J., et al., 2006. The toxicity of cadmium and resulting hazards for human health. J. Occup. Med. Toxicol. 1 (1), 1−6. Available from: https://doi.org/10.1186/1745-6673-1-22.

Hafshejani, L.D., Hooshmand, A., Naseri, A.A., Mohammadi, A.S., Abbasi, F., Bhatnagar, A., 2016. Removal of nitrate from aqueous solution by modified sugarcane bagasse biochar. Ecol. Eng. 95, 101−111. Available from: https://doi.org/10.1016/j.ecoleng.2016.06.035.

Heydari, A., et al., 2017. Polymerization of β-cyclodextrin in the presence of bentonite clay to produce polymer nanocomposites for removal of heavy metals from drinking water. Polym. Adv. Technol. 28 (4), 524–532. Available from: https://doi.org/10.1002/pat.3951.

Hokkanen, S., Bhatnagar, A., Sillanpää, M., 2016. A review on modification methods to cellulose-based adsorbents to improve adsorption capacity. Water Res. 91, 156–173. Available from: https://doi.org/10.1016/j.watres.2016.01.008.

Hong, M., et al., 2019. Heavy metal adsorption with zeolites: the role of hierarchical pore architecture. Chem. Eng. J. 359, 363–372. Available from: https://doi.org/10.1016/j.cej.2018.11.087.

Huang, G., et al., 2009. Adsorptive removal of copper ions from aqueous solution using cross-linked magnetic chitosan beads. Chin. J. Chem. Eng. 17 (6), 960–966. Available from: https://doi.org/10.1016/S1004-9541(08)60303-1. Chemical Industry and Engineering Society of China (CIESC) and Chemical Industry Press (CIP).

Jain, D.S.M.C.K., Yadav, A.K., 2017. Removal of heavy metals from emerging cellulosic low-cost adsorbents: a review. Appl. Water Sci. 7 (5), 2113–2136. Available from: https://doi.org/10.1007/s13201-016-0401-8.

Jiang, Y., et al., 2012. Inorganic impurity removal from waste oil and wash-down water by *Acinetobacter johnsonii*. J. Hazard. Mater. 239–240, 289–293. Available from: https://doi.org/10.1016/j.jhazmat.2012.08.076.

Jiménez-Castañeda, M.E., Medina, D.I., 2017. Use of surfactant-modified zeolites and clays for the removal of heavy metals from water. Water (Switz.) 9 (4). Available from: https://doi.org/10.3390/w9040235.

Jin, S., Chen, H., 2007. Near-infrared analysis of the chemical composition of rice straw. Ind. Crop. Prod. 26, 207–211. Available from: https://doi.org/10.1016/j.indcrop.2007.03.004.

Jin, H., et al., 2014. Biochar pyrolytically produced from municipal solid wastes for aqueous As(V) removal: adsorption property and its improvement with KOH activation. Bioresour. Technol. 169, 622–629. Available from: https://doi.org/10.1016/j.biortech.2014.06.103.

Jokar, M., et al., 2019. Preparation and characterization of novel bio ion exchanger from medicinal herb waste (chicory) for the removal of Pb^{2+} and Cd^{2+} from aqueous solutions. *J. Water Process Eng.* 28,88–99. https://doi.org/10.1016/j.jwpe.2019.01.007.

Joseph, L., et al., 2019. Removal of heavy metals from water sources in the developing world using low-cost materials: a review. Chemosphere 229, 142–159. Available from: https://doi.org/10.1016/j.chemosphere.2019.04.198.

Juang, R.S., Shao, H.J., 2002. Effect of pH on competitive adsorption of Cu(II), Ni(II), and Zn(II) from water onto chitosan beads. Adsorption 8 (1), 71–78. Available from: https://doi.org/10.1023/A:1015222607996.

Kesraoui, Sabeha, Kavannagh, Mark, 1997. Performance of natural zeolites for the treatment of mixed metal-contaminated Effluents. Waste Manage. Res. 15, 383–394.

Kim, S., et al., 2018. Aqueous removal of inorganic and organic contaminants by graphene-based nanoadsorbents: a review. Chemosphere 212, 1104–1124. Available from: https://doi.org/10.1016/j.chemosphere.2018.09.033.

Köse, T.E., Kıvanc, B., 2011. Adsorption of phosphate from aqueous solutions using calcined waste eggshell. Chem. Eng. J. 178, pp. 34–39. https://doi.org/10.1016/j.cej.2011.09.129.

Kumar, A., et al., 2013. Removal of fluoride from aqueous solution and groundwater by wheat straw, sawdust and activated bagasse carbon of sugarcane. Ecol. Eng. 52, 211–218. Available from: https://doi.org/10.1016/j.ecoleng.2012.12.069.

Kyzas, G.Z., Bikiaris, D.N., 2015. Recent modifications of chitosan for adsorption applications: a critical and systematic review. Mar. Drugs 13 (1), 312–337. Available from: https://doi.org/10.3390/md13010312.

Lazaridis, N.K., Bakoyannakis, D.N., Deliyanni, E.A., 2005. Chromium(VI) sorptive removal from aqueous solutions by nanocrystalline akaganèite. Chemosphere 58 (1), 65–73. Available from: https://doi.org/10.1016/j.chemosphere.2004.09.007.

Li, J., et al., 2013. Porous boron nitride with a high surface area: hydrogen storage and water treatment. Nanotechnology 24 (15). Available from: https://doi.org/10.1088/0957-4484/24/15/155603.

Li, J., Jin, P., Tang, C., 2014. Cr(III) adsorption by fluorinated activated boron nitride: a combined experimental and theoretical investigation. RSC Adv. 4 (29), 14815–14821. Available from: https://doi.org/10.1039/c4ra01684j.

Li, Z., Kong, Y., Ge, Y., 2015. Synthesis of porous lignin xanthate resin for Pb^{2+} removal from aqueous solution. Chem. Eng. J. 270, 229–234. Available from: https://doi.org/10.1016/j.cej.2015.01.123.

Li, W., et al., 2016. Extremely high arsenic removal capacity for mesoporous aluminium magnesium oxide composites. Environ. Sci: Nano 3 (1), 94–106. Available from: https://doi.org/10.1039/C5EN00171D.

Li, Z., et al., 2017. Determination of the best conditions for modified biochar immobilized petroleum hydrocarbon degradation microorganism by orthogonal test. IOP Conf. Series: Earth Environ. Sci. 94 (1). Available from: https://doi.org/10.1088/1755-1315/94/1/012191.

Li, Z., et al., 2019. Disinfection and removal performance for *Escherichia coli*, toxic heavy metals and arsenic by wood vinegar-modified zeolite. Ecotoxicol. Environ. Saf. 174, 129−136. Available from: https://doi.org/10.1016/j.ecoenv.2019.01.124.

Liang, S., et al., 2010. Effective removal of heavy metals from aqueous solutions by orange peel xanthate. Trans. Nonferrous Met. Soc. China (Engl. Ed.) 20 (Suppl. 1), s187−s191. Available from: https://doi.org/10.1016/S1003-6326(10)60037-4. The Nonferrous Metals Society of China.

Liu, Q., et al., 2015. Adsorptive removal of Cr(VI) from aqueous solutions by cross-linked chitosan/bentonite composite. Korean J. Chem. Eng. 32 (7), 1314−1322. Available from: https://doi.org/10.1007/s11814-014-0339-1.

Liu, C., et al., 2018. Ultrafast removal of cadmium(II) by green cyclodextrin metal−organic-framework-based nanoporous carbon: adsorption mechanism and application. Chem. - An Asian J. Available from: https://doi.org/10.1002/asia.201801431 (accepted article).

Long, H., et al., 2019. Hydrometallurgy Comparison of arsenic(V) removal with different lead-containing substances and process optimization in aqueous chloride solution. Hydrometallurgy 183, 199−206. Available from: https://doi.org/10.1016/j.hydromet.2018.12.006.

Ma, L., et al., 2015. Simultaneous adsorption of Cd(II) and phosphate on Al < inf > 13 < /inf > pillared montmorillonite. RSC Adv. 5 (94), 77227−77234. Available from: https://doi.org/10.1039/c5ra15744g.

Ma, Y., et al., 2017. Porous lignin based poly (acrylic acid)/organo-montmorillonite nanocomposites: swelling behaviors and rapid removal of Pb(II) ions. Polymer. (Guildf). 128, 12−23. Available from: https://doi.org/10.1016/j.polymer.2017.09.009.

Ma, G., et al., 2018. Highly active magnesium oxide nano materials for the removal of arsenates and phosphates from aqueous solutions. Nano-Struct. Nano-Objects 13, 74−81. Available from: https://doi.org/10.1016/j.nanoso.2017.11.006.

Mckay, G., Blair, H.S., Findon, A., 1989. Equilibrium studies for the sorption of metal ions onto chitosan. Indian J. Chem. 28, 356−360. Available at: http://nopr.niscair.res.in/bitstream/123456789/46710/1/IJCA28A%285%29356-360.pdf.

Mo, J., et al., 2018. A review on agro-industrial waste (AIW) derived adsorbents for water and wastewater treatment. J. Environ. Manage. 227, 395−405. Available from: https://doi.org/10.1016/j.jenvman.2018.08.069.

Mohan, D., Pittman, C.U., Steele, P.H., 2006. Single, binary and multi-component adsorption of copper and cadmium from aqueous solutions on Kraft lignin-a biosorbent. J. Colloid Interface Sci. 297 (2), 489−504. Available from: https://doi.org/10.1016/j.jcis.2005.11.023.

Moussavi, G., Barikbin, B., 2010. Biosorption of chromium(VI) from industrial wastewater onto pistachio hull waste biomass. Chem. Eng. J. 162 (3), 893−900. Available from: https://doi.org/10.1016/j.cej.2010.06.032.

Moyo, M., et al., 2013. Adsorption batch studies on the removal of Pb(II) using maize tassel based activated carbon. J. Chem. 2013, 1−8. Available from: https://doi.org/dx.doi.org/10.1155/2013/508934. Research.

Nabbou, N., Belhachemi, M., Boumelik, M., 2019. Removal of fluoride from groundwater using natural clay (kaolinite): Optimization of adsorption conditions. C. R. − Chim. 22 (2−3), 105−112. Available from: https://doi.org/10.1016/j.crci.2018.09.010.

Najiah, S., et al., 2014. Removal of Cu(II), Pb(II) and Zn(II) ions from aqueous solutions using selected agricultural wastes: adsorption and characterisation studies. J. Environ. Prot. 5 (5), 289. Available from: https://doi.org/10.4236/jep.2014.54032.

Nethaji, S., Sivasamy, A., Mandal, A.B., 2013. Preparation and characterization of corn cob activated carbon coated with nano-sized magnetite particles for the removal of Cr(VI). Bioresour. Technol. 134, 94−100. Available from: https://doi.org/10.1016/j.biortech.2013.02.012.

Ng, J.C.Y., Cheung, W.H., McKay, G., 2002. Equilibrium studies of the sorption of Cu(II) ions onto chitosan. J. Colloid. Interface. Sci. 255 (1), 64−74. Available from: https://doi.org/10.1006/jcis.2002.8664.

Nouri, L., et al., 2007. Batch sorption dynamics and equilibrium for the removal of cadmium ions from aqueous phase using wheat bran. J. Hazard. Mater. 149 (1), 115−125. Available from: https://doi.org/10.1016/j.jhazmat.2007.03.055.

Obaid, S.S., et al., 2018. Heavy metal ions removal from waste water by the natural zeolites. Mater. Today: Proc. 5 (9), 17930−17934. Available from: https://doi.org/10.1016/j.matpr.2018.06.122.

Park, S.M., et al., 2019. Adsorption characteristics of cesium on the clay minerals: structural change under wetting and drying condition. Geoderma 340, 49−54. Available from: https://doi.org/10.1016/j.geoderma.2018.12.002.

Pathirana, C., et al., 2019. Quantifying the influence of surface physico-chemical properties of biosorbents on heavy metal adsorption. Chemosphere 234, 488−495. Available from: https://doi.org/10.1016/j.chemosphere.2019.06.074.

Pepper, R.A., Couperthwaite, S.J., Millar, G.J., 2018. Re-use of waste red mud: production of a functional iron oxide adsorbent for removal of phosphorous. J. Water Process. Eng. 25, 138−148. Available from: https://doi.org/10.1016/j.jwpe.2018.07.006.

Qiu, H., et al., 2019. Fabrication of agricultural waste supported UiO-66 nanoparticles with high utilization in phosphate removal from water. Chem. Eng. J. 360, 621−630. Available from: https://doi.org/10.1016/j.cej.2018.12.017.

Qu, J., et al., 2017. Utilization of rice husks functionalized with xanthates as cost-effective biosorbents for optimal Cd(II) removal from aqueous solution via response surface methodology. Bioresour. Technol. 241, 1036−1042. Available from: https://doi.org/10.1016/j.biortech.2017.06.055.

Reddad, Z., et al., 2002. Adsorption of several metal ions onto a low-cost biosorbent: kinetic and equilibrium studies. Environ. Sci. Technol. 36 (9), 2067−2073. Available from: https://doi.org/10.1021/es0102989.

Reddy, D.H.K., Lee, S., 2013. Application of magnetic chitosan composites for the removal of toxic metal and dyes from aqueous solutions. Adv. Colloid Interface Sci. 201−202, 68−93. Available from: https://doi.org/10.1016/j.cis.2013.10.002.

Regkouzas, P., Diamadopoulos, E., 2019. Adsorption of selected organic micro-pollutants on sewage sludge biochar. Chemosphere 224, 840−851. Available from: https://doi.org/10.1016/j.chemosphere.2019.02.165.

Romera, E., et al., 2008. Biosorption of heavy metals by *Fucus spiralis*. Bioresour. Technol. 99 (11), 4684−4693. Available from: https://doi.org/10.1016/j.biortech.2007.09.081.

Rout, P.R., Bhunia, P., Dash, R.R., 2014. Modeling isotherms, kinetics and understanding the mechanism of phosphate adsorption onto a solid waste: ground burnt patties. J. Environ. Chem. Eng. 2 (3), 1331−1342. Available from: https://doi.org/10.1016/j.jece.2014.04.017.

Santos, S.C.R., Ungureanu, G., Volf, I., Boaventura, R.A.R., Botelho, C.M.S., 2018. 3- Macroalgae biomass as sorbent for metal ions. In: Popa, V., Volf, I. (Eds.), Biomass as renewable raw material to obtain bioproducts of high-tech value. Elsevier, pp. 69−112.

Sdiri, A., et al., 2011. Evaluating the adsorptive capacity of montmorillonitic and calcareous clays on the removal of several heavy metals in aqueous systems. Chem. Eng. J. 172 (1), 37−46. Available from: https://doi.org/10.1016/j.cej.2011.05.015.

Segovia-Sandoval, S.J., et al., 2018. Walnut shell treated with citric acid and its application as biosorbent in the removal of Zn(II). J. Water Process. Eng. 25, 45−53. Available from: https://doi.org/10.1016/j.jwpe.2018.06.007.

Shadbahr, J., Husain, T., 2019. Affordable and efficient adsorbent for arsenic removal from rural water supply systems in Newfoundland. Sci. Total. Environ. 660, 158−168. Available from: https://doi.org/10.1016/j.scitotenv.2018.12.319.

Sharma, R., Singh, B., 2013. Removal of Ni(II) ions from aqueous solutions using modified rice straw in a fixed bed column. Bioresour. Technol. 146 (10), 519−524. Available from: https://doi.org/10.1016/j.biortech.2013.07.146.

Shukla, A., Zhang, Y.-H., Dubey, P., Margrave, J.L., Shukla, S.S., 2002. The role of sawdust in the removal of unwanted materials from water. J. Hazard. Mater. B95, 137−152. Available from: https://doi.org/10.1017/S0017089504001740.

Sizmur, T., et al., 2017. Biochar modification to enhance sorption of inorganics from water. Bioresour. Technol. 246, 34−47. Available from: https://doi.org/10.1016/j.biortech.2017.07.082.

Tang, J., et al., 2015. Preparation and characterization of a novel graphene/biochar composite for aqueous phenanthrene and mercury removal. Bioresour. Technol. 196, 355−363. Available from: https://doi.org/10.1016/j.biortech.2015.07.047.

Thirumavalavan, M., et al., 2010. Cellulose-based native and surface modified fruit peels for the adsorption of heavy metal ions from aqueous solution: langmuir adsorption isotherms. J. Chem. Eng. Data 55 (3), 1186−1192. Available from: https://doi.org/10.1021/je900585t.

Uchimiya, M., Lima, I.M., Thomas, K.K., Chang, S., Wartelle, L.H., Rodgers, J.E., et al., 2010. Immobilization of heavy metal ions (Cu II, Cd II, Ni II, and Pb II) by broiler litter-derived biochars in water and soil. J. Agric. Food. Chem. 58 (9), 5538−5544. Available from: https://doi.org/10.1021/jf9044217.

Usman, A.R.A., Ahmad, M., 2015. Chemically modified biochar produced from conocarpus waste increases NO_3 removal from aqueous solutions chemically modified biochar produced from conocarpus waste increases NO_3 removal from aqueous solutions. Environ. Geochem. Health. Available from: https://doi.org/10.1007/s10653-015-9736-6.

Usman, A.R.A., et al., 2016. Chemically modified biochar produced from conocarpus waste increases NO_3 removal from aqueous solutions. Environ. Geochem. Health 38 (2), 511−521. Available from: https://doi.org/10.1007/s10653-015-9736-6.

Vardhan, K.H., Kumar, P.S., Panda, R.C., 2019. A review on heavy metal pollution, toxicity and remedial measures: Current trends and future perspectives. J. Mol. Liq. 290, 111197. Available from: https://doi.org/10.1016/j.molliq.2019.111197.

Wilson, K., et al., 2006. Select metal adsorption by activated carbon made from peanut shells. Bioresour. Technol. 97 (18), 2266−2270. Available from: https://doi.org/10.1016/j.biortech.2005.10.043.

Woolf, D., et al., 2010. Sustainable biochar to mitigate global climate change. Nat. Commun. 1 (5), 1−9. Available from: https://doi.org/10.1038/ncomms1053.

Xu, X., et al., 2013a. Removal of Cu, Zn, and Cd from aqueous solutions by the dairy manure-derived biochar. Environ. Sci. Pollut. Res. 20, 358−368. <https://doi.org/10.1007/s11356-012-0873-5>.

Xu, X., et al., 2013b. Uptake of perchlorate from aqueous solutions by amine-crosslinked cotton stalk. Carbohydr. Polym. 98 (1), 132−138. Available from: https://doi.org/10.1016/j.carbpol.2013.05.058.

Xue, L., et al., 2016. High efficiency and selectivity of MgFe-LDH modified wheat-straw biochar in the removal of nitrate from aqueous solutions. J. Taiwan Inst. Chem. Eng. 63, 312−317. Available from: https://doi.org/10.1016/j.jtice.2016.03.021.

Yakout, S.M., Rizk, M.A., 2015. Adsorption of uranium by low-cost adsorbent derived from agricultural wastes in multi-component system. Desalin. Water Treat. 53, 1917−1922. Available from: https://doi.org/10.1080/19443994.2013.860625.

Yang, K., et al., 2018. Treatment of wastewater containing Cu^{2+} using a novel macromolecular heavy metal chelating flocculant xanthated chitosan. Colloids Surf. A: Physicochem. Eng. Asp. 558, 384−391. Available from: https://doi.org/10.1016/j.colsurfa.2018.06.082.

Yang, X., Zhang, S., Ju, M., 2019. Preparation and modification of biochar materials and their application in soil remediation. Appl. Sci.; MDPI 9, 1365. Available from: https://doi.org/10.3390/app9071365.

Yu, Z., Zhang, X., Huang, Y., 2013. Magnetic chitosan-iron(III) hydrogel as a fast and reusable adsorbent for chromium(VI) removal. Ind. Eng. Chem. Res. 52 (34), 11956−11966. Available from: https://doi.org/10.1021/ie400781n.

Yu, S., et al., 2018. Boron nitride-based materials for the removal of pollutants from aqueous solutions: a review. Chem. Eng. J. 333, 343−360. Available from: https://doi.org/10.1016/j.cej.2017.09.163.

Yuwei, C., Jianlong, W., 2011. Preparation and characterization of magnetic chitosan nanoparticles and its application for Cu(II) removal. Chem. Eng. J. 168 (1), 286−292. Available from: https://doi.org/10.1016/j.cej.2011.01.006.

Zbair, M., et al., 2019. Carbon microspheres derived from walnut shell: rapid and remarkable uptake of heavy metal ions, molecular computational study and surface modeling. Chemosphere 231, 140−150. Available from: https://doi.org/10.1016/j.chemosphere.2019.05.120.

Zhang, M., Gao, B., 2013. Removal of arsenic, methylene blue, and phosphate by biochar/AlOOH nanocomposite. Chem. Eng. J. 226, 286−292. Available from: https://doi.org/10.1016/j.cej.2013.04.077.

Zhao, J., et al., 2012. Adsorptive characteristics of akaganeite and its environmental applications: a review. Environ. Technol. Rev. 1 (1), 114−126. Available from: https://doi.org/10.1080/09593330.2012.701239.

Zhu, R., et al., 2016. Adsorbents based on montmorillonite for contaminant removal from water: a review. Appl. Clay Sci. 123, 239−258. Available from: https://doi.org/10.1016/j.clay.2015.12.024.

Low-cost bio-adsorbent for emerging inorganic pollutants

Jyoti Singh, Priyanka Yadav and Vishal Mishra

School of Biochemical Engineering, IIT (BHU) Varanasi, Varanasi, India

11.1 Introduction

Nature served us with ample resources to sustain life on this earth, and among them, water is the most important one, without which life is impossible to live. Globally, more than 1 billion people are not able to access safe water, and it was reported that after a couple of decades, one-third of decrement will occur in the water supply. The possibility is minimal for providing safe water to the people due to increment in global population day by day (Amin et al., 2014). Ground water is vital for everyone's life, but it is affected by rapid industrialization and results in the release of various kinds of pollutants (Gupta et al., 2009). Some of them are inorganic that are brought about not naturally but by several forms of human activities, they are pollutants such as lead, zinc, and copper. As these are nonbiodegradable and remain persistent in the environment, which poses an adverse effect on flora, fauna, and possibly will cause various health problems to humans. Heavy metals like mercury, copper, cadmium, lead, arsenic, and selenium are toxic to biota at a concentration above the permissible limit in drinking water. It is thus urgently required to remove these inorganic pollutants using various technologies. These inorganic pollutants are very well known for a long time. But, some other pollutants that are not generally

Inorganic Pollutants in Water
DOI: https://doi.org/10.1016/B978-0-12-818965-8.00011-1

observed in nature, however, possess the potential to enter in the environment, which adversely affects human health, wildlife, and ecology. These pollutants are known as "emerging pollutants" (EPo) that are present in the natural environment from over decades but not noticed. New sources of EPo may vary from the release of new chemicals to the disposal of existing ones. They are not even involved in the national or international routine monitoring program. Their outcomes, actions, and the eco-toxicological effects are also unstated. EPo can be organic or inorganic. In this chapter, inorganic pollutants such as selenium, silver, thallium, copper, chlorinated paraffin, and perfluoroalkyl compounds are the topic of concern to discuss in detail. Sources of these contaminants may vary from urban, industrial, and agricultural through which their transportation may occur by various processes like as surface runoff, leaching, and erosion. More than 70% of the sanitation-lacked population live in Asia (Geissen et al., 2015).

Before going for any treatment process, it is essential to know about the nature of pollutants. Nowadays, adsorption is one of the most promising technique because of its effectiveness, economic nature, and also provides flexibility during the whole process. Different low-cost adsorbents are used to remove these pollutants which will be discussed in this chapter. These low-cost adsorbents are called green adsorbents as they originate from agricultural and fruit wastes (Kyzas and Kostoglou, 2014). Thus, the process is known as "green adsorbent". Being low-cost, they are preferable in comparison to other super-adsorbents like as chitosan, inorganic composite materials, etc., that possess more capacity to adsorb metals on it.

Some properties need to be well-thought-out for adsorbent selection, for instance, its eco-friendly nature followed by its low-cost and adsorption capacity. It is primarily essential to utilize adsorbent that has a lower impact on the environment by being nontoxic and biodegradable. One more advantage is that it lowers the massive quantity of solid waste.

Activated charcoal (AC) is one of the superior adsorbents that possess high surface area and removes a wide variety of contaminants from wastewater. The major drawback in using AC is that it is expensive. Thus it exaggerates interest in low-cost adsorbents (LCAds). It is categorized according to its availability (natural or byproduct of domestic/agricultural/industrial waste) and nature (organic and inorganic). Both categories are good enough to discuss, but we have taken up the first one for brief discussion (Gupta et al., 2009).

There are several emerging inorganic pollutants such as selenium, perchlorates, thallium, and silver that are discussed in this chapter along with their suitable low-cost adsorbent.

11.1.1 Selenium

Selenium is one of the essential elements of living organisms (Fig. 11.1). It is one among the most abundantly found natural components of the earth's crust. In the solid form, it exists as metallic, amorphous, and crystalline. In addition to this, it possesses five stable isotopes named as ^{74}Se, ^{76}Se, ^{77}Se, ^{78}Se, and ^{80}Se (Tan et al., 2016). Selenium enhances metabolism, energy releasing process, and helps to maintain proper functioning of the immune system. People obtain it from the diet they eat (Fig. 11.2).

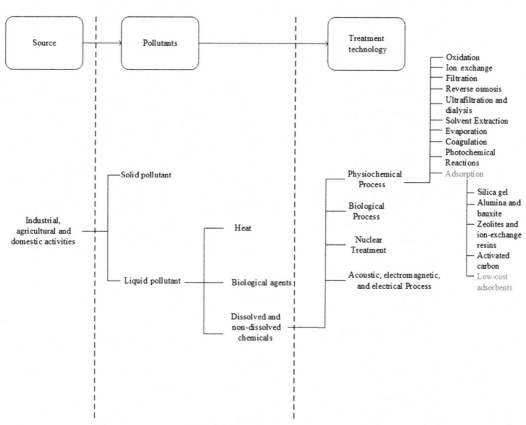

FIGURE 11.1 Schematic diagram showing sources of pollutant and its removal technologies.

One more important concept is that in animal tissues and plants, it is mainly bound to proteins and causes a number of diseases (Fig. 11.3). This is the reason that seafood, cereals, and meat are the most important sources of selenium (Zimmerman et al., 2015). Its organic form (selenomethionine) is absorbed more efficiently in comparison to the inorganic one (Selenite). Selenium finds its extensive application in electronic industries, glass industries, and rubber industries. But its chronic exposure leads to dermatitis and disturbances in CNS, and high amounts of it ultimately causes death. Thus it depends on its concentration ingested whether it gives an adverse or beneficial effect on us. Because of this reason, it shows a resemblance to fluoride (Kapoor et al., 1995). In anionic form either as selenate (SeO_4^{2-}) or selenite (SeO_3^{2-}), it is available in water that brings acidic characteristics to it. Also, it chemically resembles sulfur and physically resembles arsenic in many aspects. Selenium is adsorbed by hydroxides of aluminum and iron due to which it attaches in the soil as a complex of ferric selenite. But selenate is not affected by these types of hydroxides. Thus it enters the food chain as selenate in animals. The ingested selenium flows in the bloodstream from the intestine to the liver. At that place, it starts to reduce to selenide, and then it changes to selenoproteins such as selenocysteine.

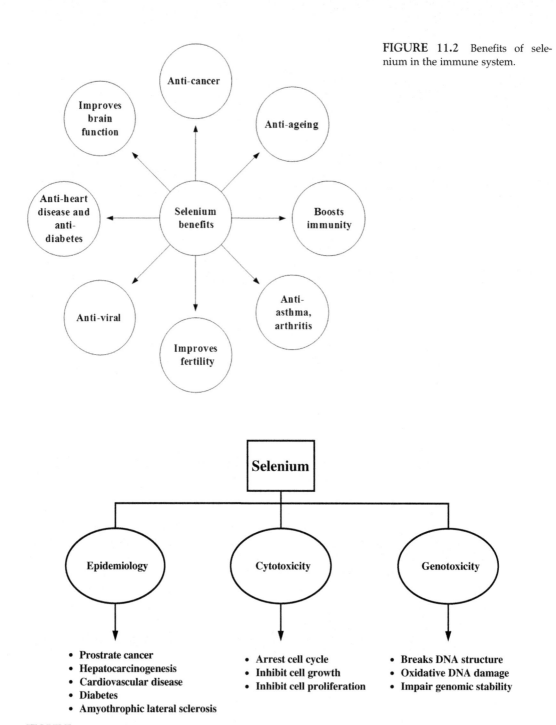

FIGURE 11.2 Benefits of selenium in the immune system.

FIGURE 11.3 Selenium toxicity in humans.

It is found that the maximum level of selenium gets deposited in RBCs, liver, spleen, heart, tooth enamel, and nails. Its excretion is mainly by urine, feces, and bile secretions. There is no information or data available for the removal of selenium in the 1990s". At that time chemical treatment processes were carried out to remove this, such as coagulation, precipitation, ion exchange, and reverse osmosis, but these methods ultimately lead to secondary contamination. Thus, this chapter focuses on the treatment process with the help of the adsorption process and using low-cost adsorbent. So, this process becomes economically viable to all. Its removal is complex as well as expensive because of generation of the high volume of wastewater and also comprises low discharge limit in μg/L. Thus removal of selenium becomes more challenging as its regulatory limit becomes sterner with above aspects. It is reported that selenium is essential as a micronutrient due to its potent antioxidant defence system and its deficiency causes liver, heart, and muscle diseases (Cardoso et al., 2015; Reis et al., 2017). Selenium is found in selenoproteins that help to scavenge free radicals and protect from cancer and many of the infectious diseases. It is bioaccumulative that is why its concentration along with the trophic chain increases (Daniel et al., 2015; Santos et al., 2015).

11.1.1.1 Low-cost adsorbent for the removal of selenium

Adsorption of selenium using bagasse fly ash was studied by Wasewar et al. (Wasewar et al., 2009). They discussed the toxicity of selenium that creates a problem for wildlife and livestock. Due to this, US EPA has endorsed a standard of 0.05 ppm for drinking water and the same rule has been suggested by CPCB for public sewers and inland surface water. They discussed different forms of selenium and its chemical properties. It is commonly found in mining wastewater with concentration from 3 to more than 12000 ppb. Several incidents related to selenium toxicity had been reported in India, such as in Karnal of NW (North West) region, severe Se poisoning in domestic buffaloes (Reilly, 1996). This is leached from the soil in the paddy field and taken up by rice plants. This is the main fodder for animals after harvesting process that leads to a cause for toxicity. A similar pattern of poisoning was seen in South India related to its poisoning in domestic animals. These reports show that selenium contamination starts to spread all over the world. There are many methods available to remove selenium, but adsorption is one of the best techniques for the low volume of water. That is why in this study, Wasewar et al. (2009) focused on reducing selenium concentration to the desired level by using sugarcane bagasse fly ash. To support this, batch studies and different characterization methods were performed. It was found that 4 g/L is the optimum dose for selenium removal at low pH values that follows pseudo-second order kinetic model and Freundlich isotherms.

Mane et al. studied the removal of selenium using pretreated biomass of algae (Jangam, 2011). Biomass can absorb metal ions from water. Algae can accumulate metal ions from their external environment using the biological mechanism as well as the physio-chemical mechanism. There is a difference between viable and nonviable biomasses that the latter one has a higher affinity for metal ions as there are no protons produced during metabolism. That is why the usage of non-living biomass is preferred over the living one. There are some advantages such as no need for growth media, nontoxic, and reusage of biomass. Uptake of metal ions is performed by three mechanisms such as binding through the surface of a cell, intracellular accumulation, and extracellular accumulation. Thus in this study, investigation of *Spirogyra* for selenium biosorption were performed. This has been initiated by physical treatment. It was found that *Spirogyra* possesses 50% selenium removal efficiency.

Mafu et al. (2014) studied the adsorption capacity of different adsorbents such as eggshell membranes, eggshells and orange peels for the removal of selenium. The effect of chemical treatment on these adsorbents by NaOH and HNO$_3$were also investigated. It has been observed that chemical treatment of eggshell membrane and orange peel increases the capacity of adsorption to 160 and 70 μg/g. No such change was observed with an eggshell adsorbent for removal of selenium. Testing these adsorbents on environmental samples demonstrated that 74.8% selenium removed with NaOH treated orange peel,and 47.3% with NaOH treated eggshell membrane (Mafu et al., 2014). This result was supported by various analyses and surface characterization techniques. In addition to this, adsorption isotherm was fitted and found that Freundlich isotherm was best suited for selenium. The effects of contact time and pH were studied and found that 100 minutes is sufficient to adsorb selenium at an optimum pH of 7 because above pH 7 the desorption process started.

Alifar et al. (2014) studied the adsorption of selenium on soil at various parameters. They studied the adsorption process at different pH and found that adsorption related experiments were performed at pH 4 and 7. But, for desorption experiments pH 7 is observed as optimum pH, as a high amount of selenium got desorbed at this pH. By using the batch technique, adsorption as well as desorption isotherms were determined. Using different isotherms, it was found that Langmuir and Freundlich's isotherms were well fitted. It is also concluded that the availability of selenium is affected by the pH of the soil, clay minerals, and organic matter. This study demonstrates that the adsorption of selenium increases with decreasing pH and increasing organic matter (Alifar et al., 2014).

Mishra et al. (2017a,b) studied the adsorption potential of rice husk for the removal of selenium. The characterization technique FTIR reveals that the presence of functional groups such as −OH, −NH, and −CO plays an important role in the adsorption process. Another characterization technique, SEM, demonstrates about the surface morphology of adsorbent and shows its porous structure. Different parameters were studied and found that optimum pH was found to be 4 and the optimum time was 120 minutes for the adsorption of selenium. Temkin isotherm was found as the best-fitted adsorption isotherm. This adsorption process followed pseudo-second-order kinetics. It was found that enthalpy and entropy were positive that suggests the endothermic nature of the operation and also, the presence of randomness at the interface of solid and solution (Mishra et al., 2017a).

Mishra et al. (2017a,b) studied the adsorption of selenium by the sweet lime peel. The FTIR technique reveals the presence of different functional groups such as −OH, −N−H, C = O, and −C−O that plays an essential role in the adsorption of selenium ions. SEM demonstrates its porous structure. Different parameters were studied and found that optimum pH was found to be 4 and the optimum time was 120 minutes for the adsorption of selenium. Dubinin Radushkevich isotherm was found as the best-fitted adsorption isotherm when the sweet lime peel is used as an adsorbent. The adsorption capacity of adsorbent was found to be 1.213. Here, also positive value enthalpy and entropy suggests the endothermic nature of the process and also, the presence of randomness at the interface of solid and solution (Mishra et al., 2017b).

Thus, in this chapter, various low-cost adsorbents are discussed for the removal of selenium such as sugarcane bagasse fly ash, pretreated biomass of algae, eggshell membrane, eggshell, orange peel, soil, rice husk, and sweet lime peel. It is observed that low-cost adsorbents have the capability to adsorb selenium with good adsorption capacity and are able to remove selenium from the environment if they are toxic to the people or affecting plants and animals at higher concentration. It is mentioned in the text above that selenium is both beneficial as well as toxic, it only depends on its concentration that the people are ingesting as it has potential to bioaccumulate that is why there is a need to remove selenium by a process that is cost-effective and common people can afford it. This is the reason for researchers to focus on low-cost adsorbent for selenium removal.

11.1.2 Perchlorate

Perchlorate is primarily used as an oxidizing agent in the aerospace and defence sector. Usage of manmade perchlorate salts in several applications such as batteries, fireworks, highway safety flares, explosives, and pyrotechnics (Bardiya and Bae, 2011; Srinivasan and Sorial, 2009). Their use and inappropriate disposal directed to its release into the environment. As perchlorate salts are emerging inorganic pollutants, they start to appear in drinking water, human saliva, surface water, and breast milk (Kannan et al., 2009). A point to remind that perchlorate ions show similarity in size with iodide ions. That is why it interferes with a natural phenomenon of uptake of iodine that ultimately inhibits the production of thyroid hormone. This leads to the inhibition of normal metabolism and causes neurological damage and anemia (Kirk et al., 2005). Because of its harm to human health, US EPA has set a limit for drinking water to 15 μg/L (Morley et al., 2009). Several technologies have been developed to remove these salts,but adsorption seems to be one of the best techniques to remove this from drinking water. Thus, here we discuss adsorbents used to remove perchlorate salts.

11.1.2.1 Low-cost adsorbent for the removal of perchlorates

The table shows the removal of perchlorates via low-cost adsorbents with the maximum efficiency of removal at different pH as well as temperature as shown in Table 11.1.

In arid Southwestern, removal of perchlorates is considered to be a major problem. Perchlorate is soluble in water and is a poor complexing agent because of that it is highly mobile, persisting contaminant in the environment (Parette and Cannon, 2006) (Fig. 11.4). Perchlorates are used in rocket fuel and are responsible for the contamination of water throughout Unites State (Parette and Cannon, 2005). With the uptake of iodine, perchlorates enter into the thyroid glands and then into the central nervous system and may also interfere with the skeleton system of infants (Ye et al., 2012).

Mainly, anthropogenic activities in the United States is responsible for widespread of perchlorates (Bardiya and Bae, 2011). The toxic nature of perchlorate with their unusual chemical as well as physical properties makes them hard to remove (Thomas et al., 2016). The use of perchlorates is mainly used in missiles, rockets, airbag inflators, electronic tubes, leather tanning, lubricating oils, paint, and enamel production as well as aluminum refining (Jothinathan and Vasudevan, 2013). Perchlorates are present in leafy vegetables, bovine milk, as well as in foodstuffs. Perchlorates are reported to be found in ground water,

TABLE 11.1   Low-cost adsorbent for the removal of perchlorates.

Biosorbents	pH	Temp.	Removal (%)/ uptake capacity	References
Wheat straw	–	20°C	91%	(Tan et al., 2012)
Granular ferric hydroxide (GFH)	6–6.5	25°C	20.0 mg/g	(Kumar et al., 2010)
Surfactant (hexadecyltrimethylammonium)-modified zeolite (SMZ)	12	–	40–47 mmol/kg	(Zhang et al., 2007)
Granular activated carbon	6–7	–	42.778 mg/g	(Thirumal and Kaliappan, 2011)
Single-walled carbon nanotubes (SWCNTs)	–	5°C–45°C	10.03–13.64 mg/g	(Lou et al., 2014)

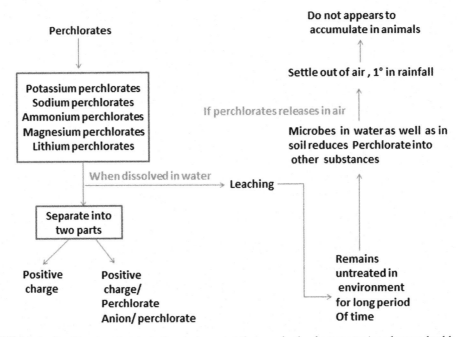

FIGURE 11.4   Perchlorate enters in to the environment that can further become toxic to human health.

drinking water, as well as surface water across the United States. The presence of perchlorates in breast milk, urine, serum, and saliva may lead to the problem of thyroid hormone disrupting-contaminants is reported from the sample from the United States population. Japan, Korea, and China are also reported for perchlorate contamination (Wu et al., 2010).

11.1.3 Thallium

Thallium is considered to be one of the most toxic pollutant for human health and leads to different diseases as shown in Fig. 11.5. The contamination of heavy metals is

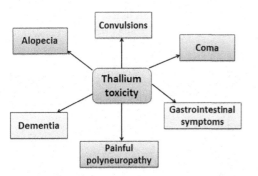

FIGURE 11.5 Toxic effects of thallium on human health.

considered to be the most hazardous as they are nondegradable (Ayangbenro and Babalola, 2017). Biosorption of heavy metals is considered to be a serious concern due to huge industrialization (Shen et al., 2017; Alluri et al., 2007). Worldwide, water pollution is considered to be a severe problem now a days. Annually, there are about 1500 km^3 of wastewater produced that is six times more than the existence of the total water in the world as estimated by the UN. Mainly in rural areas, the emerging pollutants are spread over an area that transport through runoff, air, leaching, or erosion (Benaisa et al., 2018). Low-cost biosorbents are considered to be useful for the recovery of heavy metals and are more efficient than higher cost biosorbents (Ghaedi and Mosallanejad, 2013). The discovery of thallium is done when there is the use of spectroscopy for the determination of minerals in the 1860s and is considered to be tasteless, odorless, as well as colorless heavy metals. In Germany (1920), first, thallium salts were used as a pesticide and were also used as rodenticides due to their severe toxicity, and it is poisonous, that is why thallium was banned in United Nation in 1965. In some cases, hemodialysis is beneficial in the removal of thallium toxicity (Riyaz et al., 2013).

Fig. 11.6 shows what are the possible initial as well as acute toxicity of thallium in humans as thallium is much toxic. Some of the symptoms are fever, diarrhea, stomatitis, skin eruption etc. (Fig. 11.6).

Thallium compounds can enter the human body through food, vegetables that are grown in thallium-contaminated soil, as well as inhalation. Thallium toxicity to humans can cause many problems in human health like vomiting, diarrhea, headache, bone problems, liver and lung problems, anorexia, and sometimes even death also. Therefore, the removal of thallium from wastewater is a significant concern now a days (Khavidaki et al., 2013).

11.1.3.1 Potential sources of exposure of thallium

The above figure shows the industrial as well as the nonindustrial exposure of thallium toxicity (Fig. 11.7). In the 1980s", it was said that potassium therapy is considered to be the only way to remove the poisoning of thallium (Saddique and Peterson, 1983).

11.1.3.2 Low-cost adsorbent for the removal of thallium

In 1861, thallium, a very toxic heavy metal was accidentally discovered by Sir William Crookes by burning dust from the sulfuric acid industrial plant. In our environment, mostly the exposure of thallium is mainly through dust and thallium sulfate is most common in dust (Saha, 2005). Thallium is considered to be one of the heavy metals present in the natural

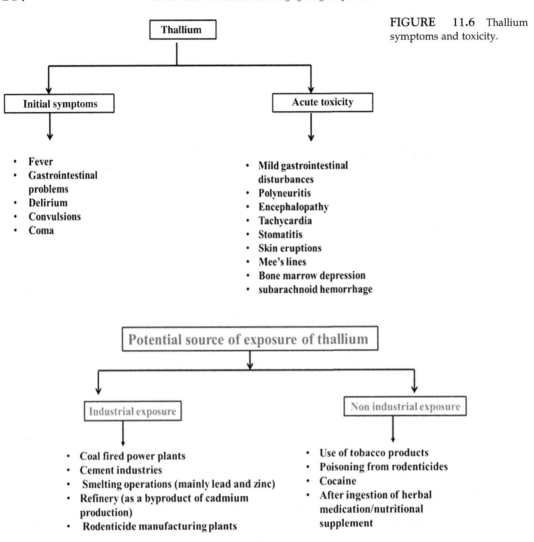

FIGURE 11.6 Thallium symptoms and toxicity.

FIGURE 11.7   Sources of exposure of thallium.

water. For the removal of Tl$^+$ions from medium (aqueous), *Eichhorniacrassipes* is considered to be an effective alternative (Martínez-Sánchez et al., 2016).

Table 11.2 shows adsorption of thallium via different low-cost adsorbents and Fig. 11.8 shows how thallium got adsorbed in a liquid bath.

11.1.4 Silver

The toxicity of silver is delivered to the human body through skin, inhalation, and eating vegetables that are contaminated with silver, etc. It was also reported (Fig. 11.9) that

TABLE 11.2 Low-cost biosorbents for the removal of thallium.

Biosorbents	Thallium ionic forms	pH	Removal (%)/uptake capacity (mg/g)	References
Rice Husk	Tl (III)	—	100%	(Alalwan et al., 2018)
Sugar beet (treated with NaOH)	Tl (I)	—	94.5%	(Zolgharnein et al., 2011)
Sugar beet (treated with NaCl)	Tl (I)	—	92.4%	(Zolgharnein et al., 2011)
Pseudomonas fluorescens strains	Tl	5	93.76 mg/g	(Long et al., 2017)
Walnut shells	Tl (I)	7.5	99%	(Karbowska et al., 2019)
Pestalotiopsis sp.	Tl (I)	—	99.80 mg/g	(Chen et al., 2018)
Modified ZnO nanoparticles	Tl (III)	6	92.2%−92.6%	(Dashti et al., 2019)

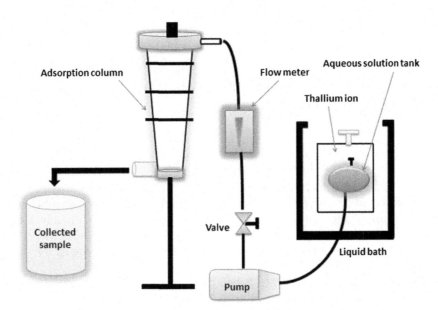

FIGURE 11.8 A Schematic diagram for showing adsorption of thallium.

the silver taken through mouth or inhalation, a little amount comes out through urine, but outrageous newsflash is that there is no release of silver given, which was transferred through the skin (ATSDR). Heavy metals are those metals whose specific gravities are greater than 5.0 g/cm³ (Salaudeen Olawale, 2019). Biosorption is considered to be an alternative to conventional methods for the recovery of heavy metals like silver from different industries and the biosorbents that are used for the recovery of these metals must be of low cost (Prabhakar and Lakkimsetty, 2011).

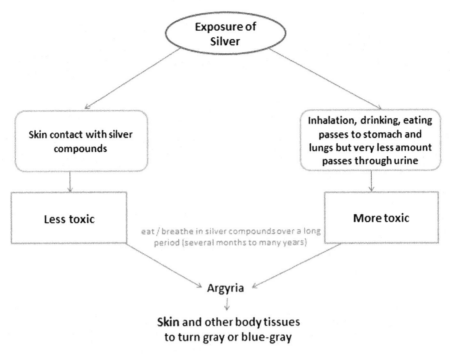

FIGURE 11.9   Exposure of silver on human body.

11.1.4.1 Low-cost adsorbent for the removal of silver

Tree bark is considered to be an abundant bioresource in the world (Şen et al., 2015). Silver is considered to be one of the most valuable and useful heavy metal due to its different industrial applications and can also elevate economic values. As silver is used for the production of batteries, mirrors, as well as photographic films. The dead and dried aquatic plants like macrophytes are considered to be low-cost biosorbents for the recovery of silver (Lech Cantuaria et al., 2014).

For many years, silver (a Nobel metal) has been used widely in the production of photographic films as well as imaging industries (Das, 2010). In 2015, it was concluded that for the bioremediation of silver the exopolysaccharide (EPS) produced by the bacteria (marine) from French Polynesia is considered to be a promising biosorbents (Deschatre et al., 2015).

The above figure shows the distribution of silver in different manufacturing industries like paper production, printing, dyeing, fibers, and many more (Fig. 11.10).

The above table shows that the removal of silver with different low-cost biosorbents and while analyzing the above table, it can be seen that the green seaweeds are more efficient low-cost biosorbents for the recovery of silver (Table 11.3) (Fig. 11.11).

As we know that silver is toxic and can lead to different problems in humans, A figure representation of different problems with exposure to silver on human health is

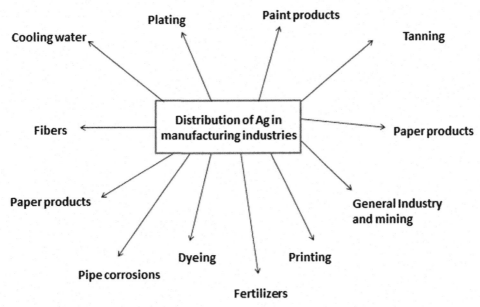

FIGURE 11.10 Distribution of Ag in manufacturing industries.

TABLE 11.3 Low-cost biosorbents for the removal of silver.

Biosorbents	pH	Temp.	Efficiency (%)/uptake capacity (mg/g)	References
Green seaweed biomass	–	–	100%	(Latinwo et al., 2015)
Grape peel	7	298K	41.7 mg/g	(Escudero et al., 2018)
Grape seed	7	298K	61.4 mg/g	(Escudero et al., 2018)
Grape stem	7	298K	46.4 mg/g	(Escudero et al., 2018)

presented, some of them are liver damage, pulmonary edema, irritation of eyes, argyria (caused due to exposure of silver or silver dust and that can lead to gray-black staining of the skin), etc.

11.2 Conclusion

The environmental system has been influenced by human, industrial, and domestic operations, leading to dramatic issues such as global warming and wastewater generation with elevated concentrations of pollutants. Because quality of water is on everyones priority but available in very less amount. Treating wastewater for pollutant removal has become highly imperative. There are some materials available that have diminutive or not

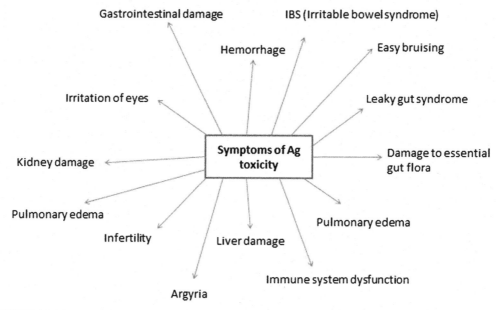

FIGURE 11.11   Symptoms of silver toxicity.

any use in nature. The use of all such products is worthwhile in the recent time. Efforts were made in this chapter to give a brief idea, in particular, to discuss and briefly highlight the low-cost adsorbents in order to use these waste / low-cost materials for the removal of emerging inorganic pollutants.

References

Alalwan, H.A., Abbas, M.N., Abudi, Z.N., Alminshid, A.H., 2018. Adsorption of thallium ion (Tl^{+3}) from aqueous solutions by rice husk in a fixed-bed column: experiment and prediction of breakthrough curves. Environ. Technol. Innov. 12, 1–13.

Alifar, N., Zaharah, A., Ishak, C., Awang, Y., Khayambashi, B., 2014. Determination of physical and chemical soil parameters on selenium adsorption, desorption by rice growing soil. Asian J. Plant Sci. 13 (4–8), 147–155.

Alluri, H., Ronda, S., Settalluri, V.S., Bondili, J., Suryanarayana, V., Venkateshwar, P., 2007. Biosorption: an eco-friendly alternative for heavy metal removal. Afr. J. Biotechnol. 6 (25), 2924–2931.

Amin, M.T., Alazba, A.A., Manzoor, U., 2014. A review of removal of pollutants from water/wastewater using different types of nanomaterials. Adv. Mater. Science Eng. 2014 (1), 1–24.

Ayangbenro, A.S., Babalola, O.O., 2017. A new strategy for heavy metal polluted environments: a review of microbial biosorbents. Int J Env. Res. Public Health 14 (1), 1–16.

Bardiya, N., Bae, J.-H., 2011. Dissimilatory perchlorate reduction: A review. Microbiological Res. 166 (2011), 237–254.

Benaisa, S., Arhoun, B., El Mail, R., Rodriguez-Maroto, J.M., 2018. Potential of brown algae biomass as new biosorbent of iron: kinetic, equilibrium and thermodynamic study. J. Mater. Environ. Sci. 9 (7), 2131–2141.

Cardoso, B.R., Roberts, B.R., Bush, A.I., Hare, D.J., 2015. Selenium, selenoproteins and neurodegenerative diseases. Metallomics 7, 1213–1228.

Chen, K., Li, H., Kong, L., Peng, Y., Chen, D., Xia, J., et al., 2018. Biosorption of thallium(I) and cadmium(II) with the dried biomass of Pestalotiopsis sp. FW-JCCW: isotherm, kinetic, thermodynamic and mechanism. Desalinaion water Treat. 111 (2018), 297–309.

Daniel, N., Subramaniyan, G., Karthik, C., Arulselvi, I., 2015. Antioxidant profiling of selenium fortified tomato (*Solanum lycopersicum*). Int. Res. J. Pharm. 6 (5), 299–304.

Das, N., 2010. Recovery of precious metals through biosorption—a review. Hydrmetallurgy 103 (1–4), 180–189.

Dashti, H., Aghaie, H., Shishehbore, M., 2019. Adsorption of thallium (III) ion from aqueous solution using modified ZnO nanopowder. J. Phys. Theor. Chem. 8 (3), 233–244.

Deschatre, M., Ghillebaert, F., Guezennec, J., Simon-Colin, C., 2015. Study of biosorption of copper and silver by marine bacterial exopolysaccharides. WIT Trans. Ecol. Environ. 196, 549–559.

Escudero, L.B., Vanni, G., Duarte, F.A., Segger, T., Dotto, G.L., 2018. Biosorption of silver from aqueous solutions using wine industry wastes. Chem. Eng. Commun. 205 (3), 325–337.

Geissen, V., Mol, H., Klumpp, E., Umlauf, G., Nadal, M., van Der ploeg, M., et al., 2015. Emerging pollutants in the environment: a challenge for water resource management. Int. Soil Water Conserv. Res. 3 (1), 57–65.

Ghaedi, M., Mosallanejad, N., 2013. Removal of heavy metal ions from polluted waters by using of low cost adsorbents: review. J. Chem. Health Risks 3 (1), 7–22.

Gupta, V.K., Carrott, P.J.M., Ribeiro Carrott, M.M.L., Suhas, 2009. Low-cost adsorbents: growing approach to wastewater treatment—a review. Crit. Rev. Environ. Science Technol. 39 (10), 783–842.

Jangam, C., 2011. Bioadsorption of selenium by pretreated algal biomass. Adv. Appl. Science. Res. 2 (2), 202–207.

Jothinathan, L., Vasudevan, S., 2013. Graphene-a promising material for removal of perchlorate (ClO4-) from water. Environ. Scienceans. Pollut. Res. 20 (8), 5114–5124.

Kannan, K., Praamsma, M.L., Oldi, J.F., Kunisue, T., Sinha, R.K., 2009. Occurrence of perchlorate in drinking water, groundwater, surface water and human saliva from India. Chemosphere 76 (1), 22–26.

Kapoor, A., Tanjore, S., Viraraghavan, T., 1995. Removal of selenium from water and wastewater. Int. J. Environ. Stud. 49 (2), 137–147.

Karbowska, B., Konował, E., Zembrzuski, W., Milczarek, G., 2019. Sorption of Thallium on Walnut Shells and its Enhancement by the Lignosulfonate-Stabilized Gold Colloid. Pol. J. Environ. Stud. 28 (3), 2151–2158.

Khavidaki, H.D., Aghaie, M., Shishehbore, M., Aghaie, H., 2013. Adsorptive removal of thallium (III) ions from aqueous solutions using eucalyptus leaves powders. Indian J. Chem. Technol. 20 (6), 380–384.

Kirk, A.B., Martinelango, P.K., Tian, K., Dutta, A., Smith, E.E., Dasgupta, P.K., 2005. Perchlorate and iodide in dairy and breast milk. Env. Sci. Technol 39 (7), 2011–2017.

Kumar, E., Bhatnagar, A., Choi, J.-A., Kumar, U., Min, B., Kim, Y., et al., 2010. Perchlorate removal from aqueous solutions by granular ferric hydroxide (GFH). Chem. Eng. J. 159 (1–3), 84–90.

Kyzas, G.Z., Kostoglou, M., 2014. Green adsorbents for wastewaters:a critical review. Materials (Basel, Switzerland) 7 (1), 333–364.

Latinwo, G.K., Jimoda, L.A., Agarry, S.E., Adeniran, J.A., 2015. Biosorption of some heavy metals from textile wastewater by green seaweed biomass. Univers. J. Environ. Res. Technol. 5 (4), 210–219.

Lech Cantuaria, M., de Almeida Neto, A.F., Vieira, M.G.A., 2014. Biosorption of silver by macrophyte salvinia cucullata. Chem. Eng. Trans. 38, 109–114.

Long, J., Chen, D., Xia, J., Luo, D., Zheng, B., Chen, Y., 2017. Equilibrium and kinetics studies on biosorption of thallium (I) by dead biomass of *Pseudomonas fluorescens*. Pol. J. Environ. Stud. 26 (4), 1591–1598.

Lou, J.C., Hsu, Y.S., Hsu, K.L., Chou, M.S., Han, J.Y., 2014. Comparing the removal of perchlorate when using single-walled carbon nanotubes (SWCNTs) or granular activated carbon: adsorption kinetics and thermodynamics. J. Environ. Science Health, Part A 49 (5), 503–513.

Mafu, L.D., Msagati, T.A.M., Mamba, B.B., 2014. Adsorption studies for the simultaneous removal of arsenic and selenium using naturally prepared adsorbent materials. Int. J. Environ. Science Technol. 11 (2014), 1723–1732.

Martínez-Sánchez, C., Torres-Rodríguez, L.M., García-de la Cruz, R.F., 2016. A kinetic, equilibrium, and thermodynamic study on the biosorption of Tl + and Cd2 + by Eichhornia crassipes roots using carbon paste electrode. J. Braz. Chem. Soc. 27 (2), 1667–1678.

Mishra, Reena, Chattree, A., Siddiqui, Shazia, 2017a. Adsorption of selenium from aqueous solution by rice husk: characterization, isotherm and kinetic studies. Int. J. Sci. Res. Science Eng. Technol. (IJSRSET) 3 (8), 495–502.

Mishra, Reena, Chattree, A., Siddiqui, Shazia, 2017b. Removal of selenium from waste water by sweet lime peel. Int. J. Chem. Stud. 5 (6), 2070–2075.

Morley, K., Brandhuber, P., Clark, S., 2009. A review of perchlorate occurrence in public drinking water systems. Am. Water Work. Assoc. 101 (11), 63–73.

Parette, R., Cannon, F.S., 2005. The removal of perchlorate from groundwater by activated carbon tailored with cationic surfactants. Water Res. 39 (16), 4020–4028.

Parette, R., Cannon, F.S., 2006. Perchlorate removal by modified activated carbon. In: Gu, B., Coates, J.D. (Eds.), Perchlorate. Springer, Boston, MA, pp. 343–372.

Prabhakar, G., Lakkimsetty, N.R., 2011. Removal of heavy metals by biosorption-an overall review. J. Eng. Res. Stud. 2 (4), 17–22.

Reilly, C., 1996. Selenium in food and health. Springer Science & Business Media, New York.

Reis, A.R., El-Ramady, H., Santos, E.F., Gratão, P.L., Schomburg, L., 2017. Overview of selenium deficiency and toxicity worldwide: affected areas, selenium-related health issues, and case studies. In: Pilon-Smits, E., Winkel, L., Lin, Z.Q. (Eds.), Selenium in plants. Plant Ecophysiology, vol. 11. Springer, Cham, pp. 209–230.

Riyaz, R., Pandalai, S., Schwartz, M., Kazzi, Z., 2013. A fatal case of thallium toxicity: challenges in management. J. Med. Toxicol. 9 (1), 75–78.

Saddique, A., Peterson, C., 1983. Thallium poisoning: a review. Veterinary Hum. Toxicol. 25 (1), 16–22.

Saha, A., 2005. Thallium toxicity: a growing concern. Indian J. Occup. Environ. Med. 9 (2), 53–56.

Salaudeen Olawale, A., 2019. Biosorption of heavy metals: a mini review. Acta Sci. Agriculture 3 (2), 22–25.

Santos, S., Ungureanu, G., Boaventura, R., Botelho, C., 2015. Selenium contaminated waters: an overview of analytical methods, treatment options and recent advances in sorption methods. Science Total. Environ. 521 (1), 246–260.

Şen, A., Pereira, H., Olivella, M.A., Villaescusa, I., 2015. Heavy metals removal in aqueous environments using bark as a biosorbent. Int. J. Environ. Sci. Technol. 12 (1), 391–404.

Shen, N., Birungi, Z.S., Chirwa, E.M., 2017. Selective biosorption of precious metals by cell-surface engineered microalgae. Chem. Eng. Trans. 61, 25–30.

Srinivasan, R., Sorial, G., 2009. Treatment of perchlorate in drinking water: a critical review. Sep. Purif. Technol. 69 (1), 7–21.

Tan, X., Gao, B., Xu, X., Wang, Y., Ling, J., Yue, Q., et al., 2012. Perchlorate uptake by wheat straw based adsorbent from aqueous solution and its subsequent biological regeneration. Chem. Eng. J. 211–212, 37–45.

Tan, L.C., Nancharaiah, Y.V., Van Hullebusch, E.D., Lens, P.N.L., 2016. Selenium: environmental significance, pollution, and biological treatment technologies. Biotechnol. Adv. 34 (5), 886–907.

Thirumal, J., Kaliappan, S., 2011. Equilibrium, kinetic and thermodynamic behavior of perchlorate adsorption onto the activated carbon. Eur. J. Sci. Res. 64 (3), 365–376.

Thomas, E., Rekha, K.G., Bhuvaneswari, S., Vijayalakshmi, K.P., George, B.K., 2016. 1,3-Dialkylimidazolium modified clay sorbents for perchlorate removal from water. RSCAdv. 6 (83), 80029–80036.

Wasewar, K.L., Prasad, B., Gulipalli, S., 2009. Adsorption of selenium using bagasse fly ash. CLEAN – Soil, Air, Water 37 (7), 534–543.

Wu, Q., Zhang, T., Sun, H., Kannan, K., 2010. Perchlorate in tap water, groundwater, surface waters, and bottled water from China and its association with other inorganic anions and with disinfection byproducts. Arch. Environ. Contamination Toxicol. 58 (3), 543–550.

Ye, L., You, H., Yao, J., Su, H., 2012. Water treatment technologies for perchlorate: a review. Desalinatin 298, 1–12.

Zhang, P., Avudzega, D.M., Bowman, R.S., 2007. Removal of perchlorate from contaminated waters using surfactant-modified zeolite. J. Environ. Qual. 36 (4), 1069–1075.

Zimmerman, M.T., Bayse, C.A., Ramoutar, R.R., Brumaghim, J.L., 2015. Sulfur and selenium antioxidants:challenging radical scavenging mechanisms and developing structure-activity relationships based on metal binding. J. Inorg. Biochem. 145, 30–40.

Zolgharnein, J., Asanjarani, N., Shariatmanesh, T., 2011. Removal of thallium (I) from aqueous solution using modified sugar beet pulp. Toxicol. Environ. Chem. 93 (2), 207–214.

Application of nanoparticles for inorganic water purification

Deepak Yadav[1], Pardeep Singh[2] and Pradeep Kumar[3]

[1]Chemical Engineering Department, HBTU, Kanpur, India [2]Department of Environmental Science, PGDAV College, University of Delhi, New Delhi, India [3]Department of Chemical Engineering & Technology, IIT (BHU), Varanasi, India

O U T L I N E

12.1 Introduction

The rapidly increasing population and its civilization require more production of livelihoods exploited from limited resources creates a big gap between their supply and demand. This mismatch has been covered by agriculture and industrialization production required for the urbanization of the growing population. These changes move fast and have advanced the practices for the obligatory agriculture and industrial growth. This increases the water consumption and water pollution; same trend is expected in the near future (Alcamo et al., 2017; WHO, 2015). The situation will increase water crisis and water quality problems around the globe (WHO, 2015). Water reserves are currently facing the biggest global problems (Cosgrove and Rijsberman, 2014). We, therefore, need to improve the water crunch by improving water treatment technologies, which must be based on environment-friendly water management systems (Altenburger et al., 2015; Loucks and Van Beek, 2017). Rapid urbanization has increased the volume of municipal wastewater (Maryam and Büyükgüngör, 2018). The recovery and recycling of wastewater is the most essential step (Lyu et al., 2016). The increasing use of natural water resources by the population contributes to the composition, complexity, property, variability, and toxicity of the various types of wastewater pollutants (Li et al., 2016). The biggest challenge of effluent treatment is the complexity of the wastewater, which increases the risk of water contamination. Therefore, according to WHO guidelines, pure water can be accessed by proper water treatment technologies (Yang et al., 2011; WHO, 2015). Therefore, efficient water treatment is the demand of the present time, which can be implemented by different water treatment strategies (Bora and Dutta, 2014; Gollavelli et al., 2013). Water and wastewater pollutants can generally be divided into three groups: inorganic pollutants, toxic organic microorganisms, and pathogens. Starting with inorganic water pollutants, these sources are generally polyatomic compounds and heavy metals responsible for ∼30% of pollutants in wastewater (Bora and Dutta, 2014). These pollutants are often announced through various treatment activities in the environment and water supply in industries such as petroleum, textile, refining, pulps and steel, and agri-food (Inglezakis et al., 2002; Martin and Griswold 2009). The excessive bioavailability of polyatomic compounds and heavy metals has been implicated in many human diseases stretching from cardiovascular disease to skeletal abnormalities (Saif et al., 2012). Table 12.1 offers an outline of these

TABLE 12.1 Outline of inorganic pollutants found in water, main sources, and effects on human health (Hlongwane et al., 2019).

Category	Representative pollutants	Main sources	Toxicity	References
Heavy metals	Chloride, lead, copper, cadmium, iron, nickel and zinc	Mining, chemical refinery, petroleum, textile industries	Cardiovascular disorders, cancer, diabetes, emphysema, hypertension, mellitus, renal damage, and skeletal malformation	Martin and Griswold (2009); Saif et al. (2012)
Polyatomic compounds	Chromate, arsenate, sulfate, and nitrate	steel and agricultural industries		Inglezakis et al. (2002)

above-said different modules of inorganic water pollutants, their main sources, and their impact on human health.

Agricultural, textile, pharmaceutical, printing, and tanning industries are the major creators of organic water pollutants (Bora and Dutta, 2014). It accounts for 70% of pollutants in wastewater (Jelić et al., 2012; Bora and Dutta, 2014). These organic compounds are biodegradable, but their low degradation rate have a tendency to bioaccumulate, which are produced in industry (Gu et al., 2016). Although there are no signs of a severe health risk in the presence of low levels of these organic pollutants, their elimination, especially from drinking water, is a preventive measure against long-term contact (Jelić et al., 2012). Acquaintance to large quantities of organic pollutants (e.g., herbicides and dyes) linked to endocrine hormone metabolism, mutagenicity, and even cancer (Tang et al., 2012; Mathur and Bhatnagar, 2007).

Microbial pollutants come from wastewater. These include viruses, bacteria, and protozoa, which are often found in the water supply due to pollution of faeces and soil (Ishii et al., 2006). Microorganisms affect water quality and threaten the general well-being of the population in terms of diarrhea, bleeding colitis, and haemolytic uremic syndrome (Zhan et al., 2014; Leclerc et al., 2002).

Several studies have been successfully verified by the use of nanoparticles in desalination, dechlorination, and removal of pollutants, for example, inorganic compounds, heavy metals, pathogens, and organic compounds from water sources (Kumar et al., 2014; Bora and Dutta, 2014). A wide variety of the review articles focus on single pollutant control, whereas the figures added is frequently selective to a specific group of nanoparticles. Wu et al. (2018) proposed composite nanoparticles for the simultaneous removal of metal ion (Cu^{2+}), bacteria (*Escherichia coli*) and anionic dye (acid fuchsin). The adsorption capacities of the nanoparticles are slightly high but for magnetic graphene (Gollavelli et al., 2013), and a few examples are given in Table 12.2.

In recent years, photocatalysis processes have revealed an abundant potential for environment-friendly and sustainable technologies in the wastewater industry. Photocatalysis is a method, where an electron—hole pair is generated when exposed to the source of light of suitable energy. Consequently, the chemical reactions that occur in the presence of catalyst and light (visible or UV light) are jointly named as photocatalytic reactions.

Adsorption have been widely studied for toxic heavy metals ions removal (Hua et al., 2012), and toxic organic compounds (Gupta et al., 2013) and microorganisms (Amin et al., 2014) in water. These are usually established on a single pollutant control viewpoint, while in reality these all often coexist in wastewater (Park et al., 2010). Hence, nanoparticles retain both adsorption capabilities, and antimicrobial activity for multipollutant control in water and wastewater. But, most of the existing studies have dedicated to remove them individually, whereas very few of them focused on simultaneous removal (toxic heavy metals, organic compounds and microorganisms) together (Purwajanti et al., 2015).

Photocatalysis is an advanced oxidation procedure based on the oxidative elimination of micro pollutants and microbial pathogens from water and wastewater (Gehrke et al., 2015). Heterogeneous photocatalysts degrade inorganic pollutants and the most popular TiO_2 (readily availability), safe, and economical, so they are widely known as a validated photocatalysts (Lazar et al., 2012; Qu et al., 2013). TiO_2 is irradiated by UV light (200—390 nm), electron—hole pair is excited move to the conduction (CB) and valent (VB)

TABLE 12.2 Photocatalytic degradation of multipollutant under different light irradiations (Hlongwane et al., 2019).

Photocatalyst (light source) nanoparticles preparation	Type of pollutant	Name of pollutants	Optimum pH	Optimum catalyst weight (g/L)	Degradation %	References
Titania/activated carbon/carbonized epoxy (TiO$_2$/AC/CE) VISIBLE LIGHT nanocomposite Solvothermal process	Inorganic (metal ions) and organic pollutants	– Pb(II) –Methylene blue (MB)	6.5	Not indicated	~97% of the Pb(II) ions in 600 min ~90% of MB dye	Benjwal and Kar (2015)
Zn-Fe mixed metal oxides coprecipitation and calcination	Metal ions and organic pollutants	– Ibuprofen – Arsenic	3.0	0.5 g/L	95.7% of ibuprofen	Di et al. (2017)

bands depending on redox potential of a substrate. The present chapter starts with the introduction of inorganic pollutants, their respective toxicity, and multifunctional nanotechnology inventions. In other cases, the focus is based on nanoparticles as application of nanotechnology and efficient assessment of adsorption applications for inorganic pollutants' removal (Ihsanullah et al., 2016). The expansion of multifunctional wastewater treatment nanotechnologies for multiple coexisting pollutants treatment to real-time solutions become the focus of the current study.

12.2 Wastewater treatment technologies

Turbidity is the most noticeable indicators of water contamination that includes transparency, clarity, odor, and turbidity offer significance of multiple pollutants in water. The potential forms of pollutants present in the form of toxic heavy metals, suspended minerals, organisms, organic substances, detergents, strains, and so on affect the turbidity or alters color of water.

Water contamination is the base of most of the intolerable odors in water for example sewage, rotten organic stuff, oils, and petrochemicals are linked with intolerable smell (Wing et al., 2014). Wastewater formed by domestic, industrial and commercial sectors are treated and disinfected to decrease aesthetic problems by several traditional methods ordered as: physical separation (suspended pollutants), biodegradation (suspended and colloidal), and chemical decomposition methods (suspended solids and toxic substances) explained by (Lofrano and Brown, 2010; Oller et al., 2011; Sarkar et al., 2006). The most widely chemical treatment method used for purifying water supplies across the world is chlorination (Sedlak and von Gunten, 2011). Chlorination has its advantages and weaknesses, but the core message is that it is unable to purify the broad variety of water and useless in eliminating pollutants for example inorganic and organic and compounds (Sedlak and von Gunten, 2011). In addition, chlorination is notorious to form detrimental by-products after chlorine responds to dissolved inorganic and organic water compounds (Gopal et al., 2007) leading to dangerous diseases like cancer (Richardson et al., 2007). Generally, the water composition changed by the addition of chemicals, it requires large sum of chemical agents that experience high capital and operational costs (Unuabonah et al., 2017; Yu et al., 2017). These limitations allied with physical separation, biodegradation, chemical decomposition reduces them to sustainable development.

12.2.1 Coexistence of pollutants in water

In fact, the contaminated water sources contain a mixture of pollutants of different categories frequently occurs depending upon the type of source (Gollavelli et al., 2013). A recent study by Bradley identified thirty four out of the thirty eight streams were sited in urban areas and agricultural regions and rest four in low population density areas across USA (Bradley et al., 2017). The authors identified a mixture of 4 to 161 organic complex pollutants coining from wastewater from residential, municipal, agricultural and industrial sources (Li et al., 2016; Ota et al., 2013). The coexistence of multipollutants in water sources worldwide increased instability and toxicity leads to complexity of wastewater (Arjoon et al., 2013).

TABLE 12.3 Summary of nanoparticles used to concurrently adsorption of multipollutants (Hlongwane et al., 2019).

Nanomaterial	Nanomaterial preparation method	Type of pollutants treated	Name of pollutants	Initial pollutant concentration	Optimum pH
Nano-alumina modified with 2,4-dinitrophenylhydrazine	Chemical synthesis	Metal ions	— Pb(II) — Cd(II) — Cr(III)	50 mg/L of each pollutant	5.0
Graphene oxide-hydrated zirconium oxide nanocomposites	Hydro-thermal coprecipitation	Metal ions	— As(III) — As(V)	65 mg/L of each pollutant	5.0
Chitosan-methacrylic acid nanoparticles	Polymerization	Metal ions	— Pb(II) — Cd(II)	50 mg/L of each pollutant	5.0
Magnetite nanoparticles	Coprecipitation	Metal ions	— Cr (III) — Pb (II)	20 mg/L of each pollutant	3.8
Magnesium oxide	Sol−gel method	Metal ions	— Pb(II) — Cd(II)	100 mg/L of each pollutant	5.0
Malachite nanoparticles	Chemical synthesis	Anionic compounds (polyatomic ions)	— Chromate — Arsenate	50 mg/L of each pollutant	4.0
Nano-zirconium oxide-crosslinked-nanolayer of carboxymethyl	Chemical synthesis	Metal ions	— Cr (III) — Cr (VI)	1 M of each metal ions	7.0 for Cr (III) and 2.0 for Cr (VI)
Nanoscale zerovalent iron (nZVI) and Au doped nZVI particles	Chemical synthesis	Metal ion and polyatomic ion	— Cd(II) — Nitrate	15 mg-N/L for nitrate 40 mg/L of Cd(II)	9.0

Therefore, it is significant that practical application is the need of the time that can solve "real life" glitches in the wastewater industry, few examples are overviewed in Table 12.3.

12.3 Importance of nanotechnology

Nanotechnology offers innovative results to the water treatment where nanoparticles are invented with special features, for example, high reactivity, aspect ratio, electrostatic, hydrophilic and hydrophobic connections (Das et al., 2014). The nanotechnology-based practices are very proficient, flexible, and multifunctional, which provide high-performance and inexpensive water and wastewater treatment results. This is suitable for various operations (adsorption, catalysis, sensing, optoelectronics, etc.). Nanoparticles are durable materials having high specific surface (S_{BET}), high surface to volume ratio controls the pollutants—bacteria interaction (Qu et al., 2013). Nanotechnology-enabled water treatment processes create major challenges to existing methods as they stretched to the

TABLE 12.4 Major limitations associated with conventional water purification methods (Hlongwane et al., 2019).

Conventional methods	Limitations
Distillation	Need high volumes of water and energy, while maximum impurities left behind. Pollutants which have boiling point >100°C, are problematic to eradicate
Chemical transformation	Excess chemicals are required to process and results a low-quality product that cannot be released directly into environment
Coagulation and flocculation	These are complex and less-proficient techniques and have need of alkaline additives to attain optimal pH for the operation
Biological treatment	Microorganisms are sensitive to environmental constraints, not cost-effective and time consuming
Ultraviolet treatment	Ineffective process for heavy metals, nonliving contaminants removal and render inoperative by turbidity of water. Expensive technique
Reverse osmosis	This method cannot eliminate chemicals, volatile organics and pharmaceuticals, moreover it needs high energy
Nanofilteration	Membrane fouling will occur with limited retention for salts and univalent ions. This technique requires high energy, pretreatment, hence expensive
Ultrafiltration	Unable to remove dissolved inorganics. Vulnerable to particulate plugging and tough to clean, requires high energy
Microfiltration	Cannot take out metals, fluoride, nitrates, volatile organics, sodium, color, and susceptible to membrane fouling. Less subtle to microbes and virus
Carbon filter	Ineffective to treat metals, fluoride, nitrates, sodium, etc. Blockage arises due to undissolved solids, requires regular filter replacement

purification of unconventional water sources in an economic system as comparison to limitations of conventional water purification methods complied in Table 12.4.

Treatment of wastewater with nanoparticles is also important in handling wastewater, as they eradicate pollutants and support in the reprocessing disinfected water with decrease in labor, time, and expenses to industry resolving numerous environmental problems (Kanchi, 2014). It should be noticed that nanoparticles for refining drinking water must be environment-friendly and nontoxic.

12.4 Applications of nanoparticle in water or wastewater treatment

Nanoparticles are generally <100 nm in size, materials with new substantially modified physical, chemical and biological properties (Theron et al., 2008). The constituents in this bowl comprise new size reliant properties, different from their larger complements. Desired assets of the nanoparticles include a large surface area for absorption, high (photocatalysis) reactivity, noble disinfection antimicrobial properties and control of biological contamination must have optical and electronic properties with a good detection range (Gehrke et al., 2015). Some nanotechnological applications for water and wastewater treatment are discussed in the following sections (Table 12.3).

12.4.1 Nanoadsorption

Adsorption is a surface phenomenon in which contaminants are adsorbed on a solid surface. Adsorption is usually done by physical forces, sometimes attributed to weak chemical bonds (Faust and Aly, 1983). The effectiveness of conventional adsorbents can be limited by their low specific surface area and selectivity (Qu et al., 2013). Nanoadsorbents are generally used to remove organic and inorganic pollutants from water and wastewater. The unique properties of nanosorbents, for example, their small size, catalytic potential, high reactivity, large surface area, simplicity of separation, and the abundant active sites-contaminants interaction constitutes ideal adsorbents for wastewater treatment (Ali, 2012). Kunduru et al. (2017) has listed different nanoparticles in water and wastewater treatment in Table 12.5.

Magnetic nanoadsorbents (MNPs), for example maghemite (γ-Fe$_2$O$_3$), hematite (α-Fe$_2$O$_3$) and spinel ferrites (M^{2+}Fe$_2$O$_4$, where M^{2+}: Cu^{2+}, Cd^{2+}, Co^{2+}, Fe^{2+}, Ni^{2+}, Mn^{2+}, Mg^{2+}, Zn^{2+}) are very good adsorbing materials for the elimination of toxic elements from polluted water (Badruddoza et al., 2011; Tan et al., 2012; Muthukumaran et al., 2016; Gómez-Pastora et al., 2014; Zhang et al., 2013; Ma et al., 2017). Magnetic properties can easily separate variety of elements, such as As, Cr, Co, Cu, Pb, and Ni in their respective ionic forms (Giakisikli and Anthemidis, 2013; Huang et al., 2008; Lee et al., 2014; Guo et al., 2014; Gautam et al., 2014; Sarkar et al., 2006). Long-term reactive iron nanoparticles (10−100 nm) as reducing ingredients validate success as detoxicants of chlorine containing compounds (organic solvents, and pesticides) (Zhang, 2003).

TABLE 12.5 Overview of different nanoparticles in water and wastewater treatment (Kunduru et al., 2017).

Nanoparticles	Physical properties and applications	Limitations
Nanoadsorbents	Contain high specific surface, better adsorption capacity, used to remove organic and inorganic pollutants and bacteria	High production costs
Nanometals and nanometal oxides	Have high specific surface area and short intraparticle diffusion distance, compressible without surface area change, abrasion resistant, magnetic and photo catalytic in nature	Less recyclable
	Used in slurry reactors, filter media, powders and pellets to remove heavy metals and radionuclides	
Membranes and membrane process	Highly reliable and generally automated process applied in all arenas of water and waste treatments	High-energy requirement
Photocatalysis	Photocatalytic activity in UV and visible light range, low human toxicity, high stability, and low cost	Selective reaction
Disinfection and microbial control	Strong and wide-spectrum antimicrobial action, low toxicity to humans and ease of usage	Disinfection residue

12.4.2 Photocatalysis

Numerous studies have been explored for heavy metals removal application of nano-technology for single pollutant control based on a specific type of nanoparticles (Kumar et al., 2014; Bora and Dutta, 2014; Feng et al., 2018; Lata et al., 2015). Then the focus is centered on group of nanoparticles with their applications can treat multiple pollutants simultaneous by multifunctional nanotechnology (Ihsanullah et al., 2016). This provides a paradigm shift from single to multiple-pollutant control strategies based on adsorption. The mechanisms for pollutant—nanoparticles treatment is the pollutant removal process is beyond the scope of present work. The economical multifunctional applications of nano-particles for multiple-pollutant control will gain interest to explore future technologies among environmental scientists, chemical, environmental engineers, ecologists, and environmental health scientists.

12.5 Adsorption for water treatment

12.5.1 Adsorption by nanoparticle

Adsorption is one of the most prominent methods usually applied for the elimination of toxic organic and inorganic water compounds due to its flexibility, high efficiency, simplicity in operation for different water systems and low cost of procedure (Qu et al., 2013). Traditional adsorbents should have high adsorption capacity, easy desorption of the adsorbed pollutants, remarkable recyclability for adsorption, nontoxicity, and selectivity to pollutants (Wang et al., 2012). Till date, various nanocomposites, for example, carbon-based adsorbents, metal oxides, and magnetic metal—organic framework composites have been discovered for the simultaneous removal of several coexisting inorganic and organic pollutants.

12.5.1.1 Simultaneous adsorption

The first nanocomposite material (2,4-Dinitrophenylhydrazine modified nano-alumina) was considered for adsorption of multiple heavy-metal cations in water. Among Cd(II), Co(II), Cr(III), Mn(II), Pb(II), and Ni (II), the utmost adsorption abilities were 100.0 mg/g for Cr(III), 100.0 mg/g for Pb(II), and 83.3 mg/g for Cd(II) ions (Afkhami et al., 2010). A higher adsorption capacity was attained using magnesium oxide (MgO) nanoparticles: 2614 mg/g for Pb(II) and 2294 mg/g for Cd(II) (Xiong et al., 2015). For example, the multi-pollutant reduction of Pb(II), Cd(II), and Ni(II) ions from water solution using recyclable chitosan-methacrylic acid nanoparticles (Heidari et al., 2013). As(III) and As(V) have been simultaneously removed using nanoparticles from aqueous solution (Luo et al., 2013).

Chromate and arsenate are two major polyatomic contaminants, and their simultaneous control using malachite nanoparticles with maximum adsorption capacity of 82.2 mg/g (chromate) and 57.1 mg/g (arsenate) (Saikia et al., 2011). Nickel sulfide nanoparticles were considered in the simultaneous adsorption of organic (safranin-O and methylene blue) dyes (Ghaedi et al., 2014). The expected removal at optimum conditions was 99.9% at 5.46 minutes for each dye. With the immense selection of nanoparticles being produced, cost-effective efficient methods are being discovered. Silica adsorbents are cost-effective under acidic conditions, simultaneously offers chemical shield besides leaching (Jiang et al., 2014).

The integrative treatment of heavy metals and organic compounds in wastewater. Chen et al. (2018) found the extreme removal capacity of 89.0 and 984.5 mg/g for Congo red and Cd(II) respectively (Song et al., 2017).

12.5.2 Photocatalysis

The kinetics and extent of degradation of water pollutants can be improved significantly over photocatalysis (Prihod'ko and Soboleva, 2013). Photocatalysis is the use of a dynamic catalyst to photo-initiate the degradation of inorganic pollutants (Bora and Dutta, 2014). The primary oxidation mechanism of photocatalyst depend on the formation of reactive radicals during the photocatalytic reactions (Oturan and Aaron, 2014). There are numerous broad reviews on the application of solar energy in water treatment methods, with detailed descriptions of the basic principles and mechanisms behind photocatalytic degradation of water pollutants (Oturan and Aaron, 2014; Zhang et al., 2018; Wang et al., 2015). Therefore, these ideas will not be replicated in the present article. With currents efforts to overcome water shortage, many nanoparticle-associated AOPs have been explored for water purification (Baniamerian and Shokrollahzadeh, 2016; Wang et al., 2015; Zhang et al., 2018). Consequently, a number of nanoparticle linked AOP based pollutant removal enhanced oxidation processes (EOPs) and advanced oxidation technologies (AOTs) have been established into the following main processes (Oppenländer, 2007):

- Fenton/photo-fenton reactions,
- Photocatalytic ozonation,
- Nonphotochemical AOP methods (include sulfate based AOTs) and
- UV wastewater treatment/heterogeneous semiconductor photocatalysis.

Toxic heavy-metal ions and organic pollutants are two different pollutants types that usually concur in wastewater and TiO_2 nanoparticles as photocatalyst examined for instantaneously photo-reduce metal ions (Cu and Ag) and organic dyes (methylene blue) ultraviolet irradiation (Doong et al., 2010). It has been revealed from the study methylene blue was complete photodegraded within 60 minutes of TiO_2 nanoparticles when the metal ion and methylene blue were preserved concurrently. Though, the degradation rate reduced because of electrons of TiO_2 nanoparticles contended by the metal ions and dye, when linked to tests whereby the metal ions were added prior to methylene blue (Doong et al., 2010). The practicality of TiO_2 is reliant on ultraviolet (UV) irradiation that may bound nanotechnology with the arrangement with inconsistent power supply (Wang et al., 2015). In the visible light range, heavy-metal ion and organic pollutant (phenol) by photocatalytic degradation, using a gold-loaded graphene oxide with degradation rate phenol and Cr(VI) reached 49.4% and 77.4% in around 4 hours (Liu et al., 2018).

Although the effort of concurrent photocatalytic degradation of coexisting pollutants using nanoparticles and nanocomposites has been described, it has not been reviewed.

TABLE 12.6 Photocatalytic degradation and adsorption of pollutants, simultaneous adsorption and photocatalysis under different light irradiations (Hlongwane et al., 2019).

Photocatalyst (light source) nanoparticles preparation	Pollutant types	Name of pollutants	Optimum pH	Optimum catalyst weight (g/L)	Degradation %	References
Titania/activated carbon/carbonized epoxy (TiO₂/AC/CE) Visible Light nanocomposite solvothermal process	Inorganic (metal ions) and organic pollutants	– Pb(II) –Methylene blue (MB)	6.5	Not indicated	~97% of the Pb(II) ions in 600 min ~90% of MB dye	Benjwal and Kar (2015)
Zn-Fe mixed metal oxides coprecipitation and calcination	Metal ions and organic pollutants	– Ibuprofen – Arsenic	3.0	0.5 g/L	95.7% of ibuprofen	Di et al. (2017)

Some of the key regions that were found in open literature where photocatalysis has played a crucial role in giving coexisting contaminants in water were collected.

Photocatalysis is an environmentally feasible solution for simultaneous removal of toxic heavy metals and organic pollutants coexist in wastewater. TiO_2 photocatalyst (nanoparticles) have been examined for simultaneously photo-reduce heavy-metal ions (copper and silver) and organic dyes (methylene blue) under UV irradiation (Doong et al., 2010). Different types of toxic organic compounds in different forms photocatalytic treatment of organic dyes using nanoparticles (Ghasemi et al., 2013) (Table 12.6).

12.5.2.1 Nanoparticle used for disinfection and microbial control

Numerous studies have offered for simultaneous treatment of different categories of pollutants with nanocomposites having photocatalytic and microbial properties. Pant et al. (2013) revealed that carbon nanofiber (CNF) composite with TiO_2/ZnO to concurrently reduce toxic chemical dye, methylene blue and *Escherichia coli* (*E. coli*), etc. They showed superb antibacterial activity, simultaneous adsorption ability and fast methylene blue degradation under UV irradiation (Pant et al., 2013). Nanoparticles that exhibit both photocatalytic activity and disinfection capabilities have found use in the corresponding removal of harmful algal blooms and produce toxins (Antoniou et al., 2014).

12.6 Influencing factors

Photocatalysis is measured by many factors, for example, concentration, catalyst amount, pH, and nature of inorganic contaminants. The amount of catalysis plays a vibrant part and degradation of pollutants in the processes of catalysis (Diyauddeen et al., 2011). The pH is an important factor as different effluents need to be treated at diverse pH values and also controls the surface properties of the photocatalyst (Turolla et al., 2016). Concentration and nature of inorganic pollutants also important on the rate of photocatalytic degradation.

Adsorption is one of the best commonly applied water remediation techniques due to its wide adaptability for diverse water systems, simplicity, high proficiency, and low operational cost (Qu et al., 2013). Photocatalytic water treatment has also been identified since the 1980s to treat many classes of pollutants has been proven in a number of scientific revisions (Mazzarino, 2001; Dong et al., 2015; Chong et al., 2010; Pi et al., 2018). A sum of new, improved nanotechnology-based pioneering claims in water treatment remains to grow in large-scale water and wastewater treatment plants (Cates, 2017). Therefore, it is also feasible to modern day challenges that multiple-pollutant control nanotechnologies in its real-world execution.

The composition of wastewater is multifarious due to the presence of coexisting contaminants such as toxic inorganic, organic compounds, and pathogenic microorganisms. The physical, chemical, and biological features of these pollutants vary massively and most encounters the use of nanoparticles for multiple-pollutant control in water treatment developments.

Nanoparticles have exhibited the potential to remove multiple contaminants. However, the nanoparticles' removal competences vary considerably from one pollutant to the other. A few experimental approaches performed by Ghaedi et al. (2014), Cai et al. (2017), and Wang et al. (2014) noted that the existence of pollutant "1" may affect (i.e., improve or hinder) the deduction of pollutant "2" In certain cases the coexistence may reciprocally improve the elimination of both pollutants as observed by Su et al. (2014) using nZVI. The removal capacity of nZVI increased for Cd in the presence of nitrate and the presence of Cd was also found to increase the removal efficiency using nZVI for nitrate. Competitive interaction happens when numerous types contend for a particular adsorption site (Mansouri et al., 2015; Yu et al., 2016). Nanoparticles would then express higher adsorption rates for the particular type. The adsorption efficiencies are dependent on its thermodynamics and kinetics. The rates of adsorption and photo-degradation are influenced by the binding capacity of nanoparticle (Wang et al., 2014), binding ability of the pollutant (Chen et al., 2018), and Gibbs free energy (Araghi and Entezari, 2015).

These features basically highlight the challenges to use nanoparticles for multiple pollutants control in water, particularly in conditions where a nanoparticle is used to treat low concentration of multiple pollutants. Therefore, it is important to treat multiple pollutants in water under hands-on practices over to understand the fundamental mechanisms and then custom these mechanisms for the development of new nanocomposite materials (by several nanoparticles) selective to a specific kind of pollutants.

12.6.1 The pH of wastewater

Actually, the pH of wastewater (acidic or alkaline) due to the presence of predominant pollutants present in wastewater. Thus, the pH of the wastewater also effect on the adsorption capacity of the adsorbents (positively or adversely) to the adsorption process (Luo et al., 2013). The literature reported these three features, which are an increase in adsorption rate with an increase in pH for one pollutant with increase (Xiong et al., 2015), decrease (Deng et al., 2013) or no effect (Tang et al., 2012) as the pH increase on adsorption rate of the adjacent pollutant.

In a broad sense, the nanoparticles had an optimum pH where the maximum adsorption capacities were extended. The adsorption rates have a tendency to increase or decrease until the optimum pH is reached (Luo et al., 2013; Afkhami et al., 2010). In real applied situations, it may be hard to match the optimum conditions for all water pollutants. The simultaneous removal of pollutants can be considered in a sequential way that the degradation of one pollutant results intermediates and outcome increases the pH of the solution, thus producing optimum conditions for the next coexisting contaminant.

The dose of the nanoparticles used and its contact time with pollutants can be strongly influenced by the pollutant removal procedure (Heidari et al., 2013). The general trend for contaminants' removal rate generally rises with the increase in nanoparticle dose and contact time till a threshold value. Beyond or under the contact time threshold limit the removal efficiencies considerably decreases (Silva et al., 2017; Deng et al., 2013).

12.6.2 Recovery, management, and disposal of exhausted nanoparticles

The capacity to safely and efficiently recover nanoparticles prior to regeneration, manage and dispose exhausted nanoparticles for commercial viability and ingenuity of nanotechnology-based water treatment methods (Lata et al., 2015). Practically, powdered form nanoparticles fail to demonstrate recovery facilities with regeneration capacity (Lee and Park, 2013) as separating nanoparticles can be difficult and exclusive in real treatment plants (Gulyas, 2014). A lack of nanoparticles management and disposal strategies could result in intense secondary pollution (Lata et al., 2015). The recovery, management, and disposal of drained nanoparticles used in water treatment devices need further assessment. The inability to safely improve nanoparticles from treatment reactors confines the large-scale application of multifunctional nanotechnology developments in water treatment.

12.6.3 Regeneration and recyclability of nanoparticles

Regeneration is significant factor that administer the lucrative nanoparticles (Qu et al., 2013). After treatment nanoparticle separation was achieved by pH reversal or simple elution. In fact, total desorption of the pollutants from the surface of the nanoparticles is easy. The nanoparticles demonstrated a stable recyclability, with the nanoparticles $\sim 100\%$ of their original properties and adsorption ability even after $3-10$ regeneration cycles deliberated valuable as it finally effect in cost saving. Thus, multiple-pollutant control studies need to broadly evaluate the effects that multiple pollutants have on regeneration of nanoparticles. Management and disposal of pollutants after regeneration requires consideration to the disposal of pollutants used for regeneration. The nanoparticles recycling could possibly increase the profitable value of nanotechnology-based on treatment methods.

The application of nanoparticles is not fully environment-friendly (Pietroiusti et al., 2018). Therefore, a comprehensive and concise researchers' shift from single to multiple-pollutant control by nanoparticles, here, the biotoxicity and ecological risks of nanoparticles can be increased. The biotoxicity is related to low metal nanoparticles concentrations such as iron, gold, and silver. It can constrain bacterial growth toxic to native plants, cause liver, brain, and stem cells loss, cause human skin diseases (Hussain et al., 2006; Rana and Kalaichelvan, 2013). Although the synthesis methods for the nanoparticles are nonhazardous, ecotoxicity of reagents during nanoparticle regeneration are purely chemical nature. The nanoparticles themselves should not only be nonhazardous but should be free from secondary pollution and use nonhazardous chemicals.

12.7 Regeneration of nanoparticles

Regeneration of water purification by nanoparticles is one of the crucial features, then it controls the water treatment technology economy. The pH-dependent solvents play crucial role in nanoparticle regeneration. They can also be attained by applying separation device or arresting nanoparticles in the treatment system. Membrane filtration is capable for the regeneration and reuse of nanoparticles because of its chemical use. Ceramic membranes are always helpful, as they are sturdy to UV compared to polymer membranes (Qu et al.,

2013). Raw water pretreatment is very essential to reduce turbidity; remove suspended particles that can retain in membranes, and can decline the water treatment efficiency. Immobilization is another technique for nanoparticles and progress of simple economical method is obligatory to immobilize nanoparticles deprived of affecting their efficiency.

Magnetic separation is another choice for the separating magnetic nanoparticles by implanted them in a solid matrix have slow release until disposed of. Identification of a nanoparticles release is a major technical hurdle and different detection systems are revealed in the research literature (Qu et al., 2013; Silva et al., 2017; Tiede et al., 2008).

According to the works, nanoparticles can be restored and recycled for water treatment to make them economically viable constituents. The regeneration ability of nanoparticles may be an extra advantage for their wastewater treatment acceptance devising highly sophisticated, exclusive with many restrictions.

At present, fast, sensitive, and selective nanoparticles analytical techniques are in great demand. Management of recycled nanoparticles and recovered pollutants is one of the most important features. Everybody is conscious of pollutant hazards and nanotoxicology exposures; proper removal must be carried out by the users. The best way is to recover the nanoparticle and exhausted nanoparticles and so on. The recovered inorganic contaminants should be treated as priority pollutants (Ali, 2012).

12.8 Current limitations and future research needs

Water or wastewater treatment methods by nanotechnology display a great promise in laboratory scale studies. Some of these technologies are advertised, and others need important research before scale up. The commercialization of these technologies are challenging: it needs to overcome technical hurdles and mark them safe and cost-effective. Research is required on full-scale nanotechnology operation for treating natural and wastewaters.

Research studies should be directed under realistic conditions to judge the efficiency of available nanotechnology to authorize the nanoparticles-enabled sensing. Another research need is to measure the long-term efficiency of existing technologies, which are conducted on a laboratory scale. Cost effectiveness can be achieved by regeneration and reuse of these nanoparticles (Qu et al., 2013). Since these materials are nanoscale, risk assessment and management is a challenge. Researchers should understand the potential hazards of these materials in the treatment of water and wastewater.

Commercialization in academic research is a measure of scientific output which directly underwrite to the country's economy (Perkmann et al., 2013). Though nanotechnology recently governs research studies in water treatment, there exists no dependable info about any current commercial multipollutant governor nanotechnologies. However, patents can be used as a direct indicator of the potential commercialization of technologies (Crespi et al., 2011). There are several patents of multifunctional water and wastewater treatment nanotechnologies for simultaneous treatment of multiple-pollutant control worldwide (Hlongwane et al., 2019). Apart from extracting links between the research studies and patents in the field of multipollutant control in water, the focus was to highlight variances between the future research technologies and the patented creations.

12.9 Conclusion

In recent decades, various exertions have been put forward for the removal of multipurpose inorganic water pollutants. This chapter offers a well-structured presentation for multiple inorganic pollutants removal from waste/wastewater treatment nanoparticle technologies over adsorption, disinfection, microbial control, photocatalysis, and integration practices. A steady increase of the interesting topics that are currently faced during remediation due to multipollutant in real life applications. The nanoparticles have simple to operate, cost-effective, comply emission standards and secondary pollution-free even for prospective large-scale applications using nanoparticles.

- Nanotechnology-based adsorption is capable of lowering pollutant concentrations to the recommended drinking water. However, an inadequate studies have compared pollutant concentrations achieved after treatment to potable water, more studies need to be addressed;
- Simultaneous removal of pollutants from water are laboratory scale and are in the initial stage for pilot or field-tested. The investigation field is already full of novel materials development and needs more research directed to actual process engineering and development;
- Commercialization of nanotechnology for simultaneous multiple-pollutant control, the economical availability need to be focused in the upcoming technical reports. Advanced scientific methods, protests, modeling and their economic analysis on real water models and conditions are very essential.

The consistent modernization in nanotechnology and advances are in laboratory, research, pilot plant, and commercial methods. Besides, nanoparticles are used most promising in nanoadsorbents, nanophotocatalysts and nanomembranes have been commercialized for large-scale wastewater treatment applications. Conversely, the risk assessment of recovered pollutants after multiple applications should be explored significantly, and more sustainable waste management approaches are essential to evade toxicity and hazards. There is urgent need to combat the upcoming challenges for collective efforts of academic and industrial resources in effectively solve water contamination globally.

Acknowledgments

Deepak Yadav and Pradeep Kumar are equally acknowledge the support for the necessary facilities by IIT (BHU), Varanasi and Harcourt Butler Technical University, Kanpur, India. The contribution of Dr. Pradeep Singh from the Department of Environment Sciences, University of Delhi is also appreciated.

References

Afkhami, A., Saber-Tehrani, M., Bagheri, H., 2010. Simultaneous removal of heavy-metal ions in wastewater samples using nano-alumina modified with 2, 4-dinitrophenylhydrazine. J. Hazard. Mater. 181 (1–3), 836–844.
Alcamo, J., Henrichs, T., Rösch, T., 2017. World water in 2025: global modeling and scenario analysis for the world commission on water for the 21st century.
Altenburger, R., Ait-Aissa, S., Antczak, P., Backhaus, T., Barceló, D., Seiler, T.B., et al., 2015. Future water quality monitoring-adapting tools to deal with mixtures of pollutants in water resource management. Sci. Total Environ. 512, 540–551.

Ali, I., 2012. New generation adsorbents for water treatment. Chem.Rev. 112, 5073–5091.

Amin, M.T., Alazba, A.A., Manzoor, U., 2014. A review of removal of pollutants from water/wastewater using different types of nanomaterials. Adv. Mater. Sci. Eng. 2014.

Antoniou, M.G., De La Cruz, A.A., Pelaez, M.A., Han, C., He, X., Dionysiou, D.D., et al., 2014. Practices that prevent the for- mation of cyanobacterial blooms in water resources and remove cyanotoxins during physical treatment of drinking water. Compr. Water Qual. Purif 2, 173–195. Available from: https://doi.org/10.1016/B978-0-12-382182-9.00032-3.

Araghi, S.H., Entezari, M.H., 2015. Amino-functionalized silica magnetite nanoparticles for the simultaneous removal of pollutants from aqueous solution. Appl. Surf. Sci. 333, 68–77.

Arjoon, A., Olaniran, A.O., Pillay, B., 2013. Co-contamination of water with chlorinated hydrocarbons and heavy metals: challenges and current bioremediation strategies. Int. J. Environ. Sci. Technol. 10 (2), 395–412.

Badruddoza, A.Z.M., Tay, A.S.H., Tan, P.Y., Hidajat, K., Uddin, M.S., 2011. Carboxymethyl-β-cyclodextrin conjugated magnetic nanoparticles as nano-adsorbents for removal of copper ions: synthesis and adsorption studies. J. Hazard. Mater. 185 (2–3), 1177–1186.

Baniamerian, H., Shokrollahzadeh, S., 2016. Improvement in photocatalysts and photocatalytic reactors for water and wastewater treatment: a review. J. Part. Sci. Technol. 2 (3), 119–140.

Benjwal, P., Kar, K.K., 2015. Simultaneous photocatalysis and adsorption based removal of inorganic and organic impurities from water by titania/activated carbon/carbonized epoxy nanocomposite. J. Environ. Chem. Eng. 3 (3), 2076–2083.

Bora, T., Dutta, J., 2014. Applications of nanotechnology in wastewater treatment—a review. J. Nanosci. Nanotechnol. 14 (1), 613–626.

Bradley, P.M., Journey, C.A., Romanok, K.M., Barber, L.B., Buxton, H.T., Foreman, W.T., et al., 2017. Expanded target-chemical analysis reveals extensive mixed-organic-contaminant exposure in US streams. Environ. Sci. Technol. 51 (9), 4792–4802.

Cai, Y., Li, C., Wu, D., Wang, W., Tan, F., Wang, X., et al., 2017. Highly active MgO nanoparticles for simultaneous bacterial inactivation and heavy metal removal from aqueous solution. Chem. Eng. J. 312, 158–166. Available from: https://doi.org/10.1016/j.cej.2016.11.134.

Cates, E.L., 2017. Photocatalytic water treatment: so where are we going with this? Environ. Sci. Technol. 51, 757–758. Available from: https://doi.org/10.1021/acs.est.6b06035.

Chen, Y.Y., Yu, S.H., Jiang, H.F., Yao, Q.Z., Fu, S.Q., Zhou, G.T., 2018. Performance and mechanism of simultaneous removal of Cd (II) and Congo red from aqueous solution by hierarchical vaterite spherulites. Appl. Surf. Sci. 444, 224–234.

Chong, M.N., Jin, B., Chow, C.W., Saint, C., 2010. Recent developments in photocatalytic water treatment technology: a review. Water Res. 44 (10), 2997–3027.

Cosgrove, W.J., Rijsberman, F.R., 2014. World Water Vision: Making Water Everybody's Business. Routledge.

Crespi, G., D'Este, P., Fontana, R., Geuna, A., 2011. The impact of academic patenting on university research and its transfer. Res. Policy 40 (1), 55–68. Available from: https://doi.org/10.1016/j.respol.2010.09.010.

Das, R., Ali, M.E., Hamid, S.B.A., Ramakrishna, S., Chowdhury, Z.Z., 2014. Carbon nanotube membranes for water purification: a bright future in water desalination. Desalination 336, 97–109.

Deng, J.H., Zhang, X.R., Zeng, G.M., Gong, J.L., Niu, Q.Y., Liang, J., 2013. Simultaneous re- moval of Cd (II) and ionic dyes from aqueous solution using magnetic graphene oxide nanocomposite as an adsorbent. Chem. Eng. J. 226, 189–200. Available from: https://doi.org/10.1016/j.cej.2013.04.045.

Di, G., Zhu, Z., Zhang, H., Zhu, J., Lu, H., Zhang, W., et al., 2017. Simultaneous removal of several pharmaceuticals and arsenic on Zn-Fe mixed metal oxides: combination of photocatalysis and adsorption. Chem. Eng. J. 328, 141–151.

Diyauddeen, B.H., Daud, W.M.A.W., Abdul Aziz, A.R., 2011. Treatment technologies for petroleum refinery effluents: a review. Process Saf. Environ. Protect 89 (2), 95–105.

Dong, S., Feng, J., Fan, M., Pi, Y., Hu, L., Han, X., et al., 2015. Recent developments in heterogeneous photocatalytic water treatment using visible light- responsive photocatalysts: a review. RSC Adv. 5 (19), 14610–14630. Available from: https://doi.org/10.1039/C4RA13734E.

Doong, R.A., Hsieh, T.C., Huang, C.P., 2010. Photoassisted reduction of metal ions and organic dye by titanium dioxide nanoparticles in aqueous solution under anoxic conditions. Sci. Total. Environ. 408 (16), 3334–3341.

Faust, S.D., Aly, O.M., 1983. Chemistry of Water Treatment. Butterworth, MA, Stoneham.

Feng, M., Zhang, P., Zhou, H.C., Sharma, V.K., 2018. Water-stable metal-organic frameworks for aqueous removal of heavy metals and radionuclides: a review. Chemosphere 208, 783–800.

Gautam, R.K., Jaiswal, A., Chattopadhyaya, M.C., 2014. Functionalized magnetic nanoparticles for heavy metal removal from aqueous solutions: kinetics and equilibrium modeling. Adv. Mater. Agric. Food Environ. Saf., 291–331.

Gehrke, I., Geiser, A., Somborn-Schulz, A., 2015. Innovations in nanotechnology for water treatment. Nanotechnol. Sci. Appl. 8, 1–17.

Ghaedi, M., Pakniat, M., Mahmoudi, Z., Hajati, S., Sahraei, R., Daneshfar, A., 2014. Synthesis of nickel sulfide nanoparticles loaded on activated carbon as a novel adsorbent for the competitive removal of Methylene blue and Safranin-O. Spectrochim. Acta Part A Mol. Biomol. Spectrosc. 123, 402–409.

Ghasemi, S., Esfandiar, A., Setayesh, S.R., Habibi-Yangjeh, A., Gholami, M.R., 2013. Synthesis and characterization of TiO2−graphene nanocomposites modified with noble metals as a photocatalyst for degradation of pollutants. Appl. Catal. A: Gen. 462, 82–90.

Giakisikli, G., Anthemidis, A.N., 2013. Magnetic materials as sorbents for metal/metalloid preconcentration and/or separation. A review. Anal. Chim. Acta 789, 1–16.

Gollavelli, G., Chang, C.C., Ling, Y.C., 2013. Facile synthesis of smart magnetic graphene for safe drinking water: heavy metal removal and disinfection control. ACS Sustain. Chem. Eng. 1 (5), 462–472.

Gómez-Pastora, J., Bringas, E., Ortiz, I., 2014. Recent progress and future challenges on the use of high performance magnetic nano-adsorbents in environmental applications. Chem. Eng. J. 256, 187–204.

Gopal, K., Tripathy, S.S., Bersillon, J.L., Dubey, S.P., 2007. Chlorination byproducts, their toxicodynamics and removal from drinking water. J. Hazard. Mater. 140 (1–2), 1–6.

Gu, J., Zhou, W., Jiang, B., Wang, L., Ma, Y., Guo, H., et al., 2016. Effects of biochar on the transformation and earthworm bioaccumulation of organic pollutants in soil. Chemosphere 145, 431–437.

Gulyas, H., 2014. Solar heterogeneous photocatalytic oxidation for water and wastewater treatment: problems and challenges. J. Adv. Chem. Eng. 4 (2), 1000108. Available from: https://doi.org/10.4172/2090-4568.1000108.

Guo, X., Du, B., Wei, Q., Yang, J., Hu, L., Yan, L., et al., 2014. Synthesis of amino functionalized magnetic graphenes composite material and its application to remove Cr (VI), Pb (II), Hg (II), Cd (II) and Ni (II) from contaminated water. J. Hazard. Mater. 278, 211–220.

Gupta, V.K., Kumar, R., Nayak, A., Saleh, T.A., Barakat, M.A., 2013. Adsorptive removal of dyes from aqueous solution onto carbon nanotubes: a review. Adv. Colloid Interface Sci. 193, 24–34.

Heidari, A., Younesi, H., Mehraban, Z., Heikkinen, H., 2013. Selective adsorption of Pb (II), Cd (II), and Ni (II) ions from aqueous solution using chitosan−MAA nanoparticles. Int. J. Biol. Macromol. 61, 251–263.

Hlongwane, G.N., Sekoai, P.T., Meyyappan, M., Moothi, K., 2019. Simultaneous removal of pollutants from water using nanoparticles: a shift from single pollutant control to multiple pollutant control. Sci. Total Environ. 656, 808–833.

Hua, M., Zhang, S., Pan, B., Zhang, W., Lv, L., Zhang, Q., 2012. Heavy metal removal from water/wastewater by nanosized metal oxides: a review. J. Hazard. Mater. 211, 317–331.

Huang, W., Wang, S., Zhu, Z., Li, L., Yao, X., Rudolph, V., et al., 2008. Phosphate removal from wastewater using red mud. J. Hazard. Mater. 158, 35–42.

Hussain, S.M., Javorina, A.K., Schrand, A.M., Duhart, H.M., Ali, S.F., Schlager, J.J., 2006. The interaction of manganese nanoparticles with PC-12 cells induces dopamine deple- tion. Toxicol. Sci 92 (2), 456–463. Available from: https://doi.org/10.1093/toxsci/kfl020.

Ihsanullah, Abbas, A., Al-Amer, A.M., Laoui, T., Al-Marri, M.J., Nasser, M.S., et al., 2016. Heavy metal removal from aqueous solution by advanced carbon nanotubes: critical review of adsorption applications. Sep. Purif. Technol. 157, 141–161. Available from: https://doi.org/10.1016/j.seppur.2015.11.039.

Inglezakis, V.J., Loizidou, M.D., Grigoropoulou, H.P., 2002. Equilibrium and kinetic ion exchange studies of Pb^{2+}, Cr^{3+}, Fe^{3+} and Cu^{2+} on natural clinoptilolite. Water Res. 36 (11), 2784–2792.

Ishii, S., Ksoll, W.B., Hicks, R.E., Sadowsky, M.J., 2006. Presence and growth of naturalized *Escherichia coli* in temperate soils from Lake Superior watersheds. Appl. Environ. Microbiol. 72 (1), 612–621.

Jelić, A., Petrović, M., Barceló, D., 2012. Pharmaceuticals in drinking water. Emerging Organic Contaminants and Human Health. Springer, Berlin, Heidelberg, pp. 47–70.

Jiang, W., Cai, Q., Xu, W., Yang, M., Cai, Y., Dionysiou, D.D., et al., 2014. Cr (VI) adsorption and reduction by humic acid coated on magnetite. Environ. Sci. Technol. 48 (14), 8078–8085.

Kanchi, S., 2014. Nanotechnology for water treatment. J. Environ. Anal. Chem. 1, e102.

Kumar, S., Ahlawat, W., Bhanjana, G., Heydarifard, S., Nazhad, M.M., Dilbaghi, N., 2014. Nanotechnology-based water treatment strategies. J. Nanosci. Nanotechnol. 14 (2), 1838−1858.

Kunduru, K.R., Nazarkovsky, M., Farah, S., Pawar, R.P., Basu, A., Domb, A.J., 2017. Nanotechnology for water purification: applications of nanotechnology methods in wastewater treatment. Water Purif. 33−74.

Lata, S., Singh, P.K., Samadder, S.R., 2015. Regeneration of adsorbents and recovery of heavy metals: a review. Int. J. Environ. Sci. Technol. 12 (4), 1461−1478. Available from: https://doi.org/10.1007/s13762-014-0714-9.

Lazar, M., Varghese, S., Nair, S., 2012. Photocatalytic water treatment by titanium dioxide: recent updates. Catalysts 2 (4), 572−601.

Leclerc, H., Schwartzbrod, L., Dei-Cas, E., 2002. Microbial agents associated with waterborne diseases. Crit. Rev. Microbiol. 28 (4), 371−409.

Lee, S.Y., Park, S.J., 2013. TiO$_2$ photocatalyst for water treatment applications. J. Ind. Eng. Chem. 19 (6), 1761−1769.

Lee, W.-L.W., Lu, C.-S., Lin, H.-P., Chen, J.-Y., Chen, C.-C., 2014. Photocatalytic degradation of ethyl violet dye mediated by TiO2 under an anaerobic condition. J. Taiwan Inst. Chem. Eng. 45 (5), 2469−2479.

Li, K., Li, P., Cai, J., Xiao, S., Yang, H., Li, A., 2016. Efficient adsorption of both methyl orange and chromium from their aqueous mixtures using a quaternary ammonium salt modified chitosan magnetic composite adsorbent. Chemosphere 154, 310−318.

Liu, J., Fang, W., Wang, Y., Xing, M., Zhang, J., 2018. Gold-loaded graphene oxide/PDPB composites for the synchronous removal of Cr (VI) and phenol. Chin. J. Catal. 39 (1), 8−15.

Lofrano, G., Brown, J., 2010. Wastewater management through the ages: a history of mankind. Sci. Total. Environ. 408 (22), 5254−5264.

Loucks, D.P., Van Beek, E., 2017. Water Resource Systems Planning and Management: An Introduction to Methods, Models, and Applications. Springer.

Luo, X., Wang, C., Wang, L., Deng, F., Luo, S., Tu, X., et al., 2013. Nanocomposites of graphene oxide-hydrated zirconium oxide for simultaneous removal of As (III) and As (V) from water. Chem. Eng. J. 220, 98−106.

Lyu, S., Chen, W., Zhang, W., Fan, Y., Jiao, W., 2016. Wastewater reclamation and reuse in China: opportunities and challenges. J. Environ. Sci. 39, 86−96.

Ma, L., Islam, S.M., Liu, H., Zhao, J., Sun, G., Li, H., et al., 2017. Selective and efficient removal of toxic oxoanions of As (III), As (V), and Cr (VI) by layered double hydroxide intercalated with MoS$_4$$^{2-}$. Chem. Mater. 29 (7), 3274−3284.

Mansouri, H., Carmona, R.J., Gomis-Berenguer, A., Souissi-Najar, S., Ouederni, A., Ania, C.O., 2015. Competitive adsorption of ibuprofen and amoxicillin mixtures from aqueous solution on activated carbons. J. Colloid Interface Sci. 449, 252−260. Available from: https://doi.org/10.1016/j.jcis.2014.12.020.

Martin, S., Griswold, W., 2009. Human health effects of heavy metals. Environ. Science Technol. briefs Citiz. 15, 1−6. Retrieved from: https://www.engg.ksu.edu/CHSR/.

Maryam, B., Büyükgüngör, H., 2018. Wastewater reclamation and reuse trends in Turkey: opportunities and challenges. J. Water Process. Eng. 9 (1), 10−17.

Mathur, N., Bhatnagar, P., 2007. Mutagenicity assessment of textile dyes from Sanganer (Rajasthan). J. Environ. Biol. 28 (1), 123−126.

Mazzarino, I., 2001. Feasibility analysis of photocatalytic wastewater treatment. WIT Trans. Ecol. Environ 49, 113−121. Available from: https://doi.org/10.2495/WP010101.

Muthukumaran, C., Sivakumar, V.M., Thirumarimurugan, M., 2016. Adsorption isotherms and kinetic studies of crystal violet dye removal from aqueous solution using surfactant modified magnetic nanoadsorbent. J. Taiwan Inst. Chem. Eng. 63, 354−362.

Oller, I., Malato, S., Sánchez-Pérez, J., 2011. Combination of advanced oxidation processes and biological treatments for wastewater decontamination—a review. Sci. Total. Environ. 409 (20), 4141−4166.

Oppenländer, T., 2007. Photochemical Purification of Water and Air: Advanced Oxidation Processes (AOPs)-Principles, Reaction Mechanisms, Reactor Concepts. John Wiley & Sons, New York.

Ota, K., Amano, Y., Aikawa, M., Machida, M., 2013. Removal of nitrate ions from water by activated carbons (ACs)—Influence of surface chemistry of ACs and coexisting chloride and sulfate ions. Appl. Surf. Sci. 276, 838−842.

Oturan, M.A., Aaron, J.J., 2014. Advanced oxidation processes in water/wastewater treatment: principles and applications. A review. Crit. Rev. Environ. Sci. Technol. 44 (23), 2577–2641.

Pant, B., Pant, H.R., Barakat, N.A., Park, M., Jeon, K., Choi, Y., et al., 2013. Carbon nano- fibers decorated with binary semiconductor (TiO$_2$/ZnO) nanocomposites for the ef- fective removal of organic pollutants and the enhancement of antibacterial activities. Ceram. Int 39 (6), 7029–7035. Available from: https://doi.org/10.1016/j.ceramint.2013.02.041.

Park, D., Yun, Y.S., Park, J.M., 2010. The past, present, and future trends of biosorption. Biotechnol. Bioprocess Eng. 15 (1), 86–102.

Perkmann, M., Tartari, V., McKelvey, M., Autio, E., Broström, A., D'Este, P., et al., 2013. Academic engagement and commercialisation: a review of the literature on university–industry relations. Res. Policy 42 (2), 423–442. Available from: https://doi.org/10.1016/j.respol.2012.09.007.

Pi, Y., Li, X., Xia, Q., Wu, J., Li, Y., Xiao, J., et al., 2018. Adsorptive and photocatalytic removal of persistent organic pollutants (POPs) in water by metal-organic frameworks (MOFs). Chem. Eng. J. 337, 351–371. Available from: https://doi.org/10.1016/j.cej.2017.12.092.

Pietroiusti, A., Stockmann-Juvala, H., Lucaroni, F., Savolainen, K., 2018. Nanomaterial ex- posure, toxicity, and impact on human health. Wiley Interdiscip. Rev. Nanomed. Nanobiotechnol e1513. Available from: https://doi.org/10.1002/wnan.1513.

Prihod'ko, R.V., Soboleva, N.M., 2013. Photocatalysis: oxidative processes in water treatment. J. Chem. 1, 1–8. Available from: https://doi.org/10.1155/2013/168701.

Purwajanti, S., Zhou, L., Ahmad Nor, Y., Zhang, J., Zhang, H., Huang, X., et al., 2015. Synthesis of magnesium oxide hierarchical microspheres: a dual-functional material for water remediation. ACS Appl. Mater. & Interfaces 7 (38), 21278–21286.

Qu, X., Alvarez, P.J., Li, Q., 2013. Applications of nanotechnology in water and wastewater treatment. Water Res. 47 (12), 3931–3946.

Rana, S., Kalaichelvan, P.T., 2013. Ecotoxicity ofnanoparticles. ISRN Toxicol. 2013, 574648. Available from: https://doi.org/10.1155/2013/574648.

Richardson, S.D., Plewa, M.J., Wagner, E.D., Schoeny, R., DeMarini, D.M., 2007. Occurrence, genotoxicity, and carcinogenicity of regulated and emerging disinfection by-products in drinking water: a review and roadmap for research. Mutat. Res./Rev. Mutat. Res. 636 (1–3), 178–242.

Saif, M.M.S., Kumar, N.S., Prasad, M.N.V., 2012. Binding of cadmium to Strychnos potatorum seed proteins in aqueous solution: adsorption kinetics and relevance to water purification. Colloids Surf. B: Biointerfaces 94, 73–79.

Saikia, J., Saha, B., Das, G., 2011. Efficient removal of chromate and arsenate from individual and mixed system by malachite nanoparticles. J. Hazard. Mater. 186 (1), 575–582.

Sarkar, B., Chakrabarti, P.P., Vijaykumar, A., Kale, V., 2006. Wastewater treatment in dairy industries—possibility of reuse. Desalination 195 (1–3), 141–152. Available from: https://doi.org/10.1016/j.desal.2005.11.015.

Sedlak, D.L., von Gunten, U., 2011. The chlorine dilemma. Science 331 (6013), 42–43.

Silva, T.A., Diniz, J., Paixão, L., Vieira, B., Barrocas, B., Nunes, C.D., et al., 2017. Novel titanate nanotubes-cyanocobalamin materials: synthesis and enhanced photocatalytic properties for pollutants removal. Solid State Sci. 63, 30–41.

Song, S., Huang, S., Zhang, R., Chen, Z., Wen, T., Wang, S., et al., 2017. Simultaneous removal of U (VI) and humic acid on defective TiO2 − x investigated by batch and spectroscopy techniques. Chem. Eng. J. 325, 576–587.

Su, Y., Adeleye, A.S., Huang, Y., Sun, X., Dai, C., Zhou, X., et al., 2014. Simultaneous removal of cadmium and nitrate in aqueous media by nanoscale zerovalent iron (nZVI) and Au doped nZVI particles. Water Res. 63, 102–111.

Tan, Y., Chen, M., Hao, Y., 2012. High efficient removal of Pb (II) by amino-functionalized Fe$_3$O$_4$ magnetic nano-particles. Chem. Eng. J. 191, 104–111.

Tang, F., Li, L., Chen, D., 2012. Mesoporous silica nanoparticles: synthesis, biocompatibility and drug delivery. Adv. Mater. 24 (12), 1504–1534.

Theron, J., Walker, J.A., Cloete, T.E., 2008. Nanotechnology and water treatment: applications and emerging opportunities. Crit. Rev. Microbiol. 34 (1), 43–69.

Tiede, K., Boxall, A.B.A., Tear, S.P., Lewis, J., David, H., Hassellov, M., 2008. Detection and characterization of engineered nanoparticles in food and the environment. Food Addit. Contam. Part A 25, 795–821.

Turolla, A., Santoro, D., de Bruyn, J.R., Crapulli, F., Antonelli, M., 2016. Nanoparticle scattering characterization and mechanistic modelling of UVeTiO2 photo- catalytic reactors using computational fluid dynamics. Water Res. 88, 117–126.

Unuabonah, E.I., Ugwuja, C.G., Omorogie, M.O., Adewuyi, A., Oladoja, N.A., 2017. Clays for efficient disinfection of bacteria in water. Appl. Clay Sci. 151, 211–223. Available from: https://doi.org/10.1016/j.clay.2017.10.005.

Wang, X., Guo, Y., Yang, L., Han, M., Zhao, J., Cheng, X., 2012. Nanomaterials as sorbents to remove heavy metal ions in wastewater treatment. J. Environ. Anal. Toxicol. 2 (7), 154.

Wang, T., Jin, X., Chen, Z., Megharaj, M., Naidu, R., 2014. Simultaneous removal of Pb (II) and Cr (III) by magnetite nanoparticles using various synthesis conditions. J. Ind. Eng. Chem. 20 (5), 3543–3549.

Wang, W., Huang, G., Jimmy, C.Y., Wong, P.K., 2015. Advances in photocatalytic disinfection of bacteria: development of photocatalysts and mechanisms. J. Environ. Sci. 34, 232–247.

Wing, S., Lowman, A., Keil, A., Marshall, S.W., 2014. Odors from sewage sludge and livestock: associations with self-reported health. Public Health Rep. 129 (6), 505–515.

World Health Organization, 2015. Progress on Sanitation and Drinking Water: 2015 Up- date and MDG Assessment. World Health Organization.

Wu, Y., Chen, L., Long, X., Zhang, X., Pan, B., Qian, J., 2018. Multi-functional magnetic water purifier for disinfection and removal ofdyes andmetal ions with superior reusability. J. Hazard. Mater. 347, 160–167. Available from: https://doi.org/10.1016/j.jhazmat.2017.12.037.

Xiong, C., Wang, W., Tan, F., Luo, F., Chen, J., Qiao, X., 2015. Investigation on the efficiency and mechanism of Cd (II) and Pb (II) removal from aqueous solutions using MgO nanoparticles. J. Hazard. Mater. 299, 664–674.

Yang, S., Hu, J., Chen, C., Shao, D., Wang, X., 2011. Mutual effects of Pb (II) and humic acid adsorption on multi-walled carbon nanotubes/polyacrylamide composites from aqueous solutions. Environ. Sci. Technol. 45 (8), 3621–3627.

Yu, S., Wang, X., Ai, Y., Tan, X., Hayat, T., Hu, W., et al., 2016. Experimental and theo- retical studies on competitive adsorption of aromatic compounds on reduced graphene oxides. J. Mater. Chem. A 4 (15), 5654–5662. Available from: https://doi.org/10.1039/C6TA00890A.

Yu, L., Han, M., He, F., 2017. A review of treating oily wastewater. Arab. J. Chem. 10, S1913–S1922.

Zhan, S., Yang, Y., Shen, Z., Shan, J., Li, Y., Yang, S., et al., 2014. Efficient removal of pathogenic bacteria and viruses by multifunctional amine-modified magnetic nanoparticles. J. Hazard. Mater. 274, 115–123.

Zhang, W.X., 2003. Nanoscale iron particles for environmental remediation: an overview. J. Nanopart. Res. 5 (3–4), 323–332.

Zhang, J., Zhai, S., Li, S., Xiao, Z., Song, Y., An, Q., et al., 2013. Pb (II) removal of Fe3O4@ SiO2–NH2 core–shell nanomaterials prepared via a controllable sol–gel process. Chem. Eng. J. 215, 461–471.

Zhang, Y., Sivakumar, M., Yang, S., Enever, K., Ramezanianpour, M., 2018. Application of solar energy in water treatment processes: a review. Desalination 428, 116–145.

Further reading

Achudume, A.C., 2009. The effect of petrochemical effluent on the water quality of Ubeji creek in Niger Delta of Nigeria. Bull. Environ. Contam. Toxicol. 83 (3), 410–415.

Anjaneyulu, Y., Chary, N.S., Raj, D.S.S., 2005. Decolourization of industrial effluents–available methods and emerging technologies–a review. Rev. Environ. Sci. Bio/Technol. 4 (4), 245–273.

Arnold, B.F., Colford Jr, J.M., 2007. Treating water with chlorine at point-of-use to improve water quality and reduce child diarrhea in developing countries: a systematic review and meta-analysis. The Am. J. Trop. Med. Hyg. 76 (2), 354–364.

Arshadi, M., Foroughifard, S., Gholtash, J.E., Abbaspourrad, A., 2015. Preparation of iron nanoparticles-loaded Spondias purpurea seed waste as an excellent adsorbent for removal of phosphate from synthetic and natural waters. J. Colloid Interface Sci. 452, 69–77.

Badruddoza, A.Z.M., Shawon, Z.B.Z., Rahman, M.T., Hao, K.W., Hidajat, K., Uddin, M.S., 2013. Ionically modified magnetic nanomaterials for arsenic and chromium removal from water. Chem. Eng. J. 225, 607–615.

Brooks, B.W., Lazorchak, J.M., Howard, M.D., Johnson, M.V.V., Morton, S.L., Perkins, D.A., et al., 2016. Are harmful algal blooms becoming the greatest inland water quality threat to public health and aquatic ecosystems? Environ. Toxicol. Chem. 35 (1), 6−13.

Chaussemier, M., Pourmohtasham, E., Gelus, D., Pécoul, N., Perrot, H., Lédion, J., et al., 2015. State of art of natural inhibitors of calcium carbonate scaling. A review article. Desalination 356, 47−55.

Chen, M., Bao, C., Cun, T., Huang, Q., 2017. One-pot synthesis of ZnO/oligoaniline nanocomposites with improved removal of organic dyes in water: effect of adsorption on photocatalytic degradation. Mater. Res. Bull. 95, 459−467.

Chowdhury, S., Balasubramanian, R., 2014. Recent advances in the use of graphene-family nanoadsorbents for removal of toxic pollutants from wastewater. Adv. Colloid Interface Sci. 204, 35−56.

Christin, M.S., Menard, L., Gendron, A.D., Ruby, S., Cyr, D., Marcogliese, D.J., et al., 2004. Effects of agricultural pesticides on the immune system of Xenopus laevis and Rana pipiens. Aquat. Toxicol. 67 (1), 33−43.

Crini, G., 2006. Nonconventional low-cost adsorbents for dye removal: a review. Bioresour. Technol. 97 (9), 1061−1085.

Deborde, M., Von Gunten, U.R.S., 2008. Reactions of chlorine with inorganic and organic compounds during water treatment—kinetics and mechanisms: a critical review. Water Res. 42 (1−2), 13−51.

Edberg, S.C.L., Rice, E.W., Karlin, R.J., Allen, M.J., 2000. *Escherichia coli*: the best biological drinking water indicator for public health protection. J. Appl. Microbiol. 88 (S1), 106S−116S.

Fosso-Kankeu, E., Mishra, A.K., 2017. Photocatalytic degradation and adsorption techniques involving nanomaterials for biotoxins removal from drinking water. Water Purif. Academic Press, pp. 323−354.

Friedmann, D., Mendive, C., Bahnemann, D., 2010. TiO2 for water treatment: parameters affecting the kinetics and mechanisms of photocatalysis. Appl. Catal. B: Environ. 99 (3−4), 398−406.

Fu, X., Chen, X., Wang, J., Liu, J., 2011. Fabrication of carboxylic functionalized superparamagnetic mesoporous silica microspheres and their application for removal basic dye pollutants from water. Microporous Mesoporous Mater. 139 (1−3), 8−15.

Fujishima, A., Zhang, X., Tryk, D.A., 2008. TiO2 photocatalysis and related surface phenomena. Surf. Sci. Rep. 63 (12), 515−582.

Gao, W., Singh, N., Song, L., Liu, Z., Reddy, A.L.M., Ci, L., et al., 2011. Direct laser writing of micro-supercapacitors on hydrated graphite oxide films. Nat. Nanotechnol. 6 (8), 496.

Gaya, U.I., Abdullah, A.H., 2008. Heterogeneous photocatalytic degradation of organic contaminants over titanium dioxide: a review of fundamentals, progress and problems. J. Photochem. Photobiol. C: Photochem. Rev. 9 (1), 1−12.

Gibbons, J., Laha, S., 1999. Water purification systems: a comparative analysis based on the occurrence of disinfection by-products. Environ. Pollut. 106 (3), 425−428.

Goel, G., Kaur, S., 2012. A study on chemical contamination of water due to household laundry detergents. J. Hum. Ecol. 38 (1), 65−69.

Griggs, D., Stafford-Smith, M., Gaffney, O., Rockström, J., Öhman, M.C., Shyamsundar, P., et al., 2013. Policy: sustainable development goals for people and planet. Nature 495 (7441), 305.

Hayat, H., Mahmood, Q., Pervez, A., Bhatti, Z.A., Baig, S.A., 2015. Comparative decolorization of dyes in textile wastewater using biological and chemical treatment. Sep. Purif. Technol. 154, 149−153.

Henze, M., Harremoes, P., la Cour Jansen, J., Arvin, E., 2001. Wastewater Treatment: Biological and Chemical Processes. Springer Science & Business Media, Berlin, Germany.

Horváth, O., Szabó-Bárdos, E., Zsilák, Z., Bajnóczi, G., 2012. Application of photocatalytic procedure combined with ozonation for treatment of industrial wastewater-a case study. Period. Polytech. Chem. Eng. 56 (2), 49−54.

Huang, J., Wu, Z., Chen, L., Sun, Y., 2015. The sorption of Cd (II) and U (VI) on sepiolite: a combined experimental and modeling studies. J. Mol. Liq. 209, 706−712.

Huang, B., Qi, C., Yang, Z., Guo, Q., Chen, W., Zeng, G., et al., 2017. Pd/Fe3O4 nanocatalysts for highly effective and simultaneous removal of humic acids and Cr (VI) by electro-Fenton with H_2O_2 in situ electro-generated on the catalyst surface. J. Catal. 352, 337−350.

Jain, R., Mathur, M., Sikarwar, S., Mittal, A., 2007. Removal of the hazardous dye rhodamine B through photocatalytic and adsorption treatments. J. Environ. Manage. 85 (4), 956−964.

Johnson, D.B., Hallberg, K.B., 2005. Acid mine drainage remediation options: a review. Sci. Total Environ. 338 (1−2), 3−14.

Jurado-Sánchez, B., Sattayasamitsathit, S., Gao, W., Santos, L., Fedorak, Y., Singh, V.V., et al., 2015. Self-Propelled Activated Carbon Janus Micromotors for Efficient Water Purification. Small 11 (4), 499–506.

Khayyat Sarkar, Z., Khayyat Sarkar, V., 2018. Removal of Mercury (II) from Wastewater by Magnetic Solid Phase Extraction with Polyethylene Glycol (PEG)-Coated Fe3O4 Nanoparticles. Int. J. Nanosci. Nanotechnol. 14 (1), 65–70.

Kobielska, P.A., Howarth, A.J., Farha, O.K., Nayak, S., 2018. Metal–organic frameworks for heavy metal removal from water. Coord. Chem. Rev. 358, 92–107.

Lee, W.M., Kim, S.W., Kwak, J.I., Nam, S.H., Shin, Y.J., An, Y.J., 2010. Research trends of ecotoxicity of nanoparticles in soil environment. Toxicol. Res. 26 (4), 253–259.

Lim, J.Y., Mubarak, N.M., Abdullah, E.C., Nizamuddin, S., Khalid, M., 2018. Recent trends in the synthesis of graphene and graphene oxide based nanomaterials for removal of heavy metals-A review. J. Ind. Eng. Chem. 66, 29–44.

Mnaya, B., Mwangomo, E., Wolanski, E., 2006. The influence of wetlands, decaying organic matter, and stirring by wildlife on the dissolved oxygen concentration in eutrophicated water holes in the Seronera River, Serengeti National Park, Tanzania. Wetlands Ecol. Manage. 14 (5), 421–425.

Mu, C., Zhang, Y., Cui, W., Liang, Y., Zhu, Y., 2017. Removal of bisphenol A over a separation free 3D Ag_3PO_4-graphene hydrogel via an adsorption-photocatalysis synergy. Appl. Catal. B: Environ. 212, 41–49.

Ngomsik, A.F., Bee, A., Talbot, D., Cote, G., 2012. Magnetic solid–liquid extraction of Eu (III), La (III), Ni (II) and Co (II) with maghemite nanoparticles. Sep. Purif. Technol. 86, 1–8.

Nyamukamba, P., Tichagwa, L., Okoh, O., Petrik, L., 2018. Visible active gold/carbon co-doped titanium dioxide photocatalytic nanoparticles for the removal of dyes in water. Mater. Sci. Semicond. Process. 76, 25–30.

Qiao, J., Cui, Z., Sun, Y., Hu, Q., Guan, X., 2014. Simultaneous removal of arsenate and fluoride from water by Al-Fe (hydr) oxides. Front. Environ. Sci. Eng. 8 (2), 169–179.

Reid, G.K., Wood, R.D., 1976. Ecology of Inland Waters and Estuaries, Second Ed D. Van Nostrand Company, New York.

Savage, N., Diallo, M.S., 2005. Nanomaterials and water purification: opportunities and challenges. J. Nanopart. Res. 7 (4–5), 331–342.

Shin, M., Lee, H.J., Kim, M.S., Park, N.B., Lee, C., 2017. Control of the red tide dinoflagellate Cochlodinium polykrikoides by ozone in seawater. Water Res. 109, 237–244.

Tan, L., Huang, C., Peng, R., Tang, Y., Li, W., 2014. Development of hybrid organic–inorganic surface imprinted Mn-doped ZnS QDs and their application as a sensing material for target proteins. Biosens. Bioelectron. 61, 506–511.

Tansel, B., 2008. New technologies for water and wastewater treatment: a survey of recent patents. Recent Pat. Chem. Eng. 1 (1), 17–26.

Tibbetts, R.S., Brumbaugh, K.M., Williams, J.M., Sarkaria, J.N., Cliby, W.A., Shieh, S.Y., et al., 1999. A role for ATR in the DNA damage-induced phosphorylation. Genes Dev. 13 (2), 152–157.

Villanueva, C.M., Cantor, K.P., Cordier, S., Jaakkola, J.J., King, W.D., Lynch, C.F., et al., 2004. Disinfection byproducts and bladder cancer: a pooled analysis. Epidemiology 15 (3), 357–367.

Weiss, S., Jakobs, J., Reemtsma, T., 2006. Discharge of three benzotriazole corrosion inhibitors with municipal wastewater and improvements by membrane bioreactor treatment and ozonation. Environ. Sci. Technol. 40 (23), 7193–7199.

Yagub, M.T., Sen, T.K., Afroze, S., Ang, H.M., 2014. Dye and its removal from aqueous solution by adsorption: a review. Adv. Colloid Interface Sci. 209, 172–184.

Yang, H., Feng, Q., 2010. Direct synthesis of pore-expanded amino-functionalized mesoporous silicas with dimethyldecylamine and the effect of expander dosage on their characterization and decolorization of sulphonated azo dyes. Microporous and Mesoporous Mater. 135 (1–3), 124–130.

Zhang, J., Dong, S., Zhang, X., Zhu, S., Zhou, D., Crittenden, J.C., 2018. Photocatalytic removal organic matter and bacteria simultaneously from real WWTP effluent with power generation concomitantly: using an ErAlZnO photo-anode. Sep. Purif. Technol. 191, 101–107.

Zvizdić, S., Rodinis-Pejić, I., Avdić-Kamberović, F., Mujkić, A., Hamzić, S., Puvacić, S., et al., 2005. Viruses in water. Med. Arh. 59 (6), 378–381.

Process intensification of treatment of inorganic water pollutants

Kailas L. Wasewar[1], Surinder Singh[2] and Sushil Kumar Kansal[2]

[1]Advance Separation and Analytical Laboratory (ASAL), Department of Chemical Engineering, Visvesvaraya National Institute of Technology (VNIT), Nagpur, India [2]Dr. S. S. Bhatnagar University Institute of Chemical Engineering and Technology, Panjab University, Chandigarh, India

O U T L I N E

Inorganic Pollutants in Water
DOI: https://doi.org/10.1016/B978-0-12-818965-8.00013-5

13.1 Introduction

Water pollutants can be classified as organic pollutants, inorganic pollutants, pathogens, suspended solids, nutrients and agriculture pollutants, thermal, radioactive, and other pollutants. Organic and inorganic pollutants are mainly discharged from industrial effluents and sewage into the water bodies.

Environment Protection Agency (EPA) of the United States has listed various water quality parameters to quantify the water quality (EPA, 2001). A few inorganic materials present in that list are arsenic, antimony, boron, beryllium, barium, chloride, calcium, copper, cadmium, chromium, cobalt, lead, iron, fluoride, manganese, molybdenum, magnesium, mercury, nitrate, nickel, nitrite, phosphates, potassium, phosphorus, salmonella, selenium, silica, sodium, silver, sulfate, sulfide, tin, tellurium, thallium, titanium, uranium, tritium, vanadium, zinc, and so on.

These materials, in the form of elements or in combination with other compounds, may be considered as inorganic pollutants if their limit exceeds permissible values, which in turn harms the environment. Heavy metal and other inorganic pollutants such as trace elements, mineral acids, sulfates, inorganic salts, metals, complexes of metals with organic compounds, and cyanides of higher concentrations pollute water bodies. These inorganic impurities are nonbiodegradable and pose threats to aquatic flora and fauna and public health (Ghangrekar, 2012). Wastewater treatment technologies are mainly classified as ecotechnologies, activated sludge technologies, anaerobic technologies, biofilm technologies, advanced oxidation processes, and membrane technologies.

Chemical and allied industries comprise of basic chemical manufacturing industries like inorganic/organic chemicals, bulk petrochemicals, pharmaceutical products and their intermediates, polymers and their derivatives, agricultural chemicals, acids, alkali, dyes, paper and pulp, and fertilizers. The chemical industries have significant impact on the environment due to pollution issues. Wastewater from chemical industry contains mainly organic and inorganic pollutants. These pollutants are toxic, mutagenic, carcinogenic, and mostly nonbiodegradable (Awaleh and Soubaneh, 2014). Complete treatment of effluents generated in the various chemical industry units in effluent treatment plant is essential and the principles of process intensification (PI) can be used for the effluent treatment. Several physical, chemical, and biological processes have been considered for treatment of wastewater obtained from chemical, biological, food, pharmaceutical, pulp and paper, dye and textile industries. The choice of methods for treatment of wastewater is based on the type, nature, and concentration of contaminants. The treated effluent should be eco-friendly and reusable (Mohajerani et al., 2009).

Inorganic and organic pollutants are main contaminants in wastewater. Organic water pollutants include bacteria and other organic pollutants from sewage, fertilizers, agricultural runoffs, forestry, food processing, tree and brush debris, industrial waste, and so on. Inorganic water pollutants include inorganic salts, mineral acids, metals, trace elements, metal compounds, complexes of metals with organic and compounds, sulfates, cyanides, acid rain caused by industrial or volcanic discharges, acid mine drainage, acid pollution of lakes by runoff from acid soils, volcanic or mineral, carbon dioxide discharges and runoff, chemical waste industrial byproducts, and so on. Chemical industry is one of the major contributors for wastewater (Nasr et al., 2007).

Heavy metals are major inorganic pollutants, which include lead, mercury, chromium, vanadium, arsenic, copper, nickel, cadmium, molybdenum, and zinc having sources from industrial coolants, mining, leather tanning, chromium salts manufacturing, e-waste, paints, smelters, ceramics, bangle industry, thermal power plants, chlor-alkali plants, fluorescent lamps, electrical appliances, hospital waste (damaged thermometers, barometers, sphygmomanometers), geogenic/natural processes, fuel, electroplating, mining, sulfuric acid plant, spent catalyst, battery industry, and so on (Verma and Dwivedi, 2013).

PI is the chemical engineering approach toward smaller, safer and cost-effective development in process industry which may be at the plant level or equipment level. This approach can be successfully implemented for the treatment of inorganic pollutants from wastewater. PI is used in chemical and allied industries to increase the output/efficiency of chemical units, reduce carbon footprint and wastes, increase plant safety, and decrease capital cost and energy consumption.

13.2 Process intensification

PI comprises of migration from large-scale processing unit to small-scale processing units, with increased productivity and decreased costs and resources. PI has innovative principles, which may result in less capital and operating cost, reduction in waste and energy with increase efficiency and improved safety (USDE, 2015). Dr. Darlene Schuster (AIChE) mentioned, PI is the interconnection of many fundamental areas of chemical engineering (USDE, 2015).

PI systems or plants follow the concept of "much in little" or the advent of miniaturization in process industry. Hence miniaturization has been identified as the prime end result of PI. Some of the applications of PI are miniature (tiny) reactors and power turbines, fuel handling and processing devices and high-volume processing reactors and other equipment. PI can be achieved by assembling process units and integrating various conventional unit operations, for example, mixing, heating, mass transfer, and reactions conducted in single, small-scale equipment. PI principles are scale-independent and their effective application to commercial processes has led to the fruitful applications, which can in future limit the use of conventional equipment.

The driving forces for the PI are mainly costs, safety, technical feasibility, market time, and company image (Rani and Sumuna, 2015). PI has been defined by various scientists/engineers such as Ramshaw (1983), Heggs (1983), Stankiewicz and Moulijn (2000, 2004), Stankiewicz (2000), Grünewald and Agar (2004), Charpentier (2007a,b), Becht et al. (2009) etc. and few of these are summarized in Table 13.1. These definitions can be summarized as technically feasible, eco-friendly, and cost-effective development in process industry.

PI has many motivations for processes like increased productivity, increased capacity building, enhanced safety, flexibility, and decreased energy usage, operational costs and less generation of waste with simplified processing (Simon et al., 2008; Lutze et al., 2010; Rong et al., 2000). Like any other technology or new development, PI also faced various barriers of reliability, risk, cost, safety, control, know-how, validation, and evaluation

TABLE 13.1 Various statements for defining process intensification.

	Reference	Statement
1.	Ramshaw (1983)	Devising an exceedingly compact plant, which reduces both the main plant as well as installation costs
2.	Heggs (1983)	PI is concerned with order-of-magnitude reduction in process plant and equipment
3.	Ramshaw (1995)	A strategy for making dramatic reductions in the size of a chemical plant so as to reach a given production objective
4.	Stankiewicz and Moulijn (2000, 2004)	PI comprises novel equipment, processing techniques, and process development methods that, compared to conventional ones, offer substantial improvements in (bio) chemical manufacturing and processing
		The development of novel and sustainable equipment that compared to the existing state-of-the-art, produces dramatic process improvements related to equipment sizes, waste production, and other factors
		Any chemical engineering development that leads to a substantially smaller, cleaner, and more energy efficient technology
5.	Stankiewicz (2000)	Consisting of development of innovative apparatuses and techniques that offer drastic improvements in chemical manufacturing and processing, substantially decreasing equipment volume, energy consumption, or waste formation and ultimately leading to cheaper, safer and sustainable technologies
6.	Dautzenberg and Mukherjee (2001)	The strategy of making significant reductions in the size of a chemical plant in order to achieve a given production objective. Innovations in catalytic reactors, which constitute the heart of such process technologies, are often the preferred starting point
7.	Tsouris and Porcelli (2003)	Technologies that replace large, expensive, energy-intensive equipment or process with ones that are smaller, less costly, more efficient or that combine multiple operations into fewer devices (or a single apparatus)
8.	Grünewald and Agar (2004)	They used PI and process integration for reaction engineering with range of techniques to get high performance of reactor
		They defined PI as a path for the future of chemical and process engineering demands in the context of globalization and sustainability
9.	Criscuoli and Drioli (2007)	Performed in terms of the following ratios: productivity/size; productivity/weight; flexibility; and modularity
10.	Becht et al. (2009)	They considered PI as an integrated approach for process and product innovation
		An integrated approach for process and product innovation in chemical research and development, and chemical engineering in order to sustain profitability even in the presence of increasing uncertainties
11.	Lutze et al. (2010)	Achieved by adding/enhancing phenomena in a process through the integration of operations, functions, phenomena or alternatively through the targeted enhancement of phenomena in an operation
		Potential for process improvement, to meet the increasing demands for sustainable production

(Continued)

TABLE 13.1 (Continued)

Reference	Statement
12. Ponce-Ortega et al. (2012)	An activity characterized by five principles—reduced size of equipment, increased throughput of process, reduced equipment holdup or inventory, reduced usage of utilities and raw materials, and, increase efficiency of process equipment
13. Boodhoo and Harvey (2013)	A promising field which can effectively tackle the challenges of significant process enhancement, whilst also offering the potential to diminish the environmental impact presented by the chemical industry
14. Lutze et al. (2013)	The targeted improvement of a process at the unit operations scale, the task scale, and/or the phenomena scale
15. Reay et al. (2013)	Process development that involves reduction in equipment (unit operation) sizes that lead to improvements in reaction kinetics, better energy efficiency, reduction in capital cost, and improvement in process safety
	Significantly improves the transport rates, it gives every molecule the same process experience. It could be achieved through improved control of reactor kinetics giving the higher selectivity/reduced wastes products, higher energy efficiency, reduced capital costs, and reduced inventory/improved intrinsic safety
16. Portha et al. (2014)	Holistic overall process-based intensification (i.e., global process intensification) in contrast to the classical approach of process intensification based on the use of techniques and methods for the drastic improvement of the efficiency of a single unit or device
17. Baldea (2015)	Any chemical engineering development that leads to substantially smaller, cleaner, safer and more energy efficient technology or that combine(s) multiple operations into fewer devices (or a single apparatus)
18. Mohunta (2015)	The process for improving profitability and remaining competitive

criteria (Rong et al., 2000; Babi et al., 2016; Stankiewicz, 2003a,b; Lutze et al., 2010; Charpentier, 2005; Harmsen, 2007; Becht et al., 2009). PI can be achieved by using one or more of the associated principles of PI. Lutze et al. (2010) have given four principles: integration of various operations (mass/heat/reaction/mechanical/other), integration of functions (separation/mixing/reaction/heat transfer), integration of phenomena and/or, targeted enhancement (conversion/selectivity/yield/separation factor) of a phenomenon of a given operation. These principles have been discussed exhaustively with suitable examples by Lutze et al. (2010).

Stankiewicz and Moulijn (2004) have classified PI mainly into two components, that is, equipment and methods. Rani and Sumuna (2015) classified further in a boarder way. It has multifunctional operations (with and without reaction), novel devices, and alternative configurations (resources and modes of operations). More details of its components are available elsewhere (Stankiewicz and Moulijn, 2004; Rani and Sumuna, 2015). Broad classification of PI is summarized in Fig. 13.1.

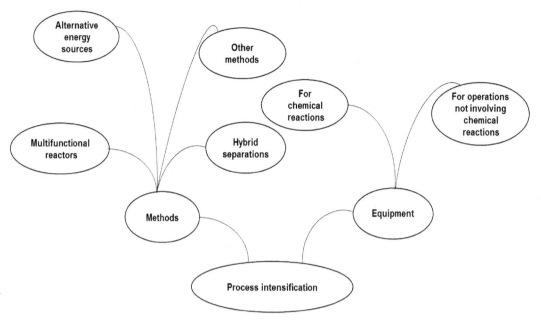

FIGURE 13.1 Broad classification of process intensification.

13.3 Process intensification for wastewater treatment

Understanding the quantitative and qualitative nature and characteristics of pollutants in water is essential for discussing and deciding suitable wastewater treatment methods and equipment. Further, the characteristics of water pollutants depend on source of pollution such as industrial, municipal, agriculture, and others. Based on nature of water pollutants, mainly it can be categorized as inorganic, organic, and biological. Heavy metals are the most common carcinogenic and toxic inorganic water pollutants. Moreover, some serious hazardous effects are also observed for sulfates, nitrates, phosphates, chlorides, fluorides, and oxalates as inorganic pollutants (Gupta et al., 2012).

Wastewater treatment technologies for inorganic water pollutants and also for organic pollutant treatment include stages of primary, secondary, and tertiary treatments. The technologies involved in these stages are started with screening and ended with recycling and reuse of water such as: centrifugal separation, filtration, gravity separation, sedimentation, coagulation, flotation, flocculation, aerobic/anaerobic processes, membrane separation, distillation, evaporation, crystallization, extraction, precipitation, oxidation, ion exchange, reverse osmosis, adsorption, electrolysis, and electrodialysis (Gupta et al., 2012).

Lots of novel opportunities are available in these technologies for PI which can be successfully implemented for wastewater treatment. The prominent advantages of PI in the domain of wastewater treatment comprises of micro/minireactors with enhanced output, process-oriented flow patterns, higher dispersion number, relatively less power inputs, compactness, recyclability, fever pollutant emissions and higher mass transfer effects. There is dire need of comprehensive approach toward wastewater treatment methodologies to enable these techniques an effective, vigorous, eco-friendly, and economically viable application for various water treatment requirements. This chapter addresses the key advancements in wastewater treatment domain, explaining the system design and component based view point, which cardinally affects the absolute solution and capability to address the overall requirements.

PI can be applied using intensified equipment and methods in chemical manufacturing, biofuel production, power generation, petroleum refining, oil and gas extraction, water treatment and recycling (Kim et al, 2017). In water treatment, it can be used as anaerobic membrane digesters, desalination equipment, water and sewage treatment, metal recycling, electronics recycling, and so on (Kim et al., 2017). Water scarcity is the main issue in many countries due to water pollution. PI can be applied to various water and wastewater treatment including membrane digester, desalination, and other methods to enhancing the treatment rate with greater efficiency. This may solve the water problem to certain extent with reliable water treatment. (Kim et al., 2017)

13.4 Intensifying approaches for inorganic water pollutants treatment

Various intensifying approaches have been used for treatment of inorganic water pollutants such as membrane-based methods, membrane-based hybrid methods, emulsion liquid membranes, sonochemical reactor, reactive extraction, reactive crystallization/precipitation, membrane distillation, membrane extraction, wet air oxidation with sonication, electrochemical and photochemical hybrid method, supercritical method, use of algae, enhanced solar evaporation, inverse fluidized bed, capacitive deionization, and use of nanotechnology.

13.4.1 Membrane-based approaches

Membrane-based operations involves membrane (bio)reactors, emulsion liquid membranes and hybrid separations (membrane distillation, membrane absorption, membrane adsorption, membrane extraction, membrane crystallization), including basic operations like nanofiltration, microfiltration, ultrafiltration, and reverse osmosis (Bartels et al., 2005). The typical membrane separation methods are summarized in Fig. 13.2.

Various functions of membrane-based operation has been listed by Sirkar et al. (1999) include product separation from reaction mixture, reactant separation from a mixed stream to feed, reactant(s) added in controlled manner, phase contact by nondispersive path, catalyst segregation, catalyst immobilized on a membrane, membrane as a catalyst, membrane as reaction medium, and also solid electrolyte membranes.

Various membrane-based approaches as mentioned earlier have been discussed separately in further sections.

13.4.2 Reverse osmosis

Reverse osmosis has contributed for around 50% desalination capacity of the world (Górak and Stankiewicz, 2011). Reverse osmosis has inherent characteristics of high

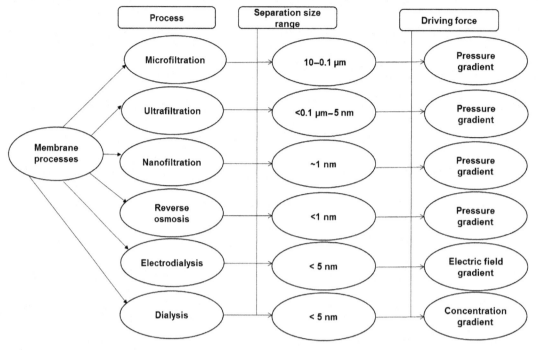

FIGURE 13.2   Membrane Separation methods.

efficiency and simplicity in operation, high permeability, and selectivity for specific solute transport among various components. It also offers compatibility between diverse and integrated membrane operations in specific systems, less energy demand, stability for wide operating conditions, compatibility with environment, ease of control and commercial scale-up, and massive flexibility.

Reverse osmosis has been successfully used for the removal of calcium, magnesium, potassium, sodium, sulfates, chlorides, fluorides, phosphorous, and nitrates with more than 98% rejection (Bartels et al., 2003; Hamoda, 2013).

13.4.3 Membrane extraction

In membrane extraction, a liquid or solid membrane is used to separate the treated aqueous phase and the organic solvent-extractant phase (Stankiewicz and Moulijn, 2004). This method can be used in wastewater treatment for pollutants removal and recovery of various trace components. Heavy metals have been recovered from wastewater by membrane extraction (Kim, 1984a,b, 1985; Degener, 1988; Jansen et al., 1992; Boyadzhyev and Lazarova, 1993; Hu and Wiencek, 1998).

The microporous membranes can be used for separation of inorganic metal pollutants. This membrane acts as an interface between organic and aqueous phase. Microporous membrane is nothing but liquid ion exchangers in organic solution. Metal ions present in aqueous phase are moved at the interface developed in the pores of membrane available in organic phase. The metal loaded organic phase is contacted with the stripping phase. The concentrated copper solution was obtained using LIX 64N. It can be successfully applied for the selective recovery of metals from various steams including metal-containing wastewater, ore leachates, and so on (Kim, 1984a,b, 1985).

Mesoporous materials can be effectively used for separation of inorganic pollutants (Kim and Yi, 2004). It has typically uniform pore channel of $1-10$ nm with 1000 m^2/g of surface area having 2D hexagonal structure or 3D interconnected pore system. These are highly thermal stable and easy to functionalized internal and external surface by self-assembled monolayer. It was used for the first time for environmental remediation for the exclusion of aqueous phase heavy toxic metal, which include mercury, lead, cadmium, and copper (Lee et al., 2000, 2002; Chah et al., 2002; Bae et al., 2000; Kim and Yi, 1999, Kim et al, 2000a,b). Metal removal efficiency was reliant on the functional groups available on the surface of the membrane. The amine-functional group having concentration of 2.43 mmol/g was used as an adsorbent obtained uptake of 0.11 mmol/g Cu ion uptake with 0.045 molar ratio of the copper ion adsorbed to the amine-functional group (Kim et al., 2003).

13.4.4 Emulsion liquid membrane

Preparation of emulsion liquid membranes involves emulsion of two immiscible phases and dispersing it in the third phase. The obtained emulsion may be oil−water−oil or water−oil−water. The liquid membrane separates the encapsulated drops in the emulsion from the external continuous phase (Chakraborty et al., 2005). It can be considered for wastewater treatment, fractionation of hydrocarbons, and other allied processes.

Emulsion liquid membrane technique can be applied for the separation of heavy metals from wastewater. Faster permeation of metals can be obtained due to high specific interfacial area in emulsion liquid membrane. It has high efficiency and is simple in operation, having extraction and stripping effects in single stage and can also be used in continuous manner (Chakraborty et al., 2005).

Generally, the emulsion consist of three phases as surfactant-extractant (carrier), internal phase (stripping solution), and membrane phase (diluent) for removal of heavy metals (Ahmad et al., 2011). Emulsion instability arises due to a variety of physical mechanisms for instance coalescence, swelling and leakage. The mechanism of swelling and leakage are the main trouble in real application of emulsion liquid membrane which breaks emulsion and decreases the efficiency. Swelling is originated through difference in ionic strength, surfactant, residence time, carrier, viscosity of diluent, temperature, pH, water volume fraction in W/O emulsion, globules size, acidity and salinity of aqueous phase, droplet size, and agitation speed. The membrane leakage is mainly exaggerated by the properties of the diluent, surfactant, internal phase and its volume fraction, membrane material, electrolyte concentration, emulsion preparation procedure, and stirring speed. To get the stable emulsion and high area for mass transfer, the small diameter of droplet of emulsion is a key decisive factor (Ahmad et al., 2011).

Emulsion liquid membrane can be applied for extraction of chromium (VI) (Hochhauser and Cussler, 1975; Fuller and Li, 1984; Salazar et al., 1990, 1992; Mori et al., 1990;

Banerjea et al., 2000; Bhowal and Dutta, 2001; Chakraborty et al., 2005), copper (II) (Strzelbicki and Charewicz, 1980; Chakraborty et al., 2003), nickel (II) and zinc (II) (Fuller and Li, 1984)

13.4.5 Membrane-based hybrid separations

Membrane-based hybrid separations is the method (Fig. 13.3) in which one or more separation techniques with membrane-based separation are integrated in a single operation like membrane distillation and membrane absorption.

In membrane distillation, distillation is combined with membrane separation. It is an integrated mass-transfer operation. In this hybrid separation, vapor pressure difference transverses due to temperature difference which allows the selective molecular transport through the membrane. It can be considered for wastewater treatment from effluents discharged in water bodies.

Integrating membrane separation with absorption gives selective removal of the components from a multicomponent gaseous feed. Membrane absorption provides bubble free gas—liquid mass transfer in separation having better application in biological systems. It may be used for removal of CO_2, H_2S, NH_3, and other compounds.

Adsorption with membrane is the membrane adsorption separation which enhances separation efficiency by functional ligand and convective flow. Membrane extraction having extraction using liquid/solid membrane is used for separation of selective components. Specific compound or pollutants can be removed from wastewater by membrane extraction.

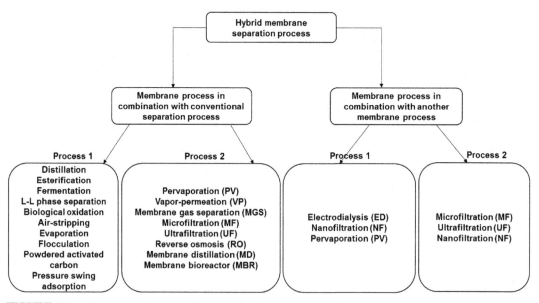

FIGURE 13.3   Membrane-based hybrid separation methods.

Gorak and Stankiewicz (2011) have suggested few novel milestones in view of the development in new/improved membrane technologies. They coined the idea of membrane crystallization and use of carbon nanotube, aquaporin channels and new protein-based membranes.

Qiu et al. (2016) have investigated the application of hybrid microfiltration-osmosis membrane bioreactor to treat nitrogen and organic matter from municipal sewage. The obtained results manifested decrease in fouling and decreased bacterial deposits. Utilizing the combination of UF/NF/RO proved beneficial to reduce the membrane fouling and removal of nitrogen and phosphorous from rendering plant wastewater (Racar et al., 2017).

13.4.6 Sonochemical reactor

Sonochemical reactors utilize the large amount of energy released from ultrasonic radiations. These reactors are acoustic cavitation based reactors having a number of advantages as compared with conventional reactors (Gogate et al., 2003). The problem of cavitation arises as the liquid pressure in the reactor varies due to acoustic waves. The liquid comes across an abrupt pressure drop at downstream resulting generation of cavities followed by subsequent growth and then collapse of the cavities, which result into release of very high energy (may be in the order of $1-10^{18}$ kW/m^3). Generation of very high temperature and pressure locally can be observed due to occurrence of cavitation simultaneously at millions of locations in the reactor. In acoustic cavitation, high-frequency ultrasound waves (16 kHz$-$100 MHz) produces pressure variation in the liquid (Gogate et al., 2003). Various applications are mentioned by Gogate et al. (2003) such as in chemical synthesis (Ando et al., 1984; Suslick, 1986; Javed et al., 1995; Lie Ken Jie and Lam, 1995), water and wastewater treatment (Gogate, 2002), polymer chemistry (Kruus et al., 1988), sono-electrochemistry, textile industry, solid$-$liquid extraction, crystallization, and in petroleum industry.

Sonochemical reactors can be one of the most important intensifying approaches for wastewater treatment. Gogate et al. (2002) have given exhaustive review on application of sonochemical reactors in wastewater treatment. Contaminants may be effectively destructed by ultrasonic irradiation due to high concentrations of oxidizing species, high temperature and pressure at local level. It can be used for the treatment of potassium iodide, sodium cyanide, and carbon tetrachloride (Gogate et al., 2003). Sonochemical reactors possess a great potential for green and sustainable chemical processing and efficient scale-up to harness the effects of sonochemistry at commercial level.

13.4.7 Reactive extraction

Stankiewicz (2003a,b) has elaborated the concept and various applications of reactive separations for PI from an industrial perspective. Reactive extraction is a liquid-liquid extraction intensified by involving a reversible reaction between the solute and extractant. Reversible reaction is the main difference between reactive extraction and solvent extraction. Extractants play the vital role in reactive extraction for reversible reaction between

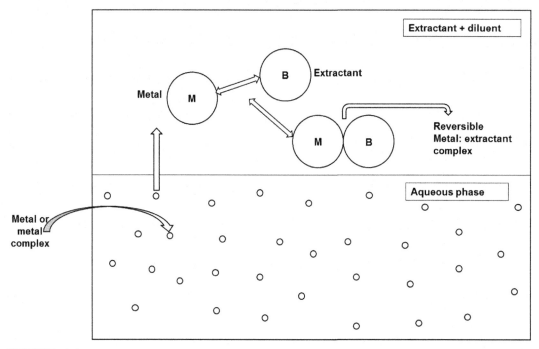

FIGURE 13.4   Typical mechanism for the removal of heavy metals by reactive extraction from aqueous phase.

solute and extractant, Further diluents also responsible for the salvation of the complex in the organic phase to attain certain level of extraction. It may be alternative for the separation of waste products or removal of waste from effluent (Minotti et al., 1998; Samant and Ng, 1999). Reactive extraction has been successfully employed for separation of toxic heavy metals from wastewater (Kueke, 2001; Elshani et al., 2001). Typical mechanism for the removal of heavy metals by reactive extraction from aqueous phase is depicted in Fig. 13.4.

13.4.8 Reactive crystallization/precipitation

In crystallization, solute gets separated in crystal forms by crystallization from liquid-solid solution. It is effective for recovery of various valuable and useful materials along with water recovery from wastewater (Lu et al., 2017). This may include (Lu et al., 2017): recovering Na_2SO_4 (Jiao and Jiao, 2015; Xu et al., 2010), NaCl (Nathoo et al., 2009), Na_2CO_3 (Spronsen et al., 2010; Himawan et al., 2006), phosphates (Rahman et al., 2014; Hutnik et al., 2013), and ammonium (Huang et al., 2014; Zhang et al., 2013); heavy metals removal (Pb^{2+}, Mn^{2+}, Ni^{2+}, Cu^{2+}, Ag^+, etc.) (Al-Tarazi, 2004; Rubio and Tessele, 1997); reclamation of pure water (Williams et al., 2015); and water softening by removing Ca^{2+} and Mg^{2+} (Rahman et al., 2011; Suzuki et al., 2002).

Reactive crystallization is the separation process by crystallization with reaction. Reactive crystallization has many industrial applications including wastewater treatment.

Reactive crystallization is employed often in pharmaceutical, fine chemicals, and related companies to formulate different intermediates and their finished products (Lu et al., 2017). In reactive crystallization, poorly soluble or insoluble substances are generated via reactions between a gas and a solution or between solutions, and thus can be separated from the system easily (Lu et al., 2017).

There are few separation technologies which may produce high purity products and reactive crystallization is one of them (Huang et al., 2014). It may be employed for the recovery and separation of: inorganic ions, such as PO_4^{3-} (Rahman et al., 2014; Hutnik et al., 2013), NH_4^{4+} (Rahman et al., 2014), heavy metals ions, Ni^{2+}, Zn^{2+}, Pb^{2+}, Co^{2+}, Cu^{2+}, Hg^{2+} (Rubio and Tessele, 1997; Janssen, 1978; Al-Tarazi, 2004), Ca^{2+} (Suzuki et al., 2002), and SO_4^{2-} (Tait et al., 2009), in wastewater treatment (Lu et al., 2017).

Reactive crystallization has been applied for the removal of ammonium and phosphate ions from the product stream of magnesium ammonium phosphate (Hirasawa, 1997). Deng et al. (2013) performed the reactive crystallization of calcium sulfate dihydrate from acidic wastewater and lime. Reactive crystallization process employed for effective separation and concentration of metal ions (Deng et al., 2013). Ammonium magnesium phosphate crystallization, calcium hydroxy phosphate crystallization, and vivianite crystallization can not only effectively remove phosphorus, but also recycle (Cheng et al., 2017).

13.4.9 Membrane distillation

Membrane distillation is an alternative method for separation of salt molecules from water and wastewater with high rejection factors. A microporous and hydrophobic membrane is used in membrane distillation to separate aqueous solutions at different compositions and temperatures (Stankiewicz, 2003a,b, 2004). At least one side of the membrane remains in contact with the liquid phase. Vapor pressure difference is developed due to temperature difference across the membrane. The molecules are transported through the pores of the membrane from the high vapor pressure side to the low vapor pressure side (Stankiewicz and Moulijn, 2004). Membrane distillation has been used for treatment of textile industry wastewater and radioactive wastewater (Calabro et al., 1991; Chmielewski et al., 1995). Membrane distillation has been utilized to remove fluoride from synthetic water (Naji et al., 2018). Results showed excellent rejection of contaminants (>98%) along with some fouling evident after approximately 25 hours of operation.

13.4.10 Membrane distillation crystallization

Membrane-distillation-crystallization is a hybrid process combines membrane distillation and crystallization. Membrane distillation is a thermally driven separation process, which is the integration of membrane and distillation technologies (Lu et al, 2017). In membrane distillation, a porous hydrophobic membrane uses which permits the water vapor through pores of the membrane which blocks the penetration of water. This leads to the separation of water from the concentrated solution (Shin and Sohn, 2016; Lu et al, 2017).

The desired supersaturation and product crystal formed after crystallization and water can be recovered via membrane distillation (Chan et al., 1997).

It has several advantages including (Lu et al, 2017): lower operating temperature and pressure, total nonvolatile solute rejection, water recovery with high purity, easy to couple with other sources of energy, for example, solar, geothermal, and low-grade heat sources (Charcosset, 2009; Lawson and Lloyd, 1997), and solute may be get up to saturation (Edwie and Chung, 2012).

It is mainly attracted for application in crystallization of protein (Profio et al., 2005). Though it can be employed to wastewater treatment, it may be considered to treat salinity of brine wastewater due to recovery of salts and water by reactive distillation crystallization (Charcosset, 2009).

The wastewater generated from taurine production, a pharmaceutical industry can be treated using membrane distillation crystallization (Wu et al., 1991). There are several challenges for its application, which need to be addressed further such as development of suitable membrane, energy consumption, temperature and concentration polarization, and overall module (Edwie and Chung, 2013; Lu et al., 2017; Guan et al., 2012).

13.4.11 Wet air oxidation with sonication

Ingale and Mahajani (1995) used the combination of acoustic sonication using ultrasonic cleaning bath followed by wet air oxidation to treat refractory waste. This is a kind of hybrid process for wastewater treatment. Sonication with wet air oxidation is better alternative compared to wet air oxidation or sonication alone. Degradation of potassium iodide and sodium cyanide from aqueous solution in the presence of carbon tetrachloride by acoustic cavitation was carried out by Shirgaonkar and Pandit (1997). Ultrasound enhanced degradation of rhodamine B studied by Sivakumar and Pandit (2001).

13.4.12 Electrochemical and photochemical hybrid method

Krystynik et al. (2015) proposed electrochemical and photochemical hybrid method for the treatment of wastewater. Combination of electro-coagulation and photo-oxidation methods leads to 100% decontamination of Fe from loading up to 50 mg/L, which resulted in process savings due to the cumulative effect of each process. The combined process showed a greater performance than the sum of each individual process.

13.4.13 Supercritical method

The wastewater treatment operations can be carried out at supercritical state (at or above critical temperature or/and pressure). At supercritical condition, water changes its polarity and becomes a nonpolar solvent which can be easily mixed with other organic part (Mishra et al., 2017). At supercritical condition, water can be mixed with the oxygen and creates homogeneous mixture to provide oxidation. This is supercritical water oxidation or hydrothermal oxidation (Mishra et al., 2017). Supercritical water oxidation or hydrothermal oxidation has steps viz. pressurizing the reagent, reaction, salt separation and depressurization. It is used for degradation of inorganic pollutants such as ammonia or cyanides which can also be converted to CO_2, H_2O, and N_2 (Mishra et al., 2017).

13.4.14 Use of algae

Algae can consume or accumulate in their cells/bodies nutrients from plants, pesticides, heavy metals, organic and inorganic toxic substances, and radioactive matters (Sahu, 2014). Hence algae can be used for wastewater treatment. It is a low-cost alternative to other complex and expensive treatment processes. Sahu (2014) observed a reduction of 67% of phosphorus, 71% total nitrogen, 51% of dissolved solids and 54% volatile solids in wastewater.

13.4.15 Enhanced solar evaporation

The solar transpired evaporation system was developed by the Solar Multiple Company (Reinsel, 2016). It has increased evaporation rates (about seven times) as compared to standard solar evaporation. In this technology, solar arrays are used to heat air and then air is bubbled through the water, till evaporated. Water vapor get away to the atmosphere and solids accumulate for further treatment of disposal in tanks (Reinsel, 2016). This technology has major disadvantage of water recovery but has several advantages including: higher efficacy than other methods, low cost (capital and operating), scalable, no carbon emission and simple design (Reinsel, 2016).

13.4.16 Inverse fluidized bed

Fluidized bed reactor has been used for the treatment and removal of sulfur from wastewater containing acidic metals. Carrier material may be sand in these cases (Kaksonen et al., 2006; Kaksonen et al., 2003). In the fluidized bed reactor, there is a biofilm developed on the particles that support bacterial growth and allows the retention of biomass within the reactor. Due to this fluidized bed reactor is operated at small hydraulic retention time (Shieh and Keenan, 1986; Marin et al., 1999).

Various types of materials are used as carrier materials such as: iron chips (Somlev and Tishkov, 1992), synthetic polymeric granules covered with iron dust (Somlev and Tishkov, 1992) and silicate minerals (Kaksonen et al., 2003).

Inverse fluidized bed reactor is the downflow fluidized bed reactor. In this reactor, floatable carrier material is fluidized downward with a downflow current of liquid. A biofilm develops over the support after inoculation that is left over at the top of the reactor. The rate of conversion of sulfate mainly depends on the adhered biomass and its formation on the material selected as carrier (Villa-Gomez, 2013; Nikolov and Karamanev, 1991).

Various materials are used as carrier materials for inverse fluidized bed having lower density than water, for example, polyethylene (Castilla et al., 2000; Celis-García et al., 2007; Celis et al., 2009; Gallegos-Garcia et al., 2009), cork (García-Calderón et al., 1998), polystyrene spheres (Nikolov and Karamanev, 1991) and extendosphere (Buffière et al.,

2000; Arnaiz et al., 2003). Celis et al. (2009) and Villa-Gomez et al. (2011, 2012) used inverse fluidized bed for sulfide and heavy metals removal.

13.4.17 Capacitive deionization

Capacitive deionization is an alternative desalination method to reverse osmosis and thermal desalination for water treatment (Kim et al., 2017). Capacitive deionization is based on electrosorption which operates at low voltage. Due to low voltage, energy-intensive electrochemical reactions are avoided (Sharma et al., 2015; Kim et al., 2017). Electrical energy can be regenerated and recycled during deionization which is stored in a capacitor. Hence capacitive deionization is more energy efficient than reverse osmosis, thermal desalination and other conventional processes (Sharma et al., 2015).

Various types of materials are developed and used as electrode including porous carbon materials, ion exchange membrane, carbon aerogel, graphite electrodes coated with mesoporous carbon (Tsouris et al., 2011; Kim et al., 2017). The salt removal efficiency can be increased by using membrane-based capacitive deionization and new type of electrodes (Kim and Choi, 2010; Jeon et al., 2013).

13.4.18 Nanotechnology

Nanotechnology has proven wide application in wastewater treatment (Amin et al., 2014). Research in the development of various nanoparticles and other nanomaterials for wastewater treatment can be made economic and process friendly (Savage and Diallo, 2005). Materials having 100 nm in one dimension are used in nanotechnology.

Due to significant change in their physical, chemical, and biological properties it is widely usable in chemistry, engineering, materials science (Masciangioli and Zhang, 2003; Eijkel and Van Den Berg, 2005) and in wastewater treatment (Rickerby and Morrison, 2007; Vaseashta et al., 2007). Nanosorbents, nanocatalysts, zeolites, dendrimers and nanostructured catalytic membranes are the type of nanomaterials which can be used to remove inorganic solutes and others from water/wastewater (Amin et al., 2014). Santhosh et al. (2016) has presented an exhaustive review on application of nanomaterials for wastewater treatment.

Carbon nanotubes can be used as an adsorbent for removal of various solutes. It was used for the removal of various heavy metals such as: Pb(II) (Li et al, 2002, 2003, 2005; Stafiej and Pyrzynska, 2008; Anitha et al., 2015), Ni(II) (Lu and Liu, 2006), Cd(II) (Li et al., 2003; Anitha et al., 2015), Zn(II) (Lu et al., 2006; Lu and Chiu, 2006), Cr(III) (Di et al., 2006; Gupta et al., 2011), Hg(II) (Anitha et al., 2015), Cu(II) (Anitha et al., 2015; Li et al., 2003), As(III) (Roy et al., 2013), As(IV) (Veličković et al., 2013) and for compounds of As (Peng et al., 2005).

Graphene is allotropic form of carbon which can be a substitute for carbon nanotube and an ideal material for water treatment (Santhosh et al., 2016). Graphene-based materials as adsorbents have distinct advantages (Sitko et al., 2013; Stafiej and Pyrzynska, 2007; Zhao et al., 2014; Kim and Yi, 2004) of two basal planes of single layered graphene for adsorption; easy for synthesis having no need for further purification. Graphene-based materials were used as adsorbents for the removal of inorganic pollutants (Huang et al., 2011; Zhao et al., 2011).

Graphene-based nanocomposites have been successfully used for the removal of heavy metals such as Cu(II) (Li et al., 2012; Nandi et al., 2013; Hu et al., 2013; Hur et al., 2015), Pb(II) (Huang et al., 2011; Lee and Yang, 2012; Madadrang et al., 2012; Hao et al., 2012; Wang et al., 2013), Cd(II) (Wang et al., 2013; Hu et al., 2014; Nandi et al., 2013; Hur et al., 2015) and Hg(II) (Sreeprasad et al., 2011).

Metal based nanomaterials such as nanosized magnetite and TiO_2 nanoparticles are used for adsorption of As(III) (Pena et al., 2005) and As(IV) (Mayo et al., 2007; Deliyanni et al., 2003). Various nanosized metal oxides as adsorbents such as manganese oxides, aluminum oxides, cerium oxides, titanium oxides, magnesium oxides, ferric oxides, etc., were used for the removal of Cu(II) (Grossl et al., 1994; Hu et al., 2006; Chen and Li, 2010), Pb (II) (Cao et al., 2010; Ma et al., 2010; Afkhami et al., 2010; Engates and Shipley, 2011), Ni(II) (Li and Zhang, 2006) and Cd(II) (Afkhami et al., 2010).

Nanocomposites can also be used to reduce/remove heavy metals. Sunlight assisted TiO_2 with graphene sheet was used for removal of chromium (Zhang et al., 2012). Similarly palladium composites were used to reduce chromium (Omole et al., 2009).

Iron oxide nanomaterials (Fe_2O_3, Fe_3O_4) as nanoiron-coated filter media, nanoiron-modified zeolite, nanomagnetic zeolites, nanocrystals were used for removal of As (Lai and Chen, 2001; Onyango et al., 2003; Oliveira et al., 2004; Yavuz et al., 2006).

Takafuji et al. (2004) used poly(1-vinylimidazole)-grafted magnetic (Fe_2O_3) nanoparticles for removal of divalent copper, nickel, and cobalt ions. Uranyl ions (UO_2^{2+}) from blood were removed by using bisphosphonate-modified magnetite nanoparticles (Wang et al., 2006).

Nanoscale zero-valent iron particles were used for transformation of Ni(II), Pb(II) (Ponder et al., 2000), As(III) (Kanel et al., 2005), Cr(VI) (Ponder et al., 2000), As(V) (Kanel et al., 2006) and metal nitrate (Yang and Lee, 2005) etc.

Zhong et al. (2006, 2007) used self-assembled 3D flowerlike iron oxide nanostructures and ceria micro/nanocomposite structure for adsorption of As(V) and Cr(VI).

Qi and Xu (2004) carried out Pb(II) adsorption from aqueous solutions on chitosan nanoparticles.

Nanotechnology can also be beneficial in membrane separations. Van der Bruggen and Vandecasteele (2003) carried out exhaustive review on separation of pollutants from surface water and groundwater by nanofiltration. It is effective for removal of arsenic and uranium (Van der Bruggen and Vandecasteele, 2003; Favre-Reguillon et al., 2003).

13.5 Discussion and conclusion

Water serves as an important ingredient for different purposes in the industry like solvent, reactant, as washing agent, stripping agent and boiler feed water etc. During and after processing the water utilized is often received as an industrial wastewater which needs to be treated in effluent treatment plants. This water contains many impurities like organic and inorganic compounds, salts, toxic and hazardous materials. Efficient treatment technologies like advanced oxidation processes and use of PI have been applied for wastewater treatment.

Wastewater treatment has four levels. Preliminary treatment removes bristly solids. Primary treatment removes settleable solids and part of organic matter. Secondary treatment removes organic matter and few nutrients. Tertiary treatment removes specific pollutants or complementary removal of pollutants. Typical conventional unit operations, unit processes and systems used for treatment of inorganic pollutants from wastewater are chemical precipitations, phytoremediation, ion exchange, reverse osmosis, membrane filtration, and electrodialysis.

The conventional treatment processes used in wastewater treatment are not efficient and sustainable due to factors like large land area usage, large number of unit operations involved, and associated problems of odor and other emissions (Keskinler and Akay, 2004). Also due to the increased demand of potable and raw water for industrial applications, the conventional treatment methods do not suffice, and hence novel treatment methods with improved treatment efficiency and throughput are required.

Hence it is vital to move beyond conventional water treatment methods (Kasher, 2009). The new approaches must have the characteristic of treating more and complex wastewater contaminants to meet environment standards, discharge norms and reusing of treated water. New approaches must be more robust, economical, and sustainable for growing requirements (Tayalia and Vijaysai, 2012).

Fraser (2019) stated that three strongest trends in wastewater treatment are nutrient removal and recovery, energy conservation and production, and PI, which involves making treatment facilities better, cheaper, faster, and in a smaller footprint. So intensification plays significant role in wastewater treatment. PI may resolve these issues, which are innovative ways having great benefits in process industry in terms of safer, efficient, flexible, cheaper, smaller, and more environment-friendly process (Charpentier, 2007a,b; Raghavan and Reddy, 2014). This approach considers all factors starting with an analysis of economic constraints to selection or development of an intensified production process. PI may also lead to improvement in performance of the process aiming to produce more from less (Tayalia and Vijaysai, 2012; Berne and Cordonnier, 1995; Ben-Guang et al., 2000; Reay et al., 2013). Tayalia and Vijaysai (2012) stated that PI in wastewater treatment can lead to flexible processes, reduced energy consumption, reduction in the waste produced, generation of value added products from waste, less market time, lower costs of lifecycle, or the combination of these factors.

Membrane-based methods, including membrane bioreactors, membrane separation, membrane hybrid separations (Howell and Noworyta, 1995) can be effectively used for the treatment of inorganic water pollutants from wastewater. Emulsion membrane separation is also a type of membrane separation having a barrier of liquid membrane for separation. Sonochemical reactor is the intensified approach at equipment level. Reactive separations (reactive extraction, reactive distillation, reactive crystallization, reactive adsorption) carried out in a single unit having the interrelation effect enhance the removal of the inorganic pollutants mainly toxic heavy metals. Other hybrid methods may include wet air oxidation with sonication and electrochemical technique combined with photochemical methods. These methods have been proved very effective for the removal of inorganic pollutants. Change of state of system may enhance the removal of pollutants, for example, using supercritical fluids. The use of supercritical fluid such as CO_2 at supercritical state has been successfully employed for the removal of contaminants. Algae consume pollutants (toxic/radioactive) from industrial effluents and hence offer low-cost treatment of wastewater. Other intensified equipment including enhanced solar evaporator, inverse fluidized bed, and capacitive deionizer has been successfully used for inorganic water pollutant treatment. Nanotechnology is a PI approach which can be considered for removal of inorganic pollutants, by employing various nanomaterials. In nutshell, the approach of PI can be efficaciously engaged for the treatment of inorganic water pollutants. There is a wide scope of PI for wastewater treatment by employing intensified approaches and equipment.

References

Afkhami, A., Saber-Tehrani, M., Bagheri, H., 2010. Simultaneous removal of heavy-metal ions in wastewater samples using nano-alumina modified with 2,4-dinitrophenylhydrazine. J. Hazard. Mater. 181 (1–3), 836–844.

Ahmad, A.L., Kusumastuti, A., Derk, C.J.C., Ooi, B.S., 2011. Emulsion liquid membrane for heavy metal removal: an overview on emulsion stabilization and destabilization. Chem. Eng. J. 171 (3), 870–882.

Al-Tarazi, M., 2004. Gas-Liquid Precipitation of Water Dissolved Heavy Metal Ions Using Hydrogen Sulfide Gas. Statistics. <http://purl.utwente.nl/publications/41412>.

Amin, M.T., Alazba, A.A., Manzoor, U., 2014. A review of removal of pollutants from water/wastewater using different types of nanomaterials. Adv. Mater. Sci. Eng. 2014.

Ando, T., Sumi, S., Kawate, T., Ichihara, J., Hanafusa, T., 1984. Sonochemical switching of reaction pathways in solid–liquid two-phase reactions. J. Chem. Soc. Chem. Commun. (7), 439–440.

Anitha, K., Namsani, S., Singh, J.K., 2015. Removal of heavy metal ions using a functionalized single-walled carbon nanotube: a molecular dynamics study. J. Phys. Chem. A 119 (30), 8349–8358.

Arnaiz, C., Buffiere, P., Elmaleh, S., Lebrato, J., Moletta, R., 2003. Anaerobic digestion of dairy wastewater by inverse fluidization: the inverse fluidized bed and the inverse turbulent bed reactors. Environ. Technol. 24 (11), 1431–1443.

Awaleh, M.O., Soubaneh, Y.D., 2014. Waste water treatment in chemical industries: the concept and current technologies. Hydrol. Curr. Res. 5 (1), 1–12.

Babi, D.K., Cruz, M.S., Gani, R., 2016. Fundamentals of process intensification: a process systems engineering view. Process Intensification in Chemical Engineering. Springer, Cham, pp. 7–33.

Bae, E., Chah, S., Yi, J., 2000. Preparation and characterization of ceramic hollow microspheres for heavy metal ion removal in wastewater. J. Colloid Interface Sci. 230 (2), 367–376.

Baldea, M., 2015. From process integration to process intensification. Comput. Chem. Eng. 81, 104–114.

Banerjea, S., Datta, S., Sanyal, S.K., 2000. Mass transfer analysis of the extraction of Cr (VI) by liquid surfactant membrane. Sep. Sci. Technol. 35 (4), 483–501.

Bartels, C., Wilf, M., Andes, K., Iong, J., 2003. Design considerations for wastewater treatment by reverse osmosis. In: Proceedings of the International Desalination and Water Reuse Conference, Tampa, FL.

Bartels, C.R., Wilf, M., Andes, K., Iong, J., 2005. Design considerations for wastewater treatment by reverse osmosis. Water Sci. Technol. 51 (6–7), 473–482.

Becht, S., Franke, R., Geißelmann, A., Hahn, H., 2009. An industrial view of process intensification. Chem. Eng. Process. Process Intensif. 48 (1), 329–332.

Ben-Guang, R.O.N.G., Fang-Yu, H., Kraslawski, A., Nyström, L., 2000. Study on the methodology for retrofitting chemical processes. Chem. Eng. Technol. Ind. Chem. Plant Equip. Process Eng. Biotechnol. 23 (6), 479–484.

Berne, F., Cordonnier, J., 1995. Industrial Water Treatment. Gulf Publishing Company.

Bhowal, A., Dutta, S., 2001. Studies on transport mechanism of Cr (VI) extraction from an acidic solution using liquid surfactant membranes. J. Membr. Sci. 188 (1), 1–8.

Boodhoo, K., Harvey, A. (Eds.), 2013. Process Intensification Technologies for GREEN Chemistry: Engineering Solutions for Sustainable Chemical Processing. John Wiley & Sons.

Boyadzhyev, L., Lazarova, Z., 1993. ChimicaOggi 11 (11–12), 29–38.

Buffière, P., Bergeon, J.P., Moletta, R., 2000. The inverse turbulent bed: a novel bioreactor for anaerobic treatment. Water Res. 34 (2), 673–677.

Calabro, V., Drioli, E., Matera, F., 1991. Membrane distillation in the textile wastewater treatment. Desalination 83 (1–3), 209–224.

Cao, C.Y., Cui, Z.M., Chen, C.Q., Song, W.G., Cai, W., 2010. Ceria hollow nanospheres produced by a template-free microwave-assisted hydrothermal method for heavy metal ion removal and catalysis. J. Phys. Chem. C 114 (21), 9865–9870.

Castilla, P., Meraz, M., Monroy, O., Noyola, A., 2000. Anaerobic treatment of low concentration waste water in an inverse fluidized bed reactor. Water Sci. Technol. 41 (4–5), 245–251.

Celis, L.B., Villa-Gómez, D., Alpuche-Solís, A.G., Ortega-Morales, B.O., Razo-Flores, E., 2009. Characterization of sulfate-reducing bacteria dominated surface communities during start-up of a down-flow fluidized bed reactor. J. Ind. Microbiol. Biotechnol. 36 (1), 111–121.

Celis-García, L.B., Razo-Flores, E., Monroy, O., 2007. Performance of a down-flow fluidized bed reactor under sulfate reduction conditions using volatile fatty acids as electron donors. Biotechnol. Bioeng. 97 (4), 771–779.

Chah, S., Kim, J.S., Yi, J., 2002. Sep. Sci. Technol. 37, 701.

Chakraborty, M., Bhattacharya, C., Datta, S., 2003. Mathematical modeling of simultaneous copper (II) and nickel (II) extraction from wastewater by emulsion liquid membranes. Sep. Sci. Technol. 38 (9), 2081–2106.

Chakraborty, M., Murthy, Z.V.P., Bhattacharya, C., Datta, S., 2005. Process intensification: extraction of chromium (VI) by emulsion liquid membrane. Sep. Sci. Technol. 40 (11), 2353–2364.

Chan, M.T., Fane, A.G., Matheickal, J.T., Sheikholeslami, R., 1997. Membrane distillation crystallization of concentrated salts—flux and crystal formation. J. Macromol. Sci. A. 34 (9), 1727–1735.

Charcosset, C., 2009. A review of membrane processes and renewable energies for desalination. Desalination 245 (1–3), 214–231.

Charpentier, J.C., 2005. Process intensification by miniaturization. Chem. Eng. Technol. Ind. Chem. Plant Equip. Process Eng. Biotechnol. 28 (3), 255–258.

Charpentier, J.C., 2007a. Modern Chemical Engineering in the Framework of Globalization, Sustainability, and Technical Innovation. Ind. Eng. Chem. Res. 46, 3465–3485.

Charpentier, J.C., 2007b. In the frame of globalization and sustainability, process intensification, a path to the future of chemical and process engineering (molecules into money). Chem. Eng. J. 134 (1–3), 84–92.

Chen, Y.H., Li, F.A., 2010. Kinetic study on removal of copper (II) using goethite and hematite nanophotocatalysts. J. Colloid Interface Sci. 347 (2), 277–281.

Cheng, R., Qiu, L., Lu, L., Hu, M., Tao, Z., 2017, Research and application of crystallization technology in treating phosphorus wastewater. In: Advances in Engineering Research (AER), vol. 143, pp. 407–410.

Chmielewski, A.G., Zakrzewska-Trznadel, G., Miljević, N.R., Van Hook, W.A., 1995. Membrane distillation employed for separation of water isotopic compounds. Sep. Sci. Technol. 30 (7–9), 1653–1667.

Criscuoli, A., Drioli, E., 2007. New metrics for evaluating the performance of membrane operations in the logic of process intensification. Ind. Eng. Chem. Res. 46 (8), 2268–2271.

Dautzenberg, F.M., Mukherjee, M., 2001. Process intensification using multifunctional reactors. Chem. Eng. Sci. 56 (2), 251–267.

Degener, W., 1988. Metall (Isernhagen, Germany); 42:817–820.

Deliyanni, E.A., Bakoyannakis, D.N., Zouboulis, A.I., Matis, K.A., 2003. Sorption of As (V) ions by akaganeite-type nanocrystals. Chemosphere 50 (1), 155–163.

Deng, L., Zhang, Y., Chen, F., Cao, S., You, S., Liu, Y., et al., 2013. Reactive crystallization of calcium sulfate dihydrate from acidic wastewater and lime. Chin. J. Chem. Eng. 21 (11), 1303–1312.

Di, Z.C., Ding, J., Peng, X.J., Li, Y.H., Luan, Z.K., Liang, J., 2006. Chromium adsorption by aligned carbon nanotubes supported ceria nanoparticles. Chemosphere 62 (5), 861–865.

Edwie, F., Chung, T.S., 2012. Development of hollow fiber membranes for water and salt recovery from highly concentrated brine via direct contact membrane distillation and crystallization. J. Membr. Sci. s421-422 (12), 111–123.

Edwie, F., Chung, T.S., 2013. Development of simultaneous membrane distillation—crystallization (SMDC) technology for treatment of saturated brine. Chem. Eng. Sci. 98 (29), 160–172.

Eijkel, J.C., Van Den Berg, A., 2005. Nanofluidics: what is it and what can we expect from it? Microfluid. Nanofluid. 1 (3), 249–267.

Elshani, S., Smart, N.G., Lin, Y., Wai, C.M., 2001. Application of supercritical fluids to the reactive extraction and analysis of toxic heavy metals from environmental matrices—system optimisation. Sep. Sci. Technol. 36 (5–6), 1197–1210.

Engates, K.E., Shipley, H.J., 2011. Adsorption of Pb, Cd, Cu, Zn, and Ni to titanium dioxide nanoparticles: effect of particle size, solid concentration, and exhaustion. Environ. Sci. Pollut. Res. 18 (3), 386–395.

EPA, Environmental Protection Agency, 2001. Parameters of Water Quality—Interpretation and Standards. Environmental Protection Agency, Ireland.

Favre-Reguillon, A., Lebuzit, G., Foos, J., Guy, A., Draye, M., Lemaire, M., 2003. Selective concentration of uranium from seawater by nanofiltration. Ind. Eng. Chem. Res. 42 (23), 5900–5904.

Fraser, J., 2019 Intensification Concept at Wastewater Plants Explained, Water online. <https://www.wateronline.com/doc/intensification-concept-at-wastewater-plants-explained-0001>.

Fuller, E.J., Li, N.N., 1984. Extraction of chromium and zinc from cooling tower blowdown by liquid membranes. J. Membr. Sci. 18, 251–271.

Gallegos-Garcia, M., Celis, L.B., Rangel-Méndez, R., Razo-Flores, E., 2009. Precipitation and recovery of metal sulfides from metal containing acidic wastewater in a sulfidogenic down-flow fluidized bed reactor. Biotechnol. Bioeng. 102 (1), 91–99.

García-Calderón, D., Buffiere, P., Moletta, R., Elmaleh, S., 1998. Influence of biomass accumulation on bed expansion characteristics of a down-flow anaerobic fluidized-bed reactor. Biotechnol. Bioeng. 57 (2), 136–144.

Ghangrekar, M.M., 2012. Classification of water pollutants and effects on environment. NPTEL 10 (12), 1–7.

Gogate, P.R., 2002. Cavitation: an auxiliary technique in wastewater treatment schemes. Adv. Environ. Res. 6 (3), 335–358.

Gogate, R., Tatake, P., Kanthale, P., Pandit, A.B., 2002. Mapping of sonochemical reactors: review, analysis, and experimental verification. AIChE J. 48 (7), 1542–1560.

Gogate, P.R., Mujumdar, S., Pandit, A.B., 2003. Large-scale sonochemical reactors for process intensification: design and experimental validation. J. Chem. Technol. Biotechnol. Int. Res. Process, Environ. Clean Technol. 78 (6), 685–693.

Gorak, A., Stankiewicz, A., 2011. Intensified reaction and separation systems. Annu. Rev. Chem. Biomol. Eng. 2 (1), 431–451.

Grossl, P.R., Sparks, D.L., Ainsworth, C.C., 1994. Rapid kinetics of Cu (II) adsorption/desorption on goethite. Environ. Sci. Technol. 28 (8), 1422–1429.

Grünewald, M., Agar, D.W., 2004. Enhanced catalyst performance using integrated structured functionalities. Chem. Eng. Sci. 59 (22–23), 5519–5526.

Guan, G., Wang, R., Wicaksana, F., Yang, X., Fane, A.G., 2012. Analysis of membrane distillation crystallization system for high salinity brine treatment with zero discharge using aspen flowsheet simulation. Ind. Eng. Chem. Res. 51 (41), 13405–13413.

Gupta, V.K., Agarwal, S., Saleh, T.A., 2011. Chromium removal by combining the magnetic properties of iron oxide with adsorption properties of carbon nanotubes. Water Res. 45 (6), 2207–2212.

Gupta, V.K., Ali, I., Saleh, T.A., Nayak, A., Agarwal, S., 2012. Chemical treatment technologies for waste-water recycling—an overview. Rsc Adv. 2 (16), 6380–6388.

Hamoda, M.F., 2013. Advances in wastewater treatment technology for water reuse. J. Eng. Res. 1, 1–27.

Hao, L., Song, H., Zhang, L., Wan, X., Tang, Y., Lv, Y., 2012. SiO2/graphene composite for highly selective adsorption of Pb (II) ion. J. Colloid Interface Sci. 369 (1), 381–387.

Harmsen, G.J., 2007. Reactive distillation: the front-runner of industrial process intensification: a full review of commercial applications, research, scale-up, design and operation. Chem. Eng. Process Process Intensif. 46 (9), 774–780.

Heggs, P.J., 1983. Experimental techniques and correlations for heat exchangers surfaces: packed beds. Chem. Engg. 394 (13), 183.

Himawan, C., Kramer, H.J.M., Witkamp, G.J., 2006. Study on the recovery of purified MgSO4 · 7H2O crystals from industrial solution by eutectic freezing. Sep. Purif. Technol. 50 (2), 240–248.

Hirasawa, I., 1997. Phosphate recovery by reactive crystallization of magnesium ammonium phosphate: application to wastewater. In: ACS Symp. Series, Separation and Purification by Crystallization, vol. 667, Honolulu, HI, pp. 267–276.

Hochhauser, A.M., Cussler, E.L., 1975. Concentrating chromium with liquid surfactant membranes. Am. Inst. Chem. Eng. 71 (152), 136–142.

Howell, J.A., Noworyta, A., 1995. Towards hybrid membrane and biotechnology solutions for polish environmental problems. Wroclaw Technical University Press, Wroclaw, Poland.

Hu, S.Y.B., Wiencek, J.M., 1998. Emulsion-liquid-membrane extraction of copper using a hollow-fiber contactor. AIChE J. 44 (3), 570–581.

Hu, J., Chen, G., Lo, I.M., 2006. Selective removal of heavy metals from industrial wastewater using maghemite nanoparticle: performance and mechanisms. J. Environ. Eng. 132 (7), 709–715.

Hu, X.J., Liu, Y.G., Wang, H., Chen, A.W., Zeng, G.M., Liu, S.M., et al., 2013. Removal of Cu (II) ions from aqueous solution using sulfonated magnetic graphene oxide composite. Sep. Purif. Technol. 108, 189–195.

Hu, X.J., Liu, Y.G., Zeng, G.M., You, S.H., Wang, H., Hu, X., et al., 2014. Effects of background electrolytes and ionic strength on enrichment of Cd (II) ions with magnetic graphene oxide–supported sulfanilic acid. J. Colloid Interface Sci. 435, 138–144.

Huang, Z.H., Zheng, X., Lv, W., Wang, M., Yang, Q.H., Kang, F., 2011. Adsorption of lead (II) ions from aqueous solution on low-temperature exfoliated graphenenanosheets. Langmuir 27 (12), 7558–7562.

Huang, H., Xiao, D., Rui, P., Han, C., Li, D., 2014. Simultaneous removal of nutrients from simulated swine wastewater by adsorption of modified zeolite combined with struvite crystallization. Chem. Eng. J. 256 (6), 431–438.

Hur, J., Shin, J., Yoo, J., Seo, Y.S., 2015. Competitive adsorption of metals onto magnetic graphene oxide: comparison with other carbonaceous adsorbents. Sci. World J. 2015.

Hutnik, N., Kozik, A., Mazienczuk, A., Piotrowski, K., Wierzbowska, B., Matynia, A., 2013. Phosphates (V) recovery from phosphorus mineral fertilizers industry wastewater by continuous struvite reaction crystallization process. Water Res. 47 (11), 3635–3643.

Ingale, M.N., Mahajani, V.V., 1995. A novel way to treat refractory waste: sonication followed by wet oxidation (SONIWO). J. Chem. Technol. Biotechnol. 64 (1), 80–86.

Jansen, A.E., Klaassen, R, van Maanen, H.C.H.J., Akkerhuis, J.J., 1992. Emulsion pertraction of heavy metals from waste water. In: Euromembrane, At Paris, France, pp. 1–6.

Janssen, C.W., 1978. Process for the removal of metals, in particular heavy metals, from waste water. US, US0279964.

Javed, T., Mason, T.J., Phull, S.S., Baker, N.R., Robertson, A., 1995. Influence of ultrasound on the Diels-Alder cyclization reaction: synthesis of some hydroquinone derivatives and lonapalene, an anti-psoriatic agent. Ultrasonicssonochemistry 2 (1), S3–S4.

Jeon, S.I., Park, H.R., Yeo, J.G., Yang, S., Cho, C.H., Han, M.H., et al., 2013. Desalination via a new membrane capacitive deionization process utilizing flow-electrodes. Energy Environ. Sci. 6 (5), 1471–1475.

Jiao, W.T., Jiao, W.X., 2015. Purifying industrial high salt waste-water involves freezing and recycling sodium sulfate waste-water, discharging brine into freezer, controlling temperature, cooling precipitated sodium sulfate, and evaporating brine. Chinese Patent No. CN105110542-A.

Kaksonen, A.H., Riekkola-Vanhanen, M.L., Puhakka, J.A., 2003. Optimization of metal sulphide precipitation in fluidized-bed treatment of acidic wastewater. Water Res. 37 (2), 255–266.

Kaksonen, A.H., Plumb, J.J., Robertson, W.J., Riekkola-Vanhanen, M., Franzmann, P.D., Puhakka, J.A., 2006. The performance, kinetics and microbiology of sulfidogenic fluidized-bed treatment of acidic metal-and sulfate-containing wastewater. Hydrometallurgy 83 (1–4), 204–213.

Kanel, S.R., Manning, B., Charlet, L., Choi, H., 2005. Removal of arsenic (III) from groundwater by nanoscale zero-valent iron. Environ. Sci. Technol. 39 (5), 1291–1298.

Kanel, S.R., Greneche, J.M., Choi, H., 2006. Arsenic (V) removal from groundwater using nano scale zero-valent iron as a colloidal reactive barrier material. Environ. Sci. Technol. 40 (6), 2045–2050.

Kasher, R., 2009. Membrane-based water treatment technologies: recent achievements, and new challenges for a chemist. Bull. Isr. Chem. Soc. (24), 10–18.

Keskinler, B., Akay, G., 2004. Process intensification in wastewater treatment: oxygen transfer characterisation of a jet loop reactor for aerobic biological wastewater treatment. Int. J. Environ. Technol. Manag. 4 (3), 220–235.

Kim, B.M., 1984a. Membrane-based solvent extraction for selective removal and recovery of metals... by liquid surfactant membranes. J. Membr. Sci. 21, 5–19.

Kim, B.M., 1984b. A membrane extraction process for selective recovery of metals from wastewater. In: Separation of Heavy Metals and Other Trace Contaminants, Houston, TX, Philadelphia, PA, San Francisco, CA, pp. 126–132.

Kim, B.M., 1985. Membrane-based solvent extraction for selective removal and recovery of metals by liquid surfactant membranes. Kagaku KogakuRonbunshu 11, 394.

Kim, J.S., Yi, J., 1999. Selective removal of copper ions from aqueous solutions using modified silica beads impregnated with LIX 84, J. Chem. Technol. Biotechnol. 74, 544–550.

Kim, Y., Yi, J., 2004. Advances in environmental technologies via the application of mesoporous materials. J. Ind. Eng. Chem. 10 (1), 41–51.

Kim, Y.J., Choi, J.H., 2010. Enhanced desalination efficiency in capacitive deionization with an ion-selective membrane. Sep. Purif. Technol. 71 (1), 70–75.

Kim, J.S., Park, J.C., Yi, J., 2000a. Zinc ion removal from aqueous solutions using modified silica impregnated with 2-ethylhexyl 2-ethylhexyl phosphonic acid, Sep. Sci. Technol. 35, 1901–1916.

Kim, J.S., Chah, S., Yi, J., 2000b. Preparation of modified silica for heavy metal removal, J. Chem. Eng. 17 (1), 118–121.

Kim, Y., Lee, B., Yi, J., 2003. Preparation of functionalized mesostructured silica containing magnetite (msm) for the removal of copper ions in aqueous solutions and its magnetic, Sep. Sci. Technol. 38 (11), 2533–2548.

Kim, Y.H., Park, L.K., Yiacoumi, S., Tsouris, C., 2017. Modular chemical process intensification: a review. Annu. Rev. Chem. Biomol. Eng. 8, 359–380.

Kruus, P., Lawrie, J.A.G., O'Neill, M.L., 1988. Polymerization and depolymerization by ultrasound. Ultrasonics 26 (6), 352–355.

Krystynik, P., Kluson, P., Tito, D.N., 2015. Water treatment process intensification by combination of electrochemical and photochemical methods. Chem. Eng. Process: Process Intensif. 94, 85–92.

Kueke F. Procedure for the reactive extraction of chromium-containing substances from aqueous feed solutions. DE 19943232, 2001.

Lai, C.H., Chen, C.Y., 2001. Removal of metal ions and humic acid from water by iron-coated filter media. Chemosphere 44 (5), 1177–1184.

Lawson, K.W., Lloyd, D.R., 1997. Membrane distillation. J. Membr. Sci. 124 (1), 1–25.

Lee, Y.C., Yang, J.W., 2012. Self-assembled flower-like TiO_2 on exfoliated graphite oxide for heavy metal removal. J. Ind. Eng. Chem. 18 (3), 1178–1185.

Lee, J.S., Kim, J.S., Yi, J., 2000. Modeling of copper ion removal from aqueous solutions using modified silica beads, Chem. Eng. Commun. 181(1), 37–55.

Lee, W., Kim, C., Yi, J., 2002. Selective recovery of silver ions from aqueous solutions using modified silica beads with Adogen 364, J. Chem. Technol. Biotechnol. 77(11), 1255–1261.

Li, X.Q., Zhang, W.X., 2006. Iron nanoparticles: the core − shell structure and unique properties for Ni (II) sequestration. Langmuir 22 (10), 4638–4642.

Li, Y.H., Wang, S., Wei, J., Zhang, X., Xu, C., Luan, Z., et al., 2002. Lead adsorption on carbon nanotubes. Chem. Phys. Lett. 357 (3–4), 263–266.

Li, Y.H., Ding, J., Luan, Z., Di, Z., Zhu, Y., Xu, C., et al., 2003. Competitive adsorption of Pb2 + , Cu2 + and Cd2 + ions from aqueous solutions by multiwalled carbon nanotubes. Carbon 41 (14), 2787–2792.

Li, Y.H., Di, Z., Ding, J., Wu, D., Luan, Z., Zhu, Y., 2005. Adsorption thermodynamic, kinetic and desorption studies of Pb2 + on carbon nanotubes. Water Res. 39 (4), 605–609.

Li, J., Zhang, S., Chen, C., Zhao, G., Yang, X., Li, J., et al., 2012. Removal of Cu (II) and fulvic acid by graphene oxide nanosheets decorated with Fe3O4 nanoparticles. ACS Appl. Mater. Interfaces 4 (9), 4991–5000.

Lie Ken Jie, M.S.F., Lam, C.K., 1995. Ultrasound-Assisted Expoxidation Reaction of Long-Chain Unsaturated Fatty Esters. Ultrason. Sonochem. 2 (1), S11–S14.

Lu, C., Chiu, H., 2006. Adsorption of zinc (II) from water with purified carbon nanotubes. Chem. Eng. Sci. 61 (4), 1138–1145.

Lu, C., Liu, C., 2006. Removal of nickel (II) from aqueous solution by carbon nanotubes. J. Chem. Technol. Biotechnol. Int. Res. Process Environ. Clean Technol. 81 (12), 1932–1940.

Lu, C., Chiu, H., Liu, C., 2006. Removal of zinc (II) from aqueous solution by purified carbon nanotubes: kinetics and equilibrium studies. Ind. Eng. Chem. Res. 45 (8), 2850–2855.

Lu, H., Wang, J., Wang, T., Wang, N., Bao, Y., Hao, H., 2017. Crystallization techniques in wastewater treatment: an overview of applications. Chemosphere . Available from: https://doi.org/10.1016/j.chemosphere.2017.01.070.

Lutze, P., Gani, R., Woodley, J.M., 2010. Process intensification: a perspective on process synthesis. Chem. Eng. Process: Process Intensif. 49 (6), 547–558.

Lutze, P., Babi, D.K., Woodley, J.M., Gani, R., 2013. Phenomena based methodology for process synthesis incorporating process intensification. Ind. Eng. Chem. Res. 52 (22), 7127–7144.

Ma, X., Wang, Y., Gao, M., Xu, H., Li, G., 2010. A novel strategy to prepare ZnO/PbS heterostructured functional nanocomposite utilizing the surface adsorption property of ZnO nanosheets. Catal. Today 158 (3–4), 459–463.

Madadrang, C.J., Kim, H.Y., Gao, G., Wang, N., Zhu, J., Feng, H., et al., 2012. Adsorption behavior of EDTA-graphene oxide for Pb (II) removal. ACS Appl. Mater. Interfaces 4 (3), 1186–1193.

Marin, P., Alkalay, D., Guerrero, L., Chamy, R., Schiappacasse, M.C., 1999. Design and startup of an anaerobic fluidized bed reactor. Water Sci. Technol. 40 (8), 63–70.

Masciangioli, T., Zhang, W.X., 2003. Peer reviewed: environmental technologies at the nanoscale. Environ. Sci. Technol. 37 (5), 102A–108A.

Mayo, J.T., Yavuz, C., Yean, S., Cong, L., Shipley, H., Yu, W., et al., 2007. The effect of nanocrystalline magnetite size on arsenic removal. Sci. Technol. Adv. Mater. 8 (1–2), 71.

Minotti, M., Doherty, M.F., Malone, M.F., 1998. Design for Simultaneous Reaction and Liquid – Liquid Extraction. Ind. & Eng. Chem. Res. 37 (12), 4748–4755.

Mishra, N.S., Reddy, R., Kuila, A., Rani, A., Mukherjee, P., Nawaz, A., et al., 2017. A review on advanced oxidation processes for effective water treatment. Curr. World Environ. 12 (3), 470.

Mohajerani, M., Mehrvar, M., Ein-Mozaffari, F., 2009. An overview of the integration of advanced oxidation technologies and other processes for water and wastewater treatment. Int. J. Eng. 3 (2), 120–146.

Mohunta, DM, 2015. Commercial, Chemical and Dev. Co. <http://www.ccdcindia.com/index2.php?act = contact>.

Mori, Y., Uemae, H., Hibino, S., Eguchi, W., 1990. Proper condition of the surfactant liquid membrane for the recovery and concentration of chromium (VI) from aqueous acid solution. Int. Chem. Eng. 30, 124–131.

Nandi, D., Basu, T., Debnath, S., Ghosh, A.K., De, A., Ghosh, U.C., 2013. Mechanistic insight for the sorption of Cd (II) and Cu (II) from aqueous solution on magnetic mn-doped Fe (III) oxide nanoparticle implanted graphene. J. Chem. Eng. Data 58 (10), 2809–2818.

Nasr, F.A., Doma, H.S., Abdel-Halim, H.S., El-Shafai, S.A., 2007. Chemical industry wastewater treatment. Environmentalist 27 (2), 275–286.

Nathoo, J., Jivanji, R., Lewis, A.E., 2009. Freezing your brines off: eutectic freeze crystallization for brine treatment. In: International Mine Water Conference, 431-437. ISBN: 9780980262353.

Nikolov, L., Karamanev, D., 1991. The inverse fluidization-a new approach to biofilm reactor design, to aerobic wastewater treatment, Studies in Environmental Science, vol. 42. Elsevier, pp. 177–182.

Naji, O., Bowtell, L., Al-juboori, R.A., Aravinthan, V., Ghaffour, N., 2018. Effect of air gap membrane distillation parameters on the removal of fluoride from synthetic water. Desalin. Water Treat. 124, 11–20.

Oliveira, L.C., Petkowicz, D.I., Smaniotto, A., Pergher, S.B., 2004. Magnetic zeolites: a new adsorbent for removal of metallic contaminants from water. Water Res. 38 (17), 3699–3704.

Omole, M.A., K'Owino, I., Sadik, O.A., 2009. Nanostructured materials for improving water quality: potentials and risks. Nanotechnology Applications for Clean Water. William Andrew Publishing, pp. 233–247.

Onyango, M.S., Kojima, Y., Matsuda, H., Ochieng, A., 2003. Adsorption kinetics of arsenic removal from groundwater by iron-modified zeolite. J. Chem. Eng. Jpn. 36 (12), 1516–1522.

Pena, M.E., Korfiatis, G.P., Patel, M., Lippincott, L., Meng, X., 2005. Adsorption of As (V) and As (III) by nanocrystalline titanium dioxide. Water Res. 39 (11), 2327–2337.

Peng, X., Luan, Z., Ding, J., Di, Z., Li, Y., Tian, B., 2005. Ceria nanoparticles supported on carbon nanotubes for the removal of arsenate from water. Mater. Lett. 59 (4), 399–403.

Ponce-Ortega, J.M., Al-Thubaiti, M.M., El-Halwagi, M.M., 2012. Process intensification: new understanding and systematic approach. Chem. Eng. Process: Process Intensif. 53, 63–75.

Ponder, S.M., Darab, J.G., Mallouk, T.E., 2000. Remediation of Cr (VI) and Pb (II) aqueous solutions using supported, nanoscale zero-valent iron. Environ. Sci. Technol. 34 (12), 2564–2569.

Portha, J.F., Falk, L., Commenge, J.M., 2014. Local and global process intensification. Chem. Eng. Process: Process Intensif. 84, 1–13.

Profio, G.D., Perrone, G., Curcio, E., Cassetta, A., Lamba, D., Drioli, E., 2005. Preparation of enzyme crystals with tunable morphology in membrane crystallizers. Ind. Eng. Chem. Res. 44 (44), 10005–10012.

Qi, L., Xu, Z., 2004. Lead sorption from aqueous solutions on chitosan nanoparticles. Colloids Surf. A: Physicochem. Eng. Asp. 251 (1–3), 183–190.

Qiu, G., Zhang, S., Raghavan, D.S.S., Das, S., Ting, Y.P., 2016. The potential of hybrid forward osmosis membrane bioreactor (FOMBR) processes in achieving high throughput treatment of municipal wastewater with enhanced phosphorus recovery. Water Res. 105, 370–382.

Racar, M., Dolar, D., Špehar, A., Košutić, K., 2017. Application of UF/NF/RO membranes for treatment and reuse of rendering plant wastewater. Process. Saf. Environ. Prot. 1 (05), 386–392.

Raghavan, K.V., Reddy, B.M., 2014. Chemical process intensification: an engineering overview. Industrial Catalysis and Separations. Apple Academic Press, pp. 28–68.

Rahman, M.M., Liu, Y.H., Kwag, J.H., Ra, C.S., 2011. Recovery of struvite from animal wastewater and its nutrient leaching loss in soil. J. Hazard. Mater. 186 (2–3), 2026–2030.

Rahman, M.M., Salleh, M.A.M., Rashid, U., Ahsan, A., Hossain, M.M., Chang, S.R., 2014. Production of slow release crystal fertilizer from wastewaters through struvite crystallization—a review. Arab. J. Chem. 7 (1), 139–155.

Ramshaw, C., 1983. Higee' distillation - an example of process intensification. Chem. Eng. 13–14.

Ramshaw, C, 1995, The incentive for process intensification. In: 1st International Conference on proceedings of Intensification for Chemical Industry, London, 1995.

Rani, Y., Sumuna, C., 2015. Chemical process intensification: an engineering overview. In: Raghavan, K.V., Reddy, B.M. (Eds.), Industrial Catalysis and Separations Innovations for Process Intensification. Taylor and Francis Group.

Reay, D., Ramshaw, C., Harvey, A., 2013. Process Intensification: Engineering for Efficiency, Sustainability and Flexibility. Butterworth-Heinemann.

Reinsel M., 2016. Industrial Water Treatment for Inorganic Contaminants: Emerging Technologies, Water Online, article.

Rickerby, D.G., Morrison, M., 2007. Nanotechnology and the environment: a European perspective. Sci. Technol. Adv. Mater. 8 (1–2), 19.

Rong, B.G., Fang-Yu, H., Kraslawski, A., Nystrom, L., 2000. Study on the methodology for retrofitting chemical processes. Chem. Eng. Technol. 23, 479–484.

Roy, P., Choudhury, M., Ali, M., 2013. As (III) and As (V) adsorption on magnetite nanoparticles: adsorption isotherms, effect of pH and phosphate, and adsorption kinetics. Int. J. 4 (1).

Rubio, J., Tessele, F., 1997. Removal of heavy metal ions by adsorptive particulate flotation. Miner. Eng. 10 (7), 671–679.

Sahu, O., 2014. Reduction of organic and inorganic pollutant from waste water by algae. Int. Lett. Nat. Sci. 8 (1), 1–8.

Salazar, E., Ortiz, M.I., Urtiaga, A.M., Irabien, J.A., 1992. Kinetics of the separation-concentration of chromium (VI) with emulsion liquid membranes. Ind. Eng. Chem. Res. 31 (6), 1523–1529.

Salazar, E., Ortiz, M.I., Irabien, J., 1990. Recovery of Cr (VI) with ELM in mechanically stirred contactors: influence of membrane composition on the yield of extraction. Inst. Chem. Engg. Symp. Ser vol. 119, 279–287.

Samant, K.D., Ng, K.M., 1999. Systematic development of extractive reaction processes. Chem. Eng. & Technol. Ind. Chem. Plant Equip. Process Eng. Biotechnol. 22 (10), 877–880.

Santhosh, C., Velmurugan, V., Jacob, G., Jeong, S.K., Grace, A.N., Bhatnagar, A., 2016. Role of nanomaterials in water treatment applications: a review. Chem. Eng. J. 306, 1116–1137.

Savage, N., Diallo, M.S., 2005. Nanomaterials and water purification: opportunities and challenges. J. Nanopart. Res. 7 (4–5), 331–342.

Sharma, K., Kim, Y.H., Gabitto, J., Mayes, R.T., Yiacoumi, S., Bilheux, H.Z., et al., 2015. Transport of ions in mesoporous carbon electrodes during capacitive deionization of high-salinity solutions. Langmuir 31 (3), 1038–1047.

Shieh, W.K., Keenan, J.D., 1986. Fluidized bed biofilm reactor for wastewater treatment. Bioproducts. Springer, Berlin, Heidelberg, pp. 131–169.

Shin, Y., Sohn, J., 2016. Mechanisms for scale formation in simultaneous membrane distillation crystallization: effect of flow rate. J. Ind. Eng. Chem. 35, 318–324.

Shirgaonkar, I.Z., Pandit, A.B., 1997. Degradation of aqueous solution of potassium iodide and sodium cyanide in the presence of carbon tetrachloride. Ultrasonicssonochemistry 4 (3), 245–253.

Simon, L.L., Osterwalder, N., Fischer, U., Hungerbühler, K., 2008. Systematic retrofit method for chemical batch processes using indicators, heuristics, and process models. Ind. Eng. Chem. Res. 47 (1), 66–80.

Sirkar, K.K., Shanbhag, P.V., Kovvali, A.S., 1999. Membrane in a reactor: a functional perspective. Ind. Eng. Chem. Res. 38 (10), 3715–3737.

Sitko, R., Zawisza, B., Malicka, E., 2013. Graphene as a new sorbent in analytical chemistry. TrAC Trends Anal. Chem. 51, 33–43.

Sivakumar, M., Pandit, A.B., 2001. Ultrasound enhanced degradation of Rhodamine B: optimization with power density. Ultrason. Sonochem. 8 (3), 233–240.

Somlev, V., Tishkov, S., 1992. Application of fluidized carrier to bacterial sulphate-reduction in industrial wastewaters purification. Biotechnol. Tech. 6 (1), 91–96.

Spronsen, J.V., Pascual, M.R., Genceli, F.E., Trambitas, D.O., Evers, H., Witkamp, G.J., 2010. Eutectic freeze crystallization from the ternary Na2CO3−NaHCO3−H2O system: a novel scraped wall crystallizer for the recovery of soda from an industrial aqueous stream. Chem. Eng. Res. Des. 88 (9), 1259–1263.

Sreeprasad, T.S., Maliyekkal, S.M., Lisha, K.P., Pradeep, T., 2011. Reduced graphene oxide−metal/metal oxide composites: facile synthesis and application in water purification. J. Hazard. Mater. 186 (1), 921–931.

Stafiej, A., Pyrzynska, K., 2007. Adsorption of heavy metal ions with carbon nanotubes. Sep. Purif. Technol. 58 (1), 49–52.

Stafiej, A., Pyrzynska, K., 2008. Solid phase extraction of metal ions using carbon nanotubes. Microchem. J. 89 (1), 29–33.

Stankiewicz, A., 2000. Process intensification in in-line monolithic reactor. Chem. Eng. Sci. 56 (2), 359–364.

Stankiewicz, A., 2003a. Reactive and Hybrid Separations: Incentives, Applications, Barriers. Re-Engineering the Chemical Processing Plant. CRC Press, pp. 258–301.

Stankiewicz, A., 2003b. Reactive separations for process intensification: an industrial perspective. Chem. Eng. Process: Process Intensif. 42 (3), 137–144.

Stankiewicz, A.I., Moulijn, J.A., 2000. Process intensification: transforming chemical engineering. Chem. Eng. Prog. 96 (1), 22–34.

Stankiewicz, A., Moulijn, J.A. (Eds.), 2004. Re-engineering the chemical processing plant: process intensification. *Marcel Dekker Inc, New York*.

Strzelbicki, J., Charewicz, W., 1980. The liquid surfactant membrane separation of copper, cobalt and nickel from multicomponent aqueous solutions. Hydrometallurgy 5 (2–3), 243–254.

Suslick, K.S., 1986. Organometallic sonochemistry, Advances in organometallic chemistry, vol. 25. Academic Press, pp. 73–119.

Suzuki, K., Tanaka, Y., Osada, T., Waki, M., 2002. Removal of phosphate, magnesium and calcium from swine wastewater through crystallization enhanced by aeration. Water Res 36 (12), 2991–2998.

Tait, S., Clarke, W.P., Keller, J., Batstone, D.J., 2009. Removal of sulfate from high-strength wastewater by crystallization. Water Res. 43 (3), 762–772.

Takafuji, M., Ide, S., Ihara, H., Xu, Z., 2004. Preparation of poly (1-vinylimidazole)-grafted magnetic nanoparticles and their application for removal of metal ions. Chem. Mater. 16 (10), 1977–1983.

Tayalia, Y., Vijaysai, P., 2012. Process intensification in water and wastewater treatment systems. In: Karimi, I.A., Srinivasan, R. (Eds.), Proceedings of the 11th International Symposium on Process Systems Engineering, 15-19. Elsevier, Singapore.

Tsouris, C., Porcelli, J.V., 2003. Process intensification—has its time finally come? Chem. Eng. Prog. 99 (10), 50–55.

Tsouris, C., Mayes, R., Kiggans, J., Sharma, K., Yiacoumi, S., DePaoli, D., et al., 2011. Mesoporous carbon for capacitive deionization of saline water. Environ. Sci. Technol. 45 (23), 10243–10249.

USDE, 2015. Process intensification, U.S. Department of Energy, Energy Efficiency and Renewable Energy, Alexandria, pp. 1–61.

Van der Bruggen, B., Vandecasteele, C., 2003. Removal of pollutants from surface water and groundwater by nanofiltration: overview of possible applications in the drinking water industry. Environ. Pollut. 122 (3), 435–445.

Vaseashta, A., Vaclavikova, M., Vaseashta, S., Gallios, G., Roy, P., Pummakarnchana, O., 2007. Nanostructures in environmental pollution detection, monitoring, and remediation. Sci. Technol. Adv. Mater. 8 (1–2), 47.

Veličković, Z.S., Bajić, Z.J., Ristić, M.Đ., Djokić, V.R., Marinković, A.D., Uskoković, P.S., et al., 2013. Modification of multi-wall carbon nanotubes for the removal of cadmium, lead and arsenic from wastewater. Dig. J. Nanomater. Biostruct. (DJNB) 8 (2).

Verma, R., Dwivedi, P., 2013. Heavy metal water pollution—a case study. Recent Res. Sci. Technol. 5 (5), 98–99. 2013.

Villa-Gomez, D., 2013. Simultaneous Sulfate Reduction and Metal Precipitation in an Inverse Fluidized Bed Reactor (Ph.D. thesis), SENSE Research School for Socio-Economic and Natural Sciences of the Environment, Wageningen University and UNESCO-IHE Institute for Water Education, Delft, The Netherlands

Villa-Gomez, D., Ababneh, H., Papirio, S., Rousseau, D.P.L., Lens, P.N.L., 2011. Effect of sulfide concentration on the location of the metal precipitates in inversed fluidized bed reactors. J. Hazard. Mater. 192 (1), 200−207.

Villa-Gomez, D.K., Papirio, S., Van Hullebusch, E.D., Farges, F., Nikitenko, S., Kramer, H., et al., 2012. Influence of sulfide concentration and macronutrients on the characteristics of metal precipitates relevant to metal recovery in bioreactors. Bioresour. Technol. 110, 26−34.

Wang, L., Yang, Z., Gao, J., Xu, K., Gu, H., Zhang, B., et al., 2006. A biocompatible method of decorporation: bisphosphonate-modified magnetite nanoparticles to remove uranyl ions from blood. J. Am. Chem. Soc. 128 (41), 13358−13359.

Wang, Y., Liang, S., Chen, B., Guo, F., Yu, S., Tang, Y., 2013. Synergistic removal of Pb (II), Cd (II) and humic acid by Fe3O4@ mesoporous silica-graphene oxide composites. PLoS One 8 (6), e65634.

Williams, P.M., Ahmad, M., Connolly, B.S., Oatley-Radcliffe, D.L., 2015. Technology for freeze concentration in the desalination industry. Desalination 356 (3), 314−327.

Wu, Y., Ying, K., Liu, J., Zhang, J., Xu, J., 1991. An experimental study on membrane distillation-crystallization for treating waste water in taurine production. Desalination 80 (2), 235−242.

Xu, H., Zhou, Q., Wang, G.R., Wang, J.F., 2010. Reclaiming solution containing phenol and nitrate comprises treating wastewater, extracting phenol using ricinoleic acid, triple-effect evaporative crystallization and recycling of sodium sulfate. Chinese Patent No. CN101654305-A.

Yang, G.C., Lee, H.L., 2005. Chemical reduction of nitrate by nanosized iron: kinetics and pathways. Water Res. 39 (5), 884−894.

Yavuz, C.T., Mayo, J.T., William, W.Y., Prakash, A., Falkner, J.C., Yean, S., et al., 2006. Low-field magnetic separation of monodisperse Fe3O4 nanocrystals. Science 314 (5801), 964−967.

Zhang, K., Kemp, K.C., Chandra, V., 2012. Homogeneous anchoring of TiO2 nanoparticles on graphene sheets for waste water treatment. Mater. Lett. 81, 127−130.

Zhang, X.M., Luan, J.Y., Peng, H.Z., 2013. High salt and high ammonia-nitrogen wastewater zero-discharge device includes film absorbing unit, filter unit, positive permeation unit cooling crystallization unit, high salt high ammonia-nitrogen wastewater and absorbing raw film. Chinese Patent No. CN104609616-A.

Zhao, G., Li, J., Ren, X., Chen, C., Wang, X., 2011. Few-layered graphene oxide nanosheets as superior sorbents for heavy metal ion pollution management. Environ. Sci. Technol. 45 (24), 10454−10462.

Zhao, J., Wang, Z., White, J.C., Xing, B., 2014. Graphene in the aquatic environment:adsorption, dispersion, toxicity and transformation. Environ. Sci. Technol. 48, 9995−10009.

Zhong, L.S., Hu, J.S., Liang, H.P., Cao, A.M., Song, W.G., Wan, L.J., 2006. Self-Assembled 3D flowerlike iron oxide nanostructures and their application in water treatment. Adv. Mater. 18 (18), 2426−2431.

Zhong, L.S., Hu, J.S., Cao, A.M., Liu, Q., Song, W.G., Wan, L.J., 2007. 3D flowerlike ceria micro/nanocomposite structure and its application for water treatment and CO removal. Chem. Mater. 19 (7), 1648−1655.

Further reading

Górak, A., Stankiewicz, A., 2012. Towards the Sustainable World of 2050: European research agenda for process intensification. Chem. IngenieurTechnik 84 (8), 1260.

Gürel, L., Altaş, L., Büyükgüngör, H., 2005. Removal of lead from wastewater using emulsion liquid membrane technique. Environ. Eng. Sci. 22 (4), 411−420.

Howard, G., 2011. Process intensification in industrial wastewater treatment. Foster. Wheeler 1−10.

Levine, A.D., Tchobanoglous, G., Asano, T., 1985. Characterization of the size distribution of contaminants in wastewater: treatment and reuse implications. J. Water Pollut. Control. Federation 805−816.

Ramesh, K.T., 2009. Nanomaterials. Nanomaterials. Springer, Boston, MA, pp. 1−20.

Stankiewicz, A., Moulijn, J.A., 2003. Re-engineering the chemical processing plant: process intensification. *CRC Press*.

Taylor, S.R., Rouse, C.M., Hoffman, A.A., 1991. Sonocatalytic Wet Oxidation for Water Purification. Taylor (Sr) and Associates, Bartlesville, OK.

Yang, W., Cicek, N., Ilg, J., 2006. State-of-the-art of membrane bioreactors: worldwide research and commercial applications in North America. J. Membr. Sci. 270 (1−2), 201−211.

Various water-treatment technologies for inorganic contaminants: current status and future aspects

Richa Soni, Shefali Bhardwaj and Dericks Praise Shukla

School of Engineering, Indian Institute of Technology, Mandi, India

14.1 Introduction

Water pollution is a grave issue affecting our lives and is expected to worsen due to rapid industrialization and population growth. Several organic and inorganic pollutants have been reported in water along with microbial populations. A number of organic and inorganic pollutants are hazardous owing to high toxicity and carcinogenic nature (Ali and Aboul-Enein, 2006). In addition, some organic and metal ions are nonbiodegradable and remain in the environment for a long time. The present study focuses on the inorganic contaminants, and therefore further discussion is based on it.

The presence of inorganic contaminants in surface and ground water continues to be one of the pervasive environmental issues of recent time. Despite various control strategies, the discharge of such contaminants is still staggering. Presence of inorganic contaminants in water may be natural, geological, or due to anthropogenic activities such as agriculture, mining, or industry. Traces of inorganic contaminants might be present in the water supplies, but their presence exceeding the maximum permissible level may lead to several health issues such as damaging the liver, kidney, nervous system, circulatory system, gastrointestinal system, bones, or skin depending upon the inorganic contaminant and level of exposure (Moore, 2012). Inorganic contaminants in water include cationic, anionic, or neutral forms of elements present in the periodic table (Clifford et al., 1986). Inorganic contaminants generally include arsenic, lead, cadmium, mercury, chromium, aluminum, nitrates, nitrites, and fluorides. A major part of inorganic contaminants is constituted by heavy metals. Heavy metals are present in a variety of waste, including fertilizers, municipal waste, mines residues, sludge, pesticides, and smelting industries (Halim et al., 2003). Mostly carcinogenic, heavy metals are a serious threat to the living population owing to their nondegradable and persistent nature (Rao and Kashifuddin, 2014). Some heavy metals have an important role in the human metabolic system but if consumed in excessive level can be harmful for the organs (Hu, 2002). Heavy metals are highly soluble in water and hence can easily enter the food chain of aquatic life and then humans causing potential health concerns (Uddin, 2017).

With deteriorating water quality and different types of contamination, it is essential to save the existing water reserves to avoid future water stress. Various methods of water treatment have been developed and used for the treatment of inorganic contaminants. This chapter proposes a general scheme of water treatment for inorganic contaminants especially heavy metals and summarizes the advantages and limitations of various techniques used.

14.2 Water-treatment technologies

A multitude of techniques are available for water treatment. Fig. 14.1 depicts a list of important techniques used for water treatment and their classification on the basis of

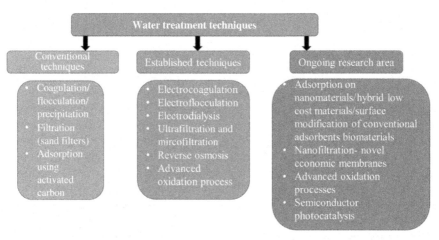

FIGURE 14.1   Classification of water-treatment techniques.

conventional techniques, established techniques, and ongoing research area. Each treatment method has its own benefits and limitations in terms of the nature of pollutant it can treat, cost, efficiency, feasibility, and environmental impact (Crini and Lichtfouse, 2018). Combination of different processes for effective and economic removal of contaminants is a major research area these days. This section discusses about various treatment technologies mentioning their recent use in the removal of metal contaminants.

14.2.1 Electro-assisted methods

Electrochemical methods are flexible techniques for water and wastewater treatment. The electrochemical method connects chemistry with electronics (Grimm et al., 1998). Various electrochemical technologies—electrocoagulation, electroflotation, electrodialysis, and electrooxidation—are used for water and wastewater treatment. Various researchers have presented the details of the electrochemical technologies (Mollah et al., 2001, 2004; Chen, 2004; Drogui et al., 2007; Martínez-Huitle and Brillas, 2008; Martinez-Huitle et al., 2015; Radjenovic and Sedlak, 2015). Electrooxidation is usually used to treat organic contaminants and hence not discussed here.

14.2.1.1 Electrocoagulation and electroflotation

Electrocoagulation is a process that finds its origin from conventional chemical coagulation (Canizares et al., 2009; An et al., 2017). Electrocoagulation process is the coagulation of colloidal particles in an aqueous medium on application of electric current as shown in Fig. 14.2. During this process suspended, emulsified or dissolved contaminants are being destabilized. In the process, electric current is applied to electrodes present in an electrolytic cell where a coagulating agent and gas bubbles are generated (Chen, 2004; Mollah

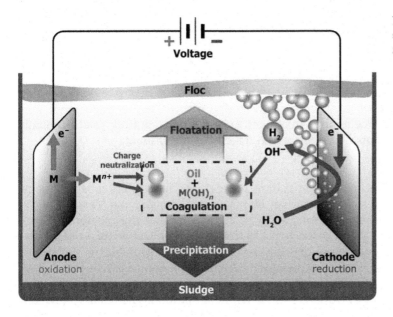

FIGURE 14.2 Principle showing process of electrocoagulation (An et al., 2017).

et al., 2004; An et al., 2017). Hydrogen is generated at the cathode and the bubbles help in additional removal of pollutants. These flocs with the characteristics of good stability, less bound water, and large particle size are easily removed by filtration (Mollah et al., 2004). It is noteworthy that during electrocoagulation treatment, numerous electrochemical, physicochemical, and chemical processes take place (Mollah et al., 2001). The interest of using electrocoagulation for water treatment is due to following: in situ generation of coagulants that reduces the cost associated with chemical transportation and storage constraints in large-scale application (Drogui et al., 2007). The effluent is not enriched with anions and salts during electrocoagulation, which otherwise happens in chemical coagulation, resulting in compact sludge production (Rajeshwar and Ibanez, 1997), effluent produced in electrocoagulation is near neutral pH, and hence posttreatment for pH neutralization is omitted (Laridi et al., 2005).

Electroflotation process removes the pollutants from water through the generation of hydrogen and oxygen gas bubbles on cathode and anode electrodes, respectively, during water electrolysis (Chen, 2004; Kyzas and Matis, 2016; An et al., 2017). Pollutants adhere onto the bubbles and move up to the surface where they are periodically skimmed off. Bubble size defines the efficiency of the process, smaller the size more efficient is the removal (Chen et al., 2000). Regulation of the current can help to adjust the size and the flow rate of gas (Drogui et al., 2007).

Akbal and Camcı (2011) reported that electrocoagulation of the metal plating wastewater using Fe−Al electrode pair showed 100% Cu, 100% Cr, and 100% Ni removal at operating parameters; current density of $10 \, mA/cm^2$, pH 3.0, and treatment time of 20 minutes. Another study by Al-Shannag et al. (2015) reported the use of electrocoagulation of metal plating wastewater for removal of Cu^{2+}, Cr^{3+}, Ni^{2+}, and Zn^{2+} with 97% of removal efficiency using current density of $4 \, mA/cm^2$, pH of 9.56, and electrocoagulation time of 45 minutes. Merzouk et al. (2009) demonstrated the separation of Fe, Ni, Cu, Zn, Pb, and Cd and achieved removal efficiency of 95%. Heidmann and Calmano (2008) explained the removal of Cr through electrocoagulation using Fe electrodes and showed that current lower than 0.1 A reduces the cost and increases the efficiency.

Although electrochemical processes are effective in removing pollutants and has its own advantages, it has some limitations such as high consumption of energy, increase of the treatment cost (Garcia-Segura et al., 2017), periodic cleaning of electrodes as scaling of electrodes reduces efficiency of the process, and sludge disposal issue (Radjenovic and Sedlak, 2015). However, electrochemical treatment may not be technically and economically feasible for some pollutants due to increased toxicity of the effluent.

14.2.1.2 Electrodialysis

Electrodialysis is an electrochemical separation process which transfers ions via membranes by using direct current as shown in Fig. 14.3 (Van der Bruggen and Vandecasteele, 2002). The feed solution consists of anions and cations which enter the membrane stack and the ions migrate toward their respective electrodes on application of voltage; the cation-exchange membranes only allow the transfer of cations and prohibit the transfer of anions (Chen, 2004). Converse is true for the anion-exchange membranes. Tzanetakis et al. (2003) used perfluorosulfonic and Nafion 117 membranes to evaluate the performance for the electrodialysis of Ni(II) and Co(II) ions from a synthetic solution. The results showed a

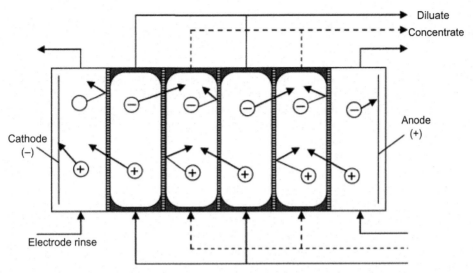

FIGURE 14.3   Principle of electrodialysis for water treatment (Van der Bruggen and Vandecasteele, 2002).

removal efficiency of Co(II) and Ni(II) as 90% and 69%, with initial metal concentrations of 0.84 and 11.72 mg/L, respectively. Jakobsen et al. (2004) reported 13% of Cd(II) removal from wastewater sludge with an initial metal concentration of 2 g/L in 120 minutes. Electrodialysis has advantages such as the production of a highly concentrated stream for recovery and the denial of unwanted impurities from water. Moreover, recovery of valuable metals such as Cr and Cu can be done (Barakat, 2011).

Electrodialysis is a membrane process, which requires clean feed, vigilant operation, and periodic maintenance to prevent damages. Electrodialysis is unable to treat effectively inorganic effluent with a metal concentration higher than 1000 mg/L, thus suggesting that the process is more appropriate for metal concentration of less than 20 mg/L (Kurniawan et al., 2006). Electrodialysis requires the development of ion-exchange membranes with higher permselectivity, lower electrical resistance, thermal and chemical stability at lower cost (Strathmann, 2010). Electro-membrane processes and components used in these processes still have technical and commercial limitations and in spite of a substantial ongoing development, need for further research to improve products and processes (Strathmann, 2010), and development of newer, economic membranes is present.

14.2.2 Adsorption-based methods

Adsorption is a surface phenomenon in which pollutants are adsorbed on the solid surface. Basically, adsorption takes place by physical forces but, sometimes, weak chemical bondings also participate in the process (Ali, 2012). Adsorption is considered as one of the suitable method for water treatment owing to ease of operation, is fast, inexpensive, and the availability of wide range of adsorbents (Lehr et al., 2005; Ali, 2012). The process of adsorption has its own limitations such as lack of high adsorption capacity adsorbents, commercial scale columns, and single adsorbent not useful for all pollutants, which prevents it from attaining the commercial importance. Some of the widely used adsorbents for pollutant removal are discussed in the following sections.

14.2.2.1 Activated carbon

Activated carbon is a popular choice among all owing to its good adsorption capacity, active free valancies, high surface area, porous structure, surface reactivity, inertness, and thermal stability (Cheremisinoff and Ellerbusch, 1978; Babel and Kurniawan, 2003). Activated carbons can be used in various forms: powdered activated carbon (PAC) has finer particle size of about 44 µm which allows faster adsorption, but it is difficult to handle in fixed adsorption beds. The granulated activated carbon (GAC), having granules of 0.6–4.0 mm in size, are hard and resistant to abrasion (Uddin, 2017). Although GAC is costlier than PAC and can be regenerated easily. The fibrous activated carbon fibers are expensive, but they can be molded easily into the shape of the adsorption system and produce low hydrodynamic resistance to flow (Uddin, 2017).

Due to the versatility, various researchers attempted to use activated carbon to work on cost reduction by generating carbon from cheap sources or by doing surface modification. Kadirvelu and Namasivayam (2003) prepared activated carbon from coir pith for adsorption of Cd(II) from aqueous solution with the adsorption capacity of 93.4 mg/g. Activated carbon modified using tetrabutyl ammonium iodide and sodium diethyl dithiocarbamate was used to remove Cu, Zn, and Cr found to be effective than plain carbon (Monser and Adhoum, 2002). Sawdust activated carbon was used to remove Cr(VI) (Karthikeyan et al., 2005). Pb(II) was removed using treated activated carbon (Goel et al., 2005). Karnib et al. (2014) performed batch experiments to evaluate removal efficiency of activated carbon for the removal of lead, cadmium, nickel, chromium, and zinc from water. Activated carbon has limitations such as cost inefficiency, expensive raw material, difficult to separate powdered form from the effluent, and costly regenerating methods. Complexing agents are required by activated carbon to improve its removal performance for inorganic matters (Babel and Kurniawan, 2003).

14.2.2.2 Clay and clay minerals

Variety of clays and clay minerals are used as adsorbent material for metal ions removal from water. Clay minerals have interlayer spaces that allow them to adsorb metal ions. In aqueous medium, clays swell and increased space between the layers accommodate the adsorbed water and ionic species (Uddin, 2017). Using clay as an adsorbent has its benefits of being cost-effective, locally available, high specific surface area, excellent adsorption properties, nontoxic nature, and large potential for ion exchange (Crini and Badot, 2010). Pretreatment is required in many cases to increase the adsorption capacity of clays (O'Connell et al., 2008). A lot of successful studies were performed, for heavy metal removal by natural clays. The removal of Cu(II) from cachaça was investigated in which 68.7% removal efficiency was achieved in 120 minutes equilibration time (Zacaroni et al., 2015). Moroccan clay materials were studied for the removal of As(V) from aqueous solution; most effective adsorbent was found to have adsorption capacity of 1.076 mg/g (Bentahar et al., 2016). Khan and Singh (2010) conducted a study to remove Cd(II), Pb(II), and Cr(VI) ions from wastewater using naturally occurring clay. In another recent work, Na-montmorillonite and Ca-montmorillonite were used to remove Pb(II), Cu(II), Co(II), Cd

(II), Zn(II), Ag(I), Hg(I), and Cr(VI) from an aqueous solution where Na-montmorillonite was more effective for heavy metal adsorption than Ca-montmorillonite (Chen et al., 2015). Modified clays were also used as they showed enhanced removal efficiency. Cellulose-clay biopolymer composite was discovered to adsorb chromium with 99.5% removal efficiency (Kumar et al., 2011). A novel titanium pillared clay impregnated with potassium iodine was made to eliminate mercury (Shen et al., 2015).

Nano-clay composites such as organically modified clays or polymer clay nanocomposites are gaining attention as hybrid organic—inorganic nanomaterials. Research works proposes that modified clay minerals demonstrate a promising class of adsorbent materials for water purification (Uddin, 2017). The efficiency of clay minerals is comparable to activated carbon, only the problem related with recovery of adsorbent from filters after use needs to be worked upon (Unuabonah et al., 2008). Nano-clay is also capturing interest of researchers due to their capacity for selective adsorption (Uddin, 2017).

14.2.2.3 Fly ash

Fly ash is a particulate material produced from the combustion of coal in thermal power plants. Huge production of fly ash is leading to its disposal problems and hence methods are being studied for its use/disposal. Many researchers have used fly ashes as adsorbents for wastewater or air pollutants control (Alinnor, 2007). Fly ash finds its potential in wastewater treatment owing to its chemical composition and its physical properties such as surface area, porosity, and particle size distribution (Cetin and Pehlivan, 2007). Moreover, fly ash acts as a good neutralizer due to its alkaline nature. A lot of research work has been carried out for effective removal of toxic metals using fly ash (Bayat, 2002a,b; Weng and Huang, 2004; Soni and Shukla, 2019). A study was performed to evaluate the effectiveness of fly ash in the removal of Zn^{2+} and Ni^{2+} by adsorption (Cetin and Pehlivan, 2007). Fly ash was also found to be effective for the removal of mercury (Kapoor and Viraraghavan, 1992), and the adsorption capacity of coal fly ash was equivalent to that of activated powdered charcoal (Sen and De, 1987).

Fly ash could be easily solidified after the heavy metals are adsorbed but the problem of leaching might persists and has to be considered and evaluated (Babel and Kurniawan, 2003). Use of fly ash may cause leaching of some elements into water, thereby creating secondary environmental pollution. Fly ash also finds its application in synthesizing zeolites, which in turn acts as an adsorbent too.

14.2.2.4 Zeolites

Zeolites are natural materials present in various deposits and can also be synthesized synthetically to tailor the properties for specific applications (Pitcher et al., 2004). Zeolites are highly crystalline aluminosilicate materials, consisting of either Si or Al atoms interconnected via oxygen bridges (Nagarjuna et al., 2015). The exchangeable ions present in the zeolites makes them suitable for removing heavy metals from wastewaters (Erdem et al., 2004). The physical and chemical properties of zeolites make them a versatile material for researchers to study and tailor the properties for removal of anionic compounds (Figueiredo and Quintelas, 2014).

Mier et al. (2001) investigated the interactions of Pb^{2+}, Cd^{2+}, and Cr^{3+} competing for ion-exchange sites in natural clinoptilolite. Inglezakis et al. (2002) studied the ion exchange of Pb^{2+}, Cu^{2+}, Fe^{3+}, and Cr^{3+} on natural clinoptilolite and demonstrated that equilibrium is unfavorable for Cu^{2+}, favorable for Pb^{2+}, and of sigmoid shape for Cr^{3+} and Fe^{3+}. Wingenfelder et al. (2005) investigated the removal of Fe, Pb, Cd, and Zn from synthetic mine waters using natural zeolite. Synthetic zeolite produced from coal fly ash was used for investigating the removal of Cr, Pb, Ni, Cu, Cd, and Zn from contaminated solution (Koukouzas et al., 2010). Synthetic zeolites was also evaluated as adsorbent in removing Cu, Pb, Zn, Ni, Cr, Fe, and As by Rios et al. (2008). El-Kamash et al. (2005) effectively removed zinc and cadmium ions using synthetic zeolite A. Soni and Shukla (2019) studied the removal of arsenic using a synthetic zeolite composite and reported the maximum adsorption capacity for 100 μg/L of initial arsenic concentration was found to be 49.23 μg/g.

14.2.2.5 Nanoparticles

Nanoparticles have gained wide attention as adsorbents, due to its properties such as high reactivity, large surface area, small size, ease of separation, catalytic potential, and large number of active sites for interaction with different contaminants (Xu et al., 2012; Ali, 2012). Nanomaterials such as nanoscale zerovalent iron (FeO), Fe_2O_3, Fe_3O_4, TiO_2, SiO_2, and Al_2O_3 are the most commonly used materials that have been applied as adsorbents (Mahdavi et al., 2012). Iron oxide nanomaterials are also extensively studied due to their novel properties and functions, nano-size range, high surface area–to-volume ratios and superparamagnetism (Afkhami et al., 2010; Xu et al., 2012). Ponder et al. (2000), Kanel et al. (2005), and Sharma et al. (2009) have used nanoparticles of zerovalent iron for sorption of arsenic, cadmium, chromium, silver, selenium, lead, and zinc metal ions. Titanium dioxide is widely investigated due to its high photocatalytic activity, nontoxicity, high stability, and excellent dielectric properties (Nagaveni et al., 2004). Nanoalumina demonstrated high resistance to chemical agents and gave brilliant performance as a catalyst in many chemical reactions (Mahdavi et al., 2012). Nanoalumina has been used for the removal of cadmium, copper, chromium, lead, and mercury metal ions (Pacheco and Rodríguez, 2001). Also, nanoparticles of alumina possess qualities of having low cost, good thermal stability, and high surface area (Alvarez-Ayuso et al., 2007). Hybrid nanoparticles have also been used in water treatment and were found potent to remove heavy metals. Zhang et al. (2008) developed a polymeric hybrid nanoparticle sorbent for sorption of lead, cadmium, and zinc ions from aqueous solution. Afkhami et al. (2010) developed 2,4-dinitrophenylhydrazine immobilized on sodium dodecylsulfate–coated nanoalumina for the removal of various metal cations. Deliyanni et al. (2007) reported high removal capacity of zinc ion on akageneite nanoparticles. Li et al. (2003) reported the removal of Cd(II) from aqueous solution on surface oxidized carbon nanotubes with H_2O_2, $KMnO_4$, and HNO_3. Graphene oxide–based composite materials are also gaining wide attention for treating water contaminants. Ultrathin graphene oxide framework layer was deposited on a modified Torlon hollow fiber support via a layer-by-layer to form composite membrane with superior nanofiltration (NF) performance to remove Pb^{2+}, Ni^{2+}, and Zn^{2+} (Zhang et al., 2016). Graphene oxide–based microbots containing nanosized multilayers of graphene oxide, nickel, and platinum removed heavy metals from water (Vilela et al., 2016). New magnetic biosorbent hydrogel beads were prepared using graphene oxide, modified biopolymer gum tragacanth, polyvinyl alcohol, to remove dyes and heavy metals (Sahraei et al., 2017).

Expert supervision is required for the removal of nanoparticles after adsorption as they have very small size and might pose health and environmental issues (Ali, 2012). Various methods employed for separation are magnetism, cross-flow filtration, and centrifugation. Though use of nanoparticles has lot of potential in water treatment but some issues related to safety are persistent. A few are nonbiodegradable and enter into the human body (Hoet et al., 2004; Singer, 2010). None is known till date which can be proclaimed to be safe completely and is nontoxic. Environmental contamination may occur during the process of synthesis, application, and disposal.

14.2.3 Bioadsorption

Water pollution as a result of inorganic pollutants and its removal using low-cost-effective techniques is a major area of concern for sustainable management of water resources. Although many physical and chemical practices such as membrane technologies, chemical oxidation, and chemical precipitation are being used for the removal of pollutants, they are not cost-effective and cannot be fully implemented, as a result of which, bioadsorbents have enticed significant attention as an effective substitute for the removal of inorganic pollutants from the water bodies. Many naturally occurring low-cost bioadsorbents are being used for the removal of heavy metals and dyes in the water bodies such as rice husk, coconut shells, native algae, peels of native fruits, leaves, roots and bark of many native plants, and recycle paper sludge. As a term bioadsorption is simply the process of adsorption of contaminants on to the surface of biological materials, it not only aid in removal of inorganic contaminants but also helps in utilization of biowaste.

14.2.3.1 Rice husk

Most widely used agricultural waste is rice husk, whose capabilities as a potential bioadsorbent has been exploited since ages. Rice husk is made up of lignin, cellulose, hemicellulose, and mineral ash, as a result of which, it is chemically and mechanically stable as well as insoluble in water. Water containing heavy metals when treated with rice husk shows the removal of cadmium, copper, zinc, and chromium to a great extent (Munaf and Zein, 1997). Various different factors can affect the adsorption rate of rice husk such as adsorbent concentration, temperature, pH, as well as size of particles. Rice husk is pretreated to increase the adsorption rate either with hydrochloric acid, tartaric acid, sodium carbonate, and sodium hydroxide (Guo et al., 2003; Kumar and Bandyopadhyay, 2006). Increase in cadmium adsorption rate is observed when rice husk is treated with HCL and Na_2CO_3. Pretreated rice husk shows best adsorption rate of copper and lead at pH 2–3 (Wong et al., 2003). Rice husk treated thermally shows high adsorption of phenolic compounds from aqueous solution (Amarasinghe and Williams, 2007). Capacity of pretreated rice husk to adsorb phosphate has been increased at pH 6 removing 89% of it. RHA (rice husk ash) produced from burning rice husk is a potential adsorbent for removing heavy metal such as Zn, Ni, and Pb. RHA main component is silica (80%–89%), as a result of which it has been used to produce low-cost precursors and value-added silica. Very recently ceramic hollow fiber membrane-based rice husk ash

(CHFM/RHA) is been used both to adsorb heavy metals from contaminated water (Ismail et al., 2016).

14.2.3.2 Sugarcane bagasse

Cellulose, lignin, and pentosan-derived sugarcane bagasse (SCB) is nothing but the waste produced from sugarcane industry which has been exploited for long as a potential adsorbent for the removal of various inorganic compounds from water bodies. Many studies show the adsorption of cadmium and lead on the silica derived from SCB (Ibrahim et al., 2006; Pehlivan et al., 2008). Activated carbon prepared from SCB adsorbs 100% cadmium and zinc on providing pH of 8 (Mohan and Singh, 2002). Very recently modified forms of SCB such as NaOH−SCB or HCL−SCB are used for the removal of mercury (Khoramzadeh et al., 2013). Pretreatment of SCB with oxalic acid, citric acid, and NaOH have shown to adsorb and remove copper ions (Cu^{2+}) from water (Altundogan et al., 2007; Akiode et al., 2015). SCB are also used to remove fluoride, nitrate, and phosphate ions from the water (Hena et al., 2015; Singh et al., 2015).

14.2.3.3 Peels of fruit and vegetable paste

Many studies have reported the adsorption of various inorganic cations such as Pb^{2+}, Zn^{2+}, CO^{2+}, Ni^{2+}, and Cu^{2+} with the help of peels of orange and banana (Annadurai et al., 2003). Various chemical groups present on peel surface such as alkenes, ester, sulfonic acid, amine, and hydroxyl are accountable for chemical adsorption. Adsorption capacity of these peels is enhanced by prior treatment of acid and alkali such as nitric acid, sodium hydroxide, calcium chloride, and citric acid in order to aggrandize the surface chemistry of the substance (Dhakal et al., 2005). Peels of some fruits such as avocado and dragon fruit have been used in water purification to remove various heavy metal and dyes such as alcian blue and methylene blue with removal rate of 95% (Mallampati et al., 2015). Various types of agro-industrial waste are used as a potential adsorbents or biosorbents in removal of heavy metals (Shakoor, 2016). Adsorption rate of Cu^{2+} are enhanced four times when peels of pineapple and banana are modified through physiochemical process (Romero-Cano et al., 2017). Most studied banana peels are found to be an effective adsorbent in the removal of anionic dyes from water. Very recently adsorbent made from banana peels are found to remove fluoride effectively from water (Mondal and Roy, 2018).

14.2.3.4 Biochar

Biochar, yet another product of biological origin and a potential adsorbent, is obtained from thermal degradation of biomass in oxygen-deprived environment. Biochar can be obtained from different materials having biological origin such as pine wood, wheat bran, oak bark, rice husk, soybean stalk, and corn straw, proved to be efficient in the removal of heavy metals from contaminated water. Biochar obtained from wood can adsorb copper and zinc to great extent (Chen et al., 2011). Eighty percent removal of mercury is reported with the help of biochar obtained from soybean stalk (Kong et al., 2011).

14.2.3.5 Miscellaneous adsorbent

Various researches are inclining toward the modifications of adsorbents surface chemically in order to increase its efficiency. In one study, magnetic adsorbent derived from walnut shells found to be efficient in the removal of lead from water (Safinejad et al., 2017). Bark obtained from *Tamarindus indica* is been used for the synthesis of BMIOP (bark-based magnetic iron oxide particle) that are found to be an efficient removal of arsenic (As^{3+}) from water. Main advantage of this is the ejection of As laden BMIOP by normal magnet (Dhoble et al., 2018). New magnetically modified hydrochar precursors, obtained from pine wood saw dust after heat treatment in air, show better removal of mercury ion from water (Wang et al., 2018). Indian bael leaves (*Aegle marmelos* tree) and ferric oxide nanoparticles were used to develop a magnetic bioadsorbent capable of removing arsenic from water (Sahu et al., 2019). A bioadsorbent obtained from spent coffee ground is found to have strong affinity toward chromium(VI) and aid in removal process (Mohan et al., 2019). A cross-linked chitosan microsphere was explored and found to be potent adsorbent for iodide ion (Zhang et al., 2019). There are various other materials that are used as bioadsorbents and have been utilized to remove various pollutants and heavy metals from water bodies. Some of them such as eggshells, almond shells, peanut shells, neem bark, black tea residues, coffee residues, and saw dust are summarized and shown in Table 14.1.

14.2.3.6 Biosorption

Biosorption is a phenomenon in which contaminants such as ions and heavy metals get bioaccumulated on to the surface of biomass (living or dead) through various physical–chemical interactions. Raw biomass of fungi, bacteria, and algae are been used to obtain biosorbents. Metals ions interact with different types of functional group present in bacteria and fungi such as carboxyl, amine, carbonyl, hydroxyl, and sulfonate that allow their sorption and accumulation on their surface. Many different species of bacteria, algae, and fungi are being used as biosorbents to remove different metals, for example, fungi (*Aspergillus, Agaricus, Trichaptum, Saccharomyces, Candida*) are used for the removal of Cd, Pb, and Cr; bacteria (*Bacillus, Escherichia, Pseudomonas, Anabaena, Synechocystis*) are used for

TABLE 14.1 List of various bioadsorbents and metals they removed.

Sr. no.	Adsorbents	Types of pollutants	References
1	Egg shell	Cd, Cr	Park et al. (2007)
2	Almond shells	Fe	Pehlivan et al. (2009)
3	Corncobs	Cu	Khan and Wahab (2007)
4	Peanut shell	Cu, Cr	Zhu et al. (2009)
5	Tea residues	Zn, Cd, and Co	Amarasinghe and Williams (2007)
6	Coffee residues	Cu, Zn, Cd, and Pb	Utomo (2007)
7	Wheat bran	Pb	Özer et al. (2004)
8	Black gram husk	Pb, Cd, Zn, and Cu	Saeed et al. (2002)
9	Walnut shell	Cr	Orhan and Büyükgüngör (1993)
10	Mango peel	Cu	Iqbal et al. (2009)

removal of Cu, Cd, Pb, and Cr; and algae (*Chlamydomonas, Enteromorpha, Codium, Sargassum, Oocystis, Gelidium, Porphyra*) are used for removal of Cd, Cu, Pb, Zn, Ni, and Cr (Abbas et al., 2014). Because of high sorption capabilities and availabilities in different environment, algae is one of the most promising biosorbents (Flouty and Estephane, 2012; Trinelli et al., 2013). Many types of different functional group are present on algae cell wall such as carboxylic group that passively can bind to metal ions through processes such as ion exchange or electrostatic attractions and aid in the removal process. Very recently high adsorption rate of Cr^{6+} is achieved by green algae *Cladophora glomerata* at pH 2 (Al-Homaidan et al., 2018). A novel strategy has been developed for removal of Pb^{2+} by using chlorella spices cultivated in phosphorous concentrations as adsorbent (Li et al., 2019). A lead-resistant bacteria (*Bacillus megaterium*) is discovered from alkaline soil contaminated with Pb, that shows highest adsorption capacity and emerge as a new biosorbent for removal of Pb from soil and water both (Li et al., 2018). Efficiency of a biosorbent to remove a metal ion may get reduced under the influence of others ion present. To overcome this, white rot fungus (*Phanerochaete chrysosporium*) was explored to observe simultaneous adsorption of Ni and Cd ions from the water (Noormohamadi et al., 2019). Very recently a nanocomposite prepared from bacterial cellulose and TiO_2 was found to be efficient in removing lead from water (Shoukat et al., 2019).

14.2.4 Membrane filtration technologies

Membrane separation is extensively used as it is convenient; it uses a semipermeable membrane for the division of stream into permeate and retentate (Mallevialle et al., 1996). The main types of membrane filtration are microfiltration (MF), ultrafiltration (UF), NF, and reverse osmosis (RO). The success of membrane-based processes depends on the membrane material. Membrane must possess qualities such as high permeate flux, high contaminant rejection, great durability, good chemical resistance, and low cost (Zhou and Smith, 2002). Inorganic membranes are brittle and costly and have less commercial value (Zhou and Smith, 2002). Organic polymers are the main commercially used membrane materials. The advantages of membrane techniques are no need of chemicals, pH adjustment, compact equipment, simple automation, and constant water quality. The major limitation of such processes lies in the cost of the membrane, membrane fouling, and regeneration.

14.2.4.1 Microfiltration and ultrafiltration

MF and UF are both low-pressure driven membrane separation process. MF membrane has a pore size of 0.1 μm or greater, whereas UF membrane is porous and allows coarsest solutes to be rejected and has a pore size of 0.002 to 0.1 μm. Applications such as cold sterilization in the pharmacy, protein separation, recovery of metals uses UF. MF can be used as a pretreatment to RO or NF to reduce fouling potential.

14.2.4.2 Nanofiltration

NF membranes have a nominal pore size of approximately 0.001 μm. A higher operation pressure is required than MF or UF to pass water from NF membrane pores. NF membranes remove the alkalinity causing the resulting water to be corrosive and hence measures have to be taken to increase alkalinity of the effluent to reduce the corrosivity. NF is also known as softening membranes as they remove hardness from water. Pretreatment is required by hard water before NF to avoid precipitation of hardness ions on the membrane. However, more energy is required for NF than MF or UF. NF also has its own limitations of membrane fouling, scaling, and limited lifetime (Van Der Bruggen et al., 2008).

14.2.4.3 Reverse osmosis

RO can effectively remove nearly all inorganic contaminants from water and can also effectively remove cysts, viruses, bacteria, and natural organic substances. The effect of RO is enhanced when used in series or in multiple units. RO unit has simple operation and lesser operator attention is required. Though it has some major limitations of high capital and operating costs, wastewater management (25%−50% of the feed), requirement of high level of pretreatment, membrane fouling.

Qdais and Moussa (2004) used both RO and NF for the treatment of wastewater containing copper and cadmium. Al-Rashdi et al. (2013) demonstrated the rejection of heavy metal ions using a commercial NF membrane (NF270); using 1000 mg/L concentration level, pH $= 1.5 \pm 0.2$, and 4 bar, the rejection was 99%, 89%, and 74% for cadmium, manganese, and lead, respectively. Nanometric graphene oxide framework membranes were developed exhibiting a high pure water permeability of 5.01 $L/m^2/h/bar$ and comparably high rejections toward Mg^{2+}, Pb^{2+}, Ni^{2+}, Cd^{2+}, and Zn^{2+} (Zhang et al., 2015). Gao et al. (2014) developed chelating polymer modified P84 NF hollow fiber membranes with rejections of heavy metals higher than 98%, also rejections to mixed ions with rejections more than 99% was demonstrated. In a study, poly(amidoamine) dendrimer was grafted on a thin film composite membrane and showed excellent rejection to Pb^{2+}, Ni^{2+}, Cd^{2+}, As^{5+}, etc. ($R > 99\%$) (Zhu et al., 2015). Another study reported the development of polybenzimidazole/polyethersulfone dual-layer NF hollow fiber membrane showing rejections to Mg^{2+} (98%) and Cd^{2+} (95%) (Zhu et al., 2014). Novel polyvinylidene fluoride NF membrane blended with functionalized halloysite nanotubes was prepared which reported a high heavy metal removal (Zeng et al., 2016).

NF and RO can treat multiple metal contaminations in wastewater. Also high rejection toward divalent cations is shown by NF as compared with monovalent cations (Al-Rashdi et al., 2011). Major challenges faced in membrane processes are control of membrane fouling and better membrane materials.

14.2.5 Advanced oxidation processes

Advanced oxidation processes (AOPs) generate in situ oxidizing agents (mainly hydroxyl radicals) to remediate the pollutants. These radicals are expected to adequately react with wastewater pollutants and render them to be less toxic, thereby providing an ultimate solution for wastewater treatment (Huang et al., 1993). All AOPs comprise of two steps, which includes the formation of oxidation species in situ and then the reaction of the produced oxidant with the contaminant (Miklos et al., 2018). AOPs are environmental friendly when compared to other chemical and biological processes as they neither cause primary pollution nor secondary pollution (Ayoub et al., 2010). Hydroxyl radicals are formed in the AOP which carries the oxidation of contaminant. AOPs are mostly used for organic contaminants, therefore it is not discussed here. But semiconductor based photocatalysis have been used by researchers for heavy metal removal and hence is discussed in the following section.

14.2.5.1 Semiconductor photocatalysis

Semiconductor photocatalysis has received considerable attention due to rapid degradation of pollutants. Semiconducting materials (TiO_2, ZnO, SnO_2, and CeO_2) act as catalysts, owing to its favorable combination of electronic structures which is characterized by a filled valence band and an empty conduction band, light absorption properties, charge transport characteristics, and excited states lifetime (Miklos et al., 2018). When photons fall on the surface of a semiconductor with greater energy than the semiconductor bandgap, formation of electron—hole pairs takes place (Barakat, 2011). The generated charge carriers migrate to the semiconductor surface and reduce or oxidize the species. TiO_2 is mostly used for such reactions due to the low costs and easy availability in various forms, sizes, nontoxicity, and photochemical stability (Miklos et al., 2018). Barakat (2005) studied the photocatalytic degradation using UV-irradiated TiO_2 suspension for destroying both cyanide ions and copper. TiO_2 thin films immobilized on glass were investigated by Kajitvichyanukul et al. (2005) which successfully removed Cr(VI). Papadam et al. (2007) and Wang et al. (2008) studied the photocatalytic reduction of Cr(VI) over TiO_2 catalysts in absence and presence of organic compounds. A novel photocatalyst, TiO_2 doped with neodymium, was prepared by the sol—gel method by Rengaraj et al. (2007) and used for the photocatalytic reduction of Cr(VI) under UV illumination.

Several constraints related to technical aspects, development of catalyst, design of reactor, and optimization process have to be addressed for the promotion of photocatalytic water-treatment technology in the near future. A lot of energy is consumed in photocatalytic processes and hence solar energy substitute is put into use to curb the cost. But for that, catalyst has to be improved to have wider solar spectra.

14.3 Summary and conclusion

During the past two decades, numerous techniques have evolved for water treatment. But the main fact remains that there is no best method of treatment. Each process has its own benefits and limitations as tabulated in Table 14.2. The treatment process that needs to be used is best decided by the type of pollutant which needs to be treated. The best treatment process is the one which is efficient, cost-effective, and has minimum environmental impact. An universal method for pollutant removal is not available and might not be cost-effective for each pollutant. Presently, hybrid techniques/materials are trending for pollutant removal as they allow to utilize the properties of several techniques/materials in a single system. Hybrid system are more efficient and allow to achieve the desired water quality in an economic way.

TABLE 14.2 Advantages and limitations of these techniques.

Technique	Advantages	Limitations
Adsorption	• Simple technique • Wide variety of adsorbents to treat various contaminants • Numerous commercial adsorbents available	• Causes secondary pollution • Regeneration is costly and not that effective • Cost of adsorbent effect the overall cost of treatment • Mostly pH-dependent processes • Pre- and posttreatment required
Electro-assisted methods	• In situ coagulant generation • Easy automation	• Energy consumption can be an issue, but these days powering by renewable energy is gaining interest • Scaling and passivation of electrodes • Sludge disposal
Membrane filtration	• Small space required • Commercial membranes available • No chemicals needed • Fast and efficient process with high quality effluent	• Investment, maintenance, and operating cost is high • Energy requirement is high • Membrane clogging
Advanced oxidation processes	• In situ radical production • Efficient for recalcitrant pollutants	• Formation of by products • Energy intensive

Various factors such as type of waste, concentration of contaminant, cost contributes toward the selection of particular treatment (Ghodbane et al., 2008). Increasing awareness in both environmental and economical aspects has led to the idea of process development which have the potential of recovery—reuse of materials to avoid stress on mineral ores and also save the manufacturing cost of materials by promoting reuse of discarded materials. The use of low-cost material is trending these days which can either be by product of some process or is abundantly available and requires less processing (Blöcher et al., 2003). Conventional adsorbents usually have limited efficiency due to the surface area or active sites, the lack of selectivity, and the adsorption kinetics. Nano-adsorbents in turn give better results as they have high specific area and tunable pore size short intraparticle diffusion distance (Qu et al., 2013). Apart from nanomaterials, doped metal oxides are also a subject of interest as they possess high surface-to-volume ratio, enhanced magnetic property, special catalytic properties, etc. (Gupta et al., 2011). The problem of high concentrations can be overcome using NF, but it suffers from its own limitations of high cost and membrane fouling. Some researchers combined different separation processes (e.g., flotation and MF membrane and flotation and NF/RO membrane) for heavy metal removal in order to minimize the limitations of using them alone (Sudilovskiy et al., 2008; Nguyen et al., 2009) combined the adsorption process (nanoscale sorbent) with the NF process to achieve very high treatment efficiency.

Nanotechnology-enabled water/wastewater treatment processes have shown great promise in laboratory studies; their readiness for commercialization requires significant research. A variety of hurdles are faced in their development such as technical aspects, cost-effectiveness, and potential environmental and human risk. Long-term performance evaluation of nanotechnology-based treatment is needed and then only a suitable comparison can be made with the existing and established techniques. Cost-effectiveness is a major parameter in which work needs to be done. Few exceptions to costly nanomaterials are nano-TiO_2, nanoscale iron oxide, and polymeric nanofibers. The cost-effectiveness can also be improved by retaining and reusing nanomaterials.

To overcome the challenges of water treatment processes a collaborative approach between research institutions, industry, government, and other stakeholders is important. It is our belief that advancing nanotechnology, oxidation processes, novel membranes while carefully weighing the opportunities, and risks on the potential environmental and health impact will be able to provide robust solutions to the challenges faced by the water-treatment techniques. While utilizing these available techniques at various stages, more robust system could be developed. A schematic diagram for that is shown in Fig. 14.4, where, through various techniques and scale of treatment, the major output is the availability of nonpotable water.

Many times the surface water from rivers, lakes, dam reservoirs are being lifted for supply as domestic water. These water tends to have higher organic contaminants that needs to be removed. Hence first stage, that is, MfHT (*multifunction high throughput filter*) could be used. Various type of micro- and nanoparticle will be tested for preparation for MfHT using adsorption technology. This filter is designed to be used as first stage for community water storages by government offices such as IPH (Irrigation and Public Health). After cleaning the first stage water will be stored in big tanks and where other heavy metal contaminants such as arsenic and fluoride will be treated using the *ECog Filter: Electrochemical Coagulation Filter*. This is the part of second stage of treatment that will be carried out at community level. After purification at community scale, lots of sludge is generated, which can be used for the preparation of bricks and for esthetic decoration purposes. This water will be supplied through pipes to domestic purposes. Over a period of time, due to lack of maintenance, some heavy metals gets in this supply and reaches households. Hence, at domestic levels, *ZrGO filter: zeolite based reduced graphene oxide filter* could be used. This falls in domain of third filtration. Also most of the houses in India have their own

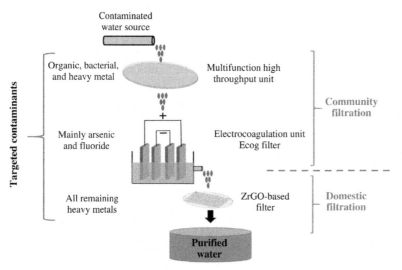

FIGURE 14.4 Scheme showing the overall concept of DWPS. *DWPS*, Drinking water purification system.

borewells for water supply. As the government supply is not 24 hours and gets erratic, many household extract ground water using their borewells. These water are generally not treated, hence the metals, heavy metals, and other trace contaminants will be purified at domestic level using ZrGO filter. These domestic filters can be used in rural and urban areas to remove heavy metals. Hence, this whole scheme displays the system where and how water can be treated and supplied so that every human can have the right to clean drinking water for various domestic purposes.

References

Abbas, S.H., Ismail, I.M., Mostafa, T.M., Sulaymon, A.H., 2014. Biosorption of heavy metals: a review. J. Chem. Sci. Technol. 3 (4), 74–102.

Afkhami, A., Saber-Tehrani, M., Bagheri, H., 2010. Simultaneous removal of heavy-metal ions in wastewater samples using nano-alumina modified with 2, 4-dinitrophenylhydrazine. J. Hazard. Mater. 181, 836–844.

Akbal, F., Camcı, S., 2011. Copper, chromium and nickel removal from metal plating wastewater by electrocoagulation. Desalination 269, 214–222.

Akiode, O., Idowu, M., Omeike, S., Akinwunm, F., 2015. Adsorption and kinetics studies of Cu(II) ions removal from aqueous solution by untreated and treated sugarcane bagasse. Global Nest 17, 583–593.

Al-Homaidan, A.A., Al-Qahtani, H.S., Al-Ghanayem, A.A., Ameen, F., Ibraheem, I.B., 2018. Potential use of green algae as a biosorbent for hexavalent chromium removal from aqueous solutions. Saudi J. Biol. Sci. 25, 1733–1738.

Ali, I., 2012. New generation adsorbents for water treatment. Chem. Rev. 112, 5073–5091.

Ali, I., Aboul-Enein, H.Y., 2006. Instrumental Methods in Metal Ion Speciation. CRC Press.

Alinnor, I., 2007. Adsorption of heavy metal ions from aqueous solution by fly ash. Fuel 86, 853–857.

Al-Rashdi, B., Somerfield, C., Hilal, N., 2011. Heavy metals removal using adsorption and nanofiltration techniques. Sep. Purif. Rev. 40, 209–259.

Al-Rashdi, B., Johnson, D., Hilal, N., 2013. Removal of heavy metal ions by nanofiltration. Desalination 315, 2–17.

Al-Shannag, M., Al-Qodah, Z., Bani-Melhem, K., Qtaishat, M.R., Alkasrawi, M., 2015. Heavy metal ions removal from metal plating wastewater using electrocoagulation: kinetic study and process performance. Chem. Eng. J. 260, 749–756.

Altundogan, H.S., Arslan, N.E., Tumen, F., 2007. Copper removal from aqueous solutions by sugar beet pulp treated by NaOH and citric acid. J. Hazard. Mater. 149, 432–439.

Alvarez-Ayuso, E., Garcia-Sanchez, A., Querol, X., 2007. Adsorption of Cr(VI) from synthetic solutions and electroplating wastewaters on amorphous aluminium oxide. J. Hazard. Mater. 142, 191–198.

Amarasinghe, B., Williams, R., 2007. Tea waste as a low cost adsorbent for the removal of Cu and Pb from wastewater. Chem. Eng. J. 132, 299–309.

An, C., Huang, G., Yao, Y., Zhao, S., 2017. Emerging usage of electrocoagulation technology for oil removal from wastewater: a review. Sci. Total Environ. 579, 537–556.

Annadurai, G., Juang, R.S., Lee, D.J., 2003. Adsorption of heavy metals from water using banana and orange peels. Water Sci. Technol. 47 (1), 185–190.

Ayoub, K., Van hullebusch, E.D., Cassir, M., Bermond, A., 2010. Application of advanced oxidation processes for TNT removal: a review. J. Hazard. Mater. 178, 10–28.

Babel, S., Kurniawan, T.A., 2003. Low-cost adsorbents for heavy metals uptake from contaminated water: a review. J. Hazard. Mater. 97, 219–243.

Barakat, M., 2005. Adsorption behavior of copper and cyanide ions at TiO$_2$–solution interface. J. Colloid Interface Sci. 291, 345–352.

Barakat, M., 2011. New trends in removing heavy metals from industrial wastewater. Arab. J. Chem. 4, 361–377.

Bayat, B., 2002a. Comparative study of adsorption properties of Turkish fly ashes: I. The case of nickel(II), copper (II) and zinc(II). J. Hazard. Mater. 95, 251–273.

Bayat, B., 2002b. Comparative study of adsorption properties of Turkish fly ashes: II. The case of chromium(VI) and cadmium(II). J. Hazard. Mater. 95, 275–290.

Bentahar, Y., Hurel, C., Draoui, K., Khairoun, S., Marmier, N., 2016. Adsorptive properties of Moroccan clays for the removal of arsenic(V) from aqueous solution. Appl. Clay Sci. 119, 385–392.

Blöcher, C., Dorda, J., Mavrov, V., Chmiel, H., Lazaridis, N., Matis, K., 2003. Hybrid flotation—membrane filtration process for the removal of heavy metal ions from wastewater. Water Res. 37, 4018–4026.

Canizares, P., Paz, R., Sáez, C., Rodrigo, M.A., 2009. Costs of the electrochemical oxidation of wastewaters: a comparison with ozonation and Fenton oxidation processes. J. Environ. Manage. 90, 410–420.

Cetin, S., Pehlivan, E., 2007. The use of fly ash as a low cost, environmentally friendly alternative to activated carbon for the removal of heavy metals from aqueous solutions. Colloids Surf. A: Physicochem. Eng. Asp. 298, 83–87.

Chen, G., 2004. Electrochemical technologies in wastewater treatment. Sep. Purif. Technol. 38, 11–41.

Chen, G., Chen, X., Yue, P.L., 2000. Electrocoagulation and electroflotation of restaurant wastewater. J. Environ. Eng. 126, 858–863.

Chen, X., Chen, G., Chen, L., Chen, Y., Lehmann, J., Mcbride, M.B., et al., 2011. Adsorption of copper and zinc by biochars produced from pyrolysis of hardwood and corn straw in aqueous solution. Bioresour. Technol. 102, 8877–8884.

Chen, C., Liu, H., Chen, T., Chen, D., Frost, R.L., 2015. An insight into the removal of Pb(II), Cu(II), Co(II), Cd(II), Zn(II), Ag(I), Hg(I), Cr(VI) by Na(I)-montmorillonite and Ca(II)-montmorillonite. Appl. Clay Sci. 118, 239–247.

Cheremisinoff, P.N., Ellerbusch, F., 1978. Carbon Adsorption Handbook. Science Publishers, Ann Arbor, MI.

Clifford, D., Subramonian, S., Sorg, T.J., 1986. Water treatment processes. III. Removing dissolved inorganic contaminants from water. Environ. Sci. Technol. 20, 1072–1080.

Crini, G., Badot, P.-M., 2010. Sorption Processes and Pollution: Conventional and Non-Conventional Sorbents for Pollutant Removal From Wastewaters. Presses Univ. Franche-Comté.

Crini, G., Lichtfouse, E., 2018. Advantages and disadvantages of techniques used for wastewater treatment. Environ. Chem. Lett. 17, 1–11.

Deliyanni, E., Peleka, E., Matis, K., 2007. Removal of zinc ion from water by sorption onto iron-based nanoadsorbent. J. Hazard. Mater. 141, 176–184.

Dhakal, R.P., Ghimire, K.N., Inoue, K., 2005. Adsorptive separation of heavy metals from an aquatic environment using orange waste. Hydrometallurgy 79 (3–4), 182–190.

Dhoble, R.M., Maddigapu, P.R., Bhole, A.G., Rayalu, S., 2018. Development of bark-based magnetic iron oxide particle (BMIOP), a bio-adsorbent for removal of arsenic(III) from water. Environ. Sci. Pollut. Res. 25, 19657–19674.

Drogui, P., Blais, J.-F., Mercier, G., 2007. Review of electrochemical technologies for environmental applications. Recent Pat. Eng. 1, 257–272.

El-Kamash, A., Zaki, A., El geleel, M.A., 2005. Modeling batch kinetics and thermodynamics of zinc and cadmium ions removal from waste solutions using synthetic zeolite A. J. Hazard. Mater. 127, 211–220.

Erdem, E., Karapinar, N., Donat, R., 2004. The removal of heavy metal cations by natural zeolites. J. Colloid Interface Sci. 280, 309–314.

Figueiredo, H., Quintelas, C., 2014. Tailored zeolites for the removal of metal oxyanions: overcoming intrinsic limitations of zeolites. J. Hazard. Mater. 274, 287–299.

Flouty, R., Estephane, G., 2012. Bioaccumulation and biosorption of copper and lead by a unicellular algae Chlamydomonas reinhardtii in single and binary metal systems: a comparative study. J. Environ. Manage. 111, 106–114.

Gao, J., Sun, S.-P., Zhu, W.-P., Chung, T.-S., 2014. Chelating polymer modified P84 nanofiltration (NF) hollow fiber membranes for high efficient heavy metal removal. Water Res. 63, 252–261.

Garcia-Segura, S., Eiband, M.M.S., De Melo, J.V., Martínez-Huitle, C.A., 2017. Electrocoagulation and advanced electrocoagulation processes: a general review about the fundamentals, emerging applications and its association with other technologies. J. Electroanalyt. Chem. 801, 267–299.

Ghodbane, I., Nouri, L., Hamdaoui, O., Chiha, M., 2008. Kinetic and equilibrium study for the sorption of cadmium(II) ions from aqueous phase by eucalyptus bark. J. Hazard. Mater. 152, 148–158.

Goel, J., Kadirvelu, K., Rajagopal, C., Garg, V.K., 2005. Removal of lead(II) by adsorption using treated granular activated carbon: batch and column studies. J. Hazard. Mater. 125, 211–220.

Grimm, J., Bessarabov, D., Sanderson, R., 1998. Review of electro-assisted methods for water purification. Desalination 115, 285–294.

Guo, Y., Yang, S., Fu, W., Qi, J., Li, R., Wang, Z., et al., 2003. Adsorption of malachite green on micro-and meso-porous rice husk-based active carbon. Dye Pigment. 56, 219—229.

Gupta, K., Bhattacharya, S., Chattopadhyay, D., Mukhopadhyay, A., Biswas, H., Dutta, J., et al., 2011. Ceria associ-ated manganese oxide nanoparticles: synthesis, characterization and arsenic(V) sorption behavior. Chem. Eng. J. 172, 219—229.

Halim, M., Conte, P., Piccolo, A., 2003. Potential availability of heavy metals to phytoextraction from contami-nated soils induced by exogenous humic substances. Chemosphere 52, 265—275.

Heidmann, I., Calmano, W., 2008. Removal of Cr(VI) from model wastewaters by electrocoagulation with Fe elec-trodes. Sep. Purif. Technol. 61, 15—21.

Hena, S., Atikah, S., Ahmad, H., 2015. Removal of phosphate ion from water using chemically modified biomass of sugarcane bagasse. Int. J. Eng. Sci. 4, 51—62.

Hoet, P.H., Brüske-Hohlfeld, I., Salata, O.V., 2004. Nanoparticles—known and unknown health risks. J. Nanobiotechnol. 2, 12.

Hu, H., 2002. Human health and heavy metals. Life Support: The Environment and Human Health. MIT Press, Cambridge, MA, p. 65.

Huang, C., Dong, C., Tang, Z., 1993. Advanced chemical oxidation: its present role and potential future in hazard-ous waste treatment. Waste Manage. 13, 361—377.

Ibrahim, S.C., Hanafiah, M.A.K.M., Yahya, M.Z.A., 2006. Removal of cadmium from aqueous solutions by adsorp-tion onto sugarcane bagasse. Am. Eurasian J. Agric. Environ. Sci. 1 (3), 179—184.

Inglezakis, V., Loizidou, M., Grigoropoulou, H., 2002. Equilibrium and kinetic ion exchange studies of Pb^{2+}, Cr^{3+}, Fe^{3+} and Cu^{2+} on natural clinoptilolite. Water Res. 36, 2784—2792.

Iqbal, M., Saeed, A., Kalim, I., 2009. Characterization of adsorptive capacity and investigation of mechanism of Cu^{2+}, Ni^{2+} and Zn^{2+} adsorption on mango peel waste from constituted metal solution and genuine electro-plating effluent. Sep. Sci. Technol. 44, 3770—3791.

Ismail, A.F., Harun, Z., Jaafar, J., Rahman, M.A., et al., 2016. A novel green ceramic hollow fiber membrane (CHFM) derived from rice husk ash as combined adsorbent-separator for efficient heavy metals removal. Ceram. Int. 43 (5), 4716—4720.

Jakobsen, M.R., Fritt-Rasmussen, J., Nielsen, S., Ottosen, L.M., 2004. Electrodialytic removal of cadmium from wastewater sludge. J. Hazard. Mater. 106, 127—132.

Kadirvelu, K., Namasivayam, C., 2003. Activated carbon from coconut coirpith as metal adsorbent: adsorption of Cd(II) from aqueous solution. Adv. Environ. Res. 7, 471—478.

Kajitvichyanukul, P., Ananpattarachai, J., Pongpom, S., 2005. Sol—gel preparation and properties study of TiO_2 thin film for photocatalytic reduction of chromium(VI) in photocatalysis process. Sci. Technol. Adv. Mater. 6, 352.

Kanel, S.R., Manning, B., Charlet, L., Choi, H., 2005. Removal of arsenic(III) from groundwater by nanoscale zero-valent iron. Environ. Sci. Technol. 39, 1291—1298.

Kapoor, A., Viraraghavan, T., 1992. Adsorption of mercury from wastewater by fly ash. Adsorpt. Sci. Technol. 9, 130—147.

Karnib, M., Kabbani, A., Holail, H., Olama, Z., 2014. Heavy metals removal using activated carbon, silica and sil-ica activated carbon composite. Energy Procedia 50, 113—120.

Karthikeyan, T., Rajgopal, S., Miranda, L.R., 2005. Chromium(VI) adsorption from aqueous solution by *Hevea brasiliensis* sawdust activated carbon. J. Hazard. Mater. 124, 192—199.

Khan, T.A., Singh, V.V., 2010. Removal of cadmium(II), lead(II), and chromium(VI) ions from aqueous solution using clay. Toxicol. Environ. Chem. 92, 1435—1446.

Khan, M.N., Wahab, M.F., 2007. Characterization of chemically modified corncobs and its application in the removal of metal ions from aqueous solution. J. Hazard. Mater. 141, 237—244.

Khoramzadeh, E., Nasernejad, B., Halladj, R., 2013. Mercury biosorption from aqueous solutions by sugarcane bagasse. J. Taiwan Inst. Chem. Eng. 44 (2), 266—269.

Kong, H., He, J., Gao, Y., Wu, H., Zhu, X., 2011. Cosorption of phenanthrene and mercury(II) from aqueous solu-tion by soybean stalk-based biochar. J. Agric. Food Chem. 59, 12116—12123.

Koukouzas, N., Vasilatos, C., Itskos, G., Mitsis, I., Moutsatsou, A., 2010. Removal of heavy metals from wastewa-ter using CFB-coal fly ash zeolitic materials. J. Hazard. Mater. 173, 581—588.

Kumar, U., Bandyopadhyay, M., 2006. Sorption of cadmium from aqueous solution using pretreated rice husk. Bioresour. Technol. 97, 104–109.

Kumar, A.S.K., Kalidhasan, S., Rajesh, V., Rajesh, N., 2011. Application of cellulose-clay composite biosorbent toward the effective adsorption and removal of chromium from industrial wastewater. Ind. Eng. Chem. Res. 51, 58–69.

Kurniawan, T.A., Chan, G.Y., Lo, W.-H., Babel, S., 2006. Physico-chemical treatment techniques for wastewater laden with heavy metals. Chem. Eng. J. 118, 83–98.

Kyzas, G.Z., Matis, K.A., 2016. Electroflotation process: a review. J. Mol. Liq. 220, 657–664.

Laridi, R., Drogui, P., Benmoussa, H., Blais, J.-F., Auclair, J.C., 2005. Removal of refractory organic compounds in liquid swine manure obtained from a biofiltration process using an electrochemical treatment. J. Environ. Eng. 131, 1302–1310.

Lehr, J.H., Keeley, J., Lehr, J., 2005. Domestic, Municipal, and Industrial Water Supply and Waste Disposal. Wiley Interscience.

Li, Y.-H., Wang, S., Luan, Z., Ding, J., Xu, C., Wu, D., 2003. Adsorption of cadmium(II) from aqueous solution by surface oxidized carbon nanotubes. Carbon 41, 1057–1062.

Li, X., Liu, X., Bao, H., Wu, T., Zhao, Y., Liu, D., et al., 2018. A novel high biosorbent of Pb-resistant bacterium isolate for the removal of hazardous lead from alkaline soil and water: biosorption isotherms in vivo and bioremediation strategy. Geomicrobiol. J. 35, 174–185.

Li, Y., Song, S., Xia, L., Yin, H., Meza, J.V.G., Ju, W., 2019. Enhanced Pb(II) removal by algal-based biosorbent cultivated in high-phosphorus cultures. Chem. Eng. J. 361, 167–179.

Mahdavi, S., Jalali, M., Afkhami, A., 2012. Removal of heavy metals from aqueous solutions using Fe_3O_4, ZnO, and CuO nanoparticles. Nanotechnology for Sustainable Development. Springer.

Mallampati, R., Xuanjun, L., Adin, A., Valiyaveettil, S., 2015. Fruit peels as efficient renewable adsorbents for removal of dissolved heavy metals and dyes from water. ACS Sustain. Chem. Eng. 3 (6), 1117–1124.

Mallevialle, J., Odendaal, P.E., Wiesner, M.R., 1996. Water Treatment Membrane Processes. American Water Works Association.

Martínez-Huitle, C.A., Brillas, E., 2008. Electrochemical alternatives for drinking water disinfection. Angew. Chem. Int. Ed. 47, 1998–2005.

Martinez-Huitle, C.A., Rodrigo, M.A., Sires, I., Scialdone, O., 2015. Single and coupled electrochemical processes and reactors for the abatement of organic water pollutants: a critical review. Chem. Rev. 115, 13362–13407.

Merzouk, B., Gourich, B., Sekki, A., Madani, K., Chibane, M., 2009. Removal turbidity and separation of heavy metals using electrocoagulation–electroflotation technique: a case study. J. Hazard. Mater. 164, 215–222.

Mier, M.V., Callejas, R.L., Gehr, R., Cisneros, B.E.J., Alvarez, P.J., 2001. Heavy metal removal with Mexican clinoptilolite: multi-component ionic exchange. Water Res. 35, 373–378.

Miklos, D.B., Remy, C., Jekel, M., Linden, K.G., Drewes, J.E., Hübner, U., 2018. Evaluation of advanced oxidation processes for water and wastewater treatment—a critical review. Water Res. 139, 118–131.

Mohan, D., Singh, K.P., 2002. Single-and multi-component adsorption of cadmium and zinc using activated carbon derived from bagasse—an agricultural waste. Water Res. 36, 2304–2318.

Mohan, G.K., Babu, A.N., Kalpana, K., Ravindhranath, K., 2019. Removal of chromium(VI) from water using adsorbent derived from spent coffee grounds. Int. J. Environ. Sci. Technol. 16, 101–112.

Mollah, M.Y.A., Schennach, R., Parga, J.R., Cocke, D.L., 2001. Electrocoagulation (EC)—science and applications. J. Hazard. Mater. 84, 29–41.

Mollah, M.Y., Morkovsky, P., Gomes, J.A., Kesmez, M., Parga, J., Cocke, D.L., 2004. Fundamentals, present and future perspectives of electrocoagulation. J. Hazard. Mater. 114, 199–210.

Mondal, N.K., Roy, A., 2018. Potentiality of a fruit peel (banana peel) toward abatement of fluoride from synthetic and underground water samples collected from fluoride affected villages of Birbhum district. Appl. Water Sci. 8, 90.

Monser, L., Adhoum, N., 2002. Modified activated carbon for the removal of copper, zinc, chromium and cyanide from wastewater. Sep. Purif. Technol. 26, 137–146.

Moore, J.W., 2012. Inorganic Contaminants of Surface Water: Research and Monitoring Priorities. Springer Science & Business Media.

Munaf, E., Zein, R., 1997. The use of rice husk for removal of toxic metals from waste water. Environ. Technol. 18, 359–362.

Nagarjuna, R., Challagulla, S., Alla, N., Ganesan, R., Roy, S., 2015. Synthesis and characterization of reduced-graphene oxide/TiO$_2$/zeolite-4A: a bifunctional nanocomposite for abatement of methylene blue. Mater. Des. 86, 621−626.

Nagaveni, K., Sivalingam, G., Hegde, M., Madras, G., 2004. Solar photocatalytic degradation of dyes: high activity of combustion synthesized nano TiO$_2$. Appl. Catal. B: Environ. 48, 83−93.

Nguyen, V., Vigneswaran, S., Ngo, H., Shon, H., Kandasamy, J., 2009. Arsenic removal by a membrane hybrid filtration system. Desalination 236, 363−369

Noormohamadi, H.R., Fat'hi, M.R., Ghaedi, M. and Ghezelbash, G.R., 2019. Potentiality of white-rot fungi in biosorption of nickel and cadmium: modeling optimization and kinetics study. Chemosphere, 216, pp. 124−130.

O'connell, D.W., Birkinshaw, C., O'dwyer, T.F., 2008. Heavy metal adsorbents prepared from the modification of cellulose: a review. Bioresour. Technol. 99, 6709−6724.

Orhan, Y., Büyükgüngör, H., 1993. The removal of heavy metals by using agricultural wastes. Water Sci. Technol. 28, 247−255.

Özer, A., Özer, D., Özer, A., 2004. The adsorption of copper(II) ions on to dehydrated wheat bran (DWB): determination of the equilibrium and thermodynamic parameters. Process. Biochem. 39, 2183−2191.

Pacheco, S., Rodríguez, R., 2001. Adsorption properties of metal ions using alumina nano-particles in aqueous and alcoholic solutions. J. Sol-Gel Sci. Technol. 20, 263−273.

Papadam, T., Xekoukoulotakis, N.P., Poulios, I., Mantzavinos, D., 2007. Photocatalytic transformation of acid orange 20 and Cr(VI) in aqueous TiO$_2$ suspensions. J. Photochem. Photobiol. A: Chem. 186, 308−315.

Park, H.J., Jeong, S.W., Yang, J.K., Kim, B.G., Lee, S.M., 2007. Removal of heavy metals using waste eggshell. J. Environ. Sci. 19, 1436−1441.

Pehlivan, E., Yanik, B.H., Ahmetli, G., Pehlivan, M., 2008. Equilibrium isotherm studies for the uptake of cadmium and lead ions onto sugar beet pulp. Bioresour. Technol. 99 (9), 3520−3527.

Pehlivan, E., Altun, T., Cetin, S., Bhanger, M.I., 2009. Lead sorption by waste biomass of hazelnut and almond shell. J. Hazard. Mater. 167, 1203−1208.

Pitcher, S., Slade, R., Ward, N., 2004. Heavy metal removal from motorway stormwater using zeolites. Sci. Total Environ. 334, 161−166.

Ponder, S.M., Darab, J.G., Mallouk, T.E., 2000. Remediation of Cr(VI) and Pb(II) aqueous solutions using supported, nanoscale zero-valent iron. Environ. Sci. Technol. 34, 2564−2569.

Qdais, H.A., Moussa, H., 2004. Removal of heavy metals from wastewater by membrane processes: a comparative study. Desalination 164, 105−110.

Qu, X., Alvarez, P.J., Li, Q., 2013. Applications of nanotechnology in water and wastewater treatment. Water Res. 47, 3931−3946.

Radjenovic, J., Sedlak, D.L., 2015. Challenges and opportunities for electrochemical processes as next-generation technologies for the treatment of contaminated water. Environ. Sci. Technol. 49, 11292−11302.

Rajeshwar, K., Ibanez, J.G., 1997. Environmental Electrochemistry: Fundamentals and Applications in Pollution Sensors and Abatement. Elsevier.

Rao, R.A.K., Kashifuddin, M., 2014. Kinetics and isotherm studies of Cd(II) adsorption from aqueous solution utilizing seeds of bottlebrush plant (*Callistemon chisholmii*). Appl. Water Sci. 4, 371−383.

Rengaraj, S., Venkataraj, S., Yeon, J.-W., Kim, Y., Li, X., Pang, G., 2007. Preparation, characterization and application of Nd−TiO$_2$ photocatalyst for the reduction of Cr(VI) under UV light illumination. Appl. Catal. B: Environ. 77, 157−165.

Rios, C.A., Williams, C.D., Roberts, C.L., 2008. Removal of heavy metals from acid mine drainage (AMD) using coal fly ash, natural clinker and synthetic zeolites. J. Hazard. Mater. 156, 23−35.

Romero-Cano, L.A., García-Rosero, H., Gonzalez-Gutierrez, L.V., Baldenegro-Pérez, L.A., et al., 2017. Functionalized adsorbents prepared from fruit peels: equilibrium, kinetic and thermodynamic studies for copper adsorption in aqueous solution. J. Clean. Prod. 162, 195−204.

Saeed, A., Iqbal, M., Akhtar, M.W., 2002. Application of biowaste materials for the sorption of heavy metals in contaminated aqueous medium. Pak. J. Sci. Ind. Res. 45, 206−211.

Safinejad, A., Chamjangali, M.A., Goudarzi, N., Bagherian, G., 2017. Synthesis and characterization of a new magnetic bio-adsorbent using walnut shell powder and its application in ultrasonic assisted removal of lead. J. Environ. Chem. Eng. 5 (2), 1429−1437.

Sahraei, R., Pour, Z.S., Ghaemy, M., 2017. Novel magnetic bio-sorbent hydrogel beads based on modified gum tragacanth/graphene oxide: removal of heavy metals and dyes from water. J. Clean. Prod. 142, 2973–2984.

Sahu, U.K., Sahu, S., Mahapatra, S.S., Patel, R.K., 2019. Synthesis and characterization of magnetic bio-adsorbent developed from *Aegle marmelos* leaves for removal of As(V) from aqueous solutions. Environ. Sci. Pollut. Res. 26, 946–958.

Sen, A.K., De, A.K., 1987. Adsorption of mercury(II) by coal fly ash. Water Res. 21, 885–888.

Shakoor, S., 2016. Removal of Dyes From Contaminated Water Using Low Cost Adsorbents. Aligarh Muslim University.

Sharma, Y.C., Srivastava, V., Singh, V., Kaul, S., Weng, C., 2009. Nano-adsorbents for the removal of metallic pollutants from water and wastewater. Environ. Technol. 30, 583–609.

Shen, B., Chen, J., Yue, S., 2015. Removal of elemental mercury by titanium pillared clay impregnated with potassium iodine. Microporous Mesoporous Mater. 203, 216–223.

Shoukat, A., Wahid, F., Khan, T., Siddique, M., Nasreen, S., Yang, G., et al., 2019. Titanium oxide-bacterial cellulose bioadsorbent for the removal of lead ions from aqueous solution. Int. J. Biol. Macromol. 129, 965–971.

Singer, P., 2010. Understanding nanotechnology safety. Solid State Technol. 53, 5–6.

Singh, K., Lataye, D.H., Wasewar, K.L., 2015. Removal of fluoride from aqueous solution by using low-cost sugarcane bagasse: kinetic study and equilibrium isotherm analyses. J. Hazard. Toxic Radioactive Waste 20 (3), 04015024.

Soni, R., Shukla, D.P., 2019. Synthesis of fly ash based zeolite-reduced graphene oxide composite and its evaluation as an adsorbent for arsenic removal. Chemosphere 219, 504–509.

Strathmann, H., 2010. Electrodialysis, a mature technology with a multitude of new applications. Desalination 264, 268–288.

Sudilovskiy, P., Kagramanov, G., Kolesnikov, V., 2008. Use of RO and NF for treatment of copper containing wastewaters in combination with flotation. Desalination 221, 192–201.

Trinelli, M.A., Areco, M.M., Dos Santos Afonso, M., 2013. Co-biosorption of copper and glyphosate by *Ulva lactuca*. Colloids Surf. B: Biointerfaces 105, 251–258.

Tzanetakis, N., Taama, W., Scott, K., Jachuck, R., Slade, R., Varcoe, J., 2003. Comparative performance of ion exchange membranes for electrodialysis of nickel and cobalt. Sep. Purif. Technol. 30, 113–127.

Uddin, M.K., 2017. A review on the adsorption of heavy metals by clay minerals, with special focus on the past decade. Chem. Eng. J. 308, 438–462.

Unuabonah, E., Olu-Owolabi, B., Adebowale, K., Yang, L., 2008. Removal of lead and cadmium ions from aqueous solution by polyvinyl alcohol-modified kaolinite clay: a novel nano-clay adsorbent. Adsorption Sci. Technol. 26, 383–405.

Utomo, H.D., 2007. The Adsorption of Heavy Metals by Waste Tea and Coffee Residues. University of Otago.

Van Der Bruggen, B., Vandecasteele, C., 2002. Distillation vs. membrane filtration: overview of process evolutions in seawater desalination. Desalination 143, 207–218.

Van Der Bruggen, B., Mänttäri, M., Nyström, M., 2008. Drawbacks of applying nanofiltration and how to avoid them: a review. Sep. Purif. Technol. 63, 251–263.

Vilela, D., Parmar, J., Zeng, Y., Zhao, Y., Sánchez, S., 2016. Graphene-based microbots for toxic heavy metal removal and recovery from water. Nano Lett. 16, 2860–2866.

Wang, L., Wang, N., Zhu, L., Yu, H., Tang, H., 2008. Photocatalytic reduction of Cr(VI) over different TiO_2 photocatalysts and the effects of dissolved organic species. J. Hazard. Mater. 152, 93–99.

Wang, H., Liu, Y., Ifthikar, J., Shi, L., Khan, A., Chen, Z., et al., 2018. Towards a better understanding on mercury adsorption by magnetic bio-adsorbents with γ-Fe_2O_3 from pinewood sawdust derived hydrochar: Influence of atmosphere in heat treatment. Bioresour. Technol. 256, 269–276.

Weng, C.-H., Huang, C., 2004. Adsorption characteristics of Zn(II) from dilute aqueous solution by fly ash. Colloids Surf. A: Physicochem. Eng. Asp. 247, 137–143.

Wingenfelder, U., Hansen, C., Furrer, G., Schulin, R., 2005. Removal of heavy metals from mine waters by natural zeolites. Environ. Sci. Technol. 39, 4606–4613.

Wong, K., Lee, C., Low, K., Haron, M., 2003. Removal of Cu and Pb by tartaric acid modified rice husk from aqueous solutions. Chemosphere 50, 23–28.

Xu, P., Zeng, G.M., Huang, D.L., Feng, C.L., Hu, S., Zhao, M.H., et al., 2012. Use of iron oxide nanomaterials in wastewater treatment: a review. Sci. Total. Environ. 424, 1–10.

Zacaroni, L.M., Magriotis, Z.M., Das Graças Cardoso, M., Santiago, W.D., Mendonça, J.G., Vieira, S.S., et al., 2015. Natural clay and commercial activated charcoal: properties and application for the removal of copper from cachaça. Food Control. 47, 536–544.

Zeng, G., He, Y., Zhan, Y., Zhang, L., Pan, Y., Zhang, C., et al., 2016. Novel polyvinylidene fluoride nanofiltration membrane blended with functionalized halloysite nanotubes for dye and heavy metal ions removal. J. Hazard. Mater. 317, 60–72.

Zhang, Q., Pan, B., Pan, B., Zhang, W., Jia, K., Zhang, Q., 2008. Selective sorption of lead, cadmium and zinc ions by a polymeric cation exchanger containing nano-Zr $(HPO_3S)_2$. Environ. Sci. Technol. 42, 4140–4145.

Zhang, Y., Zhang, S., Chung, T.-S., 2015. Nanometric graphene oxide framework membranes with enhanced heavy metal removal via nanofiltration. Environ. Sci. Technol. 49, 10235–10242.

Zhang, Y., Zhang, S., Gao, J., Chung, T.-S., 2016. Layer-by-layer construction of graphene oxide (GO) framework composite membranes for highly efficient heavy metal removal. J. Membr. Sci. 515, 230–237.

Zhang, W., Li, Q., Mao, Q., He, G., 2019. Cross-linked chitosan microspheres: an efficient and eco-friendly adsorbent for iodide removal from waste water. Carbohydr. Polym. 209, 215–222.

Zhou, H., Smith, D.W., 2002. Advanced technologies in water and wastewater treatment. J. Environ. Eng. Sci. 1, 247–264.

Zhu, C.-S., Wang, L.-P., Chen, W.-B., 2009. Removal of Cu(II) from aqueous solution by agricultural by-product: peanut hull. J. Hazard. Mater. 168, 739–746.

Zhu, W.-P., Sun, S.-P., Gao, J., Fu, F.-J., Chung, T.-S., 2014. Dual-layer polybenzimidazole/polyethersulfone (PBI/PES) nanofiltration (NF) hollow fiber membranes for heavy metals removal from wastewater. J. Membr. Sci. 456, 117–127.

Zhu, W.-P., Gao, J., Sun, S.-P., Zhang, S., Chung, T.-S., 2015. Poly(amidoamine) dendrimer (PAMAM) grafted on thin film composite (TFC) nanofiltration (NF) hollow fiber membranes for heavy metal removal. J. Membr. Sci. 487, 117–126.

Advance reduction processes for denitrification of wastewater

Rohit Chauhan and Vimal Chandra Srivastava

Department of Chemical Engineering, Indian Institute of Technology Roorkee, Roorkee, India

15.1 Introduction

Inorganic pollutants present in significant concentration (1–1000 mg/L) in water include sodium, calcium, potassium, magnesium, sulfate, chloride, and nitrate. Nitrogen is one of the most essential elements for all living beings; however, it can be a toxic for living organism if present in undesirable form or state. Nitrogen exists in various oxidation state in the water bodies in which NH_4^+ (−III), N_2 (0), NO_2^- (+III), and NO_3^- (+V) are the most common nitrogen species (Stueken et al., 2016). Among all of these, nitrate (NO_3^-) has usually highest level of concentration in comparison with other inorganic nitrogen contaminants such as nitrite (NO_2^-) and ammonium ion (NH_4^+). The permissible limit of nitrate in drinking water recommended by World Health Organization is ≈ 50 mg/L (Su et al., 2017). Permissible limit recommended by Bureau of Indian Standards is ≈ 45 mg/L (BIS, 2012). Efficient management of nitrogen cycle in water bodies during the 21st century is a grand challenge. Thus there is a need of efficient and advanced treatment technology for nitrate degradation in wastewater at large scale as well as in small point-of-use (POU) water treatment systems. An advance feature of POU systems is the need of chemical-free water treatment.

The objective of managing nitrogen cycle is to minimize the influence of anthropogenic nitrogen and produce ammonia as a usable form (i.e., fertilizer) from nitrogen cycle as shown in Fig. 15.1. However, ammonia is problematic if present in drinking water because of its toxic nature toward aquatic species in surface water. It also increases requirement of disinfectants, which further causes the biological nitrification of water system. Values of Henry's constant for nitrogen gas (N_2) with respect to water is 6.24×10^{-4} mol/(L atm), which indicates that N_2 has low solubility in water and defines its selectivity as the final product for denitrification of wastewater (Hamme and Emerson, 2004).

Several advance treatment systems have been used for the degradation of nitrate in wastewater. Biological methods (Kodera et al., 2017), catalytic methods (Murphy, 1991), reverse osmosis (Epsztein et al., 2015), ion-exchange systems (Samatya et al., 2006), and adsorption (Adeleye et al., 2016) and electrochemical (EC) methods (Martinez et al., 2017) are the conventional advance systems for denitrification process of wastewater. However, these processes have many limitations for the treatment of nitrate-contaminated wastewater. Biological and catalytic method leads to secondary pollution as well as can cause toxicity in the treated water. Reverse osmosis requires a minimum specific pressure and it is prone to biofouling. Ion-exchange process is very sensitive toward the numerous contaminants and ions present in the wastewater. Selection of adsorbent and its reusability are major issues for denitrification of wastewater (Bhatnagar and Sillanpaa, 2011; Kapoor and Viraraghavan, 1997). In all of the advance systems, EC method promises an ideal tool for treatment of nitrate-contaminated wastewater. EC reduction of nitrate wastewater produces negligible amount of sludge and secondary pollutants (Chauhan et al., 2016). Advance EC (i.e., advance electroreduction) process can reduce nitrate into nontoxic N_2 gas. It is applied for enhanced nitrate removal efficiency without scaling and fouling. Nano-catalyst-coated cathodes such as Co_3O_4/Ti (Su et al., 2017), Ti-nano cathode (Wang et al., 2016), and polypyrrole-coated copper electrode (Çirmi et al., 2015), and

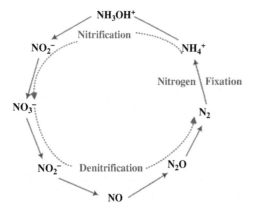

FIGURE 15.1   The inorganic compounds of nitrogen are controlled by a reaction cycle called the nitrogen cycle.

dimensionally stable anodes like boron doped diamond (BDD) (Ghazouani et al., 2017) and Ti/IrO$_2$ (Reyter et al., 2010) have been used by the researchers.

This chapter presents recent advances in EC treatment processes for denitrification of wastewater. The effect of various factors such as electrode material, pH, and current density on the nitrate reduction efficiency has been discussed. The possible reduction mechanisms for direct transfer of charge as well as indirect EC techniques involved for nitrate reduction have been discussed extensively.

15.2 Advance in the electrochemical denitrification processes

EC reactor consists of two electrodes, one is anode and the other is cathode (shown in Fig. 15.2). The electrode permits the circulation of current (transfer of electrons) over the external power supply. Therefore the exposed electrode surface assists in electron exchange with highly electroactive species within the solution. These in situ generated electro-species act as reactants in the wastewater during redox reactions through direct transfer of charge on the exposed surface of the electrodes. A cathode is that electrode where reduction processes occurs, whereas an anode is associated with oxidation reaction of electroactive species. The EC reduction of nitrate mainly aims to reduce nitrogen species into N$_2$ gas as final product. Cathodic reactions are most significant since the nitrate reduction reaction accepts electrons. The effectiveness of EC reduction is generally estimated in terms of Faradic efficiency (FE) or current efficiency (CE), which defines the consumption of electrons in an EC reaction, which is relative to the expected theoretical conversion obtained by Faraday's law (Chauhan et al., 2016).

$$\text{FE (\%) or CE (\%)} = \frac{I_{exp}}{I_{theor}} \times 100 = \frac{nFN_i}{It} \times 100 \qquad (15.1)$$

where I_{exp} is the experimental charge in Coulomb consumed during the EC reactions, I_{theor} is the total charge in Coulomb circulated, n is the number of electrons required per

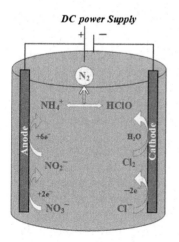

FIGURE 15.2 Schematic of EC mechanism of denitrification process in batch type reactor with anodic and cathode reactions. *EC*, Electrochemical.

mole of product in mole, F is the Faraday constant (96,487 C/mol), N_i is the mol of product produced during the EC reaction, I is the applied current intensity in A, and t is the electrolysis time in seconds. Electrooxidation and electroreduction processes occur simultaneously at the anode and cathode surface in the EC cells, respectively. In addition, some in situ reduced intermediates and/or products could be reoxidized on the anodic surface that alters the ultimate FE/CE (Macova et al., 2007). EC process is influenced by various parameters such as electrode material (metal and nonmetal), solution pH, and current density.

15.2.1 Influencing factors for advance electrochemical processes

15.2.1.1 Electrode material

Proper selection of electrode material is highly important in EC processes since electrodes directly affect the treatment efficiencies (Jiang et al., 2016). Cathode is focused more for reduction of nitrate in wastewater, and several cathodes have been used for EC denitrification process. Table 15.1 summarizes the few researches of EC reduction of nitrate by various electrode materials in synthetic and actual wastewaters. It is a fundamental understanding that d-orbital electrons promote the injection of charge in the lowest unoccupied molecular π^* orbital of nitrate molecules (Khomutov and Stamkulov, 1971). This study helped in use of advance electrodes that are made of platinoid group (Pt, Pd, Ir, Rh, and Ru) and noble metal group (Au, Ag, and Cu) materials. These metals have highly occupied d-orbitals as well as unclosed d-orbital shells. Their analogous structures (Ti, Fe, and Ni) are due to their presence in similar group of the periodic table (Garcia-Segura et al., 2018). In this way, electrode materials can be categorized in two groups, metal electrodes and nonmetal electrodes. Graphite and BDD are the well-known two nonmetallic electrodes, BDD has been considered as the most dominant electrode material for the reduction of pollutants (Ghazouani et al., 2017; Georgeaud et al., 2011).

15.2.1.1.1 Noble metals as cathode

Noble metals have excellent stability and electro-catalytic properties. Cyclic voltammetric study of different noble metals show that high current density results in enhanced electrokinetics and therefore faster reduction of nitrate on the surface of cathode. The electro-catalytic nature of these noble metals decrease in the following order: Cu > Ag > Au (Dima et al., 2003; Haque and Tariq, 2009). Fig. 15.3 shows cyclic voltammograms of Ag, Au, and Cu. Since, Cu has higher electro-catalytic properties; therefore this noble metal used extensively for research work, whereas in spite of being high cost, Au and Ag are also reported for batch treatment of nitrate-contaminated wastewater (Fedurco et al., 1999; Ohmori et al., 1999; Casella and Ritorti, 2010).

Cu cathode showed the excellent electro-catalytic reduction kinetics for nitrate to nitrite, which is also the rate-determining step of the overall process of reduction of nitrate into N_2 or NH_3 (Garcia-Segura et al., 2018). EC denitrification on Cu cathode has been applied at pH ranging from acidic to alkaline medium (Dima et al., 2003). In alkaline condition, Cu cathode deactivates gradually due to passivation of cathode (formation of amorphous Cu powder on the surface of cathode), which results in an outer shield material layer that

TABLE 15.1 Electrochemical denitrification process of synthetic and actual wastewater by several electrode materials.

Wastewater source	Electrode	Operating conditions	Results	References
Synthetic wastewater	Anode = S.S.; cathode = Cu	$I = 0.2–1$ A; $A = 11$ cm $\times$ 6 cm; 2 g adsorbent; $d = 0.5–3$ cm; pH = 3–11; $t = 200$ min	$\approx$ 70% nitrate removal	Kalaruban et al. (2017)
Synthetic wastewater	Bipolar Si/BDD cell	$j = 35.7$ mA/cm^2; $t = 2$ h; RSM methodology; $A = 70$ cm^2; pH = 5.5	$\approx$ 91% nitrate removal	Ghazouani et al. (2016)
Synthetic wastewater	Anode = Ti/TiO$_2$ nanotube array; cathode = Fe	$j = 15$ mA/cm^2; $t = 4$ h; $A = 72$ cm^2; $d = 8$ mm; Na$_2$SO$_4$ = 0.5 g/L	$\approx$ 91% nitrate removal in dual chamber cell	Li et al. (2016)
Synthetic wastewater	Conductive diamond, stainless steel, silicon carbide, graphite or lead	$j = 150–1400$ A/m^2; $A = 78$ cm^2; $d = 9$ mm; $Q = 1.5$ dm^3/min	$\approx$ 90% nitrate removal	Lacasa et al. (2012)
Synthetic wastewater	Anode = Ti/IrO$_2$–Pt; cathode = Fe, Cu, Ti	$j = 20$ mA/cm^2; $A = 75$ cm^2; $d = 8$ mm; $t = 3$ h; Na$_2$SO$_4$ = 0.5 g/L	$\approx$ 87% nitrate removal	Li et al. (2009)
Synthetic wastewater	Ti/IrO$_2$–Pt	$j = 10–60$ mA/cm^2; $A = 75$ cm^2; $d = 8$ mm; $t = 5$ h; NaCl = 0.5 g/L; pH = 6.5	$\approx$ 90% nitrate removal	Li et al. (2010)
Actual textile wastewater	Co$_3$O$_4$/Ti	$j = 10$ mA/cm^2; $A = 16$ cm^2; $d = 1.5$ cm; pH = 7; NO$_3^-$-N = 50–100 mg/L; $t = 3$ h	80% nitrate removal	Su et al. (2017)
Actual wastewater	BDD electrode	$j = 35.7$ mA/cm^2; $A = 70$ cm^2; conductivity = 2.6 and 5.2 mS/cm; $t = 3$ h	$\approx$ 85% total Kjeldahl nitrogen	Ghazouani et al. (2017)
Metal-finishing wastewater	Anode = Pt; cathode = Ti, Ni, Fe, Cu, Zn	$j = 25–100$ A/m^2; $L \times W \times H = 10$ cm $\times$ 10 cm $\times$ 20 cm; $d = 10$ mm; NO$_3$-N = 100 mg/L; pH = 2; HRT = 10–60 min; Zn:sulfamic acid: nitrate = 1:1:1	$\approx$ 90% nitrate removal	Sim et al. (2012)
Agro-industrial wastewater	BDD electrodes	$j = 450–2500$ A/m^2; $A = 68$ cm^2; $d = 10$ mm; pH = 2.5	$\approx$ 89% nitrate removal	Georgeaud et al. (2011)

BDD, Boron doped diamond; HRT, Hydraulic retention time; RSM, Response surface methodology.

prevents the electro-catalytic properties (Dima et al., 2003; Paidar et al., 1999; Bouzek et al., 2001). In alkaline environment, CE decreases from 36.5% to 8% after 6 hours of continuous electrolysis (Paidar et al., 1999). This passivation phenomena of Cu cathode can be avoided by conducting nitrate reduction process in the presence of Cu ions, which deposit on the cathode surface and create fresh active sites on the surface of cathode, and catalytic activity of cathode elongates (Paidar et al., 1999). In contrast, the passivation of Cu cathode in acidic medium is not reported. However, Cu dissolves in open-circuit environment spontaneously in solution of acidic nitrate. Hence, the continuous redeposition as well as

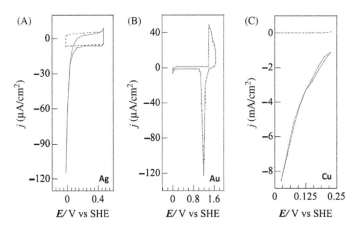

FIGURE 15.3 (A) Silver, (B) gold, and (C) copper cyclic voltammograms recorded at scan rate 20 mV/s in 0.5 M H_2SO_4 (*dashed line*) and 0.5 M H_2SO_4 in presence of 0.1 M $NaNO_3$ (*solid line*). Source: *Figure adopted from Dima, G.E., De Vooys, A.C.A., Koper, M.T.M., 2003. Electrocatalytic reduction of nitrate at low concentration on coinage and transition-metal electrodes in acid solutions. J. Electroanal. Chem. 554, 15–23.*

dissolution of Cu reinforces the surface of Cu cathode, which inhibits the passivation and poisoning of cathode (Dima et al., 2003).

Study with Ag cathode shows that nitrate reduction prior to reduction of water, although the total ammonia reduction takes place in the region of H_2 evolution reaction (Fedurco et al., 1999). Au cathode also showed effective reduction of nitrate into nitrite as well as into ammonia (Nobial et al., 2007). In the presence of 0.5 M $NaNO_3$ at 2.85 mA/cm and pH 12.5, Au electrode showed yield of 42% nitrite and 58% ammonia, after 6 hours of electrolysis time (El-Deab, 2004). Both noble metal as cathode shows minor electro-catalytic properties and sluggish electrokinetics for EC denitrification of wastewater. The corrosion as well as leaching of Cu cathode and undesired production of ammonia prohibits its long-term application for EC denitrification processes. Yet the enhanced electrocatalytic response of Cu cathode stresses it as being the most desired cathode for nitrate reduction. Silver and gold electrodes are expensive materials for the application of wastewater treatment; however, some reports are available, which show good performance toward EC reduction of nitrate with ammonia as one of the final products (Nobial et al., 2007; El-Deab, 2004).

15.2.1.1.2 Platinoid metal as cathode

Platinoids or platinum-group metals (Pt, Pd, Rh, Ir, and Ru) are that type of transition metals that have very good resistance toward corrosion and have excellent electro-catalytic properties (Kettler, 2003). Cyclic voltammetric studies reveal the order of electro-catalytic activity of platinoid metals in following order: $Rh > Ru > Ir > Pd–Pt$ (Dima et al., 2003). Fig. 15.4 shows cyclic voltammograms of Ir, Pd, Pt, Ru, and Rh. In spite of having lower electro-catalytic activity, Pd and Pt are the most researched platinoid metals for EC denitrification process as their electro-catalytic active sites are not poisoned, which is observed for other platinoid metals due to in situ formation of by-products. The poor electro-catalytic activity of Pd and Pt is related to the competitive evolution of H_2. Voltammogram studies suggest that the reduction of nitrate happens at the potentials of high coverage of H_2 (low-potential region). These results suggest that nitrate ion is

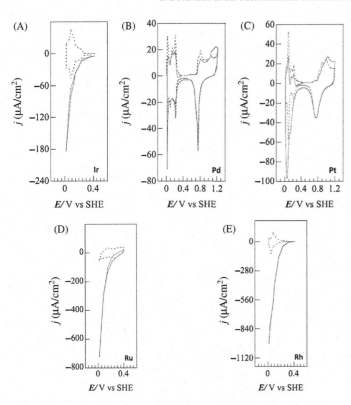

FIGURE 15.4 (A) Iridium, (B) palladium, (C) platinum, (D) ruthenium, and (E) rhodium cyclic voltammograms recorded at scan rate 20 mV/s in 0.5 M H_2SO_4 (*dashed line*) and 0.5 M H_2SO_4 in presence of 0.1 M $NaNO_3$ (*solid line*) (Dima et al., 2003).

adsorbed since nitrate sorption has low enthalpy than H_2 gas on the Pt and Pd cathode (Dima et al., 2003; Gootzen et al., 1997).

Pt and Pd also produce different types of nitrogen-based intermediates (*N*-intermediates), which recommend a nonfavored reduction pathway of denitrification reaction on the polycrystalline cathode (Da Cunha et al., 2000). Like common macroscopic metallic material electrodes, polycrystalline electrodes are that type of solid electrodes, which are composed of several crystallites of several orientations and sizes. However, single crystal with favored crystal planes of Pt electrode show strong structure sensitivity (Dima et al., 2005). Therefore N_2 gas selectivity can be enhanced. In spite of showing greater electro-catalytic activity for reduction of nitrate for other platinoid metals (Ru, Ir, and Rh), these cathodic metals are not favorable for long-period applications as their electro-catalytic activity is lowered by poisoning. This poisoning phenomenon is irreversible and occurs due to strong adsorption of in situ produced intermediates such as NO on the cathode surface (Dima et al., 2003; Da Cunha et al., 2000). Sabatier principle for poisoning explains that the interface between substrate and catalyst should be neither too weak nor too strong. The sites of catalysts must bind the ascorbates in such a way that it can enable the activation of reactants. In addition, the reaction by-products should have weak binding strength that can favor the desorption; hence, catalyst poisoning can be avoided (Hammer, Norskov, 2000; Calle-Vallejo et al., 2013). In spite of showing low negative potentials at onset potential of denitrification for Ir cathode than Pt, poisoning of catalytic activity occurs for Ir due to the association of limiting current with the adsorbed intermediates and by-products (Ureta-Zanartu and Yanez, 1997). Dima et al. (2003) reported same nature for Ru electrodes where the overall activity declines due to the very strong adsorption of highly reactive reaction intermediates, that is, poisoning of catalytic activity. The sluggish reaction kinetics for the very first charge transfer reaction is also one of the main limitations for use of Pd and Pt electrodes. However, as the rate-limiting step surpasses, the possibility of N_2 selectivity increases, which makes Pt and Pd the most promising electrode metals. In last few years, nanoparticles are being researched as electro-catalysts to enhance the reaction kinetics in enhancing the electroactive surface area as well as reducing the loss of materials (Garcia-Segura et al., 2018).

15.2.1.1.3 Other metals as cathode

There are studies reported which focus on use of the other metals (Ti, Fe, Sn, and Bi) as electrodes than noble and platinoid metals. Ti metal as cathode shows low rate of reduction of nitrate, which proves its poor activity for denitrification processes. However, TiO_2 nanotubes are used, it significantly enhances the electroactive surface area of the cathode, which shows better rate of reduction of nitrate. Moreover, Ti-based electrodes have high selectivity toward NH_3 gas (Ma et al., 2016). There are few studies available which show that Fe displays reasonable performance as cathode material for reduction of nitrate-contaminated wastewater (Li et al., 2010). The high dissolution, corrosion, and electro-chlorination of Fe in highly alkaline and acidic environment show its drawback for use as an electrode for reduction of nitrate, and selectivity of N_2 gas is very low by Fe cathode (Li et al., 2010; Kuang et al., 2016). Sn electrode accomplishes high selectivity of $\approx 80\%$ toward N_2 gas, and nitrate reduction requires cathode potential of < -2.9 V (vs Ag/AgCl electrode). However, Sn electrode starts corroding at -2.4 V (vs Ag/AgCl), which prohibits its use for nitrate reduction application (Katsounaros, Kyriacou, 2008; Katsounaros et al., 2006). Bi needs potentials under -2.0 V (vs Ag/AgCl) for efficient reduction of nitrate and exhibits lesser cathodic corrosion as compared to Sn (Katsounaros et al., 2009; Dortsiou and Kyriacou, 2009). Bi shows sluggish rate of reduction of nitrate and also shows poor selectivity toward N_2 gas than Sn cathode (Katsounaros et al., 2009).

15.2.1.1.4 Metal alloys as cathode

Alloys are mixer of two more elements, and it is necessary that one of those should be a metal. Alloys can be molded by mixing of two metals such as Cu and Sn (bronze) or by mixing one metal and other nonmetal such as Fe and C (steel). Alloy can exhibit enhanced and different properties than the individual element that combine them. The binary or ternary alloys as electrode should have the properties of enhanced selectivity toward N_2 gas as final reaction products and higher catalytic activity for nitrate reduction in wastewater.

Copper alloys (brass, bronze, and cupronickel) are used to enhance the selectivity of N_2 gas as final product, and it is used since Cu has great electro-catalytic activity for EC denitrification processes. Bronze is one of the Cu alloys, which consists of Cu and Sn metals. As discussed earlier, Sn cathodes show enhanced N_2 gas selectivity, which is the preferred outcome for EC nitrate reduction processes (Katsounaros et al., 2012, 2006). As the percentage of Sn increases up to 12 wt.% in the bronze alloy, it results in enhanced rate of reduction as well as enhanced N_2 gas selectivity at great cathodic potentials (-1.40 V vs standard hydrogen electrode (SHE)) (Macova et al., 2007). However, higher content of Sn reduces the alloy electro-catalytic activity, which has a negative effect. If bronze alloy has Sn > 49 wt.%, it shows almost similar EC behavior as Sn due to same polarization curve irrespective to tin content (El Din and El Wahab, 1977). Brass alloy shows similar characteristics like bronze though it is composed of Cu and Zn metals. The ultimate electro-catalytic activity for brass alloys happens at ≈ 41 wt.% of Cu without effects of cathode poisoning, which is generally observed for pure Cu cathode (Mácová and Bouzek, 2005; Mattarozzi et al., 2015). Cupronickel alloys are composed of mainly Cu and Ni metals, but it also contains some strengthening elements such as Fe and Mn. Cupronickel has high resistance toward the corrosion in seawater (i.e., high Cl^- content). This is the reason for use of cupronickel as cathode for the reduction of nitrate in high saline conditions. As the Ni content increases in alloy, the Faradic efficiency increases for ammonia; however, it decreases the EC reduction kinetics for nitrate (Durivault et al., 2007; Mattarozzi et al., 2013).

Platinoid alloy is used to enhance the reaction kinetics as well as keeping N_2 gas selectivity. The aim of using platinoid alloys is to improve electro-catalytic activity by the combination of catalyst-substrate with strong adsorbate interface and a catalyst-substrate with weaker interface as explained by Sabatier principle (in volcano plots) (Garcia-Segura et al., 2018). There are different types of platinoid alloys studied such as Rh–Pt (Da Cunha et al., 2000) and Ir–Pt (Chen et al., 2015), which show less cathode poisoning for nitrate reduction than pure Rh and Ir, respectively. Platinoid–platinoid alloys have not been recommended due to their poor performance, selectivity and very high cost. Platinoid alloys combine with Cu metal enhance the reduction kinetics as well as N_2 gas selectivity, and it is more economical than platinoid–platinoid alloys. Cu–Pt alloy cathode favor rapid rate of reduction of nitrate due to active sites of Cu. Therefore the alloy of Cu–Pt exhibit better electro-catalytic activity than pure Cu and Pt electrodes (Chen et al., 2015; Amertharaj et al., 2014). Pure Sn has excellent electro-catalytic activity as well as great N_2 gas selectivity; hence, it is used for composing Sn–Pt platinoid alloys. Therefore Sn–Pt (Tada and Shimazu, 2005) and Sn–Pd (Shimazu et al., 2007) alloys showed effective electro-catalytic activity toward reduction of nitrate, and the main gaseous product obtained was N_2O, which was further reduced into N_2 gas by the free active sites of Pt (Yang et al., 2011). The dominant gaseous product released was N_2O, which proposed that these electrodes can be suitable as cathode that can produce N_2 gas. However, N_2O is known as effective greenhouse gas, and it can be a big hindrance if sustainability of different systems for reduction of NO_3^- is considered.

15.2.1.1.5 Nonmetallic cathodes

In the category of nonmetal electrodes, graphite and BDD are the well-known two nonmetallic electrodes and BDD has been considered as the effective cathode material for reduction of nitrate-contaminated wastewater (Ghazouani et al., 2017; Georgeaud et al., 2011). BDD is a favorable electrode due to its exceptional EC properties such as extensive working potential, great stability even high alkaline and acidic environment, great resistance to corrosion, and poor electro-catalytic inhibition (Einaga et al., 2014; Garcia-Segura et al., 2018). Levy-Clement et al. (2003) reported that nitrate reduction by BDD electrode produces N_2 gas when cathode potential goes from -1.5 to -1.7 V, and when potentials moves from -1.7 to -2.0 V, enhanced rate of nitrate reduction was observed. BDD cathode was used for the treatment of slaughterhouse wastewater with current density of 35.7 mA/cm^2 and without using any reagent. Garcia-Gomez et al. (2014) obtained complete removal of nitrate after 180 minutes of electrolysis time. However, the application of BDD electrode at industrial level is too limited due to its very high cost.

Graphite is widely used as cathode for the EC denitrification of wastewater. Pure graphite cathode shows very poor removal efficiency of 8%. Thus modified graphite cathode such as Rh−graphite (Brylev et al., 2007) has been researched for improvement of rate of reduction of nitrate. Rh−graphite showed $\approx 60\%$ removal efficiency for nitrate removal after 96 hours, and ammonia was the key final product (Brylev et al., 2007). Hence, low rate of nitrate reduction with poor N_2 gas selectivity limits its application for denitrification process.

15.2.1.2 *Influence of pH and applied current density on denitrification processes*

The solution pH is one of the most important parameters, which influences nitrate reduction process (removal efficiency and activity of EC process) very much. For Fe and Al electrode, iron and aluminum cations with hydroxides cause deterioration of colloids, which results in active electro-coagulant species in acidic, neutral, as well as slightly alkaline pH medium. At very low pH values, hydroxide elements of dissolving iron and aluminum cannot be applied as the coagulant agents due to the ineffectiveness of coagulation activity and instability. At very high alkaline environment, $Fe(OH)^{4-}$ and $Al(OH)^{4-}$ ions get produced, and these produced ions have very low performance of coagulation (Xu et al., 2018). Few researchers report optimum nitrate removal efficiency at pH $= 5$ and minimum nitrate removal at pH $= 3$ by using Al and Fe metal electrode (Mameri et al., 1998).

Applied current density is one of the important influencing parameters for denitrification of wastewater. Enhanced current density can advance the rate of nitrate reduction; however, the polarization and passivation of the electrode gets increased, which results in high energy consumption (Sahu et al., 2014). High applied current density advances the selectivity of N_2 gas instead of any other by-products such as NH_4^+ and NO_2^-. As the current density increases the electro-energy utilization also gets affected. However, it is reported that N_2 is main gaseous product in between other gaseous products such as NO, NO_2, and N_2O (Su et al., 2017). Su et al. (2017) reported an increase in the nitrate removal efficiency, from 60% to 90%, using Co_3O_4/Ti cathode with an increase in current density from 5 mA/cm^{-2} to 25 mA/cm^{-2}.

15.3 Reaction mechanism for electrochemical denitrification process

The reaction mechanism for EC reduction of NO_3^- into N_2 gas is very complex, which includes several reactions, by-products, and intermediates (e.g., nitrite, ammonia, hydroxylamine, hydrazine, nitrous oxide, and nitric oxide) covering different types of nitrogen oxidation states from $-$ III to $+$ V. EC reduction of nitrate mainly occurs directly at cathode surface by adsorption of nitrate on the surface of cathode and gets reduced into N_2,

NH_3 and other by-products followed by different pathways. Many chemical reactions are involved during EC reduction of nitrate and formation of by-products such as nitrite, nitrous oxide on the cathode surface. Important reactions are shown by the following reactions (Xu et al., 2018; Su et al., 2017):

$$NO_3^- + H_2O + 2e^- \rightarrow NO_2^- + 2OH^- \quad (E^o = 0.01 \text{ V vs SHE}) \tag{15.i}$$

$$NO_3^- + 3H_2O + 5e^- \rightarrow 0.5N_2 + 6OH^- \quad (E^o = 0.26 \text{ V vs SHE}) \tag{15.ii}$$

$$NO_3^- + 6H_2O + 8e^- \rightarrow NH_3 + 9OH^- \quad (E^o = -0.12 \text{ V vs SHE}) \tag{15.iii}$$

$$NO_2^- + 5H_2O + 6e^- \rightarrow NH_3 + 7OH^- \quad (E^o = -0.165 \text{ V vs SHE}) \tag{15.iv}$$

$$NO_2^- + 4H_2O + 4e^- \rightarrow NH_2OH + 5OH^- \quad (E^o = -0.45 \text{ V vs SHE}) \tag{15.v}$$

$$2NO_2^- + 4H_2O + 6e^- \rightarrow N_2 + 8OH^- \quad (E^o = 0.406 \text{ V vs SHE}) \tag{15.vi}$$

$$2NO_2^- + 3H_2O + 4e^- \rightarrow N_2O + 6OH^- \quad (E^o = -0.283 \text{ V vs SHE}) \tag{15.vii}$$

$$NO_2^- + H_2O + e^- \rightarrow NO + 2OH^- \quad (E^o = -0.197 \text{ V vs SHE}) \tag{15.viii}$$

$$N_2O + 5H_2O + 4e^- \rightarrow 2NH_2OH + 4OH^- \quad (E^o = -0.387 \text{ V vs SHE}) \tag{15.ix}$$

Steps involved in the pathway are discussed hereafter.

15.3.1 Rate-limiting step (NO_3^- reduces to NO_2^-)

The main reduction barrier is the reduction of nitrate into nitrite in the aqueous solution because overall reduction kinetics is controlled by this step that is why it is called rate-limiting step for nitrate reduction process (Dima et al., 2005). Fig. 15.5 shows that first nitrate adsorbs onto the cathode surface then it gets reduced into the nitrite, but the overall reduction process is inhibited by coadsorbing ions from aqueous solution (Katsounaros et al., 2006). Therefore the limitation of adsorption of nitrate, surface-bound nitrate, and rate of EC nitrate reduction is done by mass transfer of NO_3^- ion from bulk to the cathode surface. Fick's law describes the concentration of nitrate, which decides the rate of diffusion from bulk solution to the surface of cathode (Dima et al., 2005). Reaction (15.x) shows the overall chemical reaction for EC reduction of adsorbed (ad) nitrate into nitrite. This reaction involves the formation of very short-lived nitrate dianion radicals (NO_3^{2-}) and nitrogen dioxide—free radicals ($NO_2^\bullet$), which further converts into nitrite ion adsorbed on cathode surface shown by reactions (15.x) and (15.xi) (Garcia-Segura et al., 2018).

$$NO_{3(ad)}^{2-} + H_2O \rightarrow NO_{2ad}^\bullet + 2OH^- \tag{15.x}$$

$$NO_{2ad}^\bullet + e^- \rightarrow NO_{2(ad)}^- + H_2O \quad (E^o = 1.04 \text{ V vs SHE}) \tag{15.xi}$$

The total reduction of nitrate into N_2 and NH_3 is followed by the vital quasistable intermediate, which gets produced in situ during EC reduction of nitrate. Both of the reaction shown by reactions (15.x) and (15.xi) and reaction shown by reaction (15.i) are considered as the main rate-limiting steps for the overall reduction mechanism of nitrate into N_2 and NH_3 by EC method.

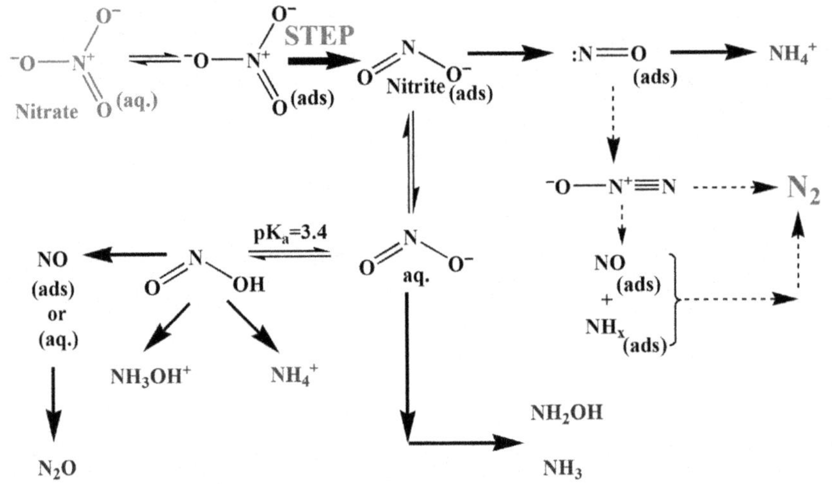

FIGURE 15.5 The major reaction pathways for nitrate and nitrite reduction on advanced metal electrodes. The "STEP" word highlights the rate-determining step. Thick lines emphasize those pathways that follow most often, while dashed lines indicate reactions taking place under more specific conditions.

15.3.1.1 Nitrite reduction into NO₂ and HNO₂

There is another pathway for the reduction of NO_3^- into nitrous acid (HNO_2) and/or nitrogen dioxide (NO_2) followed by nitrite as intermediate. This mechanism is followed at the high nitrate concentration with very low pH (acidic pH) (Garcia-Segura et al., 2018). Fig. 15.5 shows that $NO_{2(ad)}^-$ converts into NO_2 followed by HNO_2. These mechanisms are called Vetter and Schmid mechanisms, respectively. According to Vetter mechanism, NO_2 is the main product due to the in situ generated electroactive species $NO_2^\bullet$ into the bulk solution followed by the formation of HNO_2 ($pK_a = 3.4$) shown in the following reaction (Su et al., 2017):

$$NO_2^- + H^+ \Leftrightarrow HNO_2 \quad (pK_a = 3.4) \tag{15.xii}$$

In comparison, according to the Schmidt mechanism, HNO_2 is the main product followed by NO, and this reaction happens due to the formation of electroactive intermediate nitrosonium cation (NO^+) shown in the following reactions (Schmid and Delfs, 1959):

$$HNO_2 + H^+ \Leftrightarrow NO^+ + H_2O \tag{15.xiii}$$

$$NO^+ + e^- \rightarrow NO \quad (E^o = 1.28 \text{ V vs SHE}) \tag{15.xiv}$$

15.3.1.2 Nitrite reduction into N₂, N₂O, and NH₃ followed by central species NO

Nitrite is the main intermediate generated during EC reduction of nitrate. There is also a possible mechanism in which nitrite gets converted into N_2, N_2O, and NH_3 as final products. The EC reduction of nitrite ion produces the dianion radical NO_2^{2-}, which further hydrolyzes into $NO_{(ad)}$ according to the following reactions (Su et al., 2017):

$$NO_{2(ad)}^{-} + e^{-} \rightarrow NO_{2(ad)}^{2-} \quad (E^o = -0.47 \text{ V vs SHE}) \tag{15.xv}$$

$$NO_{2(ad)}^{2-} + H_2O \rightarrow NO_{(ad)} + 2OH^{-} \tag{15.xvi}$$

$NO_{(ad)}$ is the central specie for the selectivity of main and by-products by EC reduction of nitrate ion in bulk solution. According to the Duca−Feliu−Koper mechanism, nitrite gets reduced into the N_2 gas by $NO_{(ad)}$ directly and/or it may follow the path of NH_2, NH_2NO, NH_2NO_2 and finally into N_2 gas as final product (Duca et al., 2011). This pathway consists $NO_{(ad)}$ and $NH_{2(ad)}$, and both the species get reduced to the N-nitrosamide ($NONH_2$) according to Langmuir−Hinshelwood scheme and finally get converted into N_2 gas (Katsounaros and Kyriacou, 2008). Nitramide (NH_2NO_2) was also identified by few researchers which suggested that $NONH_2$ can be produced by NH_2NO_2 and get converted into N_2 gas (Katsounaros et al., 2012). There is another possibility of electrochemically reduction of NO into diazeniumdiolate (HN_2O_2) shown by the following reaction (Dutton et al., 2005):

$$NO_{(ad)} + NO_{(aq)} + H^{+} + e^{-} \rightarrow HN_2O_2 \quad (E^o = 0.0 \text{ V vs SHE}) \tag{15.xvii}$$

This electroactive species (HN_2O_2) can also take another path to produce N_2 followed by N_2O and N_2O^{-} as reported by Chumanoc group as shown in the following reactions (Zheng et al., 1999; Yang et al., 2011):

$$HN_2O_{2(ad)} + H^{+} + e^{-} \rightarrow N_2O(ad) + H_2O \quad (E^o = 1.59 \text{ V vs SHE}) \tag{15.xviii}$$

$$N_2O^{-} + e^{-} \rightarrow N_2O^{-} \quad (E^o = 1.59 \text{ V vs SHE}) \tag{15.xix}$$

In another possible pathway, NO gets reduced into nitrous oxide (N_2O) followed by azanone (HNO) and hyponitrous acid (H_2N_2O) shown in Fig. 15.5 (Katsounaros and Kyriacou, 2008; Katsounaros et al., 2012).

The final major reduced product from EC nitrate reduction is ammonia (NH_3/NH_4^{+}). NO reduced into azanone (HNO) which further reduces into H_2NO, and this step is the rate-determining step (De Vooys et al., 2001). Hydroxylamine (NH_2OH) is then produced by electron transfer from H_2NO and hydroxylamine is one of the by-products from EC reduction of nitrate in the aqueous solution (Xu et al., 2018; Su et al., 2017). Ammonia is produced quickly from the reduction of hydroxylamine and is in equilibrium with ammonium ion. Reactions (15.xx)−(15.xxiv) show all the chemical reactions involved during the ammonia production (Xu et al., 2018).

$$NO_{(ad)} + H^{+} + e^{-} \Leftrightarrow HNO_{(ad)} \quad (E^o = -0.78 \text{ V vs SHE}) \tag{15.xx}$$

$$HNO_{(ad)} + H^{+} + e^{-} \rightarrow H_2NO_{(ad)} \quad (E^o = 0.52 \text{ V vs SHE}) \tag{15.xxi}$$

$$H_2NO_{(ad)} + H^{+} + e^{-} \rightarrow NH_2OH_{(ad)} \quad (E^o = 0.90 \text{ V vs SHE}) \tag{15.xxii}$$

$$NH_2OH + 2H^{+} + 2e^{-} \rightarrow NH_3 + H_2O \quad (E^o = 0.42 \text{ V vs SHE}) \tag{15.xxiii}$$

$$NH_3 + H^{+} \Leftrightarrow NH_4^{+} \quad (pK_a = 9.25) \tag{15.xxiv}$$

However, there are several chemical reactions, which consume NH_2OH and may lead to the formation of N_2 and/or N_2O shown in Fig. 15.5.

15.3.2 Indirect anodic oxidation of by-products

Metal electrode as anode is very much capable of in situ production of active chlorine such as $HOCl/OCl^-$ (hypochlorite ion) by oxidizing the chlorine ions (Cl^-) (Chauhan et al., 2016). These active species oxidize the NH_3 and NO_2^- into N_2 and NO_3^-, respectively. Therefore N_2 selectivity toward nitrate reduction can be increased by electrochlorination (Teng et al., 2018). In the absence of chlorine, NH_3 can oxidize into N_2 directly with the help of electrochemically generated active species such as $OH^-/^\bullet OH$. Indirect oxidation of by-products such as NO_2^- and NH_3 in the absence and presence of chlorine is shown by chemical reactions (15.xxxiii)–(15.xxxvii) (Chauhan et al., 2016; Ghazouani et al., 2017).

$$M + H_2O \rightarrow M(^\bullet OH) + H^+ + e^- \tag{15.xxv}$$

$$M(^\bullet OH) \rightarrow MO + H^+ + e^- \tag{15.xxvi}$$

$$MO_{2+1} \rightarrow MO_2 + 0.5O_2 + H^+ + e^- \tag{15.xxvii}$$

$$M(^\bullet OH) \rightarrow M + 0.5O_2 + H^+ + e^- \tag{15.xxviii}$$

$$M(^\bullet OH) + Cl^- \rightarrow M(^\bullet OCl) + 0.5O_2 + H^+ + 2e^- \tag{15.xxix}$$

$$M(^\bullet OH) + Cl^- \rightarrow M(^\bullet Cl) + 0.5O_2 + H^+ + 2e^- \tag{15.xxx}$$

$$Cl^- + Cl^- \rightarrow Cl_2 + 2e^- \quad (E^o = 1.36 \text{ V vs SHE}) \tag{15.xxxi}$$

$$Cl_2 + H_2O \rightarrow HOCl + H^+ + Cl^- \tag{15.xxxii}$$

$$2NH_3 + 2OCl^- \rightarrow N_2 + 2HCl + 2H_2O + 2e^- \tag{15.xxxiii}$$

$$NH_3 + 4OCl^- \rightarrow NO_3^- + H_2O + H^+ + 4Cl^- \tag{15.xxxiv}$$

$$NO_2^- + OCl^- \rightarrow NO_3^- + Cl^- \tag{15.xxxv}$$

$$NO_3^- + 2H^+ + 5H_2 \rightarrow N_2 + 6H_2O \tag{15.xxxvi}$$

$$2NH_3 + 6OH^- \rightarrow N_2 + 6H_2O + 6e^- \tag{15.xxxvii}$$

15.4 Conclusion

The advance treatment of inorganic pollutants such as nitrate (NO_3^-) is mostly studied by EC reduction method. The EC reduction process includes several types of intermediate and final products such as NO_2^-/HNO_2, N_2O, NH_3/NH_4^+, and N_2. For denitrification of wastewater, the preferred final product is N_2 gas, since other products such as $NH_4 +$ and NO_2^- cause aesthetic, health, and operational problems. The crucial rate-limiting step is the reduction of nitrate into nitrite, which is effected by mass transfer from bulk wastewater toward the surface of cathode, the transfer of electron, and the adsorption of nitrate on the surface of cathode. The material of electrode is the key parameter that affects both the selectivity of final and by-product and the reduction kinetics during EC reduction of nitrate. The metals having greatly occupied d-orbitals such as Cu, Pt, and Ag and unclosed d-orbital shells are ideal electrode materials for EC reduction of nitrate due to the similar LUMO (lowest unoccupied molecular orbital) π* energy level of nitrate elements for assisting the first transfer of electron. In all of these cathode metals, Cu displays both the ultimate electro-catalytic kinetics as well as lowest cost. However, it is limited for wastewater treatment due to the corrosion and leaching of electrode and also selectivity toward ammonia as by-products. In addition, Pt and Pd electrodes exhibit higher selectivity of N_2 but at the cost of sluggish kinetics due to the competitive evolution of H_2 from reduction of water molecule at the surface of electrode.

The different metal alloys display prospects since privileged performance characteristics of several metals can be shared into one electrode material. Cu−Pt and Cu−Pd alloys advance the activity in comparison to pure electrodes due to the synergy between fast nitrate reduction kinetics by Cu and great selectivity of N_2 gas for Pt/Pd. Second, cupronickels show advance resistance toward corrosion and higher stability characteristics than Cu electrode. Hence, these types of electrode materials are advance and competitive for EC denitrification of wastewater.

References

Adeleye, A.S., Conway, J.R., Garner, K., Huang, Y., Su, Y., Keller, A.A., 2016. Engineered nanomaterials for water treatment and remediation: costs, benefits, and applicability. Chem. Eng. J. 286, 640−662.

Amertharaj, S., Hasnat, M.A., Mohamed, N., 2014. Electroreduction of nitrate ions at a platinum-copper electrode in an alkaline medium: influence of sodium inositol phytate. Electrochim. Acta 136, 557−564.

Bhatnagar, A., Sillanpaa, M., 2011. A review of emerging adsorbents for nitrate removal from water. Chem. Eng. J. 168 (2), 493−504.

BIS, 2012. 10500: 2012 Indian Standard Drinking Water-Specification (Second Revision). Bureau of Indian Standards (BIS), New Delhi.

Bouzek, K., Paidar, M., Sadilkova, A., Bergmann, H., 2001. Electrochemical reduction of nitrate in weakly alkaline solutions. J. Appl. Electrochem. 31 (11), 1185−1193.

Brylev, O., Sarrazin, M., Roue, L., Belanger, D., 2007. Nitrate and nitrite electrocatalytic reduction on Rh-modified pyrolytic graphite electrodes. Electrochim. Acta 52 (21), 6237−6247.

Calle-Vallejo, F., Huang, M., Henry, J.B., Koper, M.T., Bandarenka, A.S., 2013. Theoretical design and experimental implementation of Ag/Au electrodes for the electrochemical reduction of nitrate. Phys. Chem. Chem. Phys. 15 (9), 3196−3202.

Casella, I.G., Ritorti, M., 2010. Electrodeposition of silver particles from alkaline aqueous solutions and their electrocatalytic activity for the reduction of nitrate, bromate and chlorite ions. Electrochim. Acta 55 (22), 6462−6468.

Chauhan, R., Srivastava, V.C., Hiwarkar, A.D., 2016. Electrochemical mineralization of chlorophenol by ruthenium oxide coated titanium electrode. J. Taiwan Inst. Chem. Eng. 69, 106−117.

Chen, T., Li, H., Ma, H., Koper, M.T., 2015. Surface modification of Pt (100) for electrocatalytic nitrate reduction to dinitrogen in alkaline solution. Langmuir 31 (10), 3277−3281.

Çirmi, D., Aydin, R., Koleli, F., 2015. The electrochemical reduction of nitrate ion on polypyrrole coated copper electrode. J. Electroanal. Chem. 736, 101−106.

Da Cunha, M.C.P.M., De Souza, J.P.I., Nart, F.C., 2000. Reaction pathways for reduction of nitrate ions on platinum, rhodium, and platinum − rhodium alloy electrodes. Langmuir 16 (2), 771−777.

De Vooys, A.C.A., Koper, M.T.M., Van Santen, R.A., Van Veen, J.A.R., 2001. The role of adsorbates in the electrochemical oxidation of ammonia on noble and transition metal electrodes. J. Electroanal. Chem. 506 (2), 127−137.

Dima, G.E., De Vooys, A.C.A., Koper, M.T.M., 2003. Electrocatalytic reduction of nitrate at low concentration on coinage and transition-metal electrodes in acid solutions. J. Electroanal. Chem. 554, 15−23.

Dima, G.E., Beltramo, G.L., Koper, M.T.M., 2005. Nitrate reduction on single-crystal platinum electrodes. Electrochim. Acta 50 (21), 4318−4326.

Dortsiou, M., Kyriacou, G., 2009. Electrochemical reduction of nitrate on bismuth cathodes. J. Electroanal. Chem. 630 (1−2), 69−74.

Duca, M., Figueiredo, M.C., Climent, V., Rodriguez, P., Feliu, J.M., Koper, M.T., 2011. Selective catalytic reduction at quasi-perfect Pt (100) domains: a universal low-temperature pathway from nitrite to N2. J. Am. Chem. Soc. 133 (28), 10928−10939.

Durivault, L., Brylev, O., Reyter, D., Sarrazin, M., Belanger, D., Roue, L., 2007. Cu−Ni materials prepared by mechanical milling: their properties and electrocatalytic activity towards nitrate reduction in alkaline medium. J. Alloys Compd. 432 (1−2), 323−332.

Dutton, A.S., Fukuto, J.M., Houk, K.N., 2005. Theoretical reduction potentials for nitrogen oxides from CBS-QB3 energetics and (C) PCM solvation calculations. Inorg. Chem. 44 (11), 4024−4028.

Einaga, Y., Foord, J.S., Swain, G.M., 2014. Diamond electrodes: diversity and maturity. MRS Bull. 39 (6), 525−532.

El-Deab, M.S., 2004. Electrochemical reduction of nitrate to ammonia at modified gold electrodes. Electrochim. Acta 49 (9−10), 1639−1645.

El Din, A.S., El Wahab, F.A., 1977. The behaviour of copper-zinc alloys in alkaline solutions upon alternate anodic and cathodic polarization. Corros. Sci. 17 (1), 49−58.

Epsztein, R., Nir, O., Lahav, O., Green, M., 2015. Selective nitrate removal from groundwater using a hybrid nano-filtration−reverse osmosis filtration scheme. Chem. Eng. J. 279, 372−378.

Fedurco, M., Kedzierzawski, P., Augustynski, J., 1999. Effect of multivalent cations upon reduction of nitrate ions at the Ag electrode. J. Electrochem. Soc. 146 (7), 2569−2572.

Garcia-Gomez, C., Drogui, P., Zaviska, F., Seyhi, B., Gortares-Moroyoqui, P., Buelna, G., et al., 2014. Experimental design methodology applied to electrochemical oxidation of carbamazepine using Ti/PbO2 and Ti/BDD electrodes. J. Electroanal. Chem. 732, 1−10.

Garcia-Segura, S., Lanzarini-Lopes, M., Hristovski, K., Westerhoff, P., 2018. Electrocatalytic reduction of nitrate: fundamentals to full-scale water treatment applications. Appl. Catal., B: Environ. 236, 546−568.

Georgeaud, V., Diamand, A., Borrut, D., Grange, D., Coste, M., 2011. Electrochemical treatment of wastewater polluted by nitrate: selective reduction to N2 on Boron-Doped Diamond cathode. Water Sci. Technol. 63 (2), 206−212.

Ghazouani, M., Akrout, H., Jomaa, S., Jellali, S., Bousselmi, L., 2016. Enhancing removal of nitrates from highly concentrated synthetic wastewaters using bipolar Si/BDD cell: Optimization and mechanism study. J. Electroanal. Chem. 783, 28−40.

Ghazouani, M., Akrout, H., Bousselmi, L., 2017. Nitrate and carbon matter removals from real effluents using Si/BDD electrode. Environ. Sci. Pollut. Res. 24 (11), 9895−9906.

Gootzen, J.F.E., Peeters, P.G.J.M., Dukers, J.M.B., Lefferts, L., Visscher, W., Van Veen, J.A.R., 1997. The electrocatalytic reduction of NO3− on Pt, Pd and Pt + Pd electrodes activated with Ge. J. Electroanal. Chem. 434 (1−2), 171−183.

Hamme, R.C., Emerson, S.R., 2004. The solubility of neon, nitrogen and argon in distilled water and seawater. Deep Sea Res., I: Oceanogr. Res. Pap. 51 (11), 1517−1528.

Hammer, B., Norskov, J.K., 2000. Theoretical surface science and catalysis—calculations and concepts, Advances in Catalysis, vol. 45. Academic Press, pp. 71−129.

Haque, I.U., Tariq, M., 2009. Voltammetry of nitrate at solid cathodes. ECS Trans. 16 (18), 25−33.

Jiang, C., Liu, L., Crittenden, J.C., 2016. An electrochemical process that uses an Fe^0/TiO2 cathode to degrade typical dyes and antibiotics and a bio-anode that produces electricity. Front. Environ. Sci. Eng. 10 (4), 15.

Kalaruban, M., Loganathan, P., Kandasamy, J., Naidu, R., Vigneswaran, S., 2017. Enhanced removal of nitrate in an integrated electrochemical-adsorption system. Sep. Purif. Technol. 189, 260−266.

Kapoor, A., Viraraghavan, T., 1997. Nitrate removal from drinking water. J. Environ. Eng. 123 (4), 371−380.

Katsounaros, I., Kyriacou, G., 2008. Influence of nitrate concentration on its electrochemical reduction on tin cathode: identification of reaction intermediates. Electrochim. Acta 53 (17), 5477−5484.

Katsounaros, I., Ipsakis, D., Polatides, C., Kyriacou, G., 2006. Efficient electrochemical reduction of nitrate to nitrogen on tin cathode at very high cathodic potentials. Electrochim. Acta 52 (3), 1329−1338.

Katsounaros, I., Dortsiou, M., Kyriacou, G., 2009. Electrochemical reduction of nitrate and nitrite in simulated liquid nuclear wastes. J. Hazard. Mater. 171 (1−3), 323−327.

Katsounaros, I., Dortsiou, M., Polatides, C., Preston, S., Kypraios, T., Kyriacou, G., 2012. Reaction pathways in the electrochemical reduction of nitrate on tin. Electrochim. Acta 71, 270–276.

Kettler, P.B., 2003. Platinum group metals in catalysis: fabrication of catalysts and catalyst precursors. Org. Process Res. Dev. 7 (3), 342–354.

Khomutov, N.E., Stamkulov, U.S., 1971. Nitrate reduction at various metal electrodes. Sov. Electrochem. 7, 312–316.

Kodera, T., Akizuki, S., Toda, T., 2017. Formation of simultaneous denitrification and methanogenesis granules in biological wastewater treatment. Process Biochem. 58, 252–257.

Kuang, P., Feng, C., Chen, N., Hu, W., Wang, G., Peng, T., et al., 2016. Improvement on electrochemical nitrate removal by combining with the three-dimensional (3-D) perforated iron cathode and the iron net introduction. J. Electrochem. Soc. 163 (14), E397–E406.

Lacasa, E., Canizares, P., Llanos, J., Rodrigo, M.A., 2012. Effect of the cathode material on the removal of nitrates by electrolysis in non-chloride media. J. Hazard. Mater. 213, 478–484.

Levy-Clement, C., Ndao, N.A., Katty, A., Bernard, M., Deneuville, A., Comninellis, C., et al., 2003. Boron doped diamond electrodes for nitrate elimination in concentrated wastewater. Diamond Relat. Mater. 12 (3–7), 606–612.

Li, M., Feng, C., Zhang, Z., Lei, X., Chen, R., Yang, Y., et al., 2009. Simultaneous reduction of nitrate and oxidation of by-products using electrochemical method. J. Hazard. Mater. 171 (1–3), 724–730.

Li, M., Feng, C., Zhang, Z., Yang, S., Sugiura, N., 2010. Treatment of nitrate contaminated water using an electrochemical method. Bioresour. Technol. 101 (16), 6553–6557.

Li, W., Xiao, C., Zhao, Y., Zhao, Q., Fan, R., Xue, J., 2016. Electrochemical reduction of high-concentrated nitrate using Ti/TiO_2 nanotube array anode and Fe cathode in dual-chamber cell. Catal. Lett. 146 (12), 2585–2595.

Ma, X., Li, M., Feng, C., Hu, W., Wang, L., Liu, X., 2016. Development and reaction mechanism of efficient nano titanium electrode: reconstructed nanostructure and enhanced nitrate removal efficiency. J. Electroanal. Chem. 782, 270–277.

Mácová, Z., Bouzek, K., 2005. Electrocatalytic activity of copper alloys for NO_3^- reduction in a weakly alkaline solution Part 1: Copper–zinc. J. Appl. Electrochem. 35 (12), 1203–1211.

Macova, Z., Bouzek, K., Serak, J., 2007. Electrocatalytic activity of copper alloys for NO_3^- reduction in a weakly alkaline solution. J. Appl. Electrochem. 37 (5), 557–566.

Mameri, N., Yeddou, A.R., Lounici, H., Belhocine, D., Grib, H., Bariou, B., 1998. Defluoridation of septentrional Sahara water of North Africa by electrocoagulation process using bipolar aluminium electrodes. Water Res. 32 (5), 1604–1612.

Martinez, J., Ortiz, A., Ortiz, I., 2017. State-of-the-art and perspectives of the catalytic and electrocatalytic reduction of aqueous nitrates. Appl. Catal., B: Environ. 207, 42–59.

Mattarozzi, L., Cattarin, S., Comisso, N., Guerriero, P., Musiani, M., Vazquez-Gomez, L., et al., 2013. Electrochemical reduction of nitrate and nitrite in alkaline media at CuNi alloy electrodes. Electrochim. Acta 89, 488–496.

Mattarozzi, L., Cattarin, S., Comisso, N., Gerbasi, R., Guerriero, P., Musiani, M., et al., 2015. Electrodeposition of compact and porous Cu-Zn alloy electrodes and their use in the cathodic reduction of nitrate. J. Electrochem. Soc. 162 (6), D236–D241.

Murphy, A.P., 1991. Chemical removal of nitrate from water. Nature 350 (6315), 223.

Nobial, M., Devos, O., Mattos, O.R., Tribollet, B., 2007. The nitrate reduction process: a way for increasing interfacial pH. J. Electroanal. Chem. 600 (1), 87–94.

Ohmori, T., El-Deab, M.S., Osawa, M., 1999. Electroreduction of nitrate ion to nitrite and ammonia on a gold electrode in acidic and basic sodium and cesium nitrate solutions. J. Electroanal. Chem. 470 (1), 46–52.

Paidar, M., Rousar, I., Bouzek, K., 1999. Electrochemical removal of nitrate ions in waste solutions after regeneration of ion exchange columns. J. Appl. Electrochem. 29 (5), 611–617.

Reyter, D., Belanger, D., Roue, L., 2010. Nitrate removal by a paired electrolysis on copper and Ti/IrO_2 coupled electrodes—influence of the anode/cathode surface area ratio. Water Res. 44 (6), 1918–1926.

Sahu, O., Mazumdar, B., Chaudhari, P.K., 2014. Treatment of wastewater by electrocoagulation: a review. Environ. Sci. Pollut. Res. 21 (4), 2397–2413.

Samatya, S., Kabay, N., Yuksel, U., Arda, M., Yuksel, M., 2006. Removal of nitrate from aqueous solution by nitrate selective ion exchange resins. React. Funct. Polym. 66 (11), 1206–1214.

Schmid, G., Delfs, J., 1959. The autocatalytic nature of the cathodic reduction of nitric acid to nitrous acid II. The galvanostatic switch-on process. J. Electrochem., Rep. Bunsen Soc. Phys. Chem. 63 (9–10), 1192–1197.

Shimazu, K., Goto, R., Piao, S., Kayama, R., Nakata, K., Yoshinaga, Y., 2007. Reduction of nitrate ions on tin-modified palladium thin film electrodes. J. Electroanal. Chem. 601 (1–2), 161–168.

Sim, J., Seo, H., Kim, J., 2012. Electrochemical denitrification of metal-finishing wastewater: Influence of operational parameters. Korean J. Chem. Eng. 29 (4), 483–488.

Stueken, E.E., Kipp, M.A., Koehler, M.C., Buick, R., 2016. The evolution of Earth's biogeochemical nitrogen cycle. Earth Sci. Rev. 160, 220–239.

Su, L., Li, K., Zhang, H., Fan, M., Ying, D., Sun, T., et al., 2017. Electrochemical nitrate reduction by using a novel Co_3O_4/Ti cathode. Water Res. 120, 1–11.

Tada, K., Shimazu, K., 2005. Kinetic studies of reduction of nitrate ions at Sn-modified Pt electrodes using a quartz crystal microbalance. J. Electroanal. Chem. 577 (2), 303–309.

Teng, W., Bai, N., Liu, Y., Liu, Y., Fan, J., Zhang, W.X., 2018. Selective nitrate reduction to dinitrogen by electrocatalysis on nanoscale iron encapsulated in mesoporous carbon. Environ. Sci. Technol. 52 (1), 230–236.

Ureta-Zanartu, S., Yanez, C., 1997. Electroreduction of nitrate ion on Pt, Ir and on 70: 30 Pt: Ir alloy. Electrochim. Acta 42 (11), 1725–1731.

Wang, L., Li, M., Feng, C., Hu, W., Ding, G., Chen, N., et al., 2016. Ti nano electrode fabrication for electrochemical denitrification using Box–Behnken design. J. Electroanal. Chem. 773, 13–21.

Xu, D., Li, Y., Yin, L., Ji, Y., Niu, J., Yu, Y., 2018. Electrochemical removal of nitrate in industrial wastewater. Front. Environ. Sci. Eng. 12 (1), 9.

Yang, J., Duca, M., Schouten, K.J.P., Koper, M.T., 2011. Formation of volatile products during nitrate reduction on a Sn-modified Pt electrode in acid solution. J. Electroanal. Chem. 662 (1), 87–92.

Zheng, J., Lu, T., Cotton, T.M., Chumanov, G., 1999. Photoinduced electrochemical reduction of nitrite at an electrochemically roughened silver surface. J. Phys. Chem. B 103 (31), 6567–6572.

Insights on the advanced processes for treatment of inorganic water pollutants

Raj Mohan Balakrishnan, Priyanka Uddandarao, Vishnu Manirethan and Keyur Raval

Department of Chemical Engineering, National Institute of Technology Karnataka Surathkal, Surathkal, India

O U T L I N E

Inorganic Pollutants in Water
DOI: https://doi.org/10.1016/B978-0-12-818965-8.00016-0

16.1 Introduction

Water is the vital entity for the survival of all life forms in the world. Groundwater is one of the primary natural resources and it is reported that only 2.5% of world's total water is portable. With the onset of the industrial revolution, innovation grew rapidly; science became advanced, technological age came into view, and humans were able to advance further in the 21st century. This rapid industrialization and increase in population have resulted in groundwater contamination, mainly because of heavy metals found in Earth's crust, which have turned out to be an environmental hazard due to different anthropogenic activities. Metal processing in various industries, refinery processes, petroleum and coal combustions, nuclear power plants, paper processing plants, textiles, wood processing factories, plastics, electronic production centers, etc. contribute to the heavy-metal contamination of water bodies (Pacyna, 1996; Arruti et al., 2010; Sträter et al., 2010). Natural activities such as metal corrosion, soil erosion, leaching of metals to the deep soil, weathering of mineral-rich rocks, and volcanic eruption also contribute to heavy-metal contamination (Nriagu, 1989; Fergusson, 1990; Tchounwou et al., 2012). Heavy metals have a relatively high specific density of more than $5 \, g/cm^3$ and are lethal even at minor concentrations (Silwana et al., 2014).

Heavy metals cannot be disintegrated into nontoxic forms, unlike the organic pollutants even by biological as well as chemical agents (Chapman, 1996; Bolan et al., 2003; Sun et al., 2019). Moreover, these heavy metals have long half-lives and exhibit high chemical toxicity after dissolving in water. Mankind is also affected by propagation and biomagnification of heavy metals through the food chain through human ingestion or contaminated soil contact. There are various studies that aim at removing divalent mercury, divalent lead, divalent copper, and hexavalent chromium from groundwater. Most of the heavy metals present in groundwater have pH around acidic or near neutral range and they precipitate at pH close or higher than neutral, which are indistinguishable from solutions at lower concentrations. For example, mercury, copper, and lead precipitates at above pH 6.5, 7, and 5.5, respectively (Bradl, 2004; Albrecht et al., 2011; Sulaymon et al., 2013), whereas chromium precipitates around pH 8.5 (Dixit et al., 2015).

Conventional methods such as hydroponics (Haddad and Mizyed, 2011) are employed for the removal of inorganic pollutants, including heavy metals using plants. Numerous technologies are in practice such as chemical precipitation, ion exchange, membrane filtration, electrochemical treatment methodologies, coagulation, and floatation (Fu and Wang, 2011; Kobielska et al., 2018). The heavy-metal concentration is detected by various methods such as atomic absorption spectrophotometry, atomic emission spectroscopy, inductively coupled plasma—optical emission spectroscopy, and other spectroscopic techniques (Pandey et al., 2003; Zhu et al., 2008; Feng et al., 2013). Over the years various absorbents are used for the removal of heavy metals. However, the efficiency of the absorbents depends upon conditions such as pH, initial metal concentration, adsorbents dosage, and temperature of the system. In this scenario, nanosorbents came into picture, which are relatively simpler devices and requiring minimal sample preparation for heavy-metal removal (Kim and Grate, 2003). Owing to various advantages, this chapter majorly focuses on the removal of inorganic pollutants using nanoadsorbents of biological origin.

16.2 Major inorganic heavy-metal pollutants

The following section encompasses the review of the major heavy metals such as cadmium (Cd), manganese (Mn), nickel (Ni), selenium (Se), mercury (Hg), arsenic (As), chromium (Cr), lead (Pb), and copper (Cu) and annotates the incidence of these heavy metals.

16.2.1 Selenium

Se contamination in the wastewater stream could be to a great extent credited to mining, farming, petrochemical, and modern assembling activities. Se exists as Se(IV) and Se(VI). These are predominantly found in traces existing in the forms SeO_3^{2-} and SeO_2^{4-}. Their excess presence in the environment is causing tremendous loss greatly affecting human and aquatic health (Hamilton, 2004). According to World Health Organization (WHO), $10\,\mu g/L$ is the maximum Se concentration in drinking water recommended.

16.2.2 Copper

Industrial and agricultural provisions are the sources of copper contamination of water and soil. Cu(II) ion is a micronutrient essential for the body metabolism and immune system, but excess concentration will impart toxic effects leading to various health problems. Cu concentration in the body above a certain limit can damage biological systems and can cause inflammatory disorders, hepatic cell damage, and even Alzheimer's disease (Wang et al., 2015; Hu et al., 2015).

16.2.3 Manganese

Apart from the industrial sources, contamination of water and soil is also caused by erosion of crustal rocks, which are taken up by plants and enters the food chain. The most abundant forms of Mn in nature are its unsolvable oxides of MnO_2 and Mn_3O_4 among 30 Mn oxide minerals in a wide variety of geological settings. Certain regions in Bangladesh reported the Mn(II) concentration in drinking water as four times higher than what is stipulated as permissible standard for drinking water by the WHO. However, drinking water commonly has $100\,g$ Mn/L concentration. High concentrations of Mn cause neurological impacts, for example, Parkinson disorder. Likewise, contamination of foodstuffs with the Mn-containing fungicides in certain nations are viewed as a genuine ecological risk.

16.2.4 Cadmium

In cities of Texas, New York, California, and Ohio, Cds are found frequently in different children stores containing jewellery, battery-operated toys, and paints. Its toxicity is well documented due to its widespread exposure arising from cigarette smoking posing a serious threat to both environmental and occupational health. Long-term exposure of Cd may also lead to damage in DNA, heart diseases, kidney diseases, high blood pressure, and cancer. Out of 275 hazardous substances in the environment, Cd ranks seventh according to the Centers for Disease Control and Prevention.

16.2.5 Mercury

Groundwater samples at various parts in the states of Orissa, Haryana, Gujarat, Jharkhand, Andhra Pradesh, and West Bengal are reported to have Hg concentrations above the allowable limit of 0.001 mg/L stipulated by the Bureau of Indian Standards (BIS). Mercury exists in -2, 0, and $+2$, oxidation states that are toxic to all life forms (Anderson, 1986; Bernhoft, 2012). Metallic, organic, and inorganic forms of mercury are naturally found in different kinds of rocks and coal environment (Yard et al., 2012). Hg reaches the human body mainly through the food chain or even by inhalation route; can even impair the genetic material, which leads to Down syndrome; affect the reproductive system, may even led to critical cases of mental disorders and memory loss (Bernhoft, 2012; Carocci et al., 2014).

16.2.6 Arsenic

Reports states the incidence of As in drinking waters of nine districts of West Bengal affecting around 42.7 million people where the water is contaminated with arsenic concentration above the permissible limit of 0.05 mg/L as per the WHO standards. Some tube wells in West Bengal have As concentration of 3400 μg/L, where the source of As is geologic origin (Ratnaike, 2003). As exists in different oxidation states -3, 0, 4, 5 and mostly presents in as As_2O_3 (Smith, 1995). It is predominantly present in oxidation state 5 in protonated form H_3AsO_4, $H_2AsO_4^-$, $HAsO_4^{2-}$, AsO_4^3, which has a high metal ion chelating and precipitating property (Bodek, 1988). As exists as As(III) in anoxic conditions and predominantly in the protonated state of H_3AsO_3, $H_2AsO_3^-$, and $HAsO_3^{2-}$ which can chelate with the metal sulfides (Wuana and Okieimen, 2011). The toxicity of inorganic As changes with the oxidation state, that is, As(III) is 60 times more poisonous than that of As(V) form. As enters the human body mainly through inhalation and (Benramdane et al., 1999) it interferes with the enzyme functioning, energy pathway, cell mechanisms leading to lipid peroxidation, DNA functioning by replacing phosphate groups in ATP energy molecules (Cobo and Castiñeira, 1997). Chronic effects of As toxicity include Mee's lines in nails, hyperpigmentation, palmar, and solar keratosis (Lien et al., 1999). Chronic As toxicity cannot be cured and dosage of 100–300 mg can be fatal (Ratnaike, 2003).

16.2.7 Lead

Pb has been used in diverse industries such as batteries, paints, weapons, pipes, paper, pigments, and shipping (Ab Latif Wani et al., 2015). Pb is regarded as a cumulative toxicant, which affects almost all organs in the body according to WHO. It is extremely harmful to children and it directly affects brain, liver, kidney, and bones. Pb gets accumulated in tooth/bones and is subsequently released to the bloodstream and makes life at risk. The permissible limit for Pb in drinking water should not exceed 0.01 mg/L as per the BIS.

16.2.8 Chromium

In India, several locations are reported to have hexavalent Cr in groundwater, the majority of which are caused by the improper industrial effluent expulsion. Tanneries are one of the prime Cr contaminators of groundwater and other water bodies. Central Pollution Control Board of India (CPCB) reported that Kanpur in Uttar Pradesh has groundwater with Cr(VI) concentration up to 250 times that of the WHO standard of 0.05 mg/L (Sharma et al., 2012). Vellore district in Tamil Nadu is reported to have remarkably higher Cr concentration in the soil as well as the groundwater due to the tannery effluents. Cr exists in different oxidation states of −2, 0, 3, 6; out of which Cr(VI) is the most inimical one based on toxicity (Guertin, 2004). 0.05 mg/L is the maximum permissible limit of Cr in drinking water as per the Indian Standards stipulated by the CPCB and the Central Groundwater Commission. Pulmonary fibrosis, chronic bronchitis emphysema, bronchial asthma owing to hypersensitivity, pneumoconiosis, etc. are some of the pulmonary impacts of Cr poisoning. Inhalation of Cr fumes or aerosols is reported to develop cancer due to the direct mixing of Cr(VI) with blood. Cr(III) above certain concentrations has an inhibitory effect on cell growth and multiplication, while Cr(VI) imparts chromosomal aberrations leading to mutagenicity (Baruthio, 1992; Wilbur et al., 2012). The lethal dose is above 300 mg, which highly deteriorates kidneys.

16.2.9 Copper

Cu industries are one of the flourished sectors in India and there are different processes to extract the pure Cu from the ore, which leads to the production of waste and effluent consisting of Cu. These are either being discharged to the landfills or the aquifers (Dutta, 1987; Sinha et al., 1998). Cu exists mainly in oxidation states of +1 and +2, out of which cuprous form of Cu shows higher toxicity compared to the cupric form (Beswick et al., 1976). Exposure to Cu can cause irritation in eyes, nose, and mouth; but higher intake of this can damage liver and kidney, which can even be fatal. Chronic exposure results in Wilson's syndrome, which is featured by corneal Cu deposition, hepatic cirrhosis, demyelination, brain damage, renal diseases, etc.

16.3 Adsorbents for heavy-metal removal

During the recent years, various adsorbents have been developed such as activated carbon (Pattanayak et al., 2000), activated alumina (Singh and Pant, 2006), iron hydroxide (Thirunavukkarasu et al., 2001), hydrous zirconium oxide (Suzuki et al., 1997), iron oxide—coated polymeric materials (Kundu and Gupta, 2006), zerovalent iron

(Bang et al., 2005), titanium dioxide (Meng et al., 2003), chitosan (Hena, 2010), algal bloom reside (Zhang et al., 2010), sawdust activated carbon (Karthikeyan et al., 2005), and sulfonated lignite (Khezami and Capart, 2005; Demiral et al., 2008; Zhang et al., 2010). Table 16.1 depicts the utility of various heavy metals and distinct types of adsorbents used along with their adsorption capacity.

TABLE 16.1 Overview of various adsorbents for heavy-metal removal and their adsorption capacity at various operating conditions.

Sl. no.	Heavy metal	Adsorbent	Adsorption capacity (mg/g)	Optimum parameters for adsorption and Isotherm and nature of adsorption	References
1.	Hg(II)	Dry biomass of *Chlorella vulgaris*	32.6	$w = 0.5$ g/L, $C_o = 11.0–90.6$ mg/L pH = 5; Langmuir	Solisio et al. (2019)
2.	Cr(VI)	Novel cross-linked chitosan	325.2	pH = 2, $T = 30°C$, $C_o = 20$ mg/L; Freundlich	Vakili et al. (2018)
3.	Hg(II)	ECW	31.75	$w = 0.4$ g/L, $Co = 77.98$ mg/L, pH = 7, $t = 192.40$ min, $T = 33°C$, and 375 rpm; Langmuir, endothermic	Alvarez et al. (2018)
4.	Hg(II)	Thioether-functionalized corn oil biosorbent	8.15	$T = 20$, $t = 231$; Freundlich	Dunn et al. (2018)
5.	As(III) and As(V)	Chitosan coated with Fe−Mn binary oxide	3.91 [As(III)] and 3.89 [As(V)]	pH = 3−8, $w = 0.5$ g/L, 180 rpm, $T = 22 ± 1°C$, $C_o = 0.5$ mg/L; Freundlich, chemisorption	Nikić et al. (2019)
6.	As(III)	Iron hydroxide/manganese dioxide−doped straw activated carbon	75.82	pH = 3, $T = 30°C$, $w = 10$ mg, $Co = 20$ mg/L, $t = 8$ h; Langmuir, chemisorption	Xiong et al. (2017)
7.	Cu(II)	*Jatropha* biomass	22.910	pH = 5, $t = 60$ min, $T = 25°C$, 200 rpm; Langmuir, chemisorption	Nacke et al. (2016)
8.	Cu(II)	Activated carbon prepared from grape bagasse	43.47	$T = 45°C$, pH = 5; Langmuir and Dubinin−Radushkevich, chemisorption	Demiral and Güngör (2016)
9.	As(III)	NOP and COP	32.7 (NOP) and 60.9 (COP)	pH = 6.5, $w = 4$ g/L, $T = 20 ± 2°C$, $t = 120$ min, $Co = 200$ mg/L; Langmuir, chemisorption	Abid et al. (2016)
10.	As(III)	Zeolite modified with copper oxide and iron oxide	44.8 (zeolite/CuO NCs) and 47.4 (zeolite/Fe$_3$O$_4$ NCs)	$Co = 100$ mg/L, $t = 40$ min, $w = 0.15$ g, pH = 4−6; Langmuir, chemisorption	Alswat et al. (2016)
11.	Cr(VI)	Bone char	4.8	pH = 1, $t = 2$ h; Langmuir, endothermic	Hyder et al. (2015)

(Continued)

TABLE 16.1 (Continued)

Sl. no.	Heavy metal	Adsorbent	Adsorption capacity (mg/g)	Optimum parameters for adsorption and Isotherm and nature of adsorption	References
12.	Hg(II)	*Phragmites karka*	2.27	pH = 4, 100 rpm, $T = 40°C$, $t = 40$ min; Langmuir, Freundlich, endothermic	Raza et al. (2015)
13.	As(III)	Iron-impregnated biochar	2.16	$T = 20 \pm 2°C$, $w = 0.1$ g, $Co = 5$ mg/L, pH = 5.8 ± 0.2; Freundlich, chemisorption	Hu et al. (2015)
14.	Pb(II) and Cu(II)	Agricultural wastes	75% (pH 5.0–7.0), 98% (pH 5.0)	Cu ($w = 2.5$ g, $Co = 40.7$ mg/L, pH = 4.4, and $t = 64$ min), Pb ($w = 2.5$ g, $Co = 196.1$ mg/L, pH = 5.6, $t = 60$ min)	Janyasuthiwong et al. (2015)
15.	Pb(II)	Activated carbon from Algerian dates stones of *Phoenix dactylifera*	9.91	$Co = 50$ mg/L, $w = 0.1$ g, pH = 6, and $T = 25°C$; Freundlich, exothermic	Chaouch et al. (2013)
16.	As(III)	Iron coated rice husk	2.5	$t = 6$ h, pH = 4, $Co = 5$ mg/L, $w = 4$ g/L, $T = 23°C$; Langmuir	Pehlivan et al. (2013)
17.	Cu(II)	Banana peel	27.78	pH = 6, $t = 60$ min, 120 rpm, $T = 20°C$: Langmuir, exothermic	Hossain et al. (2012)
19.	Pb(II)	Pine cone activated carbon	27.53	$Co = 100$ mg/L, $w = 0.1$ g, pH = 6, $T = 25°C$; Langmuir, chemisorption	Momčilović et al. (2011)
20.	Cr(VI)	Lignin	60.4	pH = 2, $t = 24$ h; Freundlich	Albadarin et al. (2011)
21.	As(III)	*Staphylococcus xylosus* biomass pretreated with Fe(III)	54.35	pH = 7, $Co = 1$ g/L, and $t = 30$ min; Langmuir, physisorption	Aryal et al. (2010)
22.	Cr(VI)	Carbon slurry	15.24	pH = 2, $t = 70$ min, $T = 303K$; Langmuir, Freundlich	Gupta et al. (2010)

Notations: w is weight of adsorbent, C_o is initial concentration, t is time, and T is temperature. *COP*, Charred orange peel; *ECW*, exhausted coffee waste; *NOP*, natural orange peel.

16.3.1 Melanin nanoparticles as a biosorbent for heavy-metal removal

Melanin nanoparticles showed a remarkable high level of adsorption capacity at lower heavy-metal concentrations when compared with the activated carbon making it as an excellent adsorbent. Melanin has a great deal of attention due to its various biological properties such as antioxidant activity, metal ion chelation, photoprotection, and free radical scavenging behavior. The functional groups phenolic/hydroxyl (OH), carboxyl (COOH), and amine groups (NH) in squid melanin bind effectively to heavy metals. Melanin is a biopigment, which gives a brown to black coloration to the widely distributed animal kingdom and gives gray coloration as in human. It is a polymer formed from the monomeric units of 5,6-dihydroxyindole and 5,6-dihydroxy indole-2-carboxylic acid (Solano, 2014). Therefore numerous functional groups with high metal ion scavenging property can easily remove different heavy metals present in groundwater even at very low concentration at a range of 2−5 ppm and render the water fit for drinking. The term *Melanin* is derived from the Greek word *Melanos*, meaning dark. In 1840 a Swedish Chemist Berzelius extracted it from human eye membrane. Apart from its function of photoprotection and thermoregulation, it is regarded as an excellent free radical scavenger, cation chelator and has antimicrobial property. There are different types of melanin reported till date, namely, eumelanin, pheomelanin, allomelanins, and neuromelanin. Eumelanin is a dark/brown color, pheomelanin is usually red color, whereas allomelanin is a devoid nitrogen form of melanin with dark to totally black coloration, and neuromelanin is mainly found in the brain (Butler and Day, 1998; Hung et al., 2003).

Removal of uranium from aqueous solution achieved by melanin synthesized using tyrosinase enzyme was reported by Saini and Melo (2013). High adsorption capacity of 588.24 mg of uranium per gram of melanin was obtained from Langmuir plot at a broad range of pH within a very short equilibrium time of 2 hours. The adsorption process was found to be thermodynamically favorable at different temperature conditions and the experimental data of the time study showed better fit with Lagergren's pseudo-second-order model. Chen et al., 2007 reported the adsorption of Cd(II) and Pb(II) ions by squid melanin and the results reveal that the maximum content of bound ions on the surface of melanin is 0.93 mM/g for Cd(II) and 0.65 mM/g for Pb(II). Further, adsorption yield was high in the pH range of 4.0−7.0. Adsorption of Pb(II) is not affected by temperature, whereas Cd(II) binding is affected signifying the characteristics of functional groups in melanin to which the heavy metals are binding. Sono et al. (2012) effectively coated synthetic and hair melanin on hydrophobic polyvinylidene difluoride (PVDF) disks for Pb(II) removal. Synthetic eumelanin−coated PVDF disks showed higher adsorption rate toward Pb(II) than other metal ions such as Cd(II), Cu(II), and Zn(II). Maximum adsorption capacity was about 138 μg Pb(II) per disk for synthetic melanin and in the case of hair melanin, 126 μg Pb(II) per disk. Sajjan et al. (2013) conducted experiments to remove Cu(II) and Pb(II) using *Klebsiella* sp. GSK melanin immobilized in sodium alginate. The adsorption capacity reported was about 169 mg/g for Cu(II) and 280 mg/g for Pb(II). Adsorption of Pb by squid melanin, Cu and Pb by *Klebsiella* melanin immobilized alginate beads, and uranium (U) by synthetic melanin is reported to be 134, 280 [Pb(II)], 169 [Cu(II)], and 588.24 mg/g, respectively (Chen et al., 2007; Sajjan et al., 2013; Saini and Melo, 2013), whereas adsorption capacity of Pb, Cu, and U using other absorbents is 3.92−43.47, 2.89−240.06, and 3.54−98 mg/g, respectively (Tsunashima et al., 1981; Goel et al., 2005; Gerçel and Gerçel, 2007).

Melanin synthesized from the bacteria *Pseudomonas stuteri* HMGM-7, obtained from the marine sediment is used as an adsorbent for the removal of heavy metals such as Hg(II), Pb(II), Cr(VI), and Cu(II) (Thaira et al., 2019). The synthesized melanin is structurally similar to eumelanin and transmission electron microscope (TEM) analysis revealed them as nanosized spherical particles of size 32 nm. 0.2 g/L melanin attained a maximum adsorption within 90 minutes. There was an initial fast uptake of heavy metals, which was followed by a uniform increment in the rate of adsorption. The underlying stage compares to the accessibility of progressively active sites for heavy-metal binding and furthermore the higher concentration gradient between the adsorbate and adsorbent. However, as time progress, the concentration of heavy metals in melanin increased to a value greater than in solution leading to the decline of the concentration gradient; hence, the adsorption rate decreased. Based on kinetic mechanism, the Webber and Morris intraparticle diffusion model suggested that the adsorption of heavy metals is not controlled solely by intraparticular diffusion but by more than one rate-limiting step, and based on Lagergren's pseudo-second-order model kinetic data adsorption procedure is proven to be chemisorption.

Effect of pH on heavy-metal removal was studied in the range of pH 1–7. pH of the 10 mg/L concentration of heavy-metal aqueous solutions employing 0.2 g/L melanin is altered using 1 N HCl/NaOH and was agitated orbitally at 150 rpm, 318K. Maximum adsorption of Hg(II), Pb(II), and Cu(II) were shown at pH 5 while maximum at pH 3 for Cr(VI) adsorption. The heavy metals adsorption depends upon the speciation of heavy metals and also the ionization of functional groups based on the solution pH (Zhang et al., 2015; Abdel-Raouf and Abdul-Raheim, 2017). At lower pH the hydronium ions (H_3O^+) available in water protonates the functional groups of melanin and at pH greater than 4, there is a dip in the concentration of hydronium ions leading to the development of negative charge in the functional groups of melanin. Hg exists in aqueous solution as Hg^0 at pH less than 4, while it exists as $HgOH^+$ from pH 4 (Arias et al., 2017; Zhang et al., 2018). Hence, Hg(II) strongly binds to melanin at pH range of 4–6 and maximum adsorption is at pH 5.

Similarly, Pb(II) and Cu(II) exits in +2 form till pH 5 above which it starts to precipitating as metal hydroxides $PbOH^+$ and $CuOH^+$, therefore the maximum adsorption of Pb(II) and Cu(II) was observed at pH 5. Cr exists mainly as $HCrO_4^-$ and $Cr_2O_7^{2-}$ in the pH range of 2–6 and it tends to form CrO_4^{2-} as the pH increases. At pH below 4 a positive charge is induced to the functional groups of melanin; hence, the anionic form of Cr(VI) binds to melanin. However, as the pH decreases below 2, Cr(VI) forms stable $H_2Cr_2O_4$, which can only weakly bind to melanin so the maximum adsorption of Cr(VI) occurs at pH 3 (Gupta et al., 2010; Karthikeyan et al., 2005; Yang et al., 2014). The ionization energy, atomic radius, and electronegativity also play a major role in the adsorption of heavy metals to melanin. Cu has high electronegativity and lower ionization energy compared to other metals of study and binds more efficiently to melanin, even while Hg(II) has comparable electronegativity to Cu(II) and Pb(II), the high values of ionization energy and atomic radius decrease its binding efficiency to melanin. For this reason the adsorptive removal is higher for Cu(II), followed by Pb(II), Cr(VI), and Hg(II). pH studies revealed that maximum adsorption of each heavy metals occurred at a particular pH and considerable adsorption was also shown at neutral pH. Even at neutral pH the heavy metals were adsorbed to the functional groups in melanin for extensive contact time (Saini and Melo, 2013).

The maximum adsorptions took place at 328K and increase in system temperature increases the kinetic energy of heavy metals, which aids in its quick diffusion from bulk solution phase to the active sites of the adsorbent (Acharya et al., 2009; Saini and Melo, 2013). The change of temperature greatly affects the functional groups leading to the breakdown of the surface components attached to the functional groups where a large number of active sites are available for binding of heavy metals (Akpomie et al., 2015). The positive standard enthalpy change ($\Delta H°$) values revealed that the adsorption process is endothermic and the standard entropy ($\Delta S°$) values were more than zero indicating increased randomness at the solution and adsorbent interface. The Gibb's free energy values in the range of -20 to 0 kJ/mol indicate physisorption while -80 to -400 kJ/mol indicate chemisorption. In the case of heavy-metal adsorption to melanin, the $\Delta G°$ values were in between both the ranges suggesting that the adsorption might be physical adsorption enhanced by chemical effect (Yu et al., 2001). The activation energy was calculated applying Arrhenius equation and the activation energies for Hg(II), Pb(II), Cr(VI), and Cu(II) are 16.3, 19.3, 18.3, and 14.8 kJ/mol, respectively (Kara and Demirbel, 2012). The activation energy of more than 4.2 kJ/mol in the adsorption of heavy metals to melanin confirmed that the heavy metals are adsorbed to melanin is by chemisorption.

Adsorption experiments were conducted to model the isotherms by varying the adsorbate concentration from 5 to 25 mg/L with a fixed quantity of melanin (0.2 g/L). Isotherm studies showed higher coefficient of determination value for Langmuir isotherm when compared to the Freundlich isotherm inferring the binding of heavy-metal ions onto melanin to be monolayer in nature. Further it is inferred that the adsorption energy needed for heavy-metal ions to bind to the functional groups in melanin is equivalent for all active sites and the bound adjoining adsorbate entities do not interfere with each other even the occupancy condition of a functional group is not affected by another (Can et al., 2016). Maximum adsorption capacity was 82.4, 147.5, 126.9, and 167.8 mg/g for Hg(II), Pb(II), Cr(VI), and Cu(II), respectively, obtained from Langmuir adsorption isotherm models.

Functional groups in melanin were studied using the Fourier-transform infrared (FTIR) spectroscopy and the wavenumbers 1246.89, 1609.52, 1708.09, 3224.62, and 3330.25 cm^{-1} correspond to C$-$N group, amino group N$-$H, C$=$O stretching, and $-$OH group. The binding of the heavy metals Hg(II), Pb(II), Cr(VI), and Cu(II) to melanin by chemisorption phenomenon is confirmed by increase in the intensity of transmittance and the shifts in the position of wavenumbers. The analysis has shown that Hg is in the form of Hg^{2+} or as $HgOH^{+}$ after adsorption to melanin. Similarly, Pb(II) also exists in the same form as in solution after adsorption, that is, Pb^{2+}. Cr(VI) on adsorption to melanin has to undergo partial reduction to form Cr(III) and the adsorbed chromium is in the state of Cr(VI) as well as Cr(III). Cu also undergoes reduction after binding to melanin. Cu(II) ion after adsorption to melanin has turned to Cu^{+} and Cu^{0}. The optimum conditions maintained pH 5 for metals Hg(II), Pb(II), Cu(II) and pH 3 for Cr(VI) metal; temperature of 318K; orbital shaking of 150 rpm; shaking diameter of 25 mm; contact time of 2 hours showing 96% removal of Cu(II), 92% removal of Pb(II), 88% removal of Cr(VI), and 85% removal of Hg(II) using 0.2 g/L melanin and 10 mg/L heavy-metal solution. 100% removal of all metals from aqueous solution was possible at melanin concentration greater than 0.5 g/L (Manirethan et al., 2018). Fig. 16.1 postulates the mechanism of biosorption of heavy metals onto melanin nanoparticles.

16.3.2 Fungal mediated Se nanoparticles as adsorbents for Se adsorption

The industrial areas release varied heavy metals, which are often found in marine waters, and sediments, which are higher than the permissible limits, resulting in

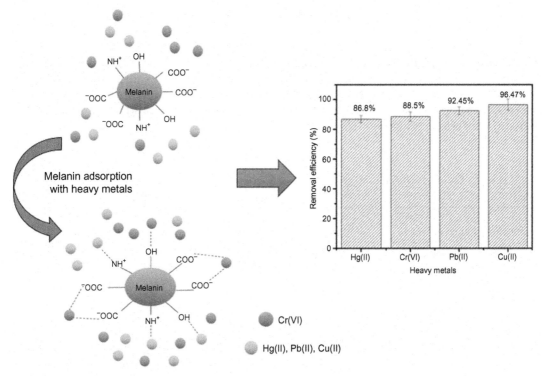

FIGURE 16.1 Schematic representation of mechanism of melanin adsorption.

modifications in the metabolic activities of the microbial communities therein. It was reported that marine water and sediments in proximity to industrial areas in Mangalore, India had higher Se content (Jacob et al., 2014). State-of-the art reveals that the biologically synthesized nanoparticles produced from Se-respiring bacteria, for example, *Sulfurospirillum barnesii*, *Bacillus selenitireducens*, and *Selnihalanaero bactershriftii*, are structurally unique. They are known to have biological activity and good adsorptive ability due to the interaction between the nanoparticles and functional groups of proteins such as $-NH-$, $-C=O-$, $COO-$, and $-C-N-$ (Zhang et al., 2004). In addition, antioxidant properties of Se hollow spherical nanoparticles, biogenesis of Se nanoparticles under anaerobic conditions, Se nanoparticles by soil bacteria *Pseudomonas aeruginosa* and *Bacillus* sp. under aerobic conditions of Se nanospheres have been demonstrated. An experiment conducted by Raj et al. (2015) explains about the biosorption of Se by the fungal biomass in upflow bioreactor. This study reported biosorption of around 85%−87% at pH 6.0−7.0 and contact time of 5 days where a marine fungus *Aspergillus terreus* was used to induce metal stress conditions and the reactor was monitored regularly confirming it as a potential biosorbant.

16.4 Nanosorbents for detection of heavy metals

Nanomaterials are the potential entities which perform dual role in the detection and removal of the heavy metals. As the nanoparticles compared to the bulk gets truncated to smaller sizes by reducing their surface to volume ratios, they exhibit rapid sensitivity, selectivity, multiplexed detection capability, response, and portability. These materials serve as optical sensors based on colorimetric, fluorescence, and surface-enhanced Raman scattering principles (Nisa et al., 2015; Wang et al., 2015; La et al., 2016) (Table 16.2).

TABLE 16.2 Overview of various nanomaterials as adsorbents for detection of metals and their detection limit.

Sl. no.	Nanosorbents	Method of detection	Detection range	Metal	References
1.	Graphitic carbon nitride quantum dots impregnated biocompatible agarose cartridge	Density functional theory	24.63 mg for 10 mg of quantum dots	Hg(II)	
2.	Fiber-optic sensor of bacteria functionalized gold nanoparticles matrix	LSPR	0.5 ppb	Hg(II) and Cd(II)	Halkare et al. (2019)
3.	Gold nanostars	Electrochemical	0.8 [As(III)], 0.5 [Hg(II)], and 4.3 ppb [Pb(II)]	As(III), Hg(II), and Pb(II)	Dutta et al. (2019)
4.	Epicatechin-capped silver nanoparticles	Colorimetric	1.52 μM	Pb(II)	Ikram et al. (2019)
5.	Mercaptobenzoheterocyclic compounds functionalized silver nanoparticle	Colorimetric	1.8 ppt	Hg(II)	Bhattacharjee et al. (2018)
6.	Gold nanoparticles	Colorimetric	23.66 [Cr(III)] and 11.21 nM [Fe(II)]	Cr(III) and Fe(II)	
7.	Fe_3O_4@catechol nanoparticles	Magnetic solid phase extraction	0.2–0.9 μg/L	Heavy metals	
8.	Carbon nanoparticles	Anodic stripping voltammetry	98.2% and 96.7%	Pb(II) and Cu(II)	
9.	Carboxymethylcellulose-stabilized silver	Colorimetric	30 nM	Hg(II)	
10.	Graphene and gold nanoparticles	Fluorescence	16.7 nM	Pb(II)	
11.	Carbon nanoparticles	Fluorescence	10–1000 ppb	Se(IV)	Devi et al. (2017)
12.	Biosynthesized gold nanoparticles	Colorimetric	1 μM to 200 μM	Hg(II)	Zohora et al. (2017)

(Continued)

TABLE 16.2 (Continued)

Sl. no.	Nanosorbents	Method of detection	Detection range	Metal	References
13.	Gold nanoparticle–embedded Nafion	Electrochemical	0.047 µg/L	Ar(III)	Zhang et al. (2016)
14.	Silver nanoparticles	Colorimetric	µM level	Cu(II) and Hg(II)	Maiti et al. (2016)
15.	Iron oxide/grapheme nanocomposite	Anodic stripping voltammetry	0.08 µg/L for Cu (II) and 0.07 µg/L for Pb(II)	Cu(II) and Pb(II)	Lee et al. (2016)
16.	Carbon nanoparticles	Fluorescent	16.5 nM	Hg(II)	Roshni and Ottoor (2015)
17.	Graphene and gold nanoparticles	Electrochemical	0.02 [Hg(II)] and 0.05 nM [Cu(II)]	Hg(II) and Cu(II)	Ting et al. (2015)
18.	Gold nanoparticles	Colorimetric	8 nM	Hg(II)	Chen et al. (2015)
19.	SiO$_2$@Au nanoparticles	Selective absoprtion then ultrasensitive immunoassay	0.5 ng/L As(III) and 0.001–100.0 µg/L Hg(II)	As(III) and Hg(II)	Tian-Hua et al. (2014)
20.	Gold nanoparticles	Fluorescence	10^{-7} M (5 ppb)	Cr(III) and Cr(VI)	Elavarasi et al. (2014)
21.	Bismuth nanoparticles	Anodic stripping square wave voltammetry	0.2 µg/L for Pb(II) and 0.6 µg/L for Cd(II)	Pb(II) and Cd(II)	Yang et al. (2014)
22.	Gold nanoparticles	Colorimetric	2.5 µg/L	As(III)	Dominguez-Medina et al. (2013)
23.	Zirconium polyacrylamide hybrid material	Adsorption in batch reactor	N/D	As(III)	Mandal et al. (2013)
24.	Cupric oxide nanoparticles	Adsorption in flow through reactor	0.001 mg/L	As(III)	Reddy et al. (2013)

LSPR, Localized surface plasmon resonance.

16.4.1 Endophytic fungus synthesized nanoadsorbents for detection

The endophytic fungi provide a broad variety of bioactive secondary metabolites with unique structures that could be explored for their ability to the biosynthesis of nanoparticles to develop an efficient environment-friendly process. De Bary (1866) introduced the term *Endophyte*, which means microorganisms that reside in the intercellular/intracellular regions of healthy plants at a particular time of their life cycle (Pimentel et al., 2011; Devi and Joshi, 2015). Usually, endophytes have a symbiotic relationship with their plant host and act as induced systemic resistance defense for the plants against foreign phytopathogens. The plant host is likewise profited by the endophytes by their natural resistance from soil contaminants, their ability to degrade xenobiotics, or their action as vectors to present degradative qualities with plants, which significantly assist in phytoremediation (Pimentel et al., 2011; Kaul et al., 2012). The endophytic protection from metals/antimicrobials might be because of their introduction to various associates in the plant/soil species and these help in the natural degradation process. It is reported that various species such as *Bjerkandera* sp., *Ceratobasidium stevensii*, *Neotyphodium coenophialum*, and *Neotyphodium uncinatum* were subjected for wood and polyaromatic compounds biodegradation and bioaugmentation of total petroleum hydrocarbons and polycyclic aromatic compounds removal.

The ZnS nanocolloids synthesized via endophytic fungus *Aspergillus flavus* are developed as a sensor for the detection of heavy metals Cu(II) and Mn(II) by colorimetric detection where the outcomes reveal that the limit of detection (LOD) for Cu(II) and Mn(II) are 1.24 and 2.14 μM (Uddandarao and Balakrishnan, 2016; Uddandarao and Balakrishnan, 2017). PbS nanocolloids for sensing capability of As(III) is studied and the results reveal that 4.5 μg/L LOD at 20 ppb concentration (Uddandarao et al., 2017). Similar studies were reported by Ni(II) detection using ZnS nanosensor and the conceivable component administering the detecting procedure is credited to the association between metal particles with the surface proteins covered on the ZnS nanoparticles. The surface of the ZnS nanoparticles consists of residual cysteine containing thiol groups $-SH$, which has an affinity to bind with heavy-metal ions (Jalilehvand et al., 2006; Hynek et al., 2012; Belmonte et al., 2016). Moreover, the scanning electron microscope (SEM) results reveal the morphology of nanoparticles before and after the addition of Ni(II) ions (Fig. 16.2) (AL-Thabaiti et al., 2016; Ma et al., 2016; Maiti et al., 2016). The Ni(II) adsorption is investigated using FTIR spectroscopy and the peaks at 3186.79, 3404.71, 3507.88 cm^{-1} correspond to $-O-H-$, weaker peaks at 2361.41, 2077.92 cm^{-1} correspond to $-C=C-$ and the peaks at 1625.7, 1230.36, 1374.03, 1073.19, 804, 778.136, and 676.89 cm^{-1} correspond to $-N-H-$, $-C-O-$, $-C-H-$, $-C-N-$, $-C-Cl-$, and $-C-S-$, respectively. The shift and disappearance of peaks were observed after the addition of the Ni(II) ions to the ZnS nanoparticles, which can be attributed to adsorption of Ni(II) ions on the ZnS nanoparticles surfaces and

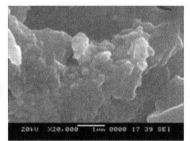

Ni(II)

ZnS nanoparticles with cysteine
and phytochelatin residues before adsorption

ZnS nanoparticles after nickel ions adsorption

FIGURE 16.2   Scanning electron microscope image before and after adsorption.

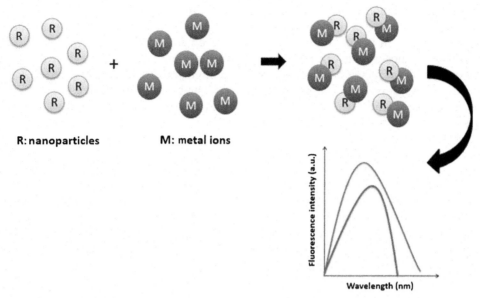

FIGURE 16.3 Pictorial representation of the mechanism of fluorescence detection.

interaction with the functional groups. After adsorption, the peak is observed at 3351.68 cm^{-1} corresponding to O–H stretching and peak at 2927.41 cm^{-1} relate to –C–H– stretching. The FTIR spectrum after adsorption with Ni(II) shows a peak at 2360.44 cm^{-1}, the reason might be the binding of thiol group residues of cysteine proteins to the binding of Ni(II) ions. The peaks observed at 2360.44, 1632.45, 1453.1, 1373.93, 1237.11, 1042.34, and 809.956 cm^{-1} correspond to –S–H–, –N–H–, –C–H–, –C–H–, –C–O–, –C–N–, and –C–Cl–, respectively (Divsar et al., 2015).

A comparative study was conducted for detection of metals such as Pb(II), Cd(II), Hg(II), Cu(II), and Ni(II) with ZnS and gadolinium (Gd)-doped ZnS nanoparticles (Uddandarao et al., 2019). Fluorescence results for intensity were recorded for Ni(II) and the studies reveal that the qualitative sensing of metal ions. Furthermore, the intensity was enhanced in the case of Pb(II), Cd(II), Hg(II), and Cu(II) ions. Gd-doped ZnS nanoparticles showed enhancement of the photoluminescence spectra for Pb(II) and Cd(II), quenched for Hg(II), Cu(II), and Ni(II) ions. The substitution of Zn(II) ions with Gd(III) ions might augment the defect sites and produce new radiation centers, which enhances the fluorescence efficiency. Fig. 16.3 postulates the pictorial representation of fluorescence detection.

16.5 Conclusion and future perspectives

Seeking lessons from nature's tools in assembling miniature functional materials in biological systems in elegant and ingenious ways, material scientists have turned their focus to harness these potent sources as biological factories to synthesize nanomaterials for adsorption of heavy metals. A vast array of biological resources available in nature, including plants, bacteria, fungi, algae, and yeast could all be employed for synthesis of nanoparticles. Research on environmental sensors has gained widespread interest because of their application in manufacturing novel nanosized sensors for pollutant detection. The utility of these nanosorbents can be improved by extending it to polymer/membrane technology, which has attracted a significant research interest in the past few decades. The sorbents can be modified by encapsulation in silica shell and immobilized by a polymer matrix, composites, and rare earth metal—doped materials, which can be used as efficient entities for removal of these inorganic pollutants.

The scale-up is a procedure for designing and building a large-scale system on the basis of the results of experiments with small-scale models. To improve the cost of production the configuration and the design parameters can be optimized based on the simulation and computational methods. Melanin being an efficient adsorbent for cations can also be used for the removal of anions such as fluorine and also ionic and nonionic surfactants from the water. The large-scale production of melanin synthesis in bioreactors can render melanin for commercial applications. These can be packed in a cartridge for household water purifier. Thus the future studies can explore the potential of these biological nanocatalysts as a novel catalytic and biological model in the field of material, energy, and environmental science.

References

Abdel-Raouf, M.S., Abdul-Raheim, A.R.M., 2017. Removal of heavy metals from industrial waste water by biomass-based materials: a review. J. Pollut. Eff. Cont. 5, 180.

Abid, M., Niazi, N.K., Bibi, I., Farooqi, A., Ok, Y.S., Kunhikrishnan, A., et al., 2016. Arsenic(V) biosorption by charred orange peel in aqueous environments. Int. J. Phytorem. 18 (5), 442–449.

Ab Latif Wani, A.A., Usmani, J.A., 2015. Lead toxicity: a review. Interdiscip. Toxicol. 8 (2), 55. 2015.

Acharya, J., Sahu, J.N., Sahoo, B.K., Mohanty, C.R., Meikap, B.C., 2009. Removal of chromium(VI) from wastewater by activated carbon developed from Tamarind wood activated with zinc chloride. Chem. Eng. J. 150 (1), 25–39.

Akpomie, K.G., Dawodu, F.A., Adebowale, K.O., 2015. Mechanism on the sorption of heavy metals from binary-solution by a low cost montmorillonite and its desorption potential. Alex. Eng. J. 54 (3), 757–767.

Albadarin, A.B., Ala'a, H., Al-Laqtah, N.A., Walker, G.M., Allen, S.J., Ahmad, M.N., 2011. Biosorption of toxic chromium from aqueous phase by lignin: mechanism, effect of other metal ions and salts. Chem. Eng. J. 169 (1–3), 20–30.

Albrecht, T.W.J., Addai-Mensah, J., Fornasiero, D., 2011. Effect of pH, concentration and temperature on copper and zinc hydroxide formation/precipitation in solution. In: Chemeca 2011: Engineering a Better World: Sydney Hilton Hotel. NSW, Australia, 18–21 September 2011, p. 2100.

Alswat, A.A., Ahmad, M.B., Saleh, T.A., 2016. Zeolite modified with copper oxide and iron oxide for lead and arsenic adsorption from aqueous solutions. J. Water Supply: Res. Technol.—AQUA 65 (6), 465–479.

AL-Thabaiti, S.A., Aazam, E.S., Khan, Z., Bashir, O., 2016. Aggregation of Congo red with surfactants and Ag-nanoparticles in an aqueous solution. Spectrochim. Acta, A: Mol. Biomol. Spectrosc. 156, 28–35.

Alvarez, N.M.M., Pastrana, J.M., Lagos, Y., Lozada, J.J., 2018. Evaluation of mercury (Hg^{2+}) adsorption capacity using exhausted coffee waste. Sustain. Chem. Pharm. 10, 60–70.

Anderson, D., 1986. IPCS international programme on chemical safety environmental health criteria 46. Br. J. Ind. Med. 43 (9), 647.

Arias, F.E.A., Beneduci, A., Chidichimo, F., Furia, E., Straface, S., 2017. Study of the adsorption of mercury(II) on lignocellulosic materials under static and dynamic conditions. Chemosphere 180, 11–23.

Arruti, A., Fernández-Olmo, I., Irabien, Á., 2010. Evaluation of the contribution of local sources to trace metals levels in urban PM2.5 and PM10 in the Cantabria region (Northern Spain). J. Environ. Monit. 12 (7), 1451–1458.

Aryal, M., Ziagova, M., Liakopoulou-Kyriakides, M., 2010. Study on arsenic biosorption using Fe(III)-treated biomass of *Staphylococcus xylosus*. Chem. Eng. J. 162 (1), 178—185.

Bang, S., Korfiatis, G.P., Meng, X., 2005. Removal of arsenic from water by zero-valent iron. J. Hazard. Mater. 121 (1—3), 61—67.

Baruthio, F., 1992. Toxic effects of chromium and its compounds. Biol. Trace Elem. Res. 32 (1—3), 145—153.

Belmonte, L., Rossetto, D., Forlin, M., Scintilla, S., Bonfio, C., Mansy, S.S., 2016. Cysteine containing dipeptides show a metal specificity that matches the composition of seawater. Phys. Chem. Chem. Phys. 18 (30), 20104—20108.

Benramdane, L., Accominotti, M., Fanton, L., Malicier, D., Vallon, J.J., 1999. Arsenic speciation in human organs following fatal arsenic trioxide poisoning—a case report. Clin. Chem. 45 (2), 301—306.

Bernhoft, R.A., 2012. Mercury toxicity and treatment: a review of the literature. J. Environ. Public Health 2012, 460508.

Beswick, P.H., Hall, G.H., Hook, A.J., Little, K., McBrien, D.C.H., Lott, K.A.K., 1976. Copper toxicity: evidence for the conversion of cupric to cuprous copper in vivo under anaerobic conditions. Chem.-Biol. Interact. 14 (3—4), 347—356.

Bhattacharjee, Y., Chatterjee, D., Chakraborty, A., 2018. Mercaptobenzo heterocyclic compounds functionalized silver nanoparticle, an ultrasensitive colorimetric probe for Hg(II) detection in water with picomolar precision: a correlation between sensitivity and binding affinity. Sens. Actuators, B: Chem. 255, 210—216.

Bodek, I., 1988. Environmental Inorganic Chemistry: Properties, Processes, and Estimation Methods. Pergamon.

Bolan, N.S., Adriano, D.C., Curtin, D., 2003. Soil acidification and liming interactions with nutrient and heavy metal transformation and bioavailability. Adv. Agron. 78 (21), 5—272.

Bradl, H.B., 2004. Adsorption of heavy metal ions on soils and soils constituents. J. Colloid Interface Sci. 277 (1), 1—18.

Butler, M.J., Day, A.W., 1998. Fungal melanins: a review. Can. J. Microbiol. 44 (12), 1115—1136.

Campbell, J.P., Joseph, A.A., 1989. Acute arsenic intoxication. Am. Fam. physician 40 (6), 93—97.

Can, N., Ömür, B.C., Altındal, A., 2016. Modeling of heavy metal ion adsorption isotherms onto metallophthalocyanine film. Sens. Actuators, B: Chem. 237, 953—961.

Carocci, A., Rovito, N., Sinicropi, M.S., Genchi, G., 2014. Mercury toxicity and neurodegenerative effects. Reviews of Environmental Contamination and Toxicology. Springer, Cham, pp. 1—18.

Chaouch, N., Ouahrani, M.R., Chaouch, S., Gherraf, N., 2013. Adsorption of cadmium(II) from aqueous solutions by activated carbon produced from Algerian dates stones of *Phoenix dactylifera* by H_3PO_4 activation. Desalin. Water Treat. 51 (10—12), 2087—2092.

Chapman, Deborah V., (ed.), Water quality assessments: a guide to the use of biota, sediments and water in environmental monitoring, CRC Press, 1996.

Chen, S.G., Xue, C.H., Xue, Y., Li, Z.J., Gao, X., Ma, Q., 2007. Studies on the free radical scavenging activities of melanin from squid ink. Chin. J. Mar. Drugs 26 (1), 24.

Chen, H., Hu, W., Li, C.M., 2015. Colorimetric detection of mercury(II) based on 2,2'-bipyridyl induced quasi-linear aggregation of gold nanoparticles. Sens. Actuators, B: Chem. 215, 421—427.

Cobo, J., Castiñeira, M., 1997. Oxidative stress, mitochondrial respiration, and glycemic control: clues from chronic supplementation with Cr^{3+} or As^{3+} to male Wistar rats. Nutrition 13 (11—12), 965—970.

Dutta, M.C., 1987. Pollution and its control in copper industry, pollution through metallurgical operation. Proceedings of the National Seminar-NSPMOP 64—75.

De Bary, A., 1866. Morphologie und Physiologie Pilze, Flechten, und myxomyceten, Hofmeister's Handbook of Physiological Botany, vol. 2. Verlag Von Wilhelm Engelmann, Leipzig.

Demiral, H., Güngör, C., 2016. Adsorption of copper(II) from aqueous solutions on activated carbon prepared from grape bagasse. J. Cleaner Prod. 124, 103—113.

Demiral, H., Demiral, I., Tümsek, F., Karabacakoğlu, B., 2008. Adsorption of chromium(VI) from aqueous solution by activated carbon derived from olive bagasse and applicability of different adsorption models. Chem. Eng. J. 144 (2), 188—196.

Devi, L.S., Joshi, S.R., 2015. Ultrastructures of silver nanoparticles biosynthesized using endophytic fungi. J. Microsc. Ultrastruct. 3 (1), 29—37.

Devi, P., Jain, R., Thakur, A., Kumar, M., Labhsetwar, N.K., Nayak, M., et al., 2017. A systematic review and meta-analysis of voltammetric and optical techniques for inorganic selenium determination in water. TrAC Trends Anal. Chem. 95, 69—85.

Divsar, F., Habibzadeh, K., Shariati, S., Shahriarinour, M., 2015. Aptamer conjugated silver nanoparticles for the colorimetric detection of arsenic ions using response surface methodology. Anal. Methods 7 (11), 4568–4576.

Dixit, R., Malaviya, D., Pandiyan, K., Singh, U., Sahu, A., Shukla, R., et al., 2015. Bioremediation of heavy metals from soil and aquatic environment: an overview of principles and criteria of fundamental processes. Sustainability 7 (2), 2189–2212.

Dominguez-Medina, S., Blankenburg, J., Olson, J., Landes, C.F., Link, S., 2013. Adsorption of a protein monolayer via hydrophobic interactions prevents nanoparticle aggregation under harsh environmental conditions. ACS Sustain. Chem. Eng. 1 (7), 833–842.

Dunn, R.O., Bantchev, G.B., Doll, K.M., Ascherl, K.L., Lansing, J.C., Murray, R.E., 2018. Thioether-functionalized corn oil biosorbents for the removal of mercury and silver ions from aqueous solutions. J. Am. Oil Chem. Soc. 95 (9), 1189–1200.

Dutta, S., Strack, G., Kurup, P., 2019. Gold nanostar electrodes for heavy metal detection. Sens. Actuators, B: Chem. 281, 383–391.

Elavarasi, M., Alex, S.A., Chandrasekaran, N., Mukherjee, A., 2014. Simple fluorescence-based detection of Cr(III) and Cr(VI) using unmodified gold nanoparticles. Anal. Methods 6 (24), 9554–9560.

Feng, J., Wang, Z., Li, L., Li, Z., Ni, W., 2013. A nonlinearized multivariate dominant factor-based partial least squares (PLS) model for coal analysis by using laser-induced breakdown spectroscopy. Appl. Spectrosc. 67 (3), 291–300.

Fergusson, J.E., 1990. Heavy Elements: Chemistry, Environmental Impact and Health Effects. Pergamon.

Fu, F., Wang, Q., 2011. Removal of heavy metal ions from wastewaters: a review. J. Environ. Manage. 92 (3), 407–418.

Gerçel, Ö., Gerçel, H.F., 2007. Adsorption of lead (II) ions from aqueous solutions by activated carbon prepared from biomass plant material of Euphorbia rigida. Chem. Eng. J. 132 (1-3), 289–297.

Goel, J., Kadirvelu, K., Rajagopal, C., Garg, V.K., 2005. Removal of lead (II) by adsorption using treated granular activated carbon: batch and column studies. J. Hazard. Mater 125 (1-3), 211–220. 2005.

Guertin, J., 2004. Toxicity and health effects of chromium (all oxidation states). Chromium (VI) Handbook 215–230.

Gupta, V.K., Rastogi, A., Nayak, A., 2010. Adsorption studies on the removal of hexavalent chromium from aqueous solution using a low cost fertilizer industry waste material. J. Colloid Interface Sci. 342 (1), 135–141.

Haddad, M., Mizyed, N., 2011. Evaluation of various hydroponic techniques as decentralised wastewater treatment and reuse systems. Int. J. Environ. Stud. 68 (4), 461–476.

Halkare, P., Punjabi, N., Wangchuk, J., Nair, A., Kondabagil, K., Mukherji, S., 2019. Bacteria functionalized gold nanoparticle matrix based fiber-optic sensor for monitoring heavy metal pollution in water. Sens. Actuators, B: Chem. 281, 643–651.

Hamilton, S.J., 2004. Review of selenium toxicity in the aquatic food chain. Sci. Total Environ. 326 (1–3), 1–31.

Hena, S., 2010. Removal of chromium hexavalent ion from aqueous solutions using biopolymer chitosan coated with poly 3-methyl thiophene polymer. J. Hazard. Mater. 181 (1–3), 474–479.

Hossain, M.A., Ngo, H.H., Guo, W.S., Nguyen, T.V., 2012. Removal of copper from water by adsorption onto banana peel as bioadsorbent. Int. J. Geomate 2 (2), 227–234.

Hu, X., Ding, Z., Zimmerman, A.R., Wang, S., Gao, B., 2015. Batch and column sorption of arsenic onto iron-impregnated biochar synthesized through hydrolysis. Water Res. 68, 206–216.

Hyder, A.G., Begum, S.A., Egiebor, N.O., 2015. Adsorption isotherm and kinetic studies of hexavalent chromium removal from aqueous solution onto bone char. J. Environ. Chem. Eng. 3 (2), 1329–1336.

Hung, Y.C., Sava, V.M., Blagodarsky, V.A., Hong, M.Y., Huang, G.S., 2003. Protection of tea melanin on hydrazine-induced liver injury. Life Sci. 72 (9), 1061–1071.

Hynek, D., Krejcova, L., Sochor, J., Cernei, N., Kynicky, J., Adam, V., et al., 2012. Study of interactions between cysteine and cadmium(II) ions using automatic pipetting system off-line coupled with electrochemical analyser. Int. J. Electrochem. Sci. 7, 1802–1819.

Ikram, F., Qayoom, A., Aslam, Z., Shah, M.R., 2019. Epicatechin coated silver nanoparticles as highly selective nanosensor for the detection of Pb^{2+} in environmental samples. J. Mol. Liq. 277, 649–655.

Jacob, J.M., Balakrishnan, R.M., Bardhan, S.K., Jagadeeshbabu, P.E., Aruna, M., 2014. Selenium and lead tolerance in marine *Aspergillus terreus* for biosynthesis of nano particles—quantum dots/rods. Int. J. Adv. Chem. Eng. Biol. Sci. 1 (1), 6–11.

Janyasuthiwong, S., Phiri, S.M., Kijjanapanich, P., Rene, E.R., Esposito, G., Lens, P.N., 2015. Copper, lead and zinc removal from metal-contaminated wastewater by adsorption onto agricultural wastes. Environ. Technol. 36 (24), 3071–3083.

Jalilehvand, F., Leung, B.O., Izadifard, M., Damian, E., 2006. Mercury(II) cysteine complexes in alkaline aqueous solution. Inorg. Chem. 45 (1), 66–73.

Kara, A., Demirbel, E., 2012. Kinetic, isotherm and thermodynamic analysis on adsorption of Cr (VI) ions from aqueous solutions by synthesis and characterization of magnetic-poly (divinylbenzene-vinylimidazole) microbeads. Water, Air, & Soil Poll. 223 (5), 2387–2403.

Karthikeyan, T., Rajgopal, S., Miranda, L.R., 2005. Chromium(VI) adsorption from aqueous solution by *Hevea brasiliensis* sawdust activated carbon. J. Hazard. Mater. 124 (1–3), 192–199.

Kaul, S., Gupta, S., Ahmed, M., Dhar, M.K., 2012. Endophytic fungi from medicinal plants: a treasure hunt for bioactive metabolites. Phytochem. Rev. 11 (4), 487–505.

Khezami, L., Capart, R., 2005. Removal of chromium(VI) from aqueous solution by activated carbons: kinetic and equilibrium studies. J. Hazard. Mater. 123 (1–3), 223–231.

Kim, J., Grate, J.W., 2003. Single-enzyme nanoparticles armored by a nanometer-scale organic/inorganic network. Nano Lett. 3 (9), 1219–1222.

Kobielska, P.A., Howarth, A.J., Farha, O.K., Nayak, S., 2018. Metal–organic frameworks for heavy metal removal from water. Coord. Chem. Rev. 358, 92–107.

Kundu, S., Gupta, A.K., 2006. Arsenic adsorption onto iron oxide-coated cement (IOCC): regression analysis of equilibrium data with several isotherm models and their optimization. Chem. Eng. J. 122 (1–2), 93–106.

La, J.A., Lim, S., Park, H.J., Heo, M.J., Sang, B.I., Oh, M.K., et al., 2016. Plasmonic-based colorimetric and spectroscopic discrimination of acetic and butyric acids produced by different types of *Escherichia coli* through the different assembly structures formation of gold nanoparticles. Anal. Chim. Acta 933, 196–206.

Lien, H.C., Tsai, T.F., Lee, Y.Y., Hsiao, C.H., 1999. Merkel cell carcinoma and chronic arsenicism. J. Am. Acad. Dermatol. 41 (4), 641–643.

Lin, W.P., Lai, H.L., Liu, Y.L., Chiung, Y.M., Shiau, C.Y., Han, J.M., Yang, C.M., Liu, Y.T., 2005. Effect of melanin produced by a recombinant Escherichia coli on antibacterial activity of antibiotics. J. Microbiol. Immunol. 38 (5), 320–326.

Ma, Y., Pang, Y., Liu, F., Xu, H., Shen, X., 2016. Microwave-assisted ultrafast synthesis of silver nanoparticles for detection of Hg^{2+}. Spectrochim. Acta, A: Mol. Biomol. Spectrosc. 153, 206–211.

Maiti, S., Barman, G., Laha, J.K., 2016. Detection of heavy metals (Cu^{+2}, Hg^{+2}) by biosynthesized silver nanoparticles. Appl. Nanosci. 6 (4), 529–538.

Mandal, S., Sahu, M.K., Patel, R.K., 2013. Adsorption studies of arsenic(III) removal from water by zirconium polyacrylamide hybrid material (ZrPACM-43). Water Resour. Ind. 4, 51–67.

Manirethan, V., Raval, K., Rajan, R., Thaira, H., Balakrishnan, R.M., 2018. Kinetic and thermodynamic studies on the adsorption of heavy metals from aqueous solution by melanin nanopigment obtained from marine source: *Pseudomonas stutzeri*. J. environ. Manage. 214, 315–324.

Meng, X.G., Jing, C.Y., Pena, M.E., 2003. Adsorption of arsenic by nanocrystalline titanium dioxide. In: Abstracts of Papers of the American Chemical Society, vol. 226. 1155 16TH ST, NW, American Chemical Society, Washington, DC 20036, pp. U582–U582.

Momčilović, M., Purenović, M., Bojić, A., Zarubica, A., Ranđelović, M., 2011. Removal of lead (II) ions from aqueous solutions by adsorption onto pine cone activated carbon. Desalination 276 (1-3), 53–59.

Nacke, H., Gonçalves, A.C., Campagnolo, M.A., Coelho, G.F., Schwantes, D., dos Santos, M.G., et al., 2016. Adsorption of Cu (II) and Zn (II) from water by Jatropha curcas L. as biosorbent. Open Chem. 14 (1), 103–117.

Nikić, J., Watson, M., Tubić, A., Isakovski, M.K., Maletić, S., Mohora, E., Agbaba, J., 2019. Arsenic removal from water using a one-pot synthesized low-cost mesoporous Fe–Mn-modified biosorbent. J. Serb. Chem. Soc. 84 (3), 327–342.

Nisa, H., Kamili, A.N., Nawchoo, I.A., Shafi, S., Shameem, N., Bandh, S.A., 2015. Fungal endophytes as prolific source of phytochemicals and other bioactive natural products: a review. Microb. Pathog. 82, 50–59.

Nriagu, J.O., 1989. A global assessment of natural sources of atmospheric trace metals. Nature 338 (6210), 47.

Pacyna, J.M., 1996. Emission inventories of atmospheric mercury from anthropogenic sources. Global and Regional Mercury Cycles: Sources, Fluxes and Mass Balances. Springer, Dordrecht, pp. 161–177.

Pandey, J., Sudhakar, P., Koshy, V.J., 2003. Determination of silver at submicrogram levels by absorption spectrophotometry.

Pattanayak, J., Mondal, K., Mathew, S., Lalvani, S.B., 2000. A parametric evaluation of the removal of As(V) and As(III) by carbon-based adsorbents. Carbon 38 (4), 589–596.

Pehlivan, E., Tran, T.H., Ouédraogo, W.K.I., Schmidt, C., Zachmann, D., Bahadir, M., 2013. Removal of As (V) from aqueous solutions by iron coated rice husk. Fuel Process. Technol. 106, 511–517.

Pimentel, M.R., Molina, G., Dionísio, A.P., Maróstica Jr, M.R., Pastore, G.M., 2011. The use of endophytes to obtain bioactive compounds and their application in biotransformation process. Biotechnol. Res. Int. 2011, 576286.

Ratnaike, R.N., 2003. Acute and chronic arsenic toxicity. Postgrad. Med. J. 79 (933), 391–396.

Raza, M.H., Sadiq, A., Farooq, U., Athar, M., Hussain, T., Mujahid, A., Salman, M., 2015. Phragmites karka as a biosorbent for the removal of mercury metal ions from aqueous solution: effect of modification. J. Chem.

Reddy, K.J., McDonald, K.J., King, H., 2013. A novel arsenic removal process for water using cupric oxide nano-particles. J. Colloid Interface Sci. 397, 96–102.

Roshni, V., Ottoor, D., 2015. Synthesis of carbon nanoparticles using one step green approach and their application as mercuric ion sensor. J. Lumin. 161, 117–122.

Saini, A.S., Melo, J.S., 2013. Biosorption of uranium by melanin: kinetic, equilibrium and thermodynamic studies. Bioresource technol. 149, 155–162.

Sajjan, S.S., Anjaneya, O., Guruprasad, B.K., Anand, S.N., Suresh, B.M., Karegoudar, T.B., 2013. Properties and functions of melanin pigment from Klebsiella sp. GSK. Korean J. Microbiol. Biotechnol 41 (1), 60–69.

Silwana, B., Van Der Horst, C., Iwuoha, E., Somerset, V., 2014. Amperometric determination of cadmium, lead, and mercury metal ions using a novel polymer immobilised horseradish peroxidase biosensor system. J. Environ. Sci. Health, A 49 (13), 1501–1511.

Sinha S.N., Agrawal M.K. and Saha, R.K., Pollution and its Control in Copper Industry. 1998.

Singh, T.S., Pant, K.K., 2006. Kinetics and mass transfer studies on the adsorption of arsenic onto activated alumina and iron oxide impregnated activated alumina. Water Qual. Res. J. 41 (2), 147–156.

Smith, L.A., 1995. Remedial options for metals-contaminated sites. *Lewis Publ.*

Solano, F., 2014. Melanins: skin pigments and much more—types, structural models, biological functions, and formation routes. New J. Sci.

Solisio, C., Al Arni, S., Converti, A., 2019. Adsorption of inorganic mercury from aqueous solutions onto dry biomass of *Chlorella vulgaris*: kinetic and isotherm study. Environ. Technol. 40 (5), 664–672.

Sono, K., Lye, D., Moore, C.A., Boyd, W.C., Gorlin, T.A., Belitsky, J.M., 2012. Melanin-based coatings as lead-binding agents. Bioinorg. Chem. Appl.

Sharma, P., Bihari, V., Agarwal, S.K., Verma, V., Kesavachandran, C.N., Pangtey, B.S., et al., 2012. Groundwater contaminated with hexavalent chromium [Cr (VI)]: a health survey and clinical examination of community inhabitants (Kanpur, India). PloS one 7 (10), 47877.

Sträter, E., Westbeld, A., Klemm, O., 2010. Pollution in coastal fog at Alto Patache, northern Chile. Environ. Sci. Pollut. Res. 17 (9), 1563–1573.

Sulaymon, A.H., Mohammed, A.A., Al-Musawi, T.J., 2013. Competitive biosorption of lead, cadmium, copper, and arsenic ions using algae. Environ. Sci. Pollut. Res. 20 (5), 3011–3023.

Sun, S., Zhu, J., Zheng, Z., Li, J., Gan, M., 2019. Biosynthesis of β-cyclodextrin modified schwertmannite and the application in heavy metals adsorption. Powder Technol. 342, 181–192.

Suzuki, T.M., Bomani, J.O., Matsunaga, H., Yokoyama, T., 1997. Removal of As(III) and As(V) by a porous spherical resin loaded with monoclinic hydrous zirconium oxide. Chem. Lett. 26 (11), 1119–1120.

Tchounwou, P.B., Yedjou, C.G., Patlolla, A.K., Sutton, D.J., 2012. Heavy metal toxicity and the environment. Molecular, clinical and environmental toxicology. Springer, Basel, pp. 133–164.

Thaira, H., Raval, K., Manirethan, V., Balakrishnan, R.M., 2019. Melanin nano-pigments for heavy metal remediation from water. Sep. Sci. Technol. 54 (2), 265–274.

Thirunavukkarasu, O.S., Viraraghavan, T., Subramanian, K.S., 2001. Removal of arsenic in drinking water by iron oxide-coated sand and ferrihydrite—batch studies. Water Qual. Res. J. 36 (1), 55–70.

Tian-Hua, L.I., Ning, G.A.N., Da-Zhen, W.U., Hai-Juan, J.I.N., Yu-Ting, C.A.O., Jiang, Q.L., 2014. An ultrasensitive simultaneous multianalyte immunoassay based on arsenic and mercury ions labeled SiO₂@Au nanoparticle probes. Chin. J. Anal. Chem. 42 (6), 817–823.

Tsunashima, A., Brindley, G.W., Bastovanov, M., 1981. Adsorption of uranium from solutions by montmorillonite; compositions and properties of uranyl montmorillonites. Clay. Clay Miner 29 (1), 10–16.

Uddandarao, P., Balakrishnan, R.M., 2016. ZnS semiconductor quantum dots production by an endophytic fungus *Aspergillus flavus*. Mater. Sci. Eng., B 207, 26–32.

Uddandarao, P., Balakrishnan, R.M., 2017. Thermal and optical characterization of biologically synthesized ZnS nanoparticles synthesized from an endophytic fungus *Aspergillus flavus*: a colorimetric probe in metal detection. Spectrochim. Acta, A: Mol. Biomol. Spectrosc. 175, 200–207.

Uddandarao, P., Akshay Gowda, K.M., Elisha, M.G., Nitish, N., 2017. Biologically synthesized PbS nanoparticles for the detection of arsenic in water. Int. Biodeterior. Biodegrad. 119, 78–86.

Uddandarao, P., Balakrishnan, R.M., Ashok, A., Swarup, S., Sinha, P., 2019. Bioinspired ZnS:Gd nanoparticles synthesized from an endophytic fungi *Aspergillus flavus* for fluorescence-based metal detection. Biomimetics 4 (1), 11.

Vakili, M., Deng, S., Li, T., Wang, W., Wang, W., Yu, G., 2018. Novel crosslinked chitosan for enhanced adsorption of hexavalent chromium in acidic solution. Chem. Eng. J. 347, 782–790.

Wang, J., Yokokawa, M., Satake, T., Suzuki, H., 2015. A micro IrOx potentiometric sensor for direct determination of organophosphate pesticides. Sens. Actuators, B: Chem. 220, 859–863.

Wilbur, S., Abadin, H., Fay, M., Yu, D., Tencza, B., Ingerman, L., et al., 2012. Health effects. Toxicological Profile for Chromium. Agency for Toxic Substances and Disease Registry (US).

Wuana, R.A., Okieimen, F.E., 2011. Heavy metals in contaminated soils: a review of sources, chemistry, risks and best available strategies for remediation. Isrn Ecol .

Xiong, Y., Tong, Q., Shan, W., Xing, Z., Wang, Y., Wen, S., Lou, Z., 2017. Arsenic transformation and adsorption by iron hydroxide/manganese dioxide doped straw activated carbon. Appl. Surf. Sci. 416, 618–627.

Yang, D., Wang, L., Chen, Z., Megharaj, M., Naidu, R., 2014. Voltammetric determination of lead(II) and cadmium(II) using a bismuth film electrode modified with mesoporous silica nanoparticles. Electrochim. Acta 132, 223–229.

Yu, B., Zhang, Y., Shukla, A., Shukla, S.S., Dorris, K.L., 2001. The removal of heavy metals from aqueous solutions by sawdust adsorption—removal of lead and comparison of its adsorption with copper. J. hazard. Mater. 84 (1), 83–94.

Yard, E.E., Horton, J., Schier, J.G., Caldwell, K., Sanchez, C., Lewis, L., Gastañaga, C., 2012. Mercury exposure among artisanal gold miners in Madre de Dios, Peru: a cross-sectional study. J. Med. Toxicol. 8 (4), 441–448.

Zhang, J., Wang, H., Bao, Y., Zhang, L., 2004. Nano red elemental selenium has no size effect in the induction of seleno-enzymes in both cultured cells and mice. Life Sci. 75 (2), 237–244.

Zhang, R., Wang, B., Ma, H., 2010. Studies on chromium(VI) adsorption on sulfonated lignite. Desalination 255 (1–3), 61–66.

Zhang, Y.J., Ou, J.L., Duan, Z.K., Xing, Z.J., Wang, Y., 2015. Adsorption of Cr(VI) on bamboo bark-based activated carbon in the absence and presence of humic acid. Colloids Surf., A: Physicochem. Eng. Aspects 481, 108–116.

Zhang, Y., McKelvie, I.D., Cattrall, R.W., Kolev, S.D., 2016. Colorimetric detection based on localised surface plasmon resonance of gold nanoparticles: merits, inherent shortcomings and future prospects. Talanta 152, 410–422.

Zhang, Q., Liu, N., Cao, Y., Zhang, W., Wei, Y., Feng, L., et al., 2018. A facile method to prepare dual-functional membrane for efficient oil removal and in situ reversible mercury ions adsorption from wastewater. Appl. Surf. Sci. 434, 57–62.

Zhu, Z., Chan, G.C.Y., Ray, S.J., Zhang, X., Hieftje, G.M., 2008. Use of a solution cathode glow discharge for cold vapor generation of mercury with determination by ICP-atomic emission spectrometry. Anal. Chem. 80 (18), 7043–7050.

Zohora, N., Kumar, D., Yazdani, M., Rotello, V.M., Ramanathan, R., Bansal, V., 2017. Rapid colorimetric detection of mercury using biosynthesized gold nanoparticles. Colloids Surf., A: Physicochem. Eng. Aspects 532, 451–457.

Further reading

Abdel-Shafy, H.I., Abdel-Sabour, M.F., Aly, R.O., 1998. Adsorption of nickel and mercury from drinking water simulant by activated carbon. Environ. Manage. Health 9 (4), 170–175.

Aydın, H., Bulut, Y., Yerlikaya, Ç., 2008. Removal of copper(II) from aqueous solution by adsorption onto low-cost adsorbents. J. Environ. Manage. 87 (1), 37–45.

Bañuelos, G.S., Arroyo, I., Pickering, I.J., Yang, S.I., Freeman, J.L., 2015. Selenium biofortification of broccoli and carrots grown in soil amended with Se-enriched hyperaccumulator *Stanleya pinnata*. Food Chem. 166, 603–608.

Brechbühl, Y., Christl, I., Elzinga, E.J., Kretzschmar, R., 2012. Competitive sorption of carbonate and arsenic to hematite: combined ATR-FTIR and batch experiments. J. Colloid Interface Sci. 377 (1), 313–321.

Chen, S., Xue, C., Wang, J., Feng, H., Wang, Y., Ma, Q., et al., 2009. Adsorption of Pb(II) and Cd(II) by squid *Ommastrephes bartrami* melanin. Bioinorg. Chem. Appl. 2009, 901563.

Chen, S., Yue, Q., Gao, B., Li, Q., Xu, X., Fu, K., 2012. Adsorption of hexavalent chromium from aqueous solution by modified corn stalk: a fixed-bed column study. Bioresour. Technol. 113, 114–120.

Chiang, Y.W., Ghyselbrecht, K., Santos, R.M., Martens, J.A., Swennen, R., Cappuyns, V., et al., 2012. Adsorption of multi-heavy metals onto water treatment residuals: sorption capacities and applications. Chem. Eng. J. 200, 405–415.

Domínguez-González, R., Varela, L.G., Bermejo-Barrera, P., 2014. Functionalized gold nanoparticles for the detection of arsenic in water. Talanta 118, 262–269.

Dupont, L., Guillon, E., 2003. Removal of hexavalent chromium with a lignocellulosic substrate extracted from wheat bran. Environ. Sci. Technol. 37 (18), 4235–4241.

Fan, L.J., Zhang, Y., Murphy, C.B., Angell, S.E., Parker, M.F., Flynn, B.R., et al., 2009. Fluorescent conjugated polymer molecular wire chemosensors for transition metal ion recognition and signaling. Coord. Chem. Rev. 253 (3–4), 410–422.

Feng, Y., Gong, J.L., Zeng, G.M., Niu, Q.Y., Zhang, H.Y., Niu, C.G., et al., 2010. Adsorption of Cd(II) and Zn(II) from aqueous solutions using magnetic hydroxyapatite nanoparticles as adsorbents. Chem. Eng. J. 162 (2), 487–494.

Khlebtsov, B.N., Khlebtsov, N.G., 2011. On the measurement of gold nanoparticle sizes by the dynamic light scattering method. Colloid J. 73 (1), 118–127.

Moritz, M., Geszke-Moritz, M., 2013. The newest achievements in synthesis, immobilization and practical applications of antibacterial nanoparticles. Chem. Eng. J. 228, 596–613.

Pereiro, I., 1999. Optimization of the coupling of multicapillary GC with ICP-MS for mercury speciation analysis in biological materials. J. Anal. At. Spectrom. 14 (5), 851–857.

Singh, P., Kim, Y.J., Zhang, D., Yang, D.C., 2016. Biological synthesis of nanoparticles from plants and microorganisms. Trends Biotechnol. 34 (7), 588–599.

Sinha, T., Ahmaruzzaman, M., 2015. Biogenic synthesis of Cu nanoparticles and its degradation behavior for methyl red. Mater. Lett. 159, 168–171.

Wang, C., Yu, C., 2013. Detection of chemical pollutants in water using gold nanoparticles as sensors: a review. Rev. Anal. Chem. 32 (1), 1–14.

Zhang, J., Wang, X., Xu, T., 2007. Elemental selenium at nano size (nano-Se) as a potential chemopreventive agent with reduced risk of selenium toxicity: comparison with se-methylselenocysteine in mice. Toxicol. Sci. 101 (1), 22–31.

Zhang, N., Fu, N., Fang, Z., Feng, Y., Ke, L., 2011. Simultaneous multi-channel hydride generation atomic fluorescence spectrometry determination of arsenic, bismuth, tellurium and selenium in tea leaves. Food Chem. 124 (3), 1185–1188.

Nanoparticles as sources of inorganic water pollutants

Arindam Malakar* and Daniel D. Snow

University of Nebraska–Lincoln, Lincoln, NE, United States

O U T L I N E

17.1 Introduction

Approximately 70.9% of Earth's surface is covered with water, that is, almost two-thirds of Earth's area, entitling Earth to be called the "Blue Planet" of our solar system. Roughly 96.5% of the Earth's crust water is saline, while 1.7% of water is present under Earth's crust, 1.7% is present in glaciers and the ice caps of Antarctica, Greenland, and

* Email: amalakar2@unl.edu.

poles, and a small fraction in other large water bodies and 0.001% in the air as water vapor, clouds, as well as precipitation (Shiklomanov, 1993). Only 2.5% of this resource occurs as freshwater, and 98.8% of that freshwater is ice (excepting ice in clouds) and groundwater. 0.3% of all freshwater is present in surface (e.g., rivers and lakes) and the atmosphere (Gleick, 1993). A greater quantity of water is found in the Earth's interior, and groundwater is by far the largest source for freshwater. The significance of clean water is invaluable, considering the relative quantity available for human use. A rapidly growing population, coupled with intensive agriculture and changing climatic patterns, have increased the depletion rate of freshwater sources around the globe. In addition, pollution of water sources by anthropogenic and geogenic sources is limiting the availability of freshwater required for vital processes.

Excessive concentrations of one or more substances in water, which can interfere with safe uses for humans and other living organisms, in general, are termed as water pollution. Water pollution occurs when any contaminant/s are directly or indirectly discharged into water bodies (oceans, lakes, rivers, and other inland waters) without adequate proper treatment for removal of these harmful contaminants. Addition and occurrence of harmful chemicals can bring about change in the chemistry of water, impacting aquatic life and living beings who are directly or indirectly utilizing those water sources for various activities. Water pollution is a major global problem, and it has been a leading cause of death and disease worldwide (Ross, 2008). It is estimated that water pollution accounts for more than 14,000 deaths daily (Bashar Bhuiyan et al., 2013).

A proper evaluation of water pollution is based in large part on our knowledge of the types and origin of pollutants that interfere with the intended use. The availability of proper and efficient analytical tools or methods for quantifying the temporal and spatial variation of pollutants is another crucial factor in understanding the processes determining the transport and fate of the pollutants. The goal of monitoring is to design optimal sampling strategies and predict future developments of a given pollution causes and to develop methods to quantify or predict adverse effects on aquatic life and human health (Malakar, 2016).

Pollution of rivers, ponds, lakes, oceans, and other surface water bodies falls under surface water pollution. Surface water pollution can be divided into two broad categories based on their origin, which can be point source water pollution, released to the environment through a well-defined space such as a pipe or ditch, and diffuse nonpoint source pollution where there is no discrete source. Groundwater sources such as aquifers are susceptible to contamination from the point and nonpoint sources that may or may not directly affect surface water bodies. A spill or ongoing release of chemical or radionuclide contaminants into soil located away from a surface water body can contaminate the aquifer below, creating a chemical plume.

Contaminants, which lead to water pollution, include a broad spectrum of chemicals and pathogens, which can be distinguished by the source and receiving water (Fig. 17.1). Water pollution may be studied under several broad categories, namely, physical, chemical, and biological (Malakar et al., 2019). Water temperature may be impacted by thermal pollution, which is caused by the discharge of hot water by power plants and industrial manufacturers (Laws, 2017). This change in temperature is a physical form of pollution. Elevated water temperatures reduce oxygen solubility and decrease oxygen levels in the water, leading to fish kills, alter food chain composition, and reduce biodiversity (Dodds and Whiles, 2010).

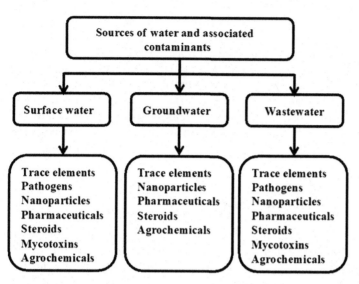

FIGURE 17.1 Main sources of water and different types of contaminants present in those water sources impacting water quality, surface water, and wastewater are subject to similar types of contamination. *Source: Taken from Malakar, A., Snow, D.D., Ray, C., 2019. Irrigation water quality—a contemporary perspective. Water 11, 1482.*

Pathogens are disease-causing microorganisms and come under the biological source of pollutants. Coliform bacteria (*Escherichia coli*, *Klebsiella*, and *Proteus*) are usually harmless or beneficial, but if present in high amount or some specific strain can cause diseases. Others pathogenic sources are *Burkholderia pseudomallei*, *Salmonella*, *Shigella*, viruses such as hepatitis A, E, Norovirus, parasitic worms include the *Schistosoma*, *Cryptosporidium parvum*, *Giardia lamblia*, *Entamoeba histolytica* (Andrade et al., 2018; Jamison et al., 2006). Pathogens levels can increase from on-site sanitation systems or improperly treated sewage discharges (Vriens et al., 2017). Antibiotic resistance gene or bacteria are another potent biological pollutant, which is a significant concern now (Zhang et al., 2009). Chemical contaminants consist of both inorganic and organic pollutant, which are a major threat to human health, and can be harmful even in trace quantities (D'Alessio et al., 2014; Malakar et al., 2016a,b). Contaminants are dynamic and evolving, and their levels are increasing in the environment due to rapid industrialization and intensification of agriculture (Malakar et al., 2019). Techniques which are applied for the treatment of contaminated water may also generate secondary pollutants which were not known earlier (Rizzo et al., 2019; Sauvé and Desrosiers, 2014). Contamination caused by organic and inorganic nanoparticles were not known until recently (Ma et al., 2016) and, generally, fall under chemical pollutants.

Nanotechnology is finding its way into every possible field from human medicine to engineering and agriculture to space science. The exclusive properties of nanomaterials can be used to fight cancer, make lightweight and durable construction materials, produce energy from solar cells, provide treatment through catalytic converters, as a few examples of their application. There is continuous research to make nanotechnology more viable and improve applications throughout the industrialized world. It is estimated that nanotechnology market will be 55 billion US dollars by the year 2022 (Inshakova and Inshakov, 2017). This exponential boost in the use of nanomaterials has increased their occurrence in water resources (González-Gálvez et al., 2017; Praetorius et al., 2012). Nanotechnology is actively being used for the treatment of groundwater, surface water, and wastewater

(Gehrke et al., 2015; Thomé et al., 2015), but the ramifications of this use are still unknown (Troester et al., 2016).

In the present chapter, the cause, study, and potential impact of the nanoparticle occurrence in the aqueous environment will be examined. Different groups and subgroups of natural and synthetic nanoparticles will be reviewed. Primary emphasis will be given to their potential impact on human health. Current analytical techniques to identify and quantify nanoparticle occurrence in water will also be discussed. This chapter will evaluate the impact of nanoparticle as a potential pollution source in water and elaborate on why the study of these particles is critical in the context of current inorganic pollutants in water.

17.2 Nanoparticles

Nanoparticles are generally classified as those materials whose size ranges between 1 and 100 nm (1 nm $= 10^{-9}$ m) (Hochella, 2002). However, others suggest that the upper limit of the size should be defined based on changes in fundamental properties from the bulk material (Banfield and Zhang, 2001). This nanoscale size regime can lead to unique properties of nanoparticles that arise due to their high specific surface area, and these properties can include rapid dissolution, high reactivity, strong sorption as well as discontinuous properties such as magnetism and the quantum confinement effect (Gehrke et al., 2015; Jeevanandam et al., 2018; Khajeh et al., 2013; Pradeep and Anshup, 2009). Nanoparticles are not only synthesized artificially but also occur throughout the environment as the Earth is opulent with nanosized mineral particles and other natural nanomaterials (Hochella, 2008; Hochella et al., 2019). Natural nanomaterials form naturally in Earth crust via different biological and geochemical processes, and it is estimated that at least 1000 Mt of nanoparticles are cycled annually through the hydrosphere (Hochella et al., 2019). Most of the environmental concerns to date have related to engineered nanoparticles. Natural and engineered nanoparticles can pass through 0.2 μm pore size membrane filters because of their sizes. Filtration through 0.2 and 0.45 μm pore size membranes is used to functionally define dissolved species in water. Thus chemical concentrations measured in filtered water may include both dissolved and nonaggregated nanoscale forms.

17.2.1 Natural nanoparticles

Natural nanoparticles are abundant in the environment, and a substantial quantity occur as water-borne nanoparticles where they are subject to transport within and between a variety of water sources (Wagner et al., 2014) (see Fig. 17.2). It is estimated that there is a flux of around 342 Mt/year of natural nanoparticles from the Earth's surface to the atmosphere (Hochella et al., 2019). The average size range of natural nanoparticle can vary between 1 and several tens of nanometers (Hochella et al., 2008). The nanoscale size of natural nanoparticles provides them with high mobility and chemical reactivity and also results in a rapid transformation in nature. The role of natural nanoparticles in the environment is a new and active area of research. Nanoparticles have been shown to serve as a buffer in environmental systems, where they both limit and control trace element

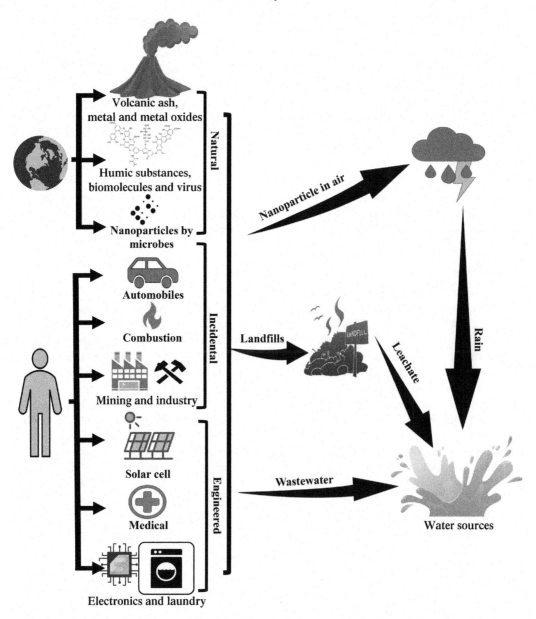

FIGURE 17.2 Sources and forms of natural and synthetic nanoparticles and the likely routes to surface and groundwater.

concentrations, and may also provide reactive surfaces and metal ions facilitating biochemical reactions (Hartland et al., 2013). Similar to engineered nanoparticles, differences in properties between the natural nanoparticle and bulk matter with similar elemental composition and crystalline structure are due to the very high specific surface area

(Lead and Wilkinson, 2006) and the spatial constraint of atom-scale properties due to their size (Wang, 2014).

Natural nanoparticles comprise a highly varied composition and morphology (Griffin et al., 2017). Environmental nanoparticles can include humic substances, organic complexes or even biomolecules (viruses), minerals generated by chemical weathering processes, volcanic eruptions, clay minerals, and metal and/or metal oxides (e.g., nano-sulfur, selenium, or uranium) nanoparticles generated by geochemical or microbial activities (Griffin et al., 2017; Mkandawire and Dudel, 2008; Suzuki et al., 2002) (Fig. 17.2). The vast majority of natural nanoparticles likely occurs mainly as oxides and oxyhydroxides of silicon, aluminum, calcium, iron, and manganese, because of their crustal abundance and chemical properties. There is a common consensus that natural nanoparticles have unique properties. Nanoscience has gained prominence in natural science and has advanced significantly to draw a framework for new understandings. Natural scientist should take a pioneering role to understand nanogeoscience phenomenon in complex ecological systems.

17.2.1.1 *Humic substances, biomolecules, and viruses*

The most studied and the most abundant group of the natural nanoparticle is the humic substances that consist of the chemically extracted fraction of total natural organic carbon pool. It is estimated that 60%−70% of total soil organic carbon is humic substances (Senesi, 2010). Humic substances generally exist as dispersed material at nano-size range (<5 nm) but can forms aggregates (Baalousha et al., 2011). Humic substances comprise relatively high to low molecular weight organic compounds, often a mixture of aliphatic and aromatic compounds, which are formed during humification (Chin et al., 1994).

The biomolecules consisting of peptides, peptidoglycans, proteins, and polysaccharides can exist in the nanoregime and is the second most abundant natural nanoparticle existing in the water. Proteins are generally globular and are less degraded compared to humic substances, while polysaccharides have mesh-like morphology. Unlike humic substances, a biomolecular nanoparticle coating promotes aggregation of metal oxides, as they are not able to reduce the surface charge of other nanoparticles as efficiently as humic substances (Buffle et al., 1998). Viruses are also termed as natural nanoparticles due to their size regime (Griffin et al., 2017; Steinmetz, 2010). New viral nanotechnology is being used for drug delivery, therapeutics, and imaging devices (Steinmetz, 2010).

17.2.1.2 *Volcanic ash, metal oxides/hydroxides, and clay minerals*

Volcanic eruptions can add millions of tons of natural nanoparticles to the Earth's surface annually (Ermolin et al., 2018). Volcanic ash nanoparticles may spread and cycle through the biosphere for many years with unknown effects on the environment and human health. The chemical composition of volcanic ash nanoparticles is mostly uncharacterized. A recent study by Ermolin et al. (2018) found that volcanic ash nanoparticles may include particles with elevated nickel, zinc, cadmium, silver, tin, selenium, tellurium, mercury, thallium, lead, and bismuth concentrations as much as 10−500 times higher than their bulk counterpart. Berlo et al. (2014) studied the sulfide and sulfur nanoparticles in mineral springs and found them to be volcanic in origin, where reactive sulfide gases rapidly form sulfur nanoparticles in water. The composition and purity of these sulfur nanoparticles are not pristine, which can be attributed to its natural origin (Faulstich et al., 2017). "Nanominerals" are those minerals that only exist in the nano-size range, such as ferrihydrite or aluminosilicate clay that are at minimum one nanoscale dimension (Hochella et al., 2008). In comparison, mineral nanoparticles are the nano-size crystalline material that can also exist as a bulk mineral. Most fresh surface and groundwater sources are full of polydisperse nano- and microscopic minerals (Griffin et al., 2017). Hydrated metal oxide/hydroxide nanominerals often have unidentified stoichiometry with variable water content. Generally, in the aqueous system, these metal oxide/hydroxide nanoparticles are associated with humic substances and other biomolecules. Natural nanoparticles can be widespread throughout the Earth's surface and control a variety of biogeochemcal cycles (Vodyanitskii and Shoba, 2016; Zhang et al., 2018). Natural metal oxide/hydroxide nanoparticles may form via various abiotic redox pathways and via natural weathering of rocks (Blanco-Andujar et al., 2012). The precipitation of carbonates, the dissolution, and recrystallization of iron minerals, and subsequent oxidation to more crystalline minerals such as hematite and maghemite are examples which generate natural nanoparticles (Guo and Barnard, 2013; Jiang et al., 2018b; Schwertmann, 1991). The natural formation of iron nanominerals or mineral nanoparticles can also be influenced by carbon dioxide and carbonate anion concentrations (Blanco-Andujar et al., 2012).

17.2.1.3 Nanoparticles from microbial activities

Natural nanoparticles formed as a product or by-product of microbial activities such as respiration are categorized as microbial nanoparticles (Hochella et al., 2015). Unlike the other natural nanoparticles, microbial nanoparticles occur more often as highly monodispersed particles with a fixed stoichiometry, often very similar to engineered nanoparticles (Gericke and Pinches, 2006). Nanotechnologists involved in synthesis are taking cues from nature, and developing methods for producing highly monodispersed nanoparticles utilizing microbes (Narayanan and Sakthivel, 2010). As properties of nanoparticles are size dependent, the production of monodispersed nanoparticles is highly desirable, so that all nanoparticles have identical property. Various nanoparticles of copper, iron, gold, selenium, silver, and uranium are known to form through microbial respiration or reduction (Das et al., 2017; Joshi et al., 2018; Suzuki et al., 2002). These metal microbial nanoparticles can be formed via plants, algae, yeast, bacteria, and fungi (Srivastava et al., 2015). Humic substances and other biomolecules can also be considered as microbial nanoparticles, as both are degradation products.

Biomineralization is nature's way of producing highly monodispersed inorganic nanoparticles such as calcium, silicon, and iron-based salts (Wu et al., 2012). Formation of these natural nanoparticles can be biologically induced and biologically controlled mineralization (Sharma et al., 2015). As the name suggests, biologically induced mineralization nanoparticles are formed as a consequence of metabolic processes, without the direct involvement of microbes. In contrast, in biologically controlled mineralization, microbes control the formation of nanoparticles, which are formed within the cell and produces highly monodispersed nanoparticles (Sharma et al., 2015), which generates highly controlled nanoparticles. Intracellular biomineralized nanoparticles often serve various tasks for the microbes, which include storage of specific metal and hardening of tissue, such as crystalline magnetite Fe_3O_4 nanoparticles are utilized for navigation by magnetotactic bacteria (Schüler and Frankel, 1999). Alternatively, redox reactions generated due to microbial metabolic processes produce Mn- and Fe-oxide, by biologically induced mechanism (Wu et al., 2012), such as iron-oxidizing bacteria are *Leptothrix* and *Gallionella* in groundwater and soils. *Geotracer sulurreducens* and *Thiobacillus denitrificans* utilize magnetite for electron shuttling and achieving nitrate reduction via acetate oxidation (Hochella et al., 2015). *Shewanella oneidensis MR-1* respires utilizing hematite nanoparticles and thereby reducing iron (Hochella et al., 2015). Copper metal nanoparticles can be produced by biomineralization processes (Wagner et al., 2014). Similarly, different eukaryotes and prokaryotes can produce natural silica nanoparticles (Wu et al., 2012).

17.2.2 Synthetic nanoparticles

Nanoparticles, which are generated due to anthropogenic activities, are categorized as synthetic nanoparticles. Currently, the major environmental concerns about nanoparticles in the aquatic environment have focused on synthetic nanoparticles. In comparison to natural nanoparticles, however, they constitute significantly smaller quantity, with roughly 10.3 Mt/year of annual flux to atmosphere (Hochella et al., 2019). Similar to natural nanoparticles, these synthetic nanoparticles can make their way to water bodies (Bundschuh et al., 2018; Frimmel and Niessner, 2010) (Fig. 17.2). Synthetic nanoparticles can be broadly classified into incidental and engineered nanoparticles.

17.2.2.1 Incidental nanoparticles

Synthetic nanoparticles, which are produced unintentionally and indirectly by human activities, are termed as incidental nanoparticles (Madl and Pinkerton, 2009). Incidental nanoparticles can come from automobile exhaust, industrial and mining wastes, wear and corrosion processes, as well as a variety of combustion processes that generate fine particulates. These can include a wide variety of carbon and metal oxide nanoparticles (Westerhoff et al., 2018). Carbon soot is a well-known example of incidental carbon nanoparticle (Tumolva et al., 2010). Another typical example is the corrosion of pipelines delivering water, which can introduce incidental nanoparticles into drinking water distribution systems (Venkatesan et al., 2018). Titanium dioxide nanoparticles, which are extensively used as white pigments in paints, paper industry, plastics, and in food, are another incidental nanoparticle that can easily get into the aquatic environment, for example, from paint peeling during heavy rainfall (Madl and Pinkerton, 2009; Westerhoff et al., 2018). Magnetite nanoparticles generated from combustion was observed in the brains of humans in Mexico and England (Maher et al., 2016), along with trace quantity of nickel, cobalt, platinum, and copper. Carbon nanotubes and fullerenes are those class of synthetic nanoparticles that are both incidental and engineered nanoparticles and can form combustion (Stern and McNeil, 2008). Generally, the stoichiometry and morphology of incidental nanoparticles can be very similar to their natural nanoparticles and engineered nanoparticles (Neal et al., 2011; von der Kammer et al., 2012; Westerhoff et al., 2018). Another form of the incidental nanoparticle is nanoplastics, which is formed by the degradation of plastic wastes in the environment. In 2018 there were critical reviews on the occurrence of nanoparticle size plastics and their increasing universal occurrence in freshwater (Alimi et al., 2018;

Lehner et al., 2019). Nanoplastics formation does include environmental degradation factor. However, it comes from anthropogenic activity from generated plastic waste, so it is suited to categorize it under synthetic nanoparticles.

17.2.2.2 Engineered nanoparticles

Nanoparticles synthesized for mass production come under the purview of engineered nanoparticles (Singh, 2015). This category includes all the nanoparticles that are used in the solar cell, cell phones, computers, chipset, agriculture, instruments, cosmetics, and household products. They can range from simple metal oxides to complex core–shell nanoparticles (Delay and Frimmel, 2012). The widespread use of engineered nanoparticles has increased their emission to aquatic environments through a variety of pathways, including wastewater (Malakar et al., 2019). Engineered nanoparticles comprise a wide variety of particle synthesized such as metals, metal oxides, carbon nanotubes, quantum dots, and semiconductors. These materials are synthesized either by top-down approach, where a bulk material is continuously grounded to its nanoscale size or bottom-up approach, where atoms are explicitly oriented to design the final nanoparticle. Given the controlled synthesis approach, engineered nanoparticles are highly monodispersed and of strict stoichiometry, which is very much needed to achieve the performance required from these nanoparticles.

Today, engineered nanoparticles are used virtually in everything, from computers, which contain semiconductors (Chen et al., 1997), to agriculture with the application of nanopesticides and nanofertilizers (Kah et al., 2018). From memory chips to medical equipment, nanotechnology is a crucial component in many sectors of the modern world. Widespread manufacture and use of nanotechnology have raised the growing concerns about their fate and toxicity in the environment (Ma et al., 2016). Nanoparticles can be highly reactive; exposure may have significant toxic effects on human health. Some engineered nanoparticles are carcinogenic and can generate reactive species in humans (Ganguly et al., 2018). Moreover, the life cycle and long-term effects of the engineered nanoparticle are understudied concerning different ecological impacts. The US Environmental Protection Agency regulates the manufacturing of new materials under the Toxic Substances Control Act (Environmental Protection Agency Federal Facilities Restoration and Reuse Office, 2017; Gehrke et al., 2015). Engineered nanoparticles can be further classified based on starting materials (e.g., carbon or metal nanoparticles) as discussed in the following subsections.

17.2.2.2.1 Carbon nanoparticles

Nanoparticles derived from carbon (e.g., carbon nanotubes) simultaneously act such as particles and high molecular weight organic compounds. Carbon nanotubes and fullerenes are allotropes of carbon, with a diameter ranging from <1 to $4\,nm$. It is one of the most vastly produced engineered nanoparticles $\sim 270\,t/year$ (Singh, 2016). The tensile strength of carbon nanotubes are 100 times that of steel, making them desirable material for high tensile structure. The thermal conductivity comparable of carbon nanotubes is similar to diamond with high electrical conductivity. Carbon nanotubes are widely used in microelectronics, medicine, hydrogen fuel cell technology. Carbon nanotubes can be multiwalled or single-walled and carbon nanowires (Harris, 2009).

17.2.2.2.2 Polymeric nanoparticles

Engineered nanoparticles based on polymeric organic molecules fall under the category of polymeric nanoparticles. These can be of varied morphology but are mainly nanospheres or nanocapsules (Khan et al., 2017). These nanoparticles are widely used in controlled release needed for drug delivery, in therapeutic and imaging (Andrieux et al., 2013). The reason for using them in a medical setup is that the organic polymers make these nanoparticles highly biodegradable and compatible.

17.2.2.2.3 Lipid-based nanoparticles

These are organic nanoparticles synthesized utilizing lipid moieties. These lipid nanoparticles are widely used in biomedical application. Although called nanoparticle, their size can vary from 100 to 1000 nm of spherical shape (Vaghasiya et al., 2013). Similar to polymeric nanoparticles, their main use is in drug delivery due to their biocompatibility (Khurana et al., 2013).

17.2.2.2.4 Metal-based nanoparticles

Engineered nanoparticles containing pure metals are metal-based nanoparticles. These nanoparticles exist in zero-valent state, and the crystal structure is generally free of defects. Metal nanoparticles are widely used in optoelectronics (gold nanoparticles) (Gad and Hegazy, 2019), and due to high reactivity, it is also used in water remediation (e.g., zero-valent iron nanoparticle) and treatment (Zou et al., 2016). Generally, the bottom-up approach is used to produce pristine metal nanoparticles of desired morphology and size. The excellent optical properties of metal nanoparticles make them good candidates for coating and using in scanning electron microscope imaging technique (e.g., gold and platinum are used to coat samples) (Khan et al., 2017). The zero-valent state of these nanoparticles makes them highly reactive, and most nanoparticles readily form an oxidized surface layer when they come in contact with air.

17.2.2.2.5 Ceramic nanoparticles

Inorganic nanoparticles based on metal and metalloid oxides, carbonates, carbides, and sulfides are ceramic nanoparticles. This group includes titanium, calcium, silicon, and zirconium and has a wide range of properties including a lower chemical inactivity and heat resistance, making them viable nanomaterials. Ceramic nanoparticles are widely used in biomedical for drug delivery, gene carriers (Thomas et al., 2015). Another aspect of ceramic nanoparticles is that their surface can be functionalized with organic molecules, giving them directional properties for efficient drug delivery (Moreno-Vega et al., 2012).

17.2.2.2.6 Semiconductor nanoparticles

Semiconductor materials are widely used in all electronic items, and they possess wide bandgaps. Nanoparticle-based semiconductors have increased the modern computers processing power and decreased its size considerably. These nanoparticles may contain both metals and metalloids and are also widely used in solar cells. The bandgap tuning in nanoparticle semiconductors, produced by different doping, make these materials imperative in the modern world where they are widely used in the energy sector, photovoltaics, photoelectron generation, photocatalysis, and hydrogen production (Khan et al., 2017). Semiconductor nanoparticles consist of zinc sulfide, zinc oxide, titanium oxide, cadmium chalcogenides-based nanoparticles. Semiconductor nanoparticles have varied morphology from zero dimension to all the way three-dimensional shape (Suresh, 2013).

17.3 Behavior of nanoparticles in water

17.3.1 Natural nanoparticles

Much of the environmental concerns and research on the occurrence of nanomaterials in the aquatic environment has been focused on engineered nanoparticles, while the fate and toxicity of natural nanoparticles are generally overlooked (Ermolin et al., 2018). A primary reason for the lack of study on natural nanoparticles has been the absence of suitable analytical tools. Moreover, natural nanoparticles have been a part of the environment long before humans evolved and thus easily overlooked in the environment (Hochella et al., 2019). However, the recent development of nanogeosciences as a field of study has brought in advanced tools to understand the fate, toxicity, and transport of natural nanoparticles in the environment.

Natural nanoparticles produced from biogeochemical processes are known to be abundant in groundwater and are generally found at low mg/L concentrations, though this baseline may increase in the presence of other groundwater contaminants (Baumann et al., 2006, 1998). It is estimated from the grain size distribution of sand and gravel in the aquifer that up to 5% of the mineral matrix is in the nanoscale range. However, the concentration and form of natural nanoparticles likely remain in equilibrium with the surrounding physical and chemical conditions.

In their review, Sharma et al. (2015) found that concentrations of natural silver and gold nanoparticles are low, and it would be difficult to predict their toxicity to aquatic organisms. Natural gold and silver nanoparticles are stabilized in water by humic substances and can be formed by reduction of respective ions in the presence of natural organic matter (Akaighe et al., 2011; Yin et al., 2014). However, under anoxic conditions, the rate of dissolution of natural nanoparticles into water likely dictates toxicity. A surface layer of sulfide limits dissolution rates of natural gold and silver nanoparticles, and there is less chance for release of metal ions and therefore result in a decrease in toxicity (Reinsch et al., 2012). Based on this phenomenon, it can be concluded that the toxicity of natural nanoparticles depends mainly on the kinetics and mechanism of dissolution of these nanoparticles under oxic/anoxic conditions.

Natural nanoparticles from volcanic eruptions can reach surface water sources, such as lakes, rivers, sea, and the ocean, via wind and through precipitation with rainwater (Hochella et al., 2019), possibly polluting these water sources. Natural nanoparticles in volcanic ash can act as a nutrient source resulting in increased phytoplankton production in marine environments (Lindenthal et al., 2013; Maters et al., 2016; Olgun et al., 2013), likely affecting carbon dioxide equilibrium (Ermolin et al., 2018). Volcanic ash nanoparticles can also contain toxic elements that may be introduced to water sources (Ermolin et al., 2018). Given the nano-size and mobility of the nanoparticles of volcanic ash, they can readily be ingested or absorbed through skin pores (Buzea et al., 2007). Raiswell et al. (2008) studied the mass flux of iron nanoparticles delivered by melting icebergs into the Southern ocean and found this source to be comparable to Aeolian inputs likely contributing to Antarctic warming.

Understanding nanoscale geochemical reactions initiated by natural nanoparticles is challenging, especially the formation of nanostructures consisting of geological materials, and their influence to dictate biogeochemical cycles (Wang, 2014). Recent efforts have shown that natural nanoparticles may not pose a direct threat as a pollutant but can dictate mobility and form of other potentially hazardous trace elements, such as arsenic, in water sources. Natural nanoparticles generally have a negative surface charge at near-neutral pH, and most of the inorganic nanoparticles have a surface layer of organic matter onto themselves, which provides a net negative charge on the surface. This surface charge provides colloidal stability by double-layer formation with counter ions. Colloidal stability can control the mobility of natural nanoparticles in different water sources, which in turn can control trace element mobility. The size, surface-charge of natural nanoparticles, and the Stokes settling velocity are the most relevant parameters for their mobility (Kretzschmar and Sticher, 1998; Wagner et al., 2014). In surface waters sources, surface charge plays a vital role in determining aggregate formation. Nanoparticle aggregation may lead to subsequent settling by gravitational forces. Natural nanoparticle mobility in groundwater is not only controlled by aggregation but is also dependent on particle deposition, size exclusion, redox potential, pH, and ionic strength of the bulk solution (Wagner et al., 2014). The significance of natural nanoparticles, especially metal hydroxides and oxides, transport lies in their capability to form complexes with trace elements.

Complexation with ion species is generally via adsorption, resulting in the formation of inner-sphere complex with natural nanoparticles and trace metal ions. Humic substances play critical roles in the complexation reactions in the aquatic environment, especially surface water (Alcacio et al., 2001; Bargar et al., 1997). Inner-sphere complexes form strong bonds and are less labile in comparison to outer-sphere complexes such as ion pairs, which have significant impact on metal-ion mobility and subsequent toxicity in water. Humic substance coatings on metal nanoparticles are affected by the presence of multiple binding sites due to the presence of metal ions. Furthermore, factors such as redox condition, pH, presence of different electrolytes, the influence of nanoparticle interaction, and concentration of humic substance compared to nanoparticles, can also impact humic substances coating on nanoparticles (Jerez, 2003). The interaction of other biomolecules with natural inorganic nanoparticles is not well known. These interactions are likely meaningful as they play a significant role in the bioavailability of trace elements in aquatic systems. Natural nanoparticles can influence the oxidation states of trace metals and other related processes at the aqueous—soil—organism interface. The general notion is that these complexes limit trace element mobility, and availability, this has been shown by Slaveykova and Wilkinson (2005) for different metal ions for the different organism.

Complexation of trace metals bound to natural nanoparticles also acts to mitigate and limit the potentially hazardous effects of trace elements in aquatic environments. The strength of complex bonds is often quite strong, but the strength is dependent on other factors such as redox and pH, which may initiate breaking of this complexation. Reductive dissolution of natural nanoparticles can initiate the release of trace elements to aquatic systems, elevating contaminant concentrations above safe levels (Erbs et al., 2010; Voegelin et al., 2019). Similar to volcanic ash natural nanoparticles formed via microbial activity or weathering of rocks can impact the phytoplankton diversity and biomass in oceans and lakes (Hochella et al., 2019). This impact of natural nanoparticles in oceans would have a profound effect on biogeochemical cycles of carbon, nitrogen, and iron. Natural nanoparticle geochemistry likely influences aquatic food chains as phytoplankton makes 40% of food production and is the base of the aqueous food chain (Hartland et al., 2013). Seawater iron mainly (80%) constitutes in the size fraction of 20−200 nm; iron is a vital nutrient in the ocean. Similarly, manganese another transition element is nearly always below 20 nm size fraction (Hochella et al., 2019).

The recent development of analytical techniques to measure and characterize nanoparticles has open doors to understand the significance of natural nanoparticles in various biogeochemical cycles, climate change, and trace element mobility. However, there is a considerable gap in knowledge of the chemical composition of natural nanoparticles and their cycles in the environment. This gap is hard to address due to the transient nature of these natural nanoparticles and also provides challenges to sample collection and characterization. The polydispersity of natural nanoparticles also adds to the complexity of this problem (Ermolin and Fedotov, 2016). The role of natural nanoparticles on the bioavailability of nutrients and toxic elements is a major feature in the progress of higher class organisms, acting as a potential buffer of the environment against modification. Lastly, it is apparent that the fate and behavior of natural nanoparticles call for further studies as these nanoparticles can add to the complexity of pollutant in different water sources both as a primary pollutant and mobilizing secondary pollutant, and these nanoparticles are by far in much more massive amounts compared to their engineering counterpart.

17.3.2 Synthetic nanoparticles

The wide application of engineered nanoparticles is intimately interconnected to the water cycle (Frimmel and Niessner, 2010). Water is a necessity for different human activities such as drinking and cleaning. Tracking engineered nanoparticles within the complex and changing composition of water is extremely difficult. Engineered nanoparticles will interact with diverse constituents of water, which can include aggregation (homo and hetero), sorption, flocculation, uptake, and biotransformation. For example, nanoparticles in the atmosphere may reach water sources via precipitation, can infiltrate to the groundwater below, and reach surface water through runoff. Treated wastewater introduces significant quantities of nanoparticle residues to water sources during discharge. Similarly, the application of municipal biosolids to soils and leachate from landfills can also infiltrate or reach surface water sources through multiple avenues (Fig. 17.2). Thus nanoparticles enter the environment during their production, use, and, finally, through discharge after disposal (Bundschuh et al., 2018). This release of nanoparticles can be directly to the environment or indirectly through wastewater treatment plants and landfills. There may be changes to the properties of nanoparticles during these release processes, especially during indirect release via wastewater treatment plant or landfill application (Bundschuh et al., 2018).

Globally, it is estimated that around 7% of total production volume of engineered nanoparticles ends up in the aquatic environment (Keller et al., 2013). Increase in outdoor applications such as the use of nanofertilizer and nanopesticide will accelerate the mass flow into the aquatic environment. Incidental nanoparticles such as titanium dioxide, extensively used to increase whiteness, and opacity in paint have been shown to reach water sources via peeling and decomposition (Kaegi et al., 2008). It is anticipated that most nanoparticles are emitted during their use and disposal phase (Keller et al., 2013). Nanoparticle release models demonstrate a growing need for environmental risk assessment for the occurrence of engineered nanomaterials (Sun et al., 2017, 2016). Studies have found that titanium dioxide nanoparticles can accumulate in sludge and biosolid treated soils and through runoff from soils to sediment and landfills. Titanium dioxide appears to be mainly emitted from wastewater, accounting for almost 85% of all titanium oxide nanoparticle release (Keller et al., 2013). Zinc oxide nanoparticles can also accumulate in soil and landfills, though these nanoparticles are mainly used in electronics, cosmetics, and medicine. Here also, the dominating pathway is through wastewater discharge (Mueller and Nowack, 2008). Unlike metal oxides, emission of carbon nanotubes is mainly during production and disposal in landfills, which is around 90% of the total emission (Sun et al., 2016). Metal nanoparticles such as silver are also released to the environment mainly during production and use and can accumulate in wastewater, and landfills (Bundschuh et al., 2018). Incidental nanoparticles such as palladium and platinum have been identified in traffic emission (Prichard and Fisher, 2012).

Engineered nanoparticles are ultimately released to the environment, and industrial and municipal wastewater plus biosolids application appears to be the largest releasing channel for engineered nanoparticle (Lazareva and Keller, 2014; Westerhoff et al., 2013). Nanoparticulate silver embedded in fabrics, which is estimated to be 20%−100% of total particle content, is released in repeated washing and end up the wastewater stream (Choi et al., 2017). Sunscreen, personal care products and cosmetics can contain titanium dioxide and are released via wastewater stream after using. Around 26%−39% of the nanoparticles that are released to wastewater stream make their way to the aquatic environment (Lazareva and Keller, 2014). A study quantified the concentration of titanium dioxide in an Arizona wastewater treatment plant, reporting an average of 843 µg/L of total Ti in the inflow (Kiser et al., 2009). Silver nanoparticles were quantified in nine wastewater treatment plants in Germany, which reported a maximum daily load of 4.4 g-Ag/day. However, primary and secondary sedimentation removed >95% of the particles (Li et al., 2013). The wastewater treatment plants play a significant role in the redistribution of engineered nanoparticles to the aquatic environment, but the fate, transport, and behavior of nanoparticles in the complex matrix of wastewater are still not clear.

In comparison to incidental and unintentional release, nanoparticles can be intentionally released to the environment for a variety of purposes such as contaminant remediation, water treatment, and agricultural application. Nanotechnology is continually developing new environmental applications such as in groundwater nanoremediation (Bardos et al., 2018). The engineered metal nanoparticle of iron, for example, nanoscale zero-valent iron particles, is widely utilized for groundwater remediation (Zou et al., 2016). Direct injection of zero-valent iron nanoparticles to aquifers are being used for in situ treatment of groundwater (Stefaniuk et al., 2016). Direct injection into groundwater can have unexpected consequences, such as unintentional interaction with microbes (Goldberg et al., 2007). Exposure to biologically active nanoparticles can affect biogeochemical cycles in the subsurface. Crampon et al. (2019) recently studied the impact on bacterial abundance from nanoparticle zero-valent iron. They assessed the denitrifying activity of the bacterial community and observed a shift in community structure. The study further concluded that remediation with nanoparticle zero-valent iron could have a long-term effect on bacterial community composition and influence groundwater ecological functions. Other studies have also found toxicity of nanoparticle zero-valent iron particles toward living organisms and observed iron inside bacteria and bacterial cell walls (Jiang et al., 2018a; O'Carroll et al., 2013; Wille et al., 2017; Xue et al., 2018). The high reactivity of zero-valent iron nanoparticles produces reactive oxygen species (ROS) in the aqueous phase (Liu et al., 2014). Elevated levels of ROS can result in toxicological effects to microbes, as DNA damage was observed in some studies (Ghosh et al., 2017). Further zero-valent iron nanoparticles utilized for the treatment of arsenic and chromium were found to enhance aggregation tendency post-treatment (Yin et al., 2012).

Use of engineered nanoparticles in water remediation is increasing due to their versatile application for treatment of organic pollutants such as pesticides, for their antimicrobial activity and toxic element such as arsenic and lead removal. Nanoparticle residues left after remediation processes are thought to be innocuous (Good et al., 2016; Simeonidis et al., 2016; Troester et al., 2016), though given their reactivity, mobility, and relative stability in water their toxicity due to drinking water consumption should be evaluated. The remediation processes must be well-monitored for the fate of nanoparticle being used; their degradation, dissolution, and interaction with trace elements need to be monitored before designing water treatment processes.

Identification of the fate of nanoparticles employed for water purification has been studied using water from five different sources, including groundwater and freshwater with spiked titanium dioxide, silver, and zinc oxide nanoparticles. The study reported that the final membrane filtration process removed nanoparticles, but there were sufficient concentrations of dissolved metals which can be hazardous to human health (Abbott Chalew et al., 2013). Exposure to nanomaterial used for adsorptive removal of water contaminants can have unknown health effects to humans and other living organisms (Hristovski et al., 2008; Simeonidis et al., 2016). Gold nanoparticles, which are known to accumulate in fish cells (Bouwmeester et al., 2011), and titanium dioxide nanoparticles are both known to incite clastogenicity, oxidative DNA damage, genotoxicity, and inflammation (Chen et al., 2014). Titania nanoparticles can enhance oxidative damage in developing fish (Fang et al., 2015). Dissolution of zinc oxide nanoparticles can generate active oxygen species. The effect of nanoparticle uptake by plants has been elaborated in a recent review by Malakar et al. (2019). The review by Malakar et al. (2019) points out that nanoparticles in different water sources and irrigation practices can lead to plant uptake, and through this process nanoparticles can be introduced into the food chain. Humans can consume bioaccumulated nanoparticles in food crops and the implication for human health is still unknown (Malakar et al., 2019).

Leaching of nanoparticles from surface soil to the groundwater below through the unsaturated zone can also occur and is controlled by recharge and capillary forces. Capillary force can influence the attachment/detachment processes of nanoparticles during wetting/drying cycles (Cheng and Saiers, 2010; Shang et al., 2008). Flow velocities are incredibly dynamic in the unsaturated zone, which imposes physical stress on attached colloids. Evaluation of the fate of nanoparticles in the unsaturated zone is dependent on the surface functionality of the nanoparticles (Rahmatpour et al., 2018). Hydrophobic particles and surfaces tend to concentrate on the air−water interface, though hydrophilic particles and surfaces orient in pore water. As with natural nanoparticles, movement and flow patterns of engineered nanoparticles are dependent on redox, pH, and ionic strength, all of which changes during wetting/drying cycles. The dynamic nature of this system emphasizes the significance of soil conditions for controlling the transport of nanosized particles.

In aqueous environment, engineered nanoparticle can also be stabilized by humic substances. Synthesized gold nanoparticles can be stabilized by citrate anions, and showed lower aggregation in surface water in the presence of humic substances (Stankus et al., 2011). Similarly, silver nanoparticles, which are used for antimicrobial effects in fabrics and treatment of water, gain stability in the presence of humic substances (Gunsolus et al., 2015). Concentrations of silver nanoparticles may reach up to 300 μg/L in river water and can form complexes of silver−sulfur−humic substances (Frimmel and Niessner, 2010). Fullerene stability was also known to increase in the presence of humic acid, which easily aggregates (Wang et al., 2012). Humic substances can also stabilize multiple-walled carbon nanotubes (Gao et al., 2018).

The long-term effects of nanoparticles as a potential pollutant in water sources is quite clear, but there are few studies of the overall occurrence and likely a growing need to regulate and monitor nanoparticles in drinking water (Westerhoff et al., 2018). Developed countries use multibarrier systems for treating drinking water (Scheurer et al., 2010), and generally surface water sources (80%) are utilized for producing drinking water. A recent review focused mainly on the presence of engineered nanoparticle in the different surface water sources, water treatment facility, and point of use sites (Westerhoff et al., 2018). The review utilized material-flow models and limited field data to predict the occurrence of nanoparticles in drinking water and concluded that there is a low risk from engineered nanoparticles from drinking water. However, another experimental study from the same group found lead, tin, iron and copper with average concentrations of 1.2 ± 1.3, 1.8 ± 3.0, 88 ± 144 and 69 ± 45 ng/L, respectively, representing a minimum of 0.4%, 18%, 16%, and 0.2% of the corresponding total dissolved concentrations in tap water collected from Phoenix, Arizona (Venkatesan et al., 2018). A 2018 report on surface water from the Dutch rivers Meuse and Ijssel showed the presence of silver and cerium-oxide nanoparticle and titanium oxide microparticles (Peters et al., 2018). The study found that concentration of silver nanoparticles at 0.8 ng/L and an average particle size of 15 nm; for cerium oxide, it was 2.7 ng/L, with an average particle size of 19 nm, and titanium oxide was found in an average concentration of 3.1 μg/L and an average particle size of 310 nm (Peters et al., 2018).

Our understanding about threats from synthetic nanoparticles contamination to water supplies is incomplete. This lack of knowledge can be attributed to a lack of understanding of their toxicity. Added to that, there is an incomplete picture of the fate and transformation of nanoparticles in the aquatic environment. The implications for the use of nanoparticles for different water treatments processes are not precise and reliable analytical tools to measure nanoparticles in complex matrices are lacking. These knowledge gaps make nanoparticle a very complex system to study with regards as a pollutant. Unlike natural nanoparticles, synthetic nanoparticles are far more studied, though comprising only a fraction of the total quantity of nanomaterials in the environment. Synthetic nanoparticles are potential pollutants in contrast to natural nanoparticles, which may be linked with trace element mobilization and geogenic contamination of water supplies.

17.4 Analytical techniques

There is a common agreement that nanoparticles have unique properties and need specialized measuring techniques. Until a few decades ago, nanoscience was virtually unknown in environmental research but has progressively gained more prominence in recent years. Analytical tools are slowly catching up and analytical nanochemistry has advanced significantly to help draw a framework for new understandings of nanoparticles in complex matrices. Environmental scientists can take a pioneering role to understand nanogeoscience phenomenon in complex ecological systems. Specialized qualitative and quantitative methods for analysis are needed for a better understanding of their fate, transformation, and impact on the environment. Analytical tools with separate enrichment, separation, and measurement technologies are needed for detecting and characterizing nanoparticles in complex matrices. High sensitivity and reproducibility are required to ensure confidence in results from these developed analytical tools.

17.4.1 Inductively coupled plasma−mass spectroscopy

Inorganic engineered nanoparticles detection in aqueous matrixes often uses inductively coupled plasma−mass spectrometry (ICP−MS) both for speciation and for quantitative analysis. ICP-MS can be coupled with a variety of separation techniques, such as field flow filtration (FFF)−based microfluidic channels, which can increase size-based or charge-based selectivity (Pornwilard and Siripinyanond, 2014). FFF−ICP−MS can be used to analyze engineered nanoparticles at the $\mu g/L$ concentration level (Bednar et al., 2013). Another hyphenated inlet for ICP−MS is a single particle (SP)−ICP−MS. SP−ICP−MS can discriminate between pH sensitive and noninert engineered nanoparticles. In SP−ICP−MS, metal nanoparticles are individually converted into a packet of ions in the plasma and detected as a SP pulse. The intensity at a specific mass is proportional to the number of atoms ionized from a single nanoparticle (Zhang et al., 2019).

In contrast, dissolved species of an element produce pulses of averaged constant intensity. SP−ICP−MS can facilitate tracing the interfacial behaviors of engineered nanoparticles and may also help in determining the fate of the nanoparticle as discrete aqueous forms. Sample size limits the sensitivity for measurement using SP−ICP−MS. The sample introduction system can influence the mass concentration of engineered nanoparticle species due to the adsorptive losses of dissolved ionic species. Polydisperse-engineered nanoparticles predominate in environmental samples, and SP−ICP−MS has difficulty in the extensively polydispersed system. Recent research has helped to address these shortcomings. Innovative tools have been utilized, for example, a monodisperse microdroplet generator may assist the use of this tool for polydispersed engineered nanoparticles (Gschwind et al., 2013). SP—ICP—MS is a sophisticated technique and requires a high level of expertise, which is a major factor limiting its use.

17.4.2 Raman spectroscopy

Surface-enhancing Raman spectrometry has been developed to measure engineered nanoparticles (Álvarez-Puebla and Liz-Marzán, 2010; Guo et al., 2017). Raman spectroscopy is sensitive for specific engineered nanoparticle and based on original Raman signatures (Zhao et al., 2018). Titanium dioxide nanoparticles, for example, were identified from a complex matrix by Zhao et al. (2018), the peak intensity of different form of titanium oxide corresponded with their concentrations. The particle size can also be obtained in Raman spectroscopy, and Zhao et al. utilized the ratio of Raman intensity of the particle and the particle-bound moiety. Raman spectroscopy has been utilized to measure gold, silver nanoparticles in complex matrices (Liou et al., 2018). Raman technology is simple to use and can be coupled with other technologies. For example, Badireddy et al. (2012) coupled Raman spectroscopy with hyperspectral imagery to measure engineered nanoparticles in the different water matrix.

17.4.3 Nanoparticle tracking analysis

Nanoparticle tracking analysis (NTA) is complementary to dynamic light scattering, which is used to analyze the size distribution of nanoparticles in the dispersed system. NTA can measure both size distribution and abundance of nanoparticles in aqueous samples (Zhang et al., 2019). NTA is widely used in water treatment and environmental monitoring. NTA can be utilized for both qualitative and quantitative detection of engineered nanoparticles. NTA measurement is easy, low-cost, and rapid technique for studying engineered nanoparticles in complex systems.

17.4.4 Electrochemical detection of nanoparticles

Electrochemical techniques utilize low-cost, user-friendly portable potentiostats and can be useful for environmental samples (Li et al., 2017). Electrochemistry has good analytical sensitivity and, in combination with disposable single-drop electrodes, can be readily adapted to in situ monitoring. Electrochemical detection cannot provide information on chemical composition but can be combined with techniques for chemical composition. Electrochemical techniques can measure nanoparticles by direct detection, utilizing redox chemistry, or through electrocatalytic reactions where detection is indirect. Surface-enhanced electrochemical techniques, the nanoparticle surface properties are utilized for detection and using a "nanoimpact" approach stochastic collisions on an electrode can detect individual nanoparticles (Martín-Yerga, 2019). Electrochemical techniques provide useful tools for observing the dissolution kinetics and speciation of assorted metallic engineered nanoparticles and has been shown to estimate percentages of dissolved species.

17.4.5 Matrix-assisted laser desorption ionization time-of-flight—MS

Matrix-assisted laser desorption ionization (MALDI) time-of-flight (TOF) mass spectroscopy (MS) is a powerful and highly sensitive technique for nanoparticles (Guan et al., 2007). A laser beam is utilized for heating and ionizing nanoparticles, causing desorption of intact nanoparticles. Sample analysis proceeds in three stages: (1) MALDI stage, (2) TOF stage, and (3) detection of ions. In the MALDI stage, particles are heated and ionized, and then enter a TOF analyzer where they are accelerated using a high voltage electrostatic field. The time of travel is directly related to particle mass-to-charge ratio. Smaller low mass nanoparticles are detected first, and high-resolution mass differences can be achieved using a TOF instrument. While quite powerful, this is a costly and highly sophisticated system for measuring nanoparticles in different matrices.

17.4.6 Other spectroscopic and microscopic methods

A variety of spectroscopic and microscopic methods has been used to characterize engineered and natural nanoparticles in environmental samples and will be briefly discussed here. Imaging instrumentation includes but not limited to transmission electron microscopy, where the electron beam is transmitted to get an image based on electron opacity, scanning electron microscopy also uses electron beam but scans the surface for imaging. The imaging techniques can be utilized to get information on the morphology of the particles and energy dispersive X-ray can be utilized to get rough estimates about the chemical composition (Bandyopadhyay et al., 2012).

X-ray diffraction (XRD) has also been used to determine nanoparticle size, based on the broadness of the diffraction pattern. XRD is also utilized to understand the crystalline phase of the nanoparticles, which can indirectly be used to understand the chemical composition. Nanoparticles size can also be determined by X-ray scattering techniques, such as small-angle X-ray scattering (SAXS), ultra-SAXS, and small-angle neutron scattering, which has gained popularity recently. These nondestructive analytical techniques can also provide structural, composition information, even in complex matrices (Bandyopadhyay et al., 2012).

17.5 Human health concerns and nanotoxicology

The study of toxicity from nanoparticles is nanotoxicology, which is a subspecialty of particle toxicology. The toxicity of nanoparticles may differ from the bulk material due to the unique size and properties of nanoparticles. The exposure pathway for nanoparticles is multifaceted, and possible hazards can be from inhalation, penetration through the skin, injected, and ingestion (Fig. 17.3). However, inhalation is considered to be the major pathway of exposure to humans. Along with size, the composition, morphology, shape, surface charge, dissolution properties, solubility, and aggregation of nanoparticles can define their potential toxicity. Currently, there are around 3000 nanoparticulate-based commercialized applications, which includes manufacturing sector, automotive sector, energy sector,

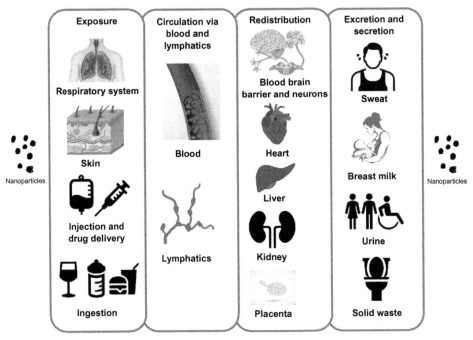

FIGURE 17.3   Exposure pathway, circulation, redistribution, and final excretion of nanoparticles inside human body.

health sector, electronics sector, agricultural sector, and food technology sector (Huang et al., 2017).

Further, engineered nanoparticles, such as titanium dioxide which are being used in personal care products, can serve as direct exposure to humans. Iron oxide nanoparticles are used in paints and polishes, zinc oxide nanoparticles are used in the food industry, textile industry for deodorizing and antibacterial products, similarly silver nanoparticles are being utilized in fabrics (Huang et al., 2017; Mantecca et al., 2017). Use of nanoparticle has also increased in biomedical applications and health sector such as drug delivery and nanoelectronics (Korin et al., 2013; Sykes et al., 2014). Nanoparticles are promising to fight diseases due to their higher therapeutic efficacy, but it can also have unwanted repercussion to humans (Huang et al., 2017).

17.5.1 Causes of nanotoxicity

There is a substantial knowledge gap in the understanding of key factors fostering toxicity from nanoparticles. It is of highest importance to evaluate these properties before predicting the effect of nanotoxicity. The fate of nanoparticle in the human body (Fig. 17.3) and the surrounding environment may be studied in the purview of direct and indirect interactions. These fundamental properties which have an impact on nanotoxicity are briefly discussed in the following subsections.

17.5.1.1 Size

Nanoparticles size is considered to be one of the most critical factors controlling toxicity to humans. The surface-to-volume ratio increases exponentially as the size is reduced, which changes chemical and physical properties of the particle, enhances reactivity and toxicity of the particle irrespective of other competing factors such as composition, morphology. Nanoscale size also enhances the capability for penetration to the human body and also penetration in different tissue and cells inside the body (Sajid et al., 2015). For examples, a nanoparticle of size <35 nm can pass through the blood—brain barrier, ~40 nm can enter nuclei and <100 nm can quickly get inside a cell (Dawson et al., 2009; Ganguly et al., 2018; Oberdörster et al., 2009, 2005). Even catalytic binding capability and adsorption rate can also be dependent on the size of the nanoparticles (Sajid et al., 2015).

Cytotoxicity in human lungs cells was induced from exposure to silver nanoparticles (Gliga et al., 2014), and it was found that the toxicity was also dependent of silver ion concentration, where lower ionic silver concentrations appear to enhance toxicity (Beer et al., 2012). In addition to being cytotoxic, silver nanoparticles are also known to cause genotoxicity and inflammation, and a wide range of effects can develop based on the size of the particle (Ning et al., 2017; Sajid et al., 2015). Cellular interaction studies have found that smaller sizes of nanoparticles can generate more ROS compared to a larger size or bulk silver (Verano-Braga et al., 2014). Sahu et al. (2014, 2016a,b) found cytotoxicity and genotoxicity of silver nanoparticles on human liver cells and confirmed that nanoparticle of same morphology, composition, surface charge but different size, 20 and 50 nm had different toxicity levels, where smaller size showed more toxicity. Gold nanoparticles, extensively used in the biomedical application, are known to impact human embryonic stem cell depending on the size (Senut et al., 2016). The study found that stem cell colonies exposed to 1.5 nm diameter gold nanoparticles showed loss of cohesiveness, rounding up, and detachment suggesting cell death but other nanoparticles of larger size (4 nm and 24 nm) did not show toxic effects in human embryonic stem cells throughout the experiments, which lasted for 19 days (Senut et al., 2016). Titanium dioxide and silicon dioxide nanoparticles can cause cytotoxicity to 3T3 fibroblasts, RAW 264.7 macrophages, and telomerase-immortalized bronchiolar epithelial cells, which was found to depend mainly on the nanoparticle size along with cell type and composition (Baranowska-Wójcik et al., 2019; Sohaebuddin et al., 2010). These studies confirm that the size of the nanoparticle has an enormous impact on deciding their toxicology; even with similar composition, the shape can show varied toxicity dependent on the size.

Wang et al. (2013) reviewed the in vivo metabolism of nanoparticles and suggested that nanoparticles can enter the bloodstream directly during their application or indirectly through inhalation, ingestion, and dermal exposure. Nanoscale particle sizes permit these to be transported through the bloodstream to different parts of the body, and where they may affect lung, liver, and kidney function. It is still challenging to generalize how nanoparticle size range affects toxicity due to the lack of standardized toxicity and analytical methods. However, a common observation among these studies is that smaller size nanoparticles are more toxic in comparison to larger size particle.

17.5.1.2 Shape

The shape of nanoparticles can also be a contributing factor in promoting their toxicity. The nanoparticles are of different shapes such as spheres, ellipsoids, cylinders, sheets, cubes, and rods. The shape of nanoparticles, however, has a less pronounced effect on nanotoxicology compared to the size of nanoparticles, but when size and surface area is kept similar for a particular type of nanoparticle, then their shape becomes more significant in the assessment of toxicity (Sajid et al., 2015). Stoehr et al. (2011) studied the toxic effects of silver wires of length 1.5–25 μm; diameter 100–160 nm, spherical silver particles of 30 nm and <45 μm on alveolar epithelial cells and concluded that silver wires strongly affect the alveolar epithelial cells, whereas spherical silver particles had no effect. In another study, it was found single-walled carbon nanotubes more effective in blocking calcium channels compared to spherical fullerenes (Park et al., 2003). Zinc oxide nanorods have more toxicity toward human lung epithelial cells compared to spherical zinc oxide nanoparticles (Hsiao and Huang, 2011). Favi et al. (2015) compared the toxicity of gold nanospheres and gold nanostars in human skin fibroblast cells and found that the gold nanospheres (~61.46 nm) showed higher toxicity with fibroblast cells in comparison to nanostars (~33.69 nm) of smaller diameter. Further, exposure to nanospheres was fatal at concentrations of 40 μg/mL, and to nanostars at 400 μg/mL.

Nanohydroxyapatite biocompatibility makes it an efficient drug delivery vehicles, a study by Zhao et al. (2013) compared needle, plate, rod, and spherical shapes of these nanoparticles on cultured human bronchial epithelial cells and found that needle and plate-like nanoparticles are more lethal. The reasoning was the pointed end of these two shapes can damage cells on contact. Hu et al. (2011) found that graphene oxide nanosheets can damage of mammalian cells. Gold nanoparticles with the same functionalization showed increased toxicity for spherical compared to rod-shaped structures on epithelial cells (Tarantola et al., 2011). Nangia and Sureshkumar (2012) simulation-based study predicted that shape and charge of nanoparticles could enhance the translocation process by 60 times through cell membranes.

17.5.1.3 Chemical composition

Size and shape of nanoparticles can influence the toxicity of nanoparticle most, but chemical composition can also have a detrimental effect on human health (Sukhanova et al., 2018). Bulk material toxicity can also vary depending on the composition, and nanoparticles toxicity will depend on the chemical form and composition. Silver nanoparticles (nominally 35 and 600–1600 nm) were found to be relatively more toxic than cerium dioxide (nominally <25 nm and 1–5 μm) nanoparticles (Gaiser et al., 2012). Yang et al. (2009) compared silicon dioxide and zinc oxide of 20 nm size and found a different mechanism of toxicity; zinc oxide caused oxidative stress and silicon dioxide altered DNA structure. However, this study was done on mouse fibroblasts.

It is not surprising that the toxic effects of nanoparticles are determined by their chemical composition, as a material known to be toxic in bulk (silver, cadmium) may replicate that toxicity in nano-size, but the impacts may vary. As nanoparticles can degrade or solubilize in complex biological matrices (pH, ionic strength), it can release metal ions which can be toxic too, which is one of the most common causes of nanoparticle toxicity

(Sukhanova et al., 2018). However, the impact of chemical composition in toxicity can be reduced by utilizing proper coating with nontoxic materials on the surface of nanoparticle that can prevent degradation and subsequently metal-ion release or leaking (Soenen et al., 2015). The chemical composition can dictate their crystallinity which can also dictate the toxicity effects, for example, titanium oxide nanoparticle of the same size, but two different crystalline structure showed different toxic effects on human bronchial epithelium cell line (Gurr et al., 2005), where anatase was nontoxic, and rutile caused oxidative damage of DNA. Further, the crystallinity of nanoparticles is dependent external environmental factors too, which can be rearranged in different matrices.

17.5.1.4 Surface chemistry and charge

Nanoparticle surface charge and surface chemistry, controlled by capping agents, determine interactions with biological systems (Schaeublin et al., 2011). Engineered nanoparticles are usually functionalized and designed with different organic and inorganic moieties for various applications in a different field. Functionalization of nanoparticles induces surface charges and different surface chemistry. The surface can have a positive charge or negative charge, which can be modified by grafting different ligands or polymers and dictate their biological interactions (Sajid et al., 2015). Engineered nanoparticle adsorption on biomolecules like proteins is dependent on the interaction between the surface groups of nanoparticle and proteins amine group (Walkey et al., 2014). Toxicity, fate, and stability of silver nanoparticles were found to be dependent on the organic capping agents (Sarma et al., 2015). Goodman et al. (2004) compared gold nanoparticles of different functionalized chains and found that cationic chains showed moderate toxicity compared to anionic chains, which were nontoxic.

Generally, cationic-charged nanoparticles show higher toxicity as they can easily enter cells due to the electrostatic attraction as cell membrane has a negative surface charge. Moreover, positively charged nanoparticles can also bind firmly to negatively charged DNA. Comparison of the cytotoxicity of negatively and positively charged polystyrene nanoparticles on HeLa and NIH/3T3 cells has shown that the latter nanoparticles are more toxic (Liu et al., 2011) but negatively charged nanoparticles have almost no effect. Positively charged nanoparticles can also bind to proteins better and enhance phagocytosis (Alexis et al., 2008). Interestingly, the adsorbed proteins, known as protein crown, can also affect the surface properties of engineered nanoparticles. The protein crown can change the surface charge, which can change aggregation characteristics and hydrodynamic diameter. Engineered nanoparticles also bring about conformational changes in the protein, affecting proteins functional activities (Sukhanova et al., 2018). This may be the cause of various diseases (e.g., amyloidosis) (Sukhanova et al., 2012). Linse et al. (2007) showed that nanoparticles could initiate fibrils formation in human $\beta 2$ microglobulin. Asati et al. (2010) estimated the cytotoxicity induced by surface-modification of cerium-oxide nanoparticles in normal and cancer cell lines and found that positively charged and neutral nanoparticle rate of absorption was similar for all cell types, whereas negatively charge accumulated in cancer cells. Thus surface modification and properties of engineered nanoparticles can dictate toxicity and interaction on different cell type, which can localize and given directional property nanoparticles.

17.5.1.5 Reactivity, mobility, stability, and agglomeration

Reactivity of engineered nanoparticles is linked to the considerable surface-to-volume ratio due to the size factor. This high reactivity of nanoparticles can initiate different catalytic reaction inside cells and can further produce ROS; this can have a chain reaction and cause cytotoxicity (Stark, 2011). The presence of charge on nanoparticles, mainly positive charge, can enhance the reactivity of nanoparticles. ROS can bring about oxidative distress in the cell (Nel et al., 2006).

The mobility of engineered nanoparticles will control toxicity through probability for interaction with humans and the environment. High mobility and size result in exposure much more quickly, and mobile nanoparticles can easily diffuse through cells. Mobility can also determine the agglomeration of nanoparticles.

The stability of engineered nanoparticles inside a living being has two-fold implications. First, it can increase the residence time of nanoparticles inside the body, and second, it can amplify the toxic effect inside the body by prolonged residence time (Gupta and Xie, 2018). pH is an essential factor in determining the stability of nanoparticles, and in biological system pH, these nanoparticles can be soluble (Sajid et al., 2015). This solubility can bring adverse effect by releasing metal ions inside the cell and can generate ROS. Therefore solubility can be a key factor in determining the toxicity of a nanoparticle.

Agglomeration of nanoparticles can bring about change in their surface properties, in comparison to dispersed nanoparticles. This change in surface property can dictate how the nanoparticles interact with biological systems regulate their adsorption and uptake (Allouni et al., 2015). The cellular uptake can determine toxicity with higher uptake, toxicity will be more. Albanese and Chan (2011) observed a difference in uptake of gold nanoparticles with a decrease of 25% in aggregated nanoparticles for HeLa and A549 cells and doubling due to agglomeration in MDA-MB 435 cells.

The full-scale toxicity of natural and engineered nanomaterials is still ambiguous in the context of complex organisms (Hochella et al., 2019). The increasing disposal of engineered nanoparticles is relevant to identify how these nanosized particles may impact the quality of different water sources (Malakar et al., 2019). Most of the present studies have been mostly focused on nanoparticles exposure through respiration. Toxicity due to exposure via ingestion of nanoparticle contaminated water is still unclear, but few studies predict the presence of nanoparticles in drinking water may have less impact (Westerhoff et al., 2018). Several recent studies have confirmed that nanoparticles occur in drinking water (Venkatesan et al., 2018), and exposure may have unidentified impacts on human health. However, studying nanomaterials in the environment is still quite challenging due to the size regime and lack of reliable, cost-effective, reproducible, and easily accessible analytical tools. The risk evaluation of nanoparticles needs to be reconsidered concerning the unique challenges involved in scrutinizing biological pathways and measuring long-term impacts on human health.

17.6 Conclusion

In the modern world presence of nanoparticles is ubiquitous; the usage of nanoparticles is going to increase in the near future. In the present, there is a vast knowledge gap about the fate, transformation, and degradation of both natural and engineered nanoparticles in complex environmental setups. The risk assessment of nanoparticles impact can only be done after addressing these prevalent knowledge gaps. The complexity to study nanoparticles is multifold. The absence of reliable, cost-effective, simple analyzing technique adds to that complexity. The size is one of the key factors that impede the understanding of nanoscale degradation products.

The toxicological impact of natural, incidental, and engineered nanoparticles is far from understanding. Similar to the aqueous environmental system, biological systems are too complex to understand with present day techniques. Current research findings have shed some light on the different interactions of nanoparticles with cell, tissue, and organs. Different exposure pathways are being studied, which in turn can also dictate the toxicity of nanoparticles on humans. More work on this front is needed, where exposure or intake of nanoparticles from varied sources are also studied.

Acknowledgments

This work is based on research that was partially supported by the Nebraska Agricultural Experiment Station with funding from the Hatch Multistate Research (Accession Number 1011588) through the USDA National Institute of Food and Agriculture. Malakar thanks the Nebraska Environmental Trust and Hasting Utilities for salary support. The authors acknowledge Dr. Sudeshna Dutta for helping with Figs. 17.2 and 17.3.

References

Abbott Chalew, T.E., Ajmani, G.S., Huang, H., Schwab, K.J., 2013. Evaluating nanoparticle breakthrough during drinking water treatment. Environ. Health Perspect. 121, 1161–1166. Available from: https://doi.org/10.1289/ehp.1306574.

Akaighe, N., MacCuspie, R.I., Navarro, D.A., Aga, D.S., Banerjee, S., Sohn, M., et al., 2011. Humic acid-induced silver nanoparticle formation under environmentally relevant conditions. Environ. Sci. Technol. 45, 3895–3901. Available from: https://doi.org/10.1021/es103946g.

Albanese, A., Chan, W.C.W., 2011. Effect of gold nanoparticle aggregation on cell uptake and toxicity. ACS Nano 5, 5478–5489.

Alcacio, T.E., Hesterberg, D., Chou, J.W., Martin, J.D., Beauchemin, S., Sayers, D.E., 2001. Molecular scale characteristics of Cu(II) bonding in goethite-humate complexes. Geochim. Cosmochim. Acta 65, 5478–5489. Available from: https://doi.org/10.1016/S0016-7037(01)00546-4.

Alexis, F., Pridgen, E., Molnar, L.K., Farokhzad, O.C., 2008. Factors affecting the clearance and biodistribution of polymeric nanoparticles. Mol. Pharm. 5, 505–515.

Alimi, O.S., Farner Budarz, J., Hernandez, L.M., Tufenkji, N., 2018. Microplastics and nanoplastics in aquatic environments: aggregation, deposition, and enhanced contaminant transport. Environ. Sci. Technol. 52, 1704–1724.

Allouni, Z.E., Gjerdet, N.R., Cimpan, M.R., Høl, P.J., 2015. The effect of blood protein adsorption on cellular uptake of anatase TiO_2 nanoparticles. Int. J. Nanomed. 10, 687–695. Available from: https://doi.org/10.2147/IJN.S72726.

Álvarez-Puebla, R.A., Liz-Marzán, L.M., 2010. Environmental applications of plasmon assisted Raman scattering. Energy Environ. Sci. 3, 1011–1017. Available from: https://doi.org/10.1039/C002437F.

Andrade, L., O'Dwyer, J., O'Neill, E., Hynds, P., 2018. Surface water flooding, groundwater contamination, and enteric disease in developed countries: a scoping review of connections and consequences. Environ. Pollut. 236, 540–549. Available from: https://doi.org/10.1016/j.envpol.2018.01.104.

Andrieux, K., Nicolas, J., Moine, L., Barratt, G., 2013. Polymeric nanoparticles for drug delivery. In: Polymeric Biomaterials: Medicinal and Pharmaceutical Applications, vol. 2. CRC Press.

Asati, A., Santra, S., Kaittanis, C., Perez, J.M., 2010. Surface-charge-dependent cell localization and cytotoxicity of cerium oxide nanoparticles. ACS Nano 4, 5321–5331.

Baalousha, M., Stolpe, B., Lead, J.R., 2011. Flow field-flow fractionation for the analysis and characterization of natural colloids and manufactured nanoparticles in environmental systems: a critical review. J. Chromatogr. A. 1218, 4078–4103. Available from: https://doi.org/10.1016/j.chroma.2011.04.063.

Badireddy, A.R., Wiesner, M.R., Liu, J., 2012. Detection, characterization, and abundance of engineered nanoparticles in complex waters by hyperspectral imagery with enhanced darkfield microscopy. Environ. Sci. Technol. 46, 10081–10088. Available from: https://doi.org/10.1021/es204140s.

Bandyopadhyay, S., Peralta-Videa, J.R., Hernandez-Viezcas, J.A., Montes, M.O., Keller, A.A., Gardea-Torresdey, J. L., 2012. Microscopic and spectroscopic methods applied to the measurements of nanoparticles in the environment. Appl. Spectrosc. Rev. 47, 180–206.

Banfield, J.F., Zhang, H., 2001. Nanoparticles in the environment. Rev. Mineral. Geochem. 44, 1–58.

Baranowska-Wójcik, E., Szwajgier, D., Oleszczuk, P., Winiarska-Mieczan, A., 2019. Effects of titanium dioxide nanoparticles exposure on human health—a review. Biol. Trace Elem. Res. 1–12. Available from: https://doi. org/10.1007/s12011-019-01706-6.

Bardos, P., Merly, C., Kvapil, P., Koschitzky, H.P., 2018. Status of nanoremediation and its potential for future deployment: risk-benefit and benchmarking appraisals. Remediation 28, 43–56. Available from: https://doi. org/10.1002/rem.21559.

Bargar, J.R., Brown, G.E., Parks, G.A., 1997. Surface complexation of Pb(II) at oxide-water interfaces: I. XAFS and bond-valence determination of mononuclear and polynuclear Pb(II) sorption products on aluminum oxides. Geochim. Cosmochim. Acta 61, 2617–2637. Available from: https://doi.org/10.1016/S0016-7037(97)00124-5.

Bashar Bhuiyan, A., Mokhtar, M.B., Ekhwan Toriman, M., Barzani Gasim, M., Choo Ta, G., Elfithri, R., et al., 2013. The environmental risk and water pollution: a review from the river basins around the world. J. Sustain. Agric. 7, 126–136.

Baumann, T., Fruhstorfer, P., Nießner, R., 1998. Sickerwassertransport von Kolloiden und Schwermetallen in Bergbaufolgelandschaften - Felduntersuchungen und Laborversuche. Grundwasser. 3, 3–13.

Baumann, T., Fruhstorfer, P., Klein, T., Niessner, R., 2006. Colloid and heavy metal transport at landfill sites in direct contact with groundwater. Water Res. 40, 2776–2786. Available from: https://doi.org/10.1016/j. watres.2006.04.049.

Bednar, A.J., Poda, A.R., Mitrano, D.M., Kennedy, A.J., Gray, E.P., Ranville, J.F., et al., 2013. Comparison of on-line detectors for field flow fractionation analysis of nanomaterials. Talanta 104, 286–292. Available from: https://doi.org/10.1016/j.talanta.2012.11.008.

Beer, C., Foldbjerg, R., Hayashi, Y., Sutherland, D.S., Autrup, H., 2012. Toxicity of silver nanoparticles—nanoparticle or silver ion? Toxicol. Lett. 208, 286–292.

Berlo, K., van Hinsberg, V.J., Vigouroux, N., Gagnon, J.E., Williams-Jones, A.E., 2014. Sulfide breakdown controls metal signature in volcanic gas at Kawah Ijen volcano, Indonesia. Chem. Geol. 371, 115–127.

Blanco-Andujar, C., Ortega, D., Pankhurst, Q.A., Thanh, N.T.K., 2012. Elucidating the morphological and structural evolution of iron oxide nanoparticles formed by sodium carbonate in aqueous medium. J. Mater. Chem. 22, 12498–12506. Available from: https://doi.org/10.1039/C2JM31295F.

Bouwmeester, H., Lynch, I., Marvin, H.J.P., Dawson, K.A., Berges, M., Braguer, D., et al., 2011. Minimal analytical characterization of engineered nanomaterials needed for hazard assessment in biological matrices. Nanotoxicology 5, 1–11. Available from: https://doi.org/10.3109/17435391003775266.

Buffle, J., Wilkinson, K.J., Stoll, S., Filella, M., Zhang, J., 1998. A generalized description of aquatic colloidal interactions: the three-colloidal component approach. Environ. Sci. Technol. 32, 2887–2899. Available from: https://doi.org/10.1021/es980217h.

Bundschuh, M., Filser, J., Lüderwald, S., McKee, M.S., Metreveli, G., Schaumann, G.E., et al., 2018. Nanoparticles in the environment: where do we come from, where do we go to? Environ. Sci. Eur. 30, 6. Available from: https://doi.org/10.1186/s12302-018-0132-6.

Buzea, C., Pacheco, I.I., Robbie, K., 2007. Nanomaterials and nanoparticles: sources and toxicity. Biointerphases 2, MR17–MR71.

Chen, C., Chen, C., Herhold, A.B., Johnson, C.S., Alivisatos, A.P., 1997. Size dependence of structural metastability in semiconductor nanocrystals. Science 398, 398–401. Available from: https://doi.org/10.1126/science.276.5311.398.

Chen, Z., Wang, Y., Ba, T., Li, Y., Pu, J., Chen, T., et al., 2014. Genotoxic evaluation of titanium dioxide nanoparticles in vivo and in vitro. Toxicol. Lett. 226, 314–319.

Cheng, T., Saiers, J.E., 2010. Colloid-facilitated transport of cesium in vadose-zone sediments: the importance of flow transients. Environ. Sci. Technol. 44, 7443–7449. Available from: https://doi.org/10.1021/es100391j.

Chin, Y.P., Alken, G., O'Loughlin, E., 1994. Molecular weight, polydispersity, and spectroscopic properties of aquatic humic substances. Environ. Sci. Technol. 28, 1853–1858. Available from: https://doi.org/10.1021/es00060a015.

Choi, S., Johnston, M.V., Wang, G.S., Huang, C.P., 2017. Looking for engineered nanoparticles (ENPs) in wastewater treatment systems: qualification and quantification aspects. Sci. Total Environ. 509, 809–817. Available from: https://doi.org/10.1016/j.scitotenv.2017.03.061.

Crampon, M., Joulian, C., Ollivier, P., Charron, M., Hellal, J., 2019. Shift in natural groundwater bacterial community structure due to zero-valent iron nanoparticles (nZVI). Front. Microbiol. 10, 533. Available from: https://doi.org/10.3389/fmicb.2019.00533.

D'Alessio, M., Vasudevan, D., Lichwa, J., Mohanty, S.K., Ray, C., 2014. Fate and transport of selected estrogen compounds in Hawaii soils: effect of soil type and macropores. J. Contam. Hydrol. 166, 1–10.

Das, R.K., Pachapur, V.L., Lonappan, L., Naghdi, M., Pulicharla, R., Maiti, S., et al., 2017. Biological synthesis of metallic nanoparticles: plants, animals and microbial aspects. Nanotechnol. Environ. Eng. 2, 18.

Dawson, K.A., Salvati, A., Lynch, I., 2009. Nanoparticles reconstruct lipids. Nat. Nanotechnol. 4, 84–85.

Delay, M., Frimmel, F.H., 2012. Nanoparticles in aquatic systems. Anal. Bioanal. Chem. 402, 583–592.

Dodds, W.K., Whiles, M.R., 2010. Chapter 16—Responses to stress, toxic chemicals, and other pollutants in aquatic ecosystems. In: Dodds, W.K., Whiles, M.R. (Eds.), Freshwater Ecology. second ed. pp. 399–436.

Environmental Protection Agency Federal Facilities Restoration and Reuse Office, 2017. Technical Fact Sheet—Nanomaterials.

Erbs, J.J., Berquó, T.S., Reinsch, B.C., Lowry, G.V., Banerjee, S.K., Penn, R.L., 2010. Reductive dissolution of arsenic-bearing ferrihydrite. Geochim. Cosmochim. Acta 74, 3382–3395.

Ermolin, M.S., Fedotov, P.S., 2016. Separation and characterization of environmental nano- and submicron particles. Rev. Anal. Chem. 35, 185–199. Available from: https://doi.org/10.1515/revac-2016-0006.

Ermolin, M.S., Fedotov, P.S., Malik, N.A., Karandashev, V.K., 2018. Nanoparticles of volcanic ash as a carrier for toxic elements on the global scale. Chemosphere 200, 16–22. Available from: https://doi.org/10.1016/j.chemosphere.2018.02.089.

Fang, Q., Shi, X., Zhang, L., Wang, Q., Wang, X., Guo, Y., et al., 2015. Effect of titanium dioxide nanoparticles on the bioavailability, metabolism, and toxicity of pentachlorophenol in zebrafish larvae. J. Hazard. Mater. 283, 897–904. Available from: https://doi.org/10.1016/j.jhazmat.2014.10.039.

Faulstich, L., Griffin, S., Nasim, M.J., Masood, M.I., Ali, W., Alhamound, S., et al., 2017. Nature's Hat-trick: can we use sulfur springs as ecological source for materials with agricultural and medical applications? Int. Biodeterior. Biodegrad. 119, 678–686. Available from: https://doi.org/10.1016/j.ibiod.2016.08.020.

Favi, P.M., Gao, M., Johana Sepúlveda Arango, L., Ospina, S.P., Morales, M., Pavon, J.J., et al., 2015. Shape and surface effects on the cytotoxicity of nanoparticles: gold nanospheres versus gold nanostars. J. Biomed. Mater. Res., A. 103, 3449–3462. Available from: https://doi.org/10.1002/jbm.a.35491.

Frimmel, F.H., Niessner, R. (Eds.), 2010. Nanoparticles in the Water Cycle. Springer, Berlin, Heidelberg.

Gad, G.M.A., Hegazy, M.A., 2019. Optoelectronic properties of gold nanoparticles synthesized by using wet chemical method. Mater. Res. Express 6, 085024.

Gaiser, B.K., Fernandes, T.F., Jepson, M.A., Lead, J.R., Tyler, C.R., Baalousha, M., et al., 2012. Interspecies comparisons on the uptake and toxicity of silver and cerium dioxide nanoparticles. Environ. Toxicol. Chem. 31, 44–154. Available from: https://doi.org/10.1002/etc.703.

Ganguly, P., Breen, A., Pillai, S.C., 2018. Toxicity of nanomaterials: exposure, pathways, assessment, and recent advances. ACS Biomater. Sci. Eng. 4, 2237–2275.

Gao, Y., Jing, H., Du, M., Chen, W., 2018. Dispersion of multi-walled carbon nanotubes stabilized by humic acid in sustainable cement composites. Nanomaterials 8, 585. Available from: https://doi.org/10.3390/nano8100858.

Gehrke, I., Geiser, A., Somborn-Schulz, A., 2015. Innovations in nanotechnology for water treatment. Nanotechnol. Sci. Appl. 8, 1–17.

Gericke, M., Pinches, A., 2006. Microbial production of gold nanoparticles. Gold Bull. 39, 22–28.

Ghosh, I., Mukherjee, A., Mukherjee, A., 2017. In planta genotoxicity of nZVI: influence of colloida; stability on uptake, DNA damage, oxidative stress and cell death. Mutagenesis 32, 371–387.

Gleick, P.H., 1993. Water in Crisis: A Guide to the World's Fresh Water Resources, first ed Oxford University Press, New York, NY.

Gliga, A.R., Skoglund, S., Odnevall Wallinder, I., Fadeel, B., Karlsson, H.L., 2014. Size-dependent cytotoxicity of silver nanoparticles in human lung cells: the role of cellular uptake, agglomeration and Ag release. Part. Fibre Toxicol. 11, 11. Available from: https://doi.org/10.1186/1743-8977-11-11.

Goldberg, S., Criscenti, L.J., Turner, D.R., Davis, J.A., Cantrell, K.J., 2007. Adsorption–desorption processes in subsurface reactive transport modeling. Vadose Zone J. 6, 407–435. Available from: https://doi.org/10.2136/vzj2006.0085.

González-Gálvez, D., Janer, G., Vilar, G., Vílchez, A., Vázquez-Campos, S., 2017. The life cycle of engineered nanoparticles. In: Tran, L., Bañares, M., Rallo, R. (Eds.), Modelling the Toxicity of Nanoparticles. Advances in Experimental Medicine and Biology, vol. 947. Springer, Cham, 41–69.

Good, K.D., Bergman, L.E., Klara, S.S., Leitch, M.E., VanBriesen, J.M., 2016. Implications of engineered nanomaterials in drinking water sources. J. Am. Water Works Assoc. 108, E1–E17. Available from: https://doi.org/10.2217/17435889.2.6.919.

Goodman, C.M., McCusker, C.D., Yilmaz, T., Rotello, V.M., 2004. Toxicity of gold nanoparticles functionalized with cationic and anionic side chains. Bioconjug. Chem. 15, 897–900. Available from: https://doi.org/10.1021/bc049951i.

Griffin, S., Masood, M., Nasim, M., Sarfraz, M., Ebokaiwe, A., Schäfer, K.-H., et al., 2017. Natural nanoparticles: a particular matter inspired by nature. Antioxidants 7, 3. Available from: https://doi.org/10.3390/antiox7010003.

Gschwind, S., Hagendorfer, H., Frick, D.A., Günther, D., 2013. Mass quantification of nanoparticles by single droplet calibration using inductively coupled plasma mass spectrometry. Anal. Chem. 85, 5875–5883. Available from: https://doi.org/10.1021/ac400608c.

Guan, B., Lu, W., Fang, J., Cole, R.B., 2007. Characterization of synthesized titanium oxide nanoclusters by MALDI-TOF mass spectrometry. J. Am. Soc. Mass Spectrom. 18, 517–524.

Gunsolus, I.L., Mousavi, M.P.S., Hussein, K., Bühlmann, P., Haynes, C.L., 2015. Effects of humic and fulvic acids on silver nanoparticle stability, dissolution, and toxicity. Environ. Sci. Technol. 49, 8078–8086. Available from: https://doi.org/10.1021/acs.est.5b01496.

Guo, H., Barnard, A.S., 2013. Naturally occurring iron oxide nanoparticles: Morphology, surface chemistry and environmental stability. J. Mater. Chem. A 1, 27–42.

Guo, H., He, L., Xing, B., 2017. Applications of surface-enhanced Raman spectroscopy in the analysis of nanoparticles in the environment. Environ. Sci. Nano 4, 2093–2107. Available from: https://doi.org/10.1039/C7EN00653E.

Gupta, R., Xie, H., 2018. Nanoparticles in daily life: applications, toxicity and regulations. J. Environ. Pathol. Toxicol. Oncol. 37. Available from: https://doi.org/10.1615/JEnvironPatholToxicolOncol.2018026009.

Gurr, J.R., Wang, A.S.S., Chen, C.H., Jan, K.Y., 2005. Ultrafine titanium dioxide particles in the absence of photoactivation can induce oxidative damage to human bronchial epithelial cells. Toxicology. 213, 66–73. Available from: https://doi.org/10.1016/j.tox.2005.05.007.

Harris, P.J.F., 2009. Ultrathin graphitic structures and carbon nanotubes in a purified synthetic graphite. J. Phys. Condens. Matter. 21, 355009. Available from: https://doi.org/10.1088/0953-8984/21/35/355009.

Hartland, A., Lead, J.R., Slaveykova, V.I., O'Carroll, D., Valsami-Jones, E., 2013. The environmental significance of natural nanoparticles. Nat. Educ. Knowl. 4, 7.

Hochella, M.F., 2002. There's plenty of room at the bottom: nanoscience in geochemistry. Geochim. Cosmochim. Acta 66, 735–743. Available from: https://doi.org/10.1016/S0016-7037(01)00868-7.

Hochella, M.F., 2008. Nanogeoscience: from origin to cutting-edge applications. Elements 4, 373–379.

Hochella, M.F., Lower, S.K., Maurice, P.A., Penn, R.L., Sahai, N., Sparks, D.L., et al., 2008. Nanominerals, mineral nanoparticles, and earth systems. Science 319, 1631–1635.

Hochella, M.F., Spencer, M.G., Jones, K.L., 2015. Nanotechnology: nature's gift or scientists' brainchild? Environ. Sci. Nano 2, 114–119.

Hochella, M.F., Mogk, D.W., Ranville, J., Allen, I.C., Luther, G.W., Marr, L.C., et al., 2019. Natural, incidental, and engineered nanomaterials and their impacts on the Earth system. Science 363, eaau8299.

Hristovski, K., Westerhoff, P., Crittenden, J., 2008. An approach for evaluating nanomaterials for use as packed bed adsorber media: a case study of arsenate removal by titanate nanofibers. J. Hazard. Mater. 156, 604–611. Available from: https://doi.org/10.1016/j.jhazmat.2007.12.073.

Hsiao, I.L., Huang, Y.J., 2011. Effects of various physicochemical characteristics on the toxicities of ZnO and TiO$_2$ nanoparticles toward human lung epithelial cells. Sci. Total Environ. 409, 1219–1228.

Hu, W., Peng, C., Lv, M., Li, X., Zhang, Y., Chen, N., et al., 2011. Protein corona-mediated mitigation of cytotoxicity of graphene oxide. ACS Nano 5, 3693–3700. Available from: https://doi.org/10.1021/nn200021j.

Huang, Y.W., Cambre, M., Lee, H.J., 2017. The toxicity of nanoparticles depends on multiple molecular and physicochemical mechanisms. Int. J. Mol. Sci. 18, 2702. Available from: https://doi.org/10.3390/ijms18122702.

Inshakova, E., Inshakov, O., 2017. World market for nanomaterials: structure and trends. MATEC Web Conf. 129, 02013.

Jamison, D.T., Breman, J.G., Measham, A.R., Alleyne, G., Claeson, M., Evans, D.B., et al., (Eds.), 2006. Disease Control Priorities in Developing Countries. second ed. Oxford University Press, Washington, DC.

Jeevanandam, J., Barhoum, A., Chan, Y.S., Dufresne, A., Danquah, M.K., 2018. Review on nanoparticles and nanostructured materials: history, sources, toxicity and regulations. Beilstein J. Nanotechnol. 9, 1050–1074.

Jerez, J., 2003. Cation binding by humic substances. Vadose Zone J. 2, 442.

Jiang, D., Zeng, G., Huang, D., Chen, M., Zhang, C., Huang, C., et al., 2018a. Remediation of contaminated soils by enhanced nanoscale zero valent iron. Environ. Res. 163, 217–227. Available from: https://doi.org/10.1016/j.envres.2018.01.030.

Jiang, Z., Liu, Q., Roberts, A.P., Barrón, V., Torrent, J., Zhang, Q., 2018b. A new model for transformation of ferrihydrite to hematite in soils and sediments. Geology 46, 987–990.

Joshi, N., Filip, J., Coker, V.S., Sadhukhan, J., Safarik, I., Bagshaw, H., et al., 2018. Microbial reduction of natural Fe(III) minerals; toward the sustainable production of functional magnetic nanoparticles. Front. Environ. Sci. 6, 127.

Kaegi, R., Ulrich, A., Sinnet, B., Vonbank, R., Wichser, A., Zuleeg, S., et al., 2008. Synthetic TiO$_2$ nanoparticle emission from exterior facades into the aquatic environment. Environ. Pollut. 156, 233–239.

Kah, M., Kookana, R.S., Gogos, A., Bucheli, T.D., 2018. A critical evaluation of nanopesticides and nanofertilizers against their conventional analogues. Nat. Nanotechnol. 13, 677–684.

Keller, A.A., McFerran, S., Lazareva, A., Suh, S., 2013. Global life cycle releases of engineered nanomaterials. J. Nanopart. Res. 15, 1692. Available from: https://doi.org/10.1007/s11051-013-1692-4.

Khajeh, M., Laurent, S., Dastafkan, K., 2013. Nanoadsorbents: classification, preparation, and applications (with emphasis on aqueous media). Chem. Rev. 113, 7728–7768. Available from: https://doi.org/10.1021/cr400086v.

Khan, I., Saeed, K., Khan, I., 2017. Nanoparticles: properties, applications and toxicities. Arab. J. Chem. Available from: https://doi.org/10.1016/j.arabjc.2017.05.011.

Khurana, S., Bedi, P.M.S., Jain, N.K., 2013. Preparation and evaluation of solid lipid nanoparticles based nanogel for dermal delivery of meloxicam. Chem. Phys. Lipids 65, 65–72. Available from: https://doi.org/10.1016/j.chemphyslip.2013.07.010.

Kiser, M.A., Westerhoff, P., Benn, T., Wang, Y., Pérez-Rivera, J., Hristovski, K., 2009. Titanium nanomaterial removal and release from wastewater treatment plants. Environ. Sci. Technol. 43, 6757–6763. Available from: https://doi.org/10.1021/es901102n.

Korin, N., Kanapathipillai, M., Ingber, D.E., 2013. Shear-responsive platelet mimetics for targeted drug delivery. Isr. J. Chem. 53, 610–615. Available from: https://doi.org/10.1002/ijch.201300052.

Kretzschmar, R., Sticher, H., 1998. Colloid transport in natural porous media: influence of surface chemistry and flow velocity. Phys. Chem. Earth 23, 133–139. Available from: https://doi.org/10.1016/S0079-1946(98)00003-2.

Laws, E.A., 2017. Aquatic Pollution: An Introductory Text. John Wiley & Sons.

Lazareva, A., Keller, A.A., 2014. Estimating potential life cycle releases of engineered nanomaterials from wastewater treatment plants. ACS Sustain. Chem. Eng. 2, 1656–1665.

Lead, J.R., Wilkinson, K.J., 2006. Aquatic colloids and nanoparticles: current knowledge and future trends. Environ. Chem. 3, 159–171. Available from: https://doi.org/10.1071/EN06025.

Lehner, R., Weder, C., Petri-Fink, A., Rothen-Rutishauser, B., 2019. Emergence of Nanoplastic in the Environment and Possible Impact on Human Health, Environ. Sci. Technol. 53, 1748–1765.

Li, L., Hartmann, G., Döblinger, M., Schuster, M., 2013. Quantification of nanoscale silver particles removal and release from municipal wastewater treatment plants in Germany. Environ. Sci. Technol. 47, 7317–7323. Available from: https://doi.org/10.1021/es3041658.

Li, M., Li, D.W., Xiu, G., Long, Y.T., 2017. Applications of screen-printed electrodes in current environmental analysis. Curr. Opin. Electrochem. 3, 137–143. Available from: https://doi.org/10.1016/j.coelec.2017.08.016.

Lindenthal, A., Langmann, B., Pätsch, J., Lorkowski, I., Hort, M., 2013. The ocean response to volcanic iron fertilisation after the eruption of Kasatochi volcano: a regional-scale biogeochemical ocean model study. Biogeosciences 10, 3715–3729. Available from: https://doi.org/10.5194/bgd-9-9233-2012.

Linse, S., Cabaleiro-Lago, C., Xue, W.-F., Lynch, I., Lindman, S., Thulin, E., et al., 2007. Nucleation of protein fibrillation by nanoparticles. Proc. Natl. Acad. Sci. U.S.A. 104, 8691–8696. Available from: https://doi.org/10.1073/pnas.0701250104.

Liou, P., Nguyen, T.H.D., Lin, M., 2018. Measurement of engineered nanoparticles in consumer products by surface-enhanced Raman spectroscopy and neutron activation analysis. J. Food Meas. Charact. 12, 736–746. Available from: https://doi.org/10.1007/s11694-017-9687-y.

Liu, Y., Li, W., Lao, F., Liu, Y., Wang, L., Bai, R., et al., 2011. Intracellular dynamics of cationic and anionic polystyrene nanoparticles without direct interaction with mitotic spindle and chromosomes. Biomaterials 32, 8291–8303. Available from: https://doi.org/10.1016/j.biomaterials.2011.07.037.

Liu, Y., Li, S., Chen, Z., Megharaj, M., Naidu, R., 2014. Influence of zero-valent iron nanoparticles on nitrate removal by *Paracoccus* sp. Chemosphere 108, 426–432. Available from: https://doi.org/10.1016/j.chemosphere.2014.02.045.

Ma, Z., Yin, X., Ji, X., Yue, J.Q., Zhang, L., Qin, J.J., et al., 2016. Evaluation and removal of emerging nanoparticle contaminants in water treatment: a review. Desalin. Water Treat. 57, 11221–11232.

Madl, A.K., Pinkerton, K.E., 2009. Health effects of inhaled engineered and incidental nanoparticles. Crit. Rev. Toxicol. 39, 629–658.

Maher, B.A., Ahmed, I.A.M., Karloukovski, V., MacLaren, D.A., Foulds, P.G., Allsop, D., et al., 2016. Magnetite pollution nanoparticles in the human brain. Proc. Natl. Acad. Sci. U.S.A. 113, 10797–10801. Available from: https://doi.org/10.1073/pnas.1605941113.

Malakar, A., 2016. Using Nanotechnology and Membrane Engineering for Effective Desalination of Water. University of Calcutta, Indian Association for the Cultivation of Science.

Malakar, A., Das, B., Islam, S., Meneghini, C., De Giudici, G., Merlini, M., et al., 2016a. Efficient artificial mineralization route to decontaminate Arsenic(III) polluted water—the Tooeleite Way. Sci. Rep. 6, 26031.

Malakar, A., Islam, S., Ali, M.A., Ray, S., 2016b. Rapid decadal evolution in the groundwater arsenic content of Kolkata, India and its correlation with the practices of her dwellers. Environ. Monit. Assess. 188. Available from: https://doi.org/10.1007/s10661-016-5592-9.

Malakar, A., Snow, D.D., Ray, C., 2019. Irrigation water quality—a contemporary perspective. Water 11, 1482.

Mantecca, P., Kasemets, K., Deokar, A., Perelshtein, I., Gedanken, A., Bahk, Y.K., et al., 2017. Airborne nanoparticle release and toxicological risk from metal-oxide-coated textiles: toward a multiscale safe-by-design approach. Environ. Sci. Technol. 51, 9305–9317.

Martín-Yerga, D., 2019. Electrochemical detection and characterization of nanoparticles with printed devices. Biosensors 9, 47.

Maters, E.C., Delmelle, P., Bonneville, S., 2016. Atmospheric processing of volcanic glass: effects on iron solubility and redox speciation. Environ. Sci. Technol. 50, 5033–5040. Available from: https://doi.org/10.1021/acs.est.5b06281.

Mkandawire, M., Dudel, E.G., 2008. Natural occurring uranium nanoparticles and the implication in bioremediation of surface mine waters. In: Uranium, Mining and Hydrogeology. Springer, Berlin, Heidelberg, pp. 487–496. Available from: https://doi.org/10.1007/978-3-540-87746-2_60.

Moreno-Vega, A.I., Gómez-Quintero, T., Nuñez-Anita, R.E., Acosta-Torres, L.S., Castaño, V., 2012. Polymeric and ceramic nanoparticles in biomedical applications. J. Nanotechnol. Available from: http://dx.doi.org/10.1155/2012/936041.

Mueller, N.C., Nowack, B., 2008. Exposure modeling of engineered nanoparticles in the environment. Environ. Sci. Technol. 42, 4447–4453.

Nangia, S., Sureshkumar, R., 2012. Effects of nanoparticle charge and shape anisotropy on translocation through cell membranes. Langmuir 28, 17666–17671. Available from: https://doi.org/10.1021/la303449d.

Narayanan, K.B., Sakthivel, N., 2010. Biological synthesis of metal nanoparticles by microbes. Adv. Colloid Interface Sci. 156, 1–13. Available from: https://doi.org/10.1016/j.cis.2010.02.001.

Neal, C., Jarvie, H., Rowland, P., Lawler, A., Sleep, D., Scholefield, P., 2011. Titanium in UK rural, agricultural and urban/industrial rivers: geogenic and anthropogenic colloidal/sub-colloidal sources and the significance of within-river retention. Sci. Total Environ. 409, 1843–1853. Available from: https://doi.org/10.1016/j.scitotenv.2010.12.021.

Nel, A., Xia, T., Mädler, L., Li, N., 2006. Toxic potential of materials at the nanolevel. Science 311, 622–627. Available from: https://doi.org/10.1126/science.1114397.

Ning, H., Zhou, Y., Zhou, Z., Cheng, S., Huang, R., 2017. Challenges to improving occupational health in China. Occup. Environ. Med. 74, 924–925. Available from: https://doi.org/10.1136/oemed-2017-104656.

Oberdörster, G., Oberdörster, E., Oberdörster, J., 2005. Nanotoxicology: an emerging discipline evolving from studies of ultrafine particles. Environ. Health Perspect. 113, 823–839. Available from: https://doi.org/10.1289/ehp.7339.

Oberdörster, G., Elder, A., Rinderknecht, A., 2009. Nanoparticles and the brain: cause for concern? J. Nanosci. Nanotechnol. 9, 4996–5007.

O'Carroll, D., Sleep, B., Krol, M., Boparai, H., Kocur, C., 2013. Nanoscale zero valent iron and bimetallic particles for contaminated site remediation. Adv. Water Resour. 51, 104–122. Available from: https://doi.org/10.1016/j.advwatres.2012.02.005.

Olgun, N., Duggen, S., Andronico, D., Kutterolf, S., Croot, P.L., Giammanco, S., et al., 2013. Possible impacts of volcanic ash emissions of Mount Etna on the primary productivity in the oligotrophic Mediterranean Sea: results from nutrient-release experiments in seawater. Mar. Chem. 152, 32–42. Available from: https://doi.org/10.1016/j.marchem.2013.04.004.

Park, K.H., Chhowalla, M., Iqbal, Z., Sesti, F., 2003. Single-walled carbon nanotubes are a new class of ion channel blockers. J. Biol. Chem. 278, 50212–50216. Available from: https://doi.org/10.1074/jbc.M310216200.

Peters, R.J.B., van Bemmel, G., Milani, N.B.L., den Hertog, G.C.T., Undas, A.K., van der Lee, M., et al., 2018. Detection of nanoparticles in Dutch surface waters. Sci. Total Environ. 621, 210–218. Available from: https://doi.org/10.1016/j.scitotenv.2017.11.238.

Pornwilard, M.M., Siripinyanond, A., 2014. Field-flow fractionation with inductively coupled plasma mass spectrometry: past, present, and future. J. Anal. At. Spectrom. 29, 1739–1752.

Pradeep, T., Anshup, 2009. Noble metal nanoparticles for water purification: a critical review. Thin Solid Films 517, 6441–6478. Available from: https://doi.org/10.1016/j.tsf.2009.03.195.

Praetorius, A., Scheringer, M., Hungerbühler, K., 2012. Development of environmental fate models for engineered nanoparticles—a case study of TiO_2 nanoparticles in the Rhine River. Environ. Sci. Technol. 46, 6705–6713.

Prichard, H.M., Fisher, P.C., 2012. Identification of platinum and palladium particles emitted from vehicles and dispersed into the surface environment. Environ. Sci. Technol. 46, 3149–3154. Available from: https://doi.org/10.1021/es203666h.

Rahmatpour, S., Mosaddeghi, M.R., Shirvani, M., Šimůnek, J., 2018. Transport of silver nanoparticles in intact columns of calcareous soils: the role of flow conditions and soil texture. Geoderma 322, 89–100. Available from: https://doi.org/10.1016/j.geoderma.2018.02.016.

Raiswell, R., Benning, L.G., Tranter, M., Tulaczyk, S., 2008. Bioavailable iron in the Southern Ocean: The significance of the iceberg conveyor belt. Geochem. Trans. 9, 7. Available from: https://doi.org/10.1186/1467-4866-9-7.

Reinsch, B.C., Levard, C., Li, Z., Ma, R., Wise, A., Gregory, K.B., et al., 2012. Sulfidation of silver nanoparticles decreases *Escherichia coli* growth inhibition. Environ. Sci. Technol. 46, 6992–7000. Available from: https://doi.org/10.1021/es203732x.

Rizzo, L., Malato, S., Antakyali, D., Beretsou, V.G., Đolić, M.B., Gernjak, W., et al., 2019. Consolidated vs new advanced treatment methods for the removal of contaminants of emerging concern from urban wastewater. Sci. Total Environ. 655, 986–1008.

Ross, N., 2008. World water quality facts and statistics. Annu. Water Rev. 35, 109.

Sahu, S.C., Zheng, J., Graham, L., Chen, L., Ihrie, J., Yourick, J.J., et al., 2014. Comparative cytotoxicity of nanosilver in human liver HepG2 and colon Caco2 cells in culture. J. Appl. Toxicol. 34, 1155–1166. Available from: https://doi.org/10.1002/jat.2994.

Sahu, S.C., Njoroge, J., Bryce, S.M., Zheng, J., Ihrie, J., 2016a. Flow cytometric evaluation of the contribution of ionic silver to genotoxic potential of nanosilver in human liver HepG2 and colon Caco2 cells. J. Appl. Toxicol. 36, 521–531. Available from: https://doi.org/10.1002/jat.3276.

Sahu, S.C., Roy, S., Zheng, J., Ihrie, J., 2016b. Contribution of ionic silver to genotoxic potential of nanosilver in human liver HepG2 and colon Caco2 cells evaluated by the cytokinesis-block micronucleus assay. J. Appl. Toxicol. 36, 532–542. Available from: https://doi.org/10.1002/jat.3279.

Sajid, M., Ilyas, M., Basheer, C., Tariq, M., Daud, M., Baig, N., et al., 2015. Impact of nanoparticles on human and environment: review of toxicity factors, exposures, control strategies, and future prospects. Environ. Sci. Pollut. Res. 22, 4122–4143.

Sarma, S.J., Bhattacharya, I., Brar, S.K., Tyagi, R.D., Surampalli, R.Y., 2015. Carbon nanotube—bioaccumulation and recent advances in environmental monitoring. Crit. Rev. Environ. Sci. Technol. 45, 905–938.

Sauvé, S., Desrosiers, M., 2014. A review of what is an emerging contaminant. Chem. Cent. J. 8, 15.

Schaeublin, N.M., Braydich-Stolle, L.K., Schrand, A.M., Miller, J.M., Hutchison, J., Schlager, J.J., et al., 2011. Surface charge of gold nanoparticles mediates mechanism of toxicity. Nanoscale. 3, 410–420.

Scheurer, M., Storck, F.R., Brauch, H.J., Lange, F.T., 2010. Performance of conventional multi-barrier drinking water treatment plants for the removal of four artificial sweeteners. Water Res. 44, 3573–3584. Available from: https://doi.org/10.1016/j.watres.2010.04.005.

Schüler, D., Frankel, R.B., 1999. Bacterial magnetosomes: microbiology, biomineralization and biotechnological applications. Appl. Microbiol. Biotechnol. 52, 464–473. Available from: https://doi.org/10.1007/s002530051547.

Schwertmann, U., 1991. Solubility and dissolution of iron oxides. Plant Soil 130, 1–25. Available from: https://doi.org/10.1007/BF00011851.

Senesi, N., 2010. Humic substances as natural nanoparticles ubiquitous in the environment. In: Molecular Environmental Soil Science at the Interfaces in the Earth's Critical Zone. Springer, Berlin, Heidelberg, pp. 249–250.

Senut, M.C., Zhang, Y., Liu, F., Sen, A., Ruden, D.M., Mao, G., 2016. Size-dependent toxicity of gold nanoparticles on human embryonic stem cells and their neural derivatives. Small 12, 631–646. Available from: https://doi.org/10.1002/smll.201502346.

Shang, J., Flury, M., Chen, G., Zhuang, J., 2008. Impact of flow rate, water content, and capillary forces on in situ colloid mobilization during infiltration in unsaturated sediments. Water Resour. Res. 44. Available from: https://doi.org/10.1029/2007WR006516.

Sharma, V.K., Filip, J., Zboril, R., Varma, R.S., 2015. Natural inorganic nanoparticles-formation, fate, and toxicity in the environment. Chem. Soc. Rev. 44, 8410–8423. Available from: https://doi.org/10.1039/c5cs00236b.

Shiklomanov, I., 1993. World fresh water resources. In: Water in Crisis a Guide to the World's Fresh Water Resources, Oxford University Press, New York, NY, pp. 7–24.

Simeonidis, K., Mourdikoudis, S., Kaprara, E., Mitrakas, M., Polavarapu, L., 2016. Inorganic engineered nanoparticles in drinking water treatment: a critical review. Environ. Sci. Water Res. Technol. 2, 43–70.

Singh, A.K., 2015. Engineered Nanoparticles: Structure, Properties and Mechanisms of Toxicity. Academic Press.

Singh, A.K., 2016. Structure, synthesis, and application of nanoparticles. In: Engineered Nanoparticles-Structure, Properties and Mechanisms of Toxicity. Academic Press, pp. 19–76. Available from: https://doi.org/10.1016/C2013-0-18974-X.

Slaveykova, V.I., Wilkinson, K.J., 2005. Predicting the bioavailability of metals and metal complexes: Critical review of the biotic ligand model. Environ. Chem. 2, 9–24. Available from: https://doi.org/10.1016/B978-0-12-801406-6.00002-9.

Soenen, S.J., Parak, W.J., Rejman, J., Manshian, B., 2015. (Intra)cellular stability of inorganic nanoparticles: Effects on cytotoxicity, particle functionality, and biomedical applications. Chem. Rev. 115, 2109–2135. Available from: https://doi.org/10.1021/cr400714j.

Sohaebuddin, S.K., Thevenot, P.T., Baker, D., Eaton, J.W., Tang, L., 2010. Nanomaterial cytotoxicity is composition, size, and cell type dependent. Part. Fibre Toxicol. 7, 22. Available from: https://doi.org/10.1186/1743-8977-7-22.

Srivastava, S.K., Ogino, C., Kondo, A., 2015. Nanoparticle synthesis by biogenic approach. In: Green Processes for Nanotechnology. Springer, Cham, pp. 237–257.

Stankus, D.P., Lohse, S.E., Hutchison, J.E., Nason, J.A., 2011. Interactions between natural organic matter and gold nanoparticles stabilized with different organic capping agents. Environ. Sci. Technol. 45, 3238–3244. Available from: https://doi.org/10.1021/es102603p.

Stark, W.J., 2011. Nanoparticles in biological systems. Angew. Chem. Int. Ed. 50, 1242–1258. Available from: https://doi.org/10.1002/anie.200906684.

Stefaniuk, M., Oleszczuk, P., Ok, Y.S., 2016. Review on nano zerovalent iron (nZVI): From synthesis to environmental applications. Chem. Eng. J. 287, 618–632. Available from: https://doi.org/10.1016/j.cej.2015.11.046.

Steinmetz, N.F., 2010. Viral nanoparticles as platforms for next-generation therapeutics and imaging devices. Nanomedicine 6, 634–641.

Stern, S.T., McNeil, S.E., 2008. Nanotechnology safety concerns revisited. Toxicol. Sci. 101, 4–21. Available from: https://doi.org/10.1093/toxsci/kfm169.

Stoehr, L.C., Gonzalez, E., Stampfl, A., Casals, E., Duschl, A., Puntes, V., et al., 2011. Shape matters: effects of silver nanospheres and wires on human alveolar epithelial cells. Part. Fibre Toxicol. 8, 36. Available from: https://doi.org/10.1186/1743-8977-8-36.

Sukhanova, A., Poly, S., Shemetov, A., Nabiev, I.R., 2012. Quantum dots induce charge-specific amyloid-like fibrillation of insulin at physiological conditions. In: Proceedings Volume 8548, Nanosystems in Engineering and Medicine. 85485F. International Society for Optics and Photonics. https://doi.org/10.1117/12.946606.

Sukhanova, A., Bozrova, S., Sokolov, P., Berestovoy, M., Karaulov, A., Nabiev, I., 2018. Dependence of nanoparticle toxicity on their physical and chemical properties. Nanoscale Res. Lett. 13, 44.

Sun, T.Y., Bornhöft, N.A., Hungerbühler, K., Nowack, B., 2016. Dynamic probabilistic modeling of environmental emissions of engineered nanomaterials. Environ. Sci. Technol. 50, 4701–4711. Available from: https://doi.org/10.1021/acs.est.5b05828.

Sun, T.Y., Mitrano, D.M., Bornhöft, N.A., Scheringer, M., Hungerbühler, K., Nowack, B., 2017. Envisioning nano release dynamics in a changing world: using dynamic probabilistic modeling to assess future environmental emissions of engineered nanomaterials. Environ. Sci. Technol. 51, 2854–2863.

Suresh, S., 2013. Semiconductor nanomaterials, methods and applications: a review. Nanosci. Nanotechnol. 3, 62–74. Available from: https://doi.org/10.5923/j.nn.20130303.06.

Suzuki, Y., Kelly, S.D., Kemnert, K.M., Banfield, J.F., 2002. Nanometre-size products of uranium bioreduction. Nature 419, 3849.

Sykes, E.A., Chen, J., Zheng, G., Chan, W.C.W., 2014. Investigating the impact of nanoparticle size on active and passive tumor targeting efficiency. ACS Nano 8, 5696–5706.

Tarantola, M., Pietuch, A., Schneider, D., Rother, J., Sunnick, E., Rosman, C., et al., 2011. Toxicity of gold-nanoparticles: synergistic effects of shape and surface functionalization on micromotility of epithelial cells. Nanotoxicology 5, 254–268. Available from: https://doi.org/10.3109/17435390.2010.528847.

Thomas, S., Harshita, B.S.P., Mishra, P., Talegaonkar, S., 2015. Ceramic nanoparticles: fabrication methods and applications in drug delivery. Curr. Pharm. Des. 21, 6165–6188. Available from: https://doi.org/10.2174/1381612821666151027153246.

Thomé, A., Reddy, K.R., Reginatto, C., Cecchin, I., 2015. Review of nanotechnology for soil and groundwater remediation: Brazilian perspectives. Water Air Soil Pollut. 226, 121.

Troester, M., Brauch, H.-J., Hofmann, T., 2016. Vulnerability of drinking water supplies to engineered nanoparticles. Water Res. 96, 255–279.

Tumolva, L., Park, J.Y., Kim, J.S., Miller, A.L., Chow, J.C., Watson, J.G., et al., 2010. Morphological and elemental classification of freshly emitted soot particles and atmospheric ultrafine particles using the TEM/EDS. Aerosol Sci. Technol. 44, 202–215. Available from: https://doi.org/10.1080/02786820903518907.

Vaghasiya, H., Kumar, A., Sawant, K., 2013. Development of solid lipid nanoparticles based controlled release system for topical delivery of terbinafine hydrochloride. Eur. J. Pharm. Sci. 49, 311–322. Available from: https://doi.org/10.1016/j.ejps.2013.03.013.

Venkatesan, A.K., Rodríguez, B.T., Marcotte, A.R., Bi, X., Schoepf, J., Ranville, J.F., et al., 2018. Using single-particle ICP-MS for monitoring metal-containing particles in tap water. Environ. Sci. Water Res. Technol. 4, 1923–1932. Available from: https://doi.org/10.1039/C8EW00478A.

Verano-Braga, T., Miethling-Graff, R., Wojdyla, K., Rogowska-Wrzesinska, A., Brewer, J.R., Erdmann, H., et al., 2014. Insights into the cellular response triggered by silver nanoparticles using quantitative proteomics. ACS Nano 8, 2161–2175.

Vodyanitskii, Y.N., Shoba, S.A., 2016. Ferrihydrite in soils. Eurasian Soil Sci. 49, 796–806.

Voegelin, A., Senn, A.-C., Kaegi, R., Hug, S.J., 2019. Reductive dissolution of As(V)-bearing Fe(III)-precipitates formed by Fe(II) oxidation in aqueous solutions. Geochem. Trans. 20, 2.

von der Kammer, F., Ferguson, P.L., Holden, P.A., Masion, A., Rogers, K.R., Klaine, S.J., et al., 2012. Analysis of engineered nanomaterials in complex matrices (environment and biota): general considerations and conceptual case studies. Environ. Toxicol. Chem. 31, 32–49. Available from: https://doi.org/10.1002/etc.723.

Vriens, B., Voegelin, A., Hug, S.J., Kaegi, R., Winkel, L.H.E., Buser, A.M., et al., 2017. Quantification of element fluxes in wastewaters: a nationwide survey in Switzerland. Environ. Sci. Technol. 51, 10943–10953.

Wagner, S., Gondikas, A., Neubauer, E., Hofmann, T., Von Der Kammer, F., 2014. Spot the difference: Engineered and natural nanoparticles in the environment-release, behavior, and fate. Angew. Chem. Int. Ed. 53, 12398−12419.

Walkey, C.D., Olsen, J.B., Song, F., Liu, R., Guo, H., Olsen, D.W.H., et al., 2014. Protein corona fingerprinting predicts the cellular interaction of gold and silver nanoparticles. ACS Nano 8, 2439−2455.

Wang, Y., 2014. Nanogeochemistry: nanostructures, emergent properties and their control on geochemical reactions and mass transfers. Chem. Geol. 378, 1−23.

Wang, Y., Li, Y., Costanza, J., Abriola, L.M., Pennell, K.D., 2012. Enhanced mobility of fullerene (C60) nanoparticles in the presence of stabilizing agents. Environ. Sci. Technol. 46, 11761−11769. Available from: https://doi.org/10.1021/es302541g.

Wang, B., He, X., Zhang, Z., Zhao, Y., Feng, W., 2013. Metabolism of nanomaterials in vivo: blood circulation and organ clearance. Acc. Chem. Res. 46, 761−769.

Westerhoff, P.K., Kiser, M.A., Hristovski, K., 2013. Nanomaterial removal and transformation during biological wastewater treatment. Environ. Eng. Sci. 30, 109−117. Available from: https://doi.org/10.1089/ees.2012.0340.

Westerhoff, P., Atkinson, A., Fortner, J., Wong, M.S., Zimmerman, J., Gardea-Torresdey, J., et al., 2018. Low risk posed by engineered and incidental nanoparticles in drinking water. Nat. Nanotechnol. 13, 661. Available from: https://doi.org/10.1038/s41565-018-0217-9.

Wille, G., Hellal, J., Ollivier, P., Richard, A., Burel, A., Jolly, L., et al., 2017. Cryo-scanning electron microscopy (SEM) and scanning transmission electron microscopy (STEM)-in-SEM for bio-and organo-mineral interface characterization in the environment. Microsc. Microanal. 23, 1159−1172. Available from: https://doi.org/10.1017/S143192761701265X.

Wu, J., Yao, J., Cai, Y., 2012. Biomineralization of natural nanomaterials. In: Nature's Nanostructures. Pan Stanford Publishing Pte. Ltd., Singapore, pp. 225−247.

Xue, W., Huang, D., Zeng, G., Wan, J., Cheng, M., Zhang, C., et al., 2018. Performance and toxicity assessment of nanoscale zero valent iron particles in the remediation of contaminated soil: a review. Chemosphere 210, 1145−1156. Available from: https://doi.org/10.1016/j.chemosphere.2018.07.118.

Yang, H., Liu, C., Yang, D., Zhang, H., Xi, Z., 2009. Comparative study of cytotoxicity, oxidative stress and genotoxicity induced by four typical nanomaterials: the role of particle size, shape and composition. J. Appl. Toxicol. 29, 69−78. Available from: https://doi.org/10.1002/jat.1385.

Yin, K., Lo, I.M.C., Dong, H., Rao, P., Mak, M.S.H., 2012. Lab-scale simulation of the fate and transport of nano zero-valent iron in subsurface environments: aggregation, sedimentation, and contaminant desorption. J. Hazard. Mater. 227, 118−125. Available from: https://doi.org/10.1016/j.jhazmat.2012.05.019.

Yin, Y., Yu, S., Liu, J., Jiang, G., 2014. Thermal and photoinduced reduction of ionic Au(III) to elemental Au nanoparticles by dissolved organic matter in water: Possible source of naturally occurring Au nanoparticles. Environ. Sci. Technol. 48, 2671−2679. Available from: https://doi.org/10.1021/es404195r.

Zhang, X.-X., Zhang, T., Fang, H.H.P., 2009. Antibiotic resistance genes in water environment. Appl. Microbiol. Biotechnol. 82, 397−414.

Zhang, D., Wang, S., Wang, Y., Gomez, M.A., Duan, Y., Jia, Y., 2018. The transformation of two-line ferrihydrite into crystalline products: effect of pH and Media (sulfate versus nitrate). ACS Earth Space Chem. 2, 577−587.

Zhang, M., Yang, J., Cai, Z., Feng, Y., Wang, Y., Zhang, D., et al., 2019. Detection of engineered nanoparticles in aquatic environments: current status and challenges in enrichment, separation, and analysis. Environ. Sci. Nano 6, 709−735. Available from: https://doi.org/10.1039/C8EN01086B.

Zhao, X., Ng, S., Heng, B.C., Guo, J., Ma, L., Tan, T.T.Y., et al., 2013. Cytotoxicity of hydroxyapatite nanoparticles is shape and cell dependent. Arch. Toxicol. 87, 1037−1052. Available from: https://doi.org/10.1007/s00204-012-0827-1.

Zhao, B., Yang, T., Zhang, Z., Hickey, M.E., He, L., 2018. A triple functional approach to simultaneously determine the type, concentration, and size of titanium dioxide particles. Environ. Sci. Technol. 52, 2863−2869. Available from: https://doi.org/10.1021/acs.est.7b05403.

Zou, Y., Wang, X., Khan, A., Wang, P., Liu, Y., Alsaedi, A., et al., 2016. Environmental remediation and application of nanoscale zero-valent iron and its composites for the removal of heavy metal ions: a review. Environ. Sci. Technol. 50, 7290−7304.

Water pollutants monitoring based on Internet of Things

Bhagavan Nvs[1] and P.L. Saranya[2]

[1]Department of Physics, Govt Degree College (Men), Srikakulam, India [2]Department of Physics, Govt College for Women (Autonomous), Srikakulam, India

OUTLINE

18.1 Introduction: artificial intelligence and Internet of Things

Artificial intelligence (AI) is the intelligence exhibited by machines, which mimic humans. The major goals of AI are analytical, human-inspired, and humanized AI. Analytical AI is to incorporate cognitive intelligence to machines, which means understanding the previous experiences and applying that knowledge in future decisions. Human-inspired AI is conglomeration of two ideas, which are cognitive and emotional intelligence; besides cognitive intelligence, interpretation of human emotions and contemplating them in their decisions. Humanized AI shows features of all types of adroitness such as cognitive, emotional, and social intelligence, is able to be self-conscious and self-aware in interactions with others.

The field of AI research was born at a workshop at Dartmouth college in 1956. The research scholars, professors, and engineers who attended the workshop are Allen Newell (CMU), Herbert Simon (CMU), John McCarthy (MIT), Mervin Minsky (MIT), and Arthur Samuel (IBM) became the path makers and front runners of AI research. AI research has been divided into subfields based on technical considerations such as machine learning, natural language processing, evolutionary computation, planning, vision, robotics, and artificial neural networks. These subfields often fail to communicate between themselves due to their ideological differences and social factors.

As machine intelligence is rapidly growing, nowadays machines are capable of understanding human speech, design of robots, which can be employed in hazardous environment such as chemical industry, space applications, and military purpose in battle fields. They are used in developing modern games such as Fortnite, which is an online battle royale game in which players are dropped onto a virtual island where they had to stay alive, by eliminating others with weapons picked up on the way. The game has grown popular since its launch 2 years ago. More than 250 million people have played this game. These modern machines are also useful in realization of driverless cars, intelligent routing networks, development of new drugs, treatment of different diseases, and inventing new weapons such as guided wave and antiballistic missiles. The AI research field basically includes logical reasoning, knowledge representation, understanding different languages, sense of perception, and the ability of self-motion and interacting with different objects. The long-term goal of the AI research is to inculcate general intelligence in machines. Different approaches namely, computational intelligence, statistical approaches, probability, mathematical optimization, and artificial neural networks are used in AI. AI is a combined output of different areas of research such as computer science, information technology, mathematics, statistics, probability, physics, chemistry, medicine, mechanical engineering, biotechnology, psychology, linguistics, philosophy, and many other fields.

Generally machines with AI will understand the environment surrounding to them with the help of network of sensors and they make decisions to improve the quality of work. The quality of AI research output will depend on the strength of the algorithm written on the machine. Algorithm is a set of obvious instructions with which a smart device can execute and analyze the data for future applications. Protocol layer is based on cluster of complex algorithms, which are generally written over the simpler algorithms. The complex algorithms, which are written for AI, have a capacity to learn from previous data and they can enhance themselves by writing new algorithms, Bayesian networks, decision tree, and nearest neighbor, which are able to approximate any function. These advanced algorithms could able to learn all possible knowledge, by considering each and every hypothesis and comparing them against the data. Due to combinatorial explosion it is impossible for the machines to consider every possibility. Where the time required solving a problem grows exponentially. Most of AI research involves in finding out the number of possibilities, which are not useful in decision-making.

The main goal of AI research is to create technology that bolsters the knowledge levels of computers and automation of machine functioning in industry as well as in all other sectors. The previous researchers in initial stage designed algorithms that will work on bit by bit reasoning like humans use when they solve logical reasoning problems. During the period 1980 and 1990, AI research had introduced novel techniques for dealing with complex data, related to concepts of probability and economics. As these algorithms encountered a problem called combinatorial explosion, these are incapable of solving large reasoning problems. The functioning of these algorithms becomes drastically slow as the problem becomes complicated.

The fundamental of idea AI research is delineation of knowledge. Some intelligent devices try to gather knowledge endowed with experts in particular areas, for example, to study the scanning reports of brain tumor, bone fractures, and various sensitive organs present in the human body.

For representation of knowledge and to solve a particular problem, logic is the key element, this logic can be applied to solve other problems that are of similar nature. For example, the algorithm developed for SATPLAN is based on the logic for planning and inductive logic programming. The AI research has used different forms of logical operators such as OR, EX-OR, NOR, AND, NAND, and NOT. Most of the times agents are compelled to work with inadequate data researchers have designed numerous algorithms to overcome these hurdles, using the concepts from theory of probability and economics. Bayesian inference algorithm is a powerful tool that can be used to solve different kinds of problems by logical reasoning. Planning can be done by using decision networks and dynamic Bayesian networks are used in developing perception intelligence to machines. Probabilistic algorithms can also be used for filtering noise, futuristic prediction, tailoring the signal and finding explanations for streams of data; and developing perception intelligence to systems. The advanced algorithms such as Markov chain Monte Carlo method are used to solve the complicated graphs.

AI along with Internet of Things (IoT) has made human life easy as they have potential applications in almost all sectors. In this chapter, we are trying to present water pollutants monitoring based on IoT.

18.1.1 Internet of Things

The rapid growth of Internet usage for communication through e-mails, interacting via social networking applications such as Facebook, YouTube, Orkut, Twitter, WhatsApp, and Skype; sharing huge amount of data; transacting financial business of banking sectors via ATMs, bank-to-bank financial transactions, e-commerce (Amazon, Flipcart, etc.), e-governance, in different service sectors such as transport, food catering, medicine, old age homes, and education; playing online games; analyzing data; and summarizing it and lot of applications through Internet. There is another big area of use of Internet begins to emerge as a global platform for allowing the machines, smart devices, and electronic objects to communicate, compute, and coordinate. In the recent years, everything, such as refrigerators, cars, cameras, fans, lights, washing machines, doors of the house, overhead water tank and pump house, and sprinklers, are interconnected to create a better world for human beings; these devices are now collectively called IoT.

The IoT is the interconnection of physical devices and objects used in industry via wireless sensor networks (WSNs), integrated with microcontrollers are provided with radio frequency identification technique (RFIDs) and the ability to transfer data over Internet without requiring human-to-human or human-to-computer interaction, and they can be controlled and monitored remotely. The IoT refers to the ever-growing network of physical objects that feature an IP address for Internet connectivity, and the communication that occurs between these objects and other Internet-enabled devices and systems. The IoT extends Internet-connectivity beyond traditional devices such as desktop and laptop computers, smartphones, and tablets to a diverse range of devices and everyday things that utilize embedded technology to communicate and interact with the external environment, all via the Internet. IoT is a conglomeration of WSNs, control systems, embedded systems, automation, cloud computing, etc.

18.2 Origin and later developments in Internet of Things

In the year 1982 the first invention of smart device network was occurred in the case of coke machine, which operated through IoT at Carnegie Mellon University, which can give information about newly loaded drinks whether they are cold or not. Simultaneously the contemporary vision of the IoT was emphasized by Mark Weiser's paper on ubiquitous computing in 1991, which is the computer of the 21st century. Kevin Ashton of Procter & Gamble first coined the term "Internet of Things" in 1999, followed by MIT Auto-ID Center. He applied radio frequency identification (RFID) as a powerful tool to the IoT at that point of time, which can admit microcontrollers to control all individual objects through sensors/actuators via Internet connection.

In June 2002 the conference for Nordic researchers in Norway received a research article, which had mentioned about the term "Internet of Things," which was followed by a research paper published in January 2002 in Finnish, a communication system infrastructure for intelligently connected devices, similar to the modern machine developed by Kary Framling and his team at Helsinki University of Technology. The IoT can be redefined as "at instance the number of machines connected to the Internet is more than people." According to Cisco Systems assessment, the IOT was invented in 2008 with machine-to-human ratio increasing from 0.08 in 2003 to 1.84 in 2010. Home or building automation can be done by the IoT devices, which includes lighting, heating, air-conditioning, overhead water tank filling, driers, vacuum cleaners, media, and security systems. With these IoT devices we enjoy the benefits such as energy saving by autonomously turned off industrial heavy machinery and electronic goods. With the deployment of IoT devices in

Industry called industrial IoT (IIoT) in production of goods, which reduces manufacturing costs and improves the quality of goods, this paves the path to the fourth industrial revolution called Industry 4.0. Upon implementation of IIoT technology in industries, it may create $12 trillion gross domestic product across the globe by 2030.

In manufacturing asset predictive maintenance, industrial big data analytics will play an important role, though industrial big data can handle so many sectors. The core technology of industrial big data is cyber-physical system (CPS), and it also acts as a connectivity between people and the machines. Based on connection, conversion, cyber, cognition, and configuration architecture, CPS was designed and it will convert the picked up data into workable information and finally deals with the physical devices to optimize processes.

18.3 Use of Internet of Things in various sectors

The National Science Foundation Industry and University collaborative research center for intelligent maintenance system at the University of Cincinnati designed an IoT-enabled intelligent system in 2001 and later demonstrated in the case of bandsaw machine at IMTS 2014 in Chicago. Bandsaw machines are not much expensive, but the bandsaw belt degrades much faster; due to this, its maintenance cost enormously increasing. Without sensing and intelligent analytics by virtue of experience, it can be assessed when bandsaw belt will actually break. The developed sensor network system with Wi-Fi enabled can recognize and monitor the degradation of bandsaw belt and advises the operaters when is the optimum time to change it. This IoT-enabled bandsaw belt system is significantly improves the experience of the user and safety of the operator, and ultimately maintenance cost will be reduced.

The IoT finds its potential applications in examining the functioning of infrastructures such as bridges, windmills, and railway tracks. The detriments in structures, which conciliate safety and enhance the risk, can be monitored in real time by using IoT infrastructure. In the construction industry by using IoT, one can reduce the cost of the project, minimize the time, and improve the quality of work, paperless transaction, which in turn increases the productivity. IoT that gives real-time data analytics helps in taking quick decisions and saves money. By bringing together service providers and users, IoT facilitates in solving the common problems in an effective manner.

The energy-consuming devices such as switches, bulbs, televisions, refrigerators, air conditioners, washing machines, and all other home appliances can be integrated to Internet connectivity, so that they can communicate with utilities and optimize energy consumption and can be operated remotely by users via a cloud-based interface. The production and distribution of electricity can be managed in an efficient manner with the use of IoT devices such as advanced metering infrastructure, with which the data can be collected from end user, Internet-connected transformers will manage power distribution autonomously leading to smart grid technology.

The monitoring of water sources, environmental conditions, and weather and geographical movements of wild life can be done by different kinds of sensor-assisted IoT technology. The emergency services such as earthquake, tsunami, and thunder warning systems can be developed by using IoT devices. The enormous applications of IoT in different fields give an idea that standardization of IoT in wireless sensing is higher prospectus of future generations.

The IoT intelligence architecture consists of three tiers: IoT devices, edge gateway, and cloud computing. In IIoT equipment, generally we found devices such as sensors and actuators using the protocols such as Modbus, ZigBee, or proprietary protocols, to transfer data through an Edge Gateway by using neural networks. The edge gateway contains sensor data collection systems, which execute tasks, namely, improving, cloud security using modern equipment such as Web Sockets, Internet hub, and fog computing. The signal-to-noise ratio cloud computing for IIoT uses the microcontroller and its peripherals, which are generally multipurpose in nature and highly secured provided with HTTPS/OAuth protocols.

The autonomous control and decision-making of machines at each stage will rely on response time of the IoT application. For an instance, to avoid an accident, a driver-less vehicle's camera requires to detect an instantaneous hurdle. Sending data collected by the camera to cloud and cloud to the vehicle will hinder the decision-making speed. Therefore these all operations should be performed locally in the vehicle to increase the decision-making speed. It is possible through integrating advanced machine learning such as convolution neural networks, long short-term memory, and variational auto encoders to IoT devices.

The large burst of dataflow through Internet can be prevented by fog computing technique. The computation capability of edge devices is inadequate to evaluate and process the collected data. Because of this limited computational power of these IoT edge devices, their basic purpose is to collect data from physical objects and transfer to the cloud. As computing devices of the higher level use excessive battery power which in turn impedes the IoT's capacity to operate, with a minimal processing power. IoT device transfers the sensor collected data to the server via Internet.

The IoT is capable of interconnecting 50–100 trillion devices simultaneously and intelligent enough to track them. There are 1000–5000 smart traceable devices deployed in urban areas at various places to monitor the people. In 2015, 83 million smart objects are used in home appliances. According to latest research, the number of IoT devices across the globe has reached 10 billion by 2018. According to the Gartner's predictions, there will be 25 billion IoT devices by 2021. A total of 127 new IoT devices are connected to the Internet for every second according Mckinsey Global Institution survey. The global IoT market is forecast to be worth $1.7 trillion in 2019 according to Statista. The number of cellular IoT connections is expected to reach 3.5 billion in 2023 according to Ericsson.

18.4 Wireless sensor network

WSN is a collection of distributed and customized sensors, which are mounted on appropriate positions to monitor and record the chemical compositions and physical behavior of the environment and arrange the recorded data and send to the cloud. WSN is able to monitor different environmental and weather conditions such as air pollution, soil erosion, and pollutants in water.

In WSN system, sensor data can be transported wirelessly with spontaneous formation networks. These electronically embedded sensors synchronizingly send their data via neural networks to the cloud by using protocol algorithm. The novel networks can function in the both the directions and able to monitor the sensor activity. The improvement in WSNs was first initiated by military applications. Nowadays these advanced sensor networks find their potential application in various fields such as industries, design and development of goods, logistics, and health monitoring of bridges.

The basic building block of WSN is nodes. A node is a connection point or redistribution point or a communication end point, which is generally made of one or more number of sensors. Depending upon the application, one can choose the size of the network with required number of nodes may be hundreds or thousands. Each node consists of transmitter, receiver, antenna, and embedded electronic circuits with power source.

18.5 Water pollution/quality and Internet of Things

Most of the theories have been concluded that first living organism was evolved in water. Water is a fuel for all living organisms on the planet earth and no life can survive without it. Out of the available water, maximum percentage of it is not useful for drinking purpose because of various factors such as high salinity in sea water; rigorous chemical pollution due to industrialization, globalization, urbanization, agriculture, etc.; the presence of heavy metals in particular geographical regions because of their geological factors and solidified water in polar regions. Nowadays, availability of safe drinking water is a tremendous problem across the globe. The rapid growth of population is the major culprit of depletion of available means of water and water quality degradation. Also the quality of underground water has been infected by weed killer and fungicides. As water is used in various sectors for wide range of activities, such as consumption, as a solvent in chemical industry, coolant in radiators of vehicles, for bathing and sanitation, for irrigation of agricultural lands, all these will affect the quality of water.

Water pollution is the degradation of quality of water resources due to various man-made actions. Number of water resources is contaminated due to industrial growth, urbanization, poor sanitation, population explosion, industrial effluents, fossil fuel plant discharges, and thermal and nuclear power plants. Water resources are also polluted by chemicals coming from chemical industry, agricultural fields, sanitary cleaners, plastic wastage, etc. Two billion people across the world are using the contaminated drinking water out of which 785 million people lack even basic drinking water service. The contaminants present in the water will create severe health disorders to the people. A total of 4.85 lakh people who drinks the contaminated water are succumbed to diarrhea every year.

Water pollution is one of the biggest worries for the green globalization. In order to ensure the safe supply of drinking water for consumption of human beings, animals, and plants, the quality needs to be monitored in real time. The basic observed water parameters for water quality determination are pH level, dissolved oxygen (DO), chemical oxygen demand, biological oxygen demand, total suspended solid, and ammoniacal nitrogen (NH_3N). All these water parameters will be measured to determine the water quality before, so that it can be consumed safely.

The percentage of pollutants present in water can be quantified by analyzing water samples. Physical, chemical, and biological tests can be performed on the water samples for quantification of pollutants. By employing appropriate infrastructure and management plans, water pollution can be controlled. To protect natural water bodies such as rivers, ponds, lakes, and oceans, the infrastructures, such as waste-water treatment plants, sewage treatment plants, agricultural waste-water treatment plants, and industrial waste-water treatment plants, should be installed wherever it is required.

Nowadays, the biggest challenge faced by all the governments across the globe is day-to-day increasing levels of pollutants in water bodies. It is high time to emphasis on review of global water policy at different levels. In the 21st century there were lots of inventions but at the same time pollution and global warming are being formed, because of this there is no safe drinking water for the world's population. The monitoring of water standard is a complex process as it has several laboratory testing methods and is time-consuming. To overcome this difficulty a real-time monitoring of water goodness by using IoT can be employed. IoT together with the AI exhibited by sensor water meters will effectively govern the quality of water. For real-time monitoring of the water goodness in water sources, different sensors such as turbidity sensor, pH sensor, temperature sensor, and conductivity sensor can be employed. The controller accesses the information that is monitored by the use of sensors. The accessed data is controlled by the usage of Arduino controllers.

The central idea behind the every IoT technology is "devices are integrated with the virtual world of Internet and interact with it by tracking, sensing, and monitoring objects and their environment." The features of smart devices to act as members of IoT network are collect data by sensors/actuators and transmit the collected data via wireless/wired mode, actuate devices based on triggers and receive information from network or Internet.

To automate real-time water monitoring process, water quality monitoring sensors/actuators, Arduino UNO/Raspberry Pi IDE, XBee/ZigBee modules, and data concentrator module are physically placed in each and every water sources. The water quality monitoring sensors gather data from the water source to be monitored and forward that data to Arduino UNO/Raspberry Pi IDE, for analog to digital conversion; Arduino UNO/Raspberry Pi IDE forward that data to concentrator module through ZigBee module for remote transfer of data to the lab. The data concentrator is located in each and every water source and transfers that data to the cloud configured server, which is located in the testing laboratory. From there it monitors this data remotely and securely provides this data to the requested users, which is stored in the cloud. Water quality parameter is stored in the cloud, which will be securely provided to the requested users using the cryptographic techniques.

18.6 Proposed water pollutant monitoring system

In recent days, a water quality monitoring system based on WSN technique, using different wireless communication standards, has attracted intensive interest. The WSN in IoT-based water quality monitoring system enables the information and communication systems invisibly embedded in the water resources, this sensor network enables people to interact with real world remotely. The PC management software is developed using different software platforms few examples are Arduino UNO, Raspberry Pi, ZigBee, XBee, etc. In the proposed smart water quality monitoring system a reconfigurable smart water

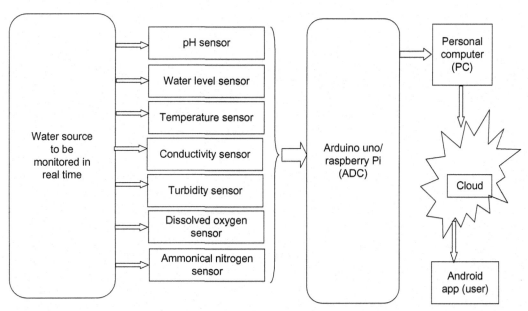

FIGURE 18.1 Schematic diagram of IoT-based water pollutants monitoring system.

sensor interface device that integrates data storage, data processing, and wireless transmission is designed. The hardware experimental setup of smart water quality monitoring system is shown in Fig. 18.1. The hardware's of water quality monitoring system comprises (1) pH sensor, (2) temperature sensor, (3) turbidity sensor, (4) conductivity sensor, (5) dissolved sensor, and (6) ammoniacal nitrogen sensors.

These sensors are connected to core controller, the core controller accesses the data from these sensors and processes it to transfer to the Internet. The sensor data can be viewed on the Internet Wi-Fi system. The overall block diagram of the proposed method is explained, each and every block of the system is discussed in detail in this section.

18.6.1 pH sensor

Acidic or alkaline nature of water can be expressed in pH units, by using a pH meter activity in water-based solutions. The basic principle involved in pH meter is to measure the electric potential difference between pH electrode and a reference electrode, based on its operating principle it can also be called "potentiometric pH meter." The difference in electrical potential relates to the acidity or pH of the solution. The pH meter is used in many applications ranging from laboratory experimentation to quality control.

It consists of three types of probes (1) glass electrode, (2) reference electrode, and (3) combination of gel electrode. pH is defined as the negative logarithm of hydrogen ion concentration in water.

$$pH = -\log(H+)$$

pH meter consists of a pair of electrodes out of which first one is glass electrode and second one is reference electrode. These electrodes are dipped into a solution for which the pH value has to be calculated. By measuring the electric potential between these two electrodes, one can find the pH value of the solution by proper calibration of the pH meter.

In developing a good quality pH meter, the key part is to design the electrodes by choosing an appropriate material and structure. The electrodes have rod-like structures preferably made with glass. For determining the pH value of any given solution, the electrode with a glass bulb, which has sensor at bottom, is exclusively designed to measure hydrogen ion concentration. When a test solution comes in contact with glass electrodes, the hydrogen ions from the test solution replace the positively charged ions on the glass bulb, which in turn creates an electrical potential difference across the electrodes. The electronic amplifier amplifies the electrical potential difference between the two electrodes and converts into equivalent pH units. According to Nernst equation, the electrical potential difference generated between two electrodes is linearly related to pH value of the solution.

In a pH meter the electrode that is used as reference is unresponsive to the hydrogen ion concentration in the solution, which is made of a metallic conductor connected to the display unit. This electrode is dipped into an electrolyte solution usually potassium chloride, which has a contact with the test solution via porous ceramic membrane.

The moment we dip the two electrodes in any solution, immediately an electric path is established, which in turn creates a potential difference between electrodes and it can be measured by the voltmeter. The electric path so established is running from metallic element of the reference electrode to the electrolyte solution, from there it is passing through porus membrane test solution, the hydrogen ion−sensitive electrode to the solution inside the glass electrode, to the silver of the glass electrode, and ultimately the output of the pH meter is connected to a smart device to communicate with Arduino UNO board.

The pH of pure water is 7. In general, water with a pH lower than 7 is considered acidic in nature, and with a pH greater than 7 is considered basic in nature. The normal range for pH in surface water system is 6.5−8.5, and the pH range for groundwater systems is in between 6 and 8.5. The pH value for apple juice is 3, orange juice 3.5, coffee 5.5, milk 6.2, baking soda 8.5, soapy water 10, and bleach 12. Alkalinity is a measure of the capacity of the water to resist a change in pH that would tend to make the water more acidic. The measurement of alkalinity and pH is needed to determine the corrosiveness of the water. The Environmental Protection Agency (EPA) warned that consuming excessively acidic or alkaline water is harmful. Drinking water must have a pH value of in between 6.5 and 8.5 to fall within EPA standards, and they further note that even within the acceptable pH range, slightly high- or low-pH water can be unappealing for several reasons.

High-pH water has a slippery feel, tastes a bit like baking soda, and may leave deposits on fixtures, according to the EPA website. Low-pH water, on the other hand, may have a bitter or metallic taste and may contribute to fixture corrosion (Fig. 18.2).

It is an analog pH meter, specially designed for *Arduino controllers* and has built-in simple, convenient, and practical connection and features. It has an LED that works as the

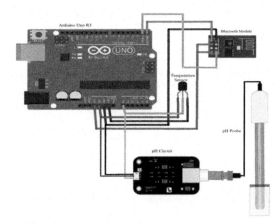

Specification:
- module power: 5.00 V,
- module size: 43 mm × 32 mm, measuring range: 0–14 pH,
- measuring temperature: 0°C–60°C, accuracy: ± 0.1 pH (25°C),
- response time: ≤ 1 minute,
- pH sensor with BNC connector, pH 2.0 interface (3 ft patch), gain adjustment potentiometer,
- power indicator LED,
- cable length from sensor to BNC connector: 660 mm.

FIGURE 18.2   The pH sensor for water based on IoT application.

power indicator, a Bayonet Neill Concelman Connector (BNC), and pH 2.0 sensor interface. To use it, just connect the pH sensor with BNC connector, and plug the pH 2.0 interface into the analog input port of any Arduino UNO controller. If it is preprogramed, you will get the pH value on LED display board. The entire system comes in compact plastic box with foams for better mobile storage. In order to ensure the accuracy of the pH probe, you need to use the standard solution to calibrate it regularly. Generally, the period is about half a year for calibration. If you measure the dirty aqueous solution, you need to increase the frequency of calibration.

18.6.2 Water-level sensor

Level sensor senses the level of the water in water sources such as reservoirs, ponds, lakes, overhead tanks, rivers, channels, water in agricultural lands, seas, and oceans. The level of the water in water sources can be obtained remotely by level sensor through Internet connection at any point of time. These level sensors are further classified into continuous level sensors and point-level sensors. Continuous level sensor determines the liquid level within given range and quantifies the amount of liquid. Point-level sensor specifies the liquid level with respect to the particular sensing point (Fig. 18.3).

Water-level sensor can be designed for detecting water level in reservoir and over head tanks. This is widely used in sensing the water leakage, water level, and the rainfall. It consists of mainly three parts: 1 MΩ resistor, an electronic brick connector, and several lines of bare conducting wires. It works by having a series of exposed traces, which are connected to ground. This is also interlaced between grounded traces and the sunstrokes. A weak pull-up resistor of 1 MΩ is present. 1 MΩ resistor pulls up the sensor value until a drop of water shorts the sensor trace to the grounded trace. This can measure the water droplets/water size by using a series of exposed parallel wires. It has characteristics such as low power consumption and high sensitivity.

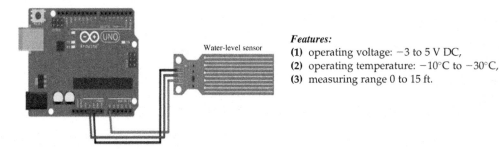

Water-level sensor

Features:
(1) operating voltage: -3 to 5 V DC,
(2) operating temperature: $-10°C$ to $-30°C$,
(3) measuring range 0 to 15 ft.

FIGURE 18.3   The water-level sensor based on IoT application.

FIGURE 18.4   Ultrasonic water-level sensor.

Ultrasonic water-level sensor

Ultrasonic level transmitter is a noncontact ultrasonic continuous level measurement product based on Arduino UNO technology. Ultrasonic level transmitter offers affordable solution for a variety of applications for water, wastewater, petrochemical, chemical, and food and beverage.

It has the following features:

extended range of 0.25–15 m (0.8–49 ft)
integral LCD display, keypad, and integral 4–20 mA output
automatic compensation for different environment
approved for installations in hazardous areas (Ex)
simple to install and configure, direct installation on tanks, vessels, and reactors (Fig. 18.4).

The basic principle of ultrasonic level sensor is based on emitting ultrasonic (20–200 kHz) acoustic waves that are reflected back due to density variations in the medium and detected by the sensor. Ultrasonic level sensors are used to measure depth of oceans, to measure the water level of the lakes, ponds, bore wells as well as sensing of cracks or fractures in bulk solids. Due to moisture, temperature, and pressure, the speed of the sound waves will be changed, which affects the functioning of ultrasonic level sensors. The sensitivity of the ultrasonic sensor is influenced by chaos, air bubbles, chemical smog, and vapor present in the water source. Chaos and air bubbles cause major hindrance to the acoustic waves while traveling toward bottom of the water source and reflected back to the transducer. Chemical smog and vapor act as absorbers of ultrasonic waves and diminish the intensity of these waves, which results in poor quality of sensing.

The optimum capture of reflected echo waves from the beneath of the water source is possible only by placing the sensor in an appropriate position. The different types of obstacles, which are present in the test water source, lead to incorrect results, even though the latest technologies are available for good quality echo signal processing. The basic purpose of ultrasonic transducer is to transmit and receive the acoustic signal. But in its operation it undergoes physical vibration called ringing. This kind of physical vibration should be nullified by proper means before the reflected signal is processed. Due to this physical vibration, the sensor is not able to sense an object. It is called the blanking zone, ranging from 15 to 100 cm, according to the detection level of the transducer.

To design a smart ultrasonic sensor, a sophisticated electronic system, which can process the collected sensor data, is required. Ultrasonic sonic device can act as point level as well as continuous monitoring sensor. The microcontroller that is present in the signal processing system has a capacity to calibrate and improve the signal-to-noise ratio, monitor the sensor collected data from a distant place by using wireless technology with a less power consumption. The ultrasonic water-level indicators attractive due to minimal cost and sophisticated functionality.

18.6.3 Temperature sensor

Temperature sensor is an integrated circuit sensor. The output voltage is linearly proportional to the centigrade temperature. The sensor shown in figure is compatible with Arduino UNO device. The applications of the temperature sensor are in microwave ovens, fridges, household devices, air conditioners, and atmosphere and water temperature monitoring. It can measure not only the hot bodies but also cold bodies. There are two types of sensors, they are noncontact temperature sensors and contact temperature sensors. Contact temperature sensors are again divided into three subtypes: electromechanical, resistive resistance temperature detectors, and semiconductor-based temperature sensors (Fig. 18.5).

18.6.4 Conductivity sensor

In general, conductivity sensors are designed by potentiometric method. Each sensor consists of four electrodes that are designed like a concentrical cylinder. Platinum metal is used for the preparation of electrode. On application of alternating current to the external electrodes, the electric potential between internal electrodes can be measured. By using Ohm's law ($J = \sigma E$, where J is current density, σ is conductivity of medium, and E is the applied electric field). The conductivity of the medium can be calculated with the help of electrodes

FIGURE 18.5   Temperature sensor.

surface area and separation between them. With electrolytes of well-known conductivity, the conductivity sensor can be calibrated to improve the accuracy of the measurement.

The conductivity probe, which operates on the principle of inductive method, has the benefit that the liquid will not have the direct contact with the electric circuit of the sensor, which is generally used in industrial purpose. In this method, two coils coupled with mutual inductance are employed in which first coil generates magnetic field upon application of calibrated emf. The second coil functions like a secondary coil in the transformer. The fluid guiding through a groove liquid passing through a channel in the sensor forms one turn in the secondary winding of the transformer. The current induces in the secondary coil due to passage of liquid through it. The amount of this current will be used for sensing purpose.

The alternative method is to use four probe conductivity sensors in which the electrode probes are made from materials sustainable to environment. The four probe conductivity sensors are capable of measuring conductivities of liquids, which are lower than $100\,\mu S/cm$, whereas inductive sensor does not have such a low sensitivity (Fig. 18.6).

In industries for process control widely used unit is conductivity sensor because it can be monitored easily, quickly and cost of maintenance is less. The value of conductivity so obtained is useful for taking crucial decisions in the process control. The conductivity measurements are essential to assess the quantity of ions present in the liquids. Based on the values of temperature and conductivity of the liquid, it is possible to compute concentration of liquid. With the help of predetermined graphs between concentration versus conductivity at a particular temperature, it is easy to assess concentration levels of unknown liquid.

18.6.5 Turbidity sensor

Turbidity in a fluid arises due to the presence of huge number of microscopic particles, which cannot be seen directly with eye. The percentage of turbidity indicates the quality of drinking water.

Features:
- 3–5 V wide operating range.
- Hardware filtered output signal, low jitter.
- AC excitation port, effectively reduce polarization.
- Gravity connector and BNC connector, plug and play, no welding.
- Software library supports two-point calibration and automatically identifies standard buffer solution and integrates temperature compensates algorithm.
- Uniform size and connector convenient for the design of mechanical structure.

FIGURE 18.6 Analog conductivity sensor.

Features:
- working voltage: DC5V; operating current: 30 mA (*max*)
- response time: <500 ms
- insulation resistance: 100 MΩ (*min*)
- works with microcontroller—cannot work alone (no display)
- the output way: analog output 0–4.5 V

FIGURE 18.7 Turbidity sensor.

The solid particles of the order of micron or submicron size, which are miscible in water, are resposible for the water to appear turbid. Most of the suspended particles come down in a stable vessel of liquid due to their heavy size. The particles that are in very small dimensions will settle very slowly. These small solid particles are responsible for turbidity in water sources. Most of these solid particles are responsible for health hazards.

Due to growth of phytoplankton in open water, the turbidity is gradually increasing. Human activities such as agriculture, mining, construction, and transportation that disturb land can lead to sediments entering into resources of water in rainy season. The turbidity levels increases rapidly because of all kinds of soil corrosion, landslides due to heavy rains, volcanic eruptions. Turbidity also increases because of man-made activities such as construction of big buildings, coal mining, and oil refineries.

If people consume high turbid drinking water, they are more prone to different kinds of diseases such as malfunctioning of kidneys and indigestion problems. In high turbid water the presence of microorganisms is also likely to be high, which in turn causes serious damage to immune system in human body. These microscopic particles, which are responsible for turbidity, hinder the different processes for eliminating microorganisms (Fig. 18.7).

18.7 Drinking water standards

The amount of turbidity presents in water can be measured by reflection of electromagnetic waves from the microscopic particles present in the water. Nephelometer with a photodetector arrangement on the side of EM beam is used to measure the level of turbidity present in water sources. Even if there are lots of small particles present in the water sources due to scattering, more light reaches the detector. The quantity of turbidity can be measured in nephelometric turbidity units (NTU) by using nephelometer. The amount of reflected light depends on the quantity and physical properties of the microscopic particles suspended in water. In general, the safe drinking water turbidity levels should not exceed 4 NTU and preferably less than 1 NTU (Fig. 18.8).

18.7.1 Dissolved oxygen sensor

The amount of DO in the water source can be estimated through an electronic sensor called oxygen sensor. The oxygen dissolves by diffusion from the surrounding air; drying of water that has tumbled over falls; and most importantly as a secondary product of photosynthesis reaction, which generally takes place in plants.

Photosynthesis equation can be represented as

$$\underset{CO_2}{Carbon\ dioxide} + \underset{H_2O}{Water} \rightarrow \underset{O_2}{Oxygen} + \underset{C_6H_{12}O_6}{Carbon\text{-}rich\ foods}$$

Only green plants and some bacteria can undergo photosynthesis and similar processes. Animals cannot split oxygen from water (H_2O) or other oxygen-containing compounds. The oxygen we are using in our day-to-day life is available only through the green plants.

The drinking water standards also depend on percentage of oxygen dissolved in it. DO levels can be measured in parts per million (ppm), number of mg/L, and percentage of saturation. Percentage of saturation is the amount of oxygen dissolved in the water sample compared to the maximum amount that could be present at the same temperature. If DO levels in water are below 5.0 mg/L, aquatic animals cannot survive; hence, it is generally named as dead zone. At room temperature standard atmospheric pressure, the maximum amount of oxygen that can dissolve in fresh water is 9 ppm. At levels of 4 ppm or less,

Specification
- range: 0–15 mg/L (or ppm)
- accuracy: ± 0.2 mg/L
- response time: 95% of final reading in 30 seconds, 98% in 45 seconds
- typical resolution: 0.014 mg/L
- temperature compensation: automatic from 5°C to 35°C
- pressure compensation: manual
- salinity compensation: manual
- minimum sample flow: 20 cm/s

FIGURE 18.8   Dissolved oxygen sensor. Source: © *Vernier Software & Technology Used with permission.*

some fish and macroinvertebrate populations will begin to decline. DO percent saturation values of 80%−120% are considered to be excellent and values less than 60% or over 125% are considered to be poor.

High levels of DO in water make it tasty. But there will be oxidation problems in pipes if the oxygen is further more. Hence, it is always better to maintain at required levels of DO. Oxygen sensors have their applications in different sectors: in medical field, for anesthesia monitors, respirators, and oxygen concentrators; in automobile industries, for fire prevention units; etc. There are different methods to measure these DO such as electrochemical, electromagnetic radiation methods, and very latest method is through lasers.

18.7.2 Ammonia nitrogen sensor

The neutral, unionized form of ammonia is a chemical compound of both nitrogen and hydrogen with the formula NH_3. It is primarily sourced by fish excrete through their gills or urine and is highly toxic to the aquatic life. Ammonia is a major food source to nitrifying bacteria, which is deadly to fish. It can cause gill and organ damage so always be aware of any symptoms such as cloudy eyes or frayed fins. This makes ammonia an important parameter to measure when cycling a new tank. Anything measuring above 0.02 mg/L (ppm) can be considered harmful. Another source of ammonia is decaying organic substances such as dead animals or uneaten food. As the substance decomposes over time nitrogen is released producing bacteria, which then produces ammonia. It can also come from domestic, industrial, or agricultural pollution, primarily from fertilizers, organic matter or fecal matter.

When measuring ammonia it is important to always measure pH and temperature as it will give a clearer idea as to the toxicity of the ammonia.

Ammonium (NH_4^+) is a positively charged ion (or cation) that is formed by the addition of a hydrogen proton to ammonia (NH_3). This ammonium ion is created when ammonia, which is a weak alkaline, reacts with a Bronsted acid. A Bronsted acid is a molecule or ion that has the ability to lose, or "donate," a hydrogen cation. Water acts as a Bronsted acid when it reacts with ammonia. This means that when ammonia dissolves in water, a tiny amount of ammonium is produced.

The formula looks like

$$H_2O + NH_3 \rightleftharpoons OH^- + NH_4^+$$

The amount of ammonium generated depends on the pH, temperature, and concentration of dissolved salts in the water. If the pH is low, a greater number of ammonium ions are created. Conversely if the pH is high then the hydroxide ion takes a proton from the ammonium ion and creates ammonia (Fig. 18.9).

18.8 Hardware and software modules in Internet of Things

18.8.1 Arduino UNO

The Arduino is a smallest, cheapest device that allows us to easily connect with some electronic devices that have made to our computer and to the Internet. And it

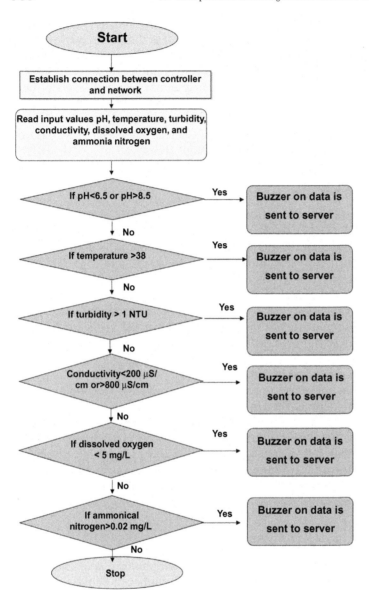

FIGURE 18.9 Flowchart of IoT-based water monitoring system.

brings all sorts of madcap invention to the IoT. It does for making connected hardware what blogging did for publishing, makes it easy and liberates ideas. Hardware hackers are the kind of people who strap ordinary cameras to weather balloons to photograph space, give them things like the Arduino and they make machines that blow bubbles when they see their own names on Twitter. Or they make pairs of lamps for lovers separated by distance—connected lamps, so if you switch one of them off, the other goes off too—a little reminder of what your love in another time zone is up to.

These are the same curious, hybrid, and inventive sort of people who built the web and pioneered social media. They are turning from mucking about with the web to mucking about with the real world because there seems to be a whole new set of interesting things to invent, unoccupied, uncolonized space.

18.8.2 Hardware

Arduino UNO is a specially designed microcontroller with peripheral devices. This hardware system can be easily integrated with IoT devices for different kinds of applications in various industries. Specifically coded software called integrated development environment (IDE), which facilitates to read and write new algorithms to control and monitor devices, is employed in the physical environment.

The key features are as follows:

- Arduino microcontroller integrated chips are capable of receive and analyze, analog or digital input signals from different sensors and able to take decisions such as triggering a motor, actuate LED on/off, connect to the cloud, and many other actions.
- One can control the board functions by giving series of commands to the microcontroller via Arduino IDE (referred to as uploading software).
- With the help of USB cable required software's can be easily loads on Arduino UNO board. This feature attracts the more number of users.
- The Arduino IDE is developed with C++ language so that it can be further modified easily.
- The Arduino UNO offers the microcontrollers with dedicated functionalities, which are highly useful in performing special tasks.

18.8.3 Arduino—board description

The Arduino UNO board is the most popular board in the Arduino board family. In addition, it is the best board to get started with electronics and coding. Some boards look a bit different from the one given in Fig. 18.10, but most Arduinos have majority of these components in common.

1 *Power USB*
Arduino UNO integrated circuit can also be charged through USB cable connects to personnel computer or laptop. We just have to insert the USB cable in to the USB port.
2 *Power (Barrel Jack)*
The alternate way to charge Arduino circuit is connecting it directly to alternating current through the Barrel Jack.
3 *Voltage regulator*
The role of voltage regulator is crucial in protecting the arduino circuit from burnt up. As it delivers only required amount of voltage to the input of the arduino board and controls the DC power supply utilized by the microcontroller and other peripheral devices.

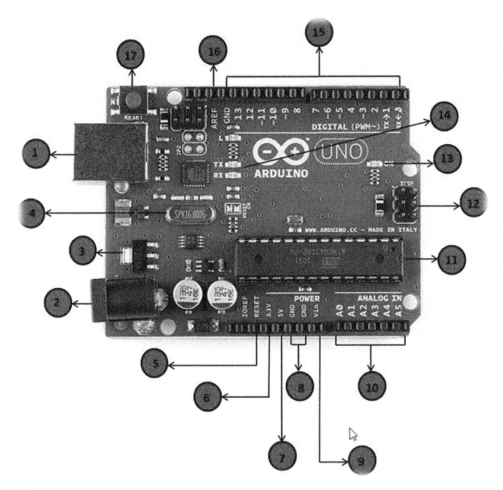

FIGURE 18.10   Arduino UNO board with detailed description.

4 *Crystal oscillator*
With the help of crystal oscillator the microcontroller is able to calculate the time. The time factor plays a vital role in receiving and sending signals to peripheral and other physical devices. Based on the oscillator frequency, the microcontroller is able to take speedy decisions. Generally arduino crystal consists of 16 MHz frequency, which is indicated by the number 16.000H9H on the board.

5,17 *Arduino reset*
Any microcontroller requires a function called reset, to delete unwanted memory or refresh the main software to ready for new task. The arduino controller can be reset by two pin functions, either by resetting the pin number17 or by external reset button number 5.

Pins (3.3, 5, GND, Vin)
- 3.3 V (6)—This pin renders 3.3 V power supply as output.
- 5 V (7)—This pin gives 5 V power supply as output.
- The devices that are compatible with Arduino UNO board are powered by either 3.3 or 5 V supply.
- GND (8)(Ground): Grounding is highly essential for any integrated circuit, to safeguard from high voltage fluctuations. The Arduino UNO board has multiple pins to ground the circuit where it is necessary.
  - Vin(9): This pin offers an additional power supply from external source.

Analog pins

The Arduino UNO board is embedded with six analog input pins A0−A5, which are capable of retrieving the analog signal data from sensors or actuators placed in remotely monitored environment and can convert this analog signals into digital signals for further processing.

Main microcontroller

Every Arduino UNO board consists of a microcontroller made of usually from ATMEL company. We can treat this microcontroller as brain of the board. Depending upon type of microcontroller required software can be installed to monitor the environment in which it is placed.

ICSP pin

ICSP (12) is a serial peripheral interface (SPI) which is an additional functionality acts like an output pin. The output devices can be connected to SPI to retrieve data from Arduino UNO board.

Power LED indicator

LED indicator will glow when Arduino UNO board is powered properly, otherwise it will turn off. If the LED glow with diminished intensity or completely off then we need to check power cable as it is properly inserted or not.

Transmitting port (TX) and receiving port (RX LEDs)

TX LED indicates whether the data is transmitted from Arduino UNO board to other peripheral devices. The data is always transmitted in a serial manner in digital form as 0 and 1. So that TX LED indicator response according to the voltage levels. The rate at which TX LED flashes depends on transmitting capacity of the microcontroller. RX LED is an indicator for receiving data which also functions similar to the TX LED in the case of receiving process.

Digital input and output (I/O)

The Arduino UNO board is embedded with 14 digital I/O pins out of which 6 pins are used as PWM (pulse width modulation) output. These pins also helpful in understanding voltage levels either 0 or 1. These pins are also used for triggering indicator LEDs, relays, etc.

AREF

Analog reference is abbreviated as AREF. The functionality of this pin is to set the upper voltage limit 5 V in the case of analog signals.

18.8.4 Arduino UNO board with I/O pin configuration

In Fig. 18.11 arduino board depicts the I/O pin configuration with number of internal circuits. The Arduino UNO is subdivided into three parts diecimila, duemilanove, and current uno, which consist of 14 digital I/O pins. Among them six pins create pulse width modulated signals, and six work as analog inputs, one USB cable connection, and one power connection.

Arduino UNO is one of the widely used microcontroller boards for beginners. The Arduino IDE software is available for multiple platforms, including Windows, Mac OSX, and Linux. C and C++ are the base programming languages used to program and customize the behavior of the Arduino microcontroller board. The Arduino IDE can be downloaded without any cost, along with the live updates of the software. Arduino can be connected with live platforms of IoT by using Wi-Fi or similar network-based chips. ESP8266 is a very popular device for communication with the Internet that can be easily interfaced with Arduino microcontroller boards. The ESP8266 Wi-Fi chip is very low cost and can be directly connected with the Arduino board to establish communication with the real cloud.

18.8.5 Integrated development environment

IDE is a java language—based application useful for different operating systems such as Windows, Mac OS, and Linux. It contains options for different editing works, which are possible in normal word document such as cut, paste, find and replace, indenting, and highlighting with a single click connected to an Arduino board. It is also provided with an area for messaging text, console for text, and a toolbar that can be operated for different instructions.

Arduino UNO board software IDE is compatible with programming languages such as C and C++ by applying unique regulations for codes. These programed codes need only two fundamental functions one is for initiating the plan and the other is to execute central program loop. The Arduino IDE makes use of *avrdude* program and decodes the main code into a normal text file in hexadecimal encoding, which is installed into the Arduino board by a program in the board's firmware.

18.9 Advances in Internet of Things for water quality monitoring

Raspberry Pi 3 model is a small-sized system on chip, which is capable of doing all the tasks that an average desktop computer can do such as spreadsheets, word processing, Internet browsing, programing, and gaming. It is a mini computer usually with a Linux operating system to run multiple programs. Raspberry Pi has the built-in Internet port through which one connects to the network. This model was built on latest 1.2 GHz quad code Broadcom BCM 2837 Armv8 64 bit processor is the third generation of Raspberry Pi and is faster and more powerful than its predecessors. With built-in wireless and Bluetooth modules it is an ideal IoT solution. It also consists of 1 GB RAM, 4 USB ports, 40

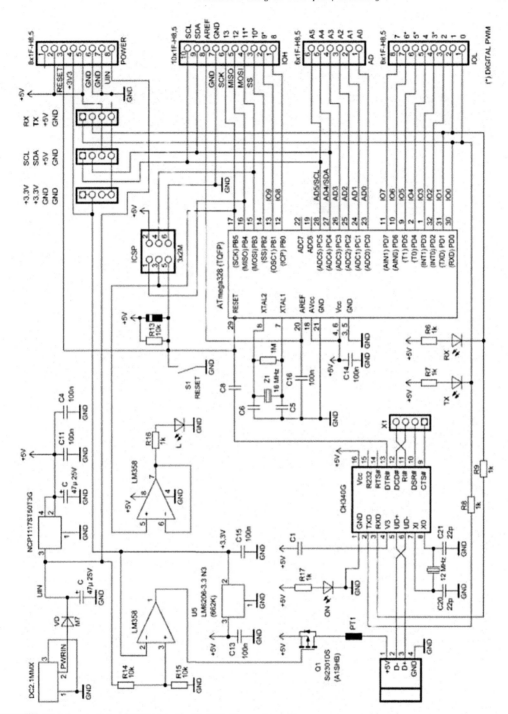

FIGURE 18.11 Arduino UNO board with I/O pin configuration.

FIGURE 18.12   NodeMCU firmware for the IoT. *IoT*, Internet of Things.

pin extended GPIO, HDMI and RCA video output, hence makes it ideal for computing the sensor data.

NodeMCU, an open-source IoT-based hardware platform with inbuilt Wi-Fi: To communicate and transfer data to the cloud in real-time, there is another chip called NodeMCU (http://www.nodemcu.com) that can directly communicate with live servers. NodeMCU integrates the firmware with inbuilt Wi-Fi connectivity so that direct interaction with the network can be done without physical jumper wires. It is an Arduino-like board that is used for interfacing with the cloud in real-time so that the live signals or data can be logged or stored for predictive analysis (Fig. 18.12).

18.10 ZigBee

ZigBee is an open global standard for wireless technology designed to use low-power digital radio signals for personal area networks. ZigBee operates on the IEEE 802.15.4 specification and is used to create networks that require a low data transfer rate, energy efficiency, and secure networking. It is employed in a number of applications such as building automation systems, heating and cooling control and in medical devices. ZigBee is designed to be simpler and less expensive than other personal area network technologies such as Bluetooth. ZigBee is a cost- and energy-efficient wireless network standard. It employs mesh network topology, allowing it provide high reliability and a reasonable range. One of ZigBee's defining features is the secure communications it is able to provide. This is accomplished through the use of 128-bit cryptographic keys. This system is based on symmetric keys, which means that both the recipient and originator of a transaction need to share the same key. These keys are either preinstalled, transported by a "trust center" designated within the network or established between the trust center and a device without being transported. Security in a personal area network is most crucial when ZigBee is used in corporate or manufacturing networks.

18.10.1 Storing sensor data in Internet of Things platforms

There are a number of IoT platforms that can be used for the storage of real-time streaming sensor data from Arduino and similar boards. Once the circuit is created using Arduino, the signals can be transmitted and stored directly onto IoT platforms. These open-source IoT platforms can be used for storage, processing, prediction, and visualization of the data recorded from the sensors on the Arduino board.

A few prominent IoT platforms that can be used for device management and the visualization of data are

- ThingSpeak: *thingspeak.com*
- Carriots: *carriots.com*
- KAA: *kaaproject.org*
- Things Board: *thingsboard.io*
- MainFlux: *mainflux.com*
- Thinger: *thinger.io*
- DeviceHive: *devicehive.com*

18.10.2 Transferring data to ThingSpeak

ThingSpeak (https://www.thingspeak.com) is a widely used open IoT platform for the collection, storage, and visualization of real-time sensor data fetched from Arduino, Raspberry Pi, BeagleBone Black, and similar boards. Live streaming data on the cloud can be preserved using channels in ThingSpeak.

To work with ThingSpeak a free account can be created. After successfully signing up on ThingSpeak, a channel is created on the dashboard and a "Write Key" is generated, which is to be used in the Arduino code.

The key features of ThingSpeak include

- configuration of devices for sending data to a cloud-based environment
- aggregation of sensor data from different devices and motherboards connected with sensors
- visualization of real-time sensor data, including historical records
- preprocessing and evaluation of sensor data for predictive analysis and knowledge discovery

The source code that can be customized and executed on Arduino IDE so that real-time data can be fetched on board and then transmitted to ThingSpeak is given next. For simplicity the execution of the code is done for NodeMCU hardware, which has inbuilt Wi-Fi for sending the data to the IoT platform.

18.11 Security framework for Internet of Things Arduino platform

The traditional security services are not directly applied on IoT due to the devices run on different platforms and use different protocols to communicate especially in the wireless or wire system need to communicate with other devices without human intervention as a machine to machine interaction. There is a need for a new framework to apply the security services and machine to machine (M2M) communication architecture. So a flexible and common security framework is need to be implemented, which deals with security threats in IoT environment.

As the IoT is rapidly growing the threats to the IoT is also increasing. Sensors, the communication between sensor and access point, M2M gateway the access point network, and backend server are potential points for attacks. In order to establish a secure M2M communication architecture, security mechanisms should achieve requirements such as confidentiality, authentication, nonrepudiation, access control, availability, and privacy are listed.

1. Confidentiality ensures that only authorized entities can read M2M sensing data.
2. Authentication allows the backend server to certify the sensory data of the M2M nodes.
3. Nonrepudiation guarantees that M2M nodes once sending data cannot deny the transmission
4. Access control represents the availability to restrict and control the application domain that is it allows only authorized M2M application systems to gain access to the backend server.
5. Availability ensures that whenever M2M application systems access the backend server is always available.
6. Privacy is very important in the case of privacy sensitive M2M communication system.

In general, cryptographic technique can be used to achieve confidentiality while digital signature and message authentication code techniques can achieve others. It should be noted that most security mechanisms only efficiently depend against external attacks that launched by attackers who are not equipped with key materials in an M2M communication system while an internal attack cause more serious damage to the M2M systems compared to the external attacks. For example the connected malicious things will cause a violation a substantial fraction of the system. The application of the traditional security services on M2M communication architecture directly represents a big challenge to deal with the internal attack issues in IoT environment due to the devices run on different platforms and uses different protocols to communicate.

To enforce the secure and robust connection for smart devices that want to connect legitimately for communicating with other devices are IoT based, the proposed secured M2M communication framework the procedure of workflow for the smart devices will be as in Fig. 18.13.

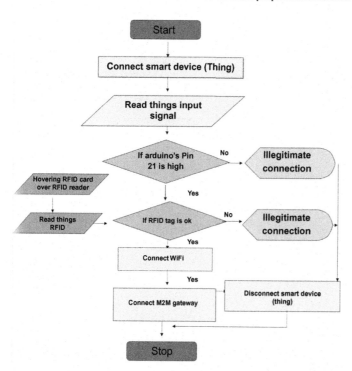

FIGURE 18.13 Certification authority workflow chart.

18.12 Future prospects and conclusion

The abovementioned frameworks are focused on two issues the first issue is to provide a common platform using Arduino/Raspberry Pi/ZigBee that allow to configuring and coding the common platform to connect smart devices that run on different platforms and uses different protocols without concern to the memory limitations and processing power of smart devices, while second issue is to enforce a level of authentication for smart devices trust to be robust enough to be connected without denial of service threats.

In future, RFID sensors can be replaced with biometric sensors to add a level of personality authentication for certification authority. Also the cryptographic techniques can be utilized to add a level of confidentiality for the implemented platform.

Paper-based microfluidic sensor devices for inorganic pollutants monitoring in water

Amit Bansiwal[1], Kirpa Ram[2] and Rashmi Dahake[1]

[1]CSIR-National Environmental Engineering Research Institute, Nagpur, India [2]Institute of Environment and Sustainable Development, Banaras Hindu University, Varanasi, India

19.1 Introduction

Water pollution and vector-borne diseases have become one of the most serious and life-threatening issues because almost all the biota is affected by pollution in some way or other. Due to overexploitation of natural water resources and unfair means of using water, the water bodies are getting contaminated and becoming unsafe for drinking purposes. The presence of several organic pollutants and heavy metals in ground as well as surface water is a serious threat not only to human health but also to living organisms

Inorganic Pollutants in Water
DOI: https://doi.org/10.1016/B978-0-12-818965-8.00019-6

(Jang et al., 2011; Cleary et al., 2010). Furthermore, the rapid industrialization and toxic chemicals released from industries and agriculture activities subsequently make their way to either surface water or groundwater. However, the groundwater is much more seriously threatened than surface water in the world today because of accumulation of pollutants due to its larger residence time. Furthermore, there are a variety of environmental contaminants, including heavy metals, which require regular monitoring in order to assess their impacts and to arrive at remedial measures. One of the effective mitigation and management of surface water and ground water requires regular and real-time monitoring of water quality as well as source identification of pollutants, both on a high spatial and temporal resolutions.

There are several conventional analytical techniques available for monitoring of heavy-metal contaminants based on spectroscopic and chromatographic techniques such as inductively coupled plasma mass spectrometry, atomic absorption and emission spectrometry, atomic fluorescence spectrometry, ion chromatography, and high-performance liquid chromatography. These techniques are highly sensitive and selective but at the same time are exorbitantly costly and not suitable for field analysis as they are bulky and require professional operators and laborious operations. In addition, sometimes measurement of heavy metals by these conventional techniques requires pretreatment of samples from the environmental matrices. Therefore big challenge that remains to the scientific community is to develop low-cost, portable, and sensitive field testing devices for selective pollutants.

Microfluidics, a transversal technology which links different scientific disciplines (such as chemistry, biology, mechanics, control systems, microscale physics and thermal/fluidic transport, simulation of micro-flows, and materials science) with engineering, has evolved as a reliable and expanding technology for developing affordable and field deployable sensors for various applications in the last two decades (Kim et al., 2013). Tailored microfluidic devices are being developed to address a variety of applications, including determination of air, water, and food quality and detection of pathogens, biofilms, environmental contaminants. However, recent efforts mainly focus on the development of integrated microfluidic devices by incorporating fluidic components, analytical, and detection techniques into a single platform. In this context, paper-based microfluidic devices have been developed recently, which are not only have reduced the cost but also have several other advantageous features such as portability, disposability, and small sample consumption. There is a rapid growth in research in the field of paper-based microfluidics since 2009. The number of publications is increasing day by day as depicted in the Fig. 19.1.

Paper-based microfluidic sensors utilize a combination of microfluidic platform and electrochemical detection techniques for detection of measurands (here pollutants) in any environmental matrices. In the development of microfluidic technique, sensor development and its integration in microfluidic devices are of great importance because they enable and act as the realization of point-of-care and lab-on-a-chip systems. A sensor includes following components: a receptor that recognizes the species to detect with high specificity and selectivity, fabrication techniques, and finally a detection technique. The fabrication techniques in paper-based microfluidics can achieved in different ways, including screen printing, polydimethyl siloxane, wax patterning/printing, flexographic printing, shaping/cutting, Inkjet etching, 3D printing, photolithography, and stamping.

Separation of solute (or measurand) in microfluidics is achieved by applying an external force based on the characteristics of solute and solution. For example, solutes of interest's

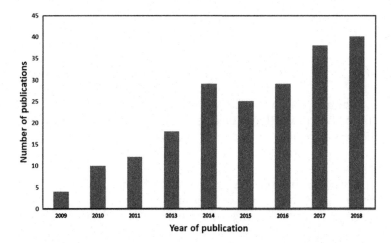

FIGURE 19.1 Growth of publication in the area of paper based microfluidics since 2009 (Source — Web of science, Keyword — Paper based microfluidic devices)

can be separated if they have different mobility (e.g., solute with larger mass will have smaller velocity) and dielectric properties, etc.

The detection methods mainly depend on the prior knowledge and identification of a specific and selective chemical or physical property of measurands (here pollutants), which is the main and tough task in development of any microfluidic-based sensors. Diverse detection approaches have been explored for paper-based microfluidic paper-based analytical devices (μPAD). Some of the very standard detection methods use colorimetric, electrochemical, fluorescence, chemiluminescence, and electrochemiluminescence sensing methods. A number of specific electrochemical techniques have also been adapted to μPAD, including impedance and resistivity (Jiang, 2014), conductivity, voltammetry (linear, cyclic, square wave, and differential pulse), and amperometry. For example, the colorimetric assays have been developed to test for amino acids, bacteria, biomarkers for cancer, lung and liver function, gases, infectious diseases, ions, heavy metals, small molecules, pH, and proteins. In electrochemical approach a transducer is used to translate the detection into a quantifiable physical signal, for example, electrical current/potential in the case of electrochemical voltammetric/amperometric sensors.

The success of microfluidic devices lies mostly in the fabrication, detection of pollutants with high sensitivity and accuracy. With the increasing need for quick and accurate analysis and handiness, screen-printed electrodes (SPEs) are reported to possess several advantages, including bulk fabrication, low cost, low sample requirement, and possibility of miniaturization. However, poor stability and reproducibility are the problems associated with SPE. The chapter offers a comprehensive insight into recent developments pertaining to materials and techniques for detection and fabrication of paper-based microfluidic devices.

19.2 Fabrication methods for paper-based microfluidics

There are various fabrication methods available for manufacturing of microfluidic paper−based devices (μPAD). The methods used by researchers are screen printing, wax printing/patterning, flexographic printing, inkjet patterning, photolithography, cutting/shaping, etc. In this chapter, we had explained fabrication process in detail.

19.2.1 Screen printing

This technique is the most common for fabrication of electrodes in electrochemical detection methods. In contracts to the conventional electrodes, this technique has reference, auxillary, and working electrodes, which can be replaced by SPEs. Reference, auxillary, and working electrodes are screen printed using Ag/AgCl, silver and carbon ink, respectively (Dungchai et al., 2009).

For the fabrication of these electrodes, first the design is finalized and exposed on the screen of polyester or various fine mesh with the help of light and coating solutions. At the time of exposure the blank areas are covered with impermeable material and the design of electrode will allow passage of inks only. Ink is forcibly passed through the mesh using squeegee onto the substrate such as paper, ceramic, and plastic. Different layers can be printed using different screens. First, the auxillary and working electrodes are printed with ink and cured at specific temperature (generally provided by ink manufacturer) then reference electrode is printed and cured. Final layer is printed using insulated materials and electrodes are ready to use. The procedure of fabrication is depicted in Fig. 19.2.

Fabrication of reference electrode is very important as accuracy of electrochemical measurements strongly depends on this electrode. Reference electrode provides a fixed potential stability, which depends on the stability of Ag/AgCl film. In the case of screen printing the fabricated reference electrode is called pseudo electrode as the dissolvation of Ag/AgCl film significantly changes the potential of the electrode. The number of steps in fabrication of such electrodes is reduced in screen printing as the ink used is composed of mixture of silver/silver chloride. In recent days, microfabrication of electrode is very popular and different automated screen printing machines are also available in the market, which allows bulk fabrication of electrode in very less time, reduced cost, and high precision.

19.2.2 Wax patterning/printing

Wax patterning/printing is the most economical and rapid method used for making channels and hydrophobic barriers. The time required for the wax printing is very less compared to the other patterning processes and the total cost for fabrication is $0.001 per device except the printer and energy cost. The channels of wax printed substrate are compatible with aqueous solvents but not with organic solvents. For instance, in the case of acetone wax will not perform as a barrier; however, this may be advantageous for the easy cleaning of the devices from the reagents and sample in some cases. Wax melts after heating and penetrates into the paper, which will create channels for flow of fluid.

In 1902 Dieterich patented a technique of wax printing for preventing the cross-reactivity between the two testing zones (Dieterich, 1902). Followed by this work, Muller et al. (1951) used paraffin paper that was impregnated on to a filter paper by heated die to create hydrophobic barrier. This fabrication method requires a multistep production and

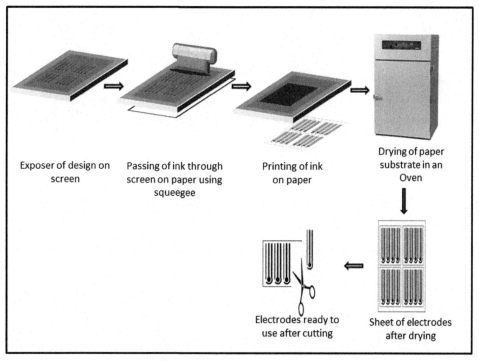

FIGURE 19.2 Steps involved in screen printing of electrodes using carbon, silver and Ag/AgCl ink.

considering the difficulty in handling, many efforts were made with other hydrophobic materials. Lu et al. in 2009 have briefly described the patterning process using wax pen and wax printer. Three different fabrication approaches were demonstrated, which include painting with wax pen, printing with inkjet printer followed by wax pen, and directly printing the barriers with wax printer (Lu et al., 2009).

For wax printing by pen the design was drawn using wax pen on paper and then paper is heated in an oven for about 5 minutes at 150°C. The melted wax will create channels for flow of fluid by penetrating into the paper. Wax dipping approach is also used to abolish the use of complicated instrumentation. It requires a mold that was created by laser cutting. The paper was inserted into the mold and conserved by using magnets and dipped into the molten wax (Songjaroen et al., 2011). Though wax pen and wax dipping methodologies do not require high-cost equipment for fabrication and are easy to handle, they suffer from low reproducibility among different batches. Nowadays, wax printers are available for the patterning of the microchannels in which complex designs can be printed very easily. As illustrated in Fig. 19.3, the design is printed directly on paper through printer and the paper is heated in an oven or on hot plate at particular temperature and time. This process can be completed within 5−10 minutes allowing preparation of more number of devices in very short period of time and the reproducibility for every batch is very high. The use of wax printer is more convenient, rapid, reproducible, and more suitable for the complex designs compared to other approaches.

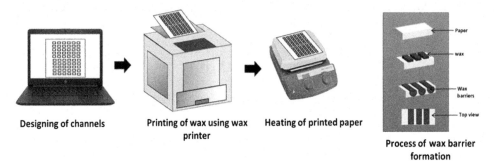

FIGURE 19.3   Steps involved in fabrication of microchannel using wax printing.

19.2.3 Flexographic printing

Paper-based microfluidics can be manufactured in bulk using a flexographic printing of polystyrene. Roll-to-roll production is possible by this method using ceramic rolls and patterned printing plates having relief pattern. Ink is filled in the ink reservoir and substrate paper is fixed with impression roll. Anilox roll contains microfluidic pattern, volume of which will decide the amount of ink to be transferred to the printing plates. Anilox roll will accelerate to the set printing speed and distribute the ink evenly. Then rotation of the plate and impression roll will print the ink ultimately onto the paper. Printing speed and ink penetration on the paper depends upon the acceleration gained by anilox and pressure between the plate roll and the impression roll, respectively.

Normally one step printing is not enough for creation of appropriate hydrophobic microfluidic channel. Hence, polystyrene mixed with xylene is normally used as the printing inks on backside and microfluidic pattern is printed on front side. This type of technique needs at least two cycles of printing for obtaining proper hydrophobic barrier (Olkkonen et al., 2010). As flexographic printing is a well-established method considering the existing manufacturing techniques and is a most widely used technique for large-scale production of paper-based devices.

19.2.4 Inkjet patterning

Inkjet technology can be of drop on demand (DOD) and continuous inkjet type depending on the application. Piezoelectric, thermal, electrostatic, and acoustic are the four modes in which DOD technique works, but among them piezoelectric and thermal mode are most commonly employed by the researchers. The ink used for printing is different than the one used for regular printing, and its physicochemical properties are the directives of the resolution, drop velocity, and reliability of printing. Generally UV-curable inks such as alkyl ketene dimer, and polyacrylate are used to create sufficiently hydrophobic barriers in the microfluidic channels. The black cartridges are filled with the inks. In this technique also the pattern is first drawn using different software in which black color represents hydrophobic layer and white as hydrophilic. The pattern is printed on paper, which is generally No. 5C filter paper having smallest particle retention size (1 μm). Immediately after printing, the paper is exposed to UV light source for curing. Thereafter, the paper is again printed on backside with ink that will cover all the area, which is patterned on topside (Maejima et al., 2013). Selection of ink is the important factor in this type of printing.

While in inkjet etching the whole filter paper is entirely covered with hydrophobic materials such as polystyrene followed by drying. The paper is then printed by inkjet printers with picoliter volume of toluene, which will create a channel (De Gans et al., 2007). The size of the channel and reproducibility of the pattern is optimized by varying the number of printing cycles. Though this method is simple to create paper-based devices, it has major drawback of coating the polystyrene on whole paper.

19.2.5 Photolithography

This technique has been explained by Whiteside for fabrication of microfluidic paper devices. The paper is soaked in a particular photoresist (e.g., SU-8-2010) used for photolithographic techniques followed by spinning at high RPM and baking to remove the solvent present in the photoresist. The paper and photoresist is then exposed to UV light through photomask, which results in cross-linking of the exposed portion of the photoresist. Soaking and washing of the paper is normally done using propylene glycol monomethyl ether acetate and propan-2-ol respectively for removal of unpolymerized photoresist as well as cleaning the pattern. To increase the hydrophilicity of the paper, entire surface of the paper is exposed to oxygen plasma (Martinez et al., 2007). High resolution is obtained by this process, but the cost of the photoresist and complexity of operation are some of the limitations of this process (Table 19.1).

19.3 Detection techniques

Selection of fabrication method and design of pattern to be created on paper is highly dependent on the detection method. Various detection methods such as colorimetric, electrochemical, chemiluminescence, fluorometric, and electrophoresis are available. In this section, commonly used techniques such as colorimetric, electrochemical detection, and chemiluminescence/electrochemiluminescence are explained.

19.3.1 Colorimetric sensing

In recent era, colorimetric detection is trending due to fast detection efficiency, visual detection, low cost, applicability in remote areas, operability, and stability. These μPADs have been used for the different analytes such as biomolecules and toxic ions present in different environmental matrices. Colorimetry is defined by the capillary action of the analytes to the zone of reaction thereby reacting with the specifically laden reagents giving a color change. The color change can be captured through cameras or smartphones or by naked eye (Martinez et al., 2008). The paper-based platforms created with different microfluidic pattern will enable the flow of fluid through different channels and mixing. The

TABLE 19.1   Advantages and limitations for different fabrication technique of paper-based microfluidic devices.

Fabrication technique	Advantages	Limitations
Screen printing	Simple and easy, uniformity in fabrication of devices, bulk fabrication, economical	Low resolution of microfluidic channels
Wax printing	Simple, fast, bulk fabrication is easy	It requires wax printers that are expensive
Photolithography	High resolution microfluidic channels	Complex process, expensive devices are needed
Flexographic printing	Roll-to-roll production in bulk	Printing quality depends on paper quality, polystyrene is required in more quantity
Inkjet printing	Rapid and bulk production of devices	Modified inkjet printers are required

mixed reagents meet at the detection zone, where analyte of interest reacts with receptors and exhibit a color change. Many times the change in color is concentration specific or it can just predict the presence of analyte. In many cases the analyte will oxidize or reduce the receptors, which corresponds to color change. In some cases, nanomaterials agglomerated after reacting with analytes giving color change. Though colorimetric detection of heavy metal is a popular method, its main drawback is the release of toxic gases during analysis.

19.3.2 Electrochemical sensing

Application of μPADs for electrochemical sensing has become one of the most popular and researched techniques that provide more stable and quantifiable signals. It deals with the interplay between electricity and chemistry. It will give the idea about the relationship between chemical parameters such as concentration by measuring electrical quantities such as current, potential, or charge. Different types of electrochemical techniques are explored by researchers, which include voltammetry, amperometry, and impedance spectroscopy (Wang, 2006). Henry et al. have demonstrated the use of electrochemical PADs for the analysis of biological, environmental, and other sample matrix for the first time (Dungchai et al., 2009).

Most commonly used μPAD for electrochemical sensing are developed by using screen printing technique. In this method, working, auxiliary, and reference electrodes play an important role in the detection of the target analyte. The working electrode is generally made of carbon or other conducting materials. Nowadays, the carbon-based electrodes are modified using the significant receptor, which will reduce or oxidize the analyte present in the sample and change in the electrical signals, which can be attributed to the concentration of analyte (Liu et al., 2014).

19.3.3 Chemiluminescence and electrochemiluminescence

This technique has arisen as a sensitive and efficient analytical technique because it can provide good sensitivity and wide dynamic range. Digital cameras and smartphones are of low cost, which will work as the readout units in this method. This method has wide application in biochemical studies such as salivary, cortisol, and genotyping of single-nucleotide polymorphisms. Excitation of molecules after reaction generates fluorescence, the signal of which directly corresponds to the concentration of analytes (Alahmad et al., 2016). On the other hand, electrochemiluminescence uses electrochemical reactions to generate luminescence signal. Electrochemiluminescence has advantages such as lower background optical signals, easier to control reagent generation through potential control, and improved selectivity through potential control at the electrodes (Miao, 2008) (Table 19.2).

19.4 Application of μPAD for the metal detection

μPADs have application in the detection and analysis of biological, biochemical, diagnostic, chemical, and environmental matrices, which includes determination of pH, blood glucose, urine analytes, heavy metals, ions, hormones, and infectious diseases. The main objective of these devices is to replace glass or plastic by paper with inbuilt microfluidic channels. Though μPADs have wide range of application in environmental monitoring, this chapter is mostly focusing on metal detection in water.

Colorimetric techniques are the fast and easy detection methods employed for the metal detection. Nanoparticles of several precious metal are very much employed in the colorimetric detection, but Ag nanoparticles are preferred and widely chosen due to its low cost and higher extinction coefficient compared to Au and Pt. For example, hexadecyltrimethylammonium bromide–modified Ag nanoparticles were used for the detection of Cu^{2+} ion in the presence of thiosulfate. Modified Ag nanoparticles were coated on test zone of paper. Once the sample containing Cu^{2+} ion contacted the paper, a change in color from violet-red to colorless is observed due to oxidation of Ag nanoparticles/thiosulfate by Cu^{2+} ions. The detection limit of the device is 1.0×10^{-3} ppm (Chaiyo et al., 2015). The higher extinction coefficient of Ag nanoparticles shows superior visibility of color. Ratnarathorn et al. fabricated μPAD by knife cutting on 180 μm thick filter paper and used homocysteine- and dithiothreitol-modified Ag nanoparticles in the detection zone. In the presence of Cu^{2+} ions, oxidation of homocysteine and aggregation of Ag nanoparticles will give yellow to orange color change. If the concentration of Cu^{2+} ions increases the change is from orange to greenish brown due to reduced surface Plasmon resonance coupling (Ratnarathorn et al., 2012).

Feng et al. have developed the enrichment-based μPAD by employing wax printing technique on a 215 μm thick paper and colorimetric detection of heavy metal was successfully completed. In the enrichment-based μPADs, water adsorbent bed is attached at the end of each channel for enhancing the sensitivity as the channels can hold up to 800 μL of sample, which is more than the closed end–type μPADs, which can hold maximum 20 μL of sample (Feng et al., 2013).

TABLE 19.2 Performance of various paper-based microfluidic devices for the detection of metals.

Analyte	Detection platform	Sample matrix	Transduction	Detection limits, range	References
Cu^{2+}	HCFI reagent	Aqueous	Colorimetric	$10^{-2}-10^1\,\mu g/mL$	Mujawar and El-Shahawi (2019)
Hg^{2+}	AuNCs/MIL-68(In)-NH$_2$/Cys		Fluorometric	6.7 pM, 5 nM to 50 M	Wu et al. (2019)
Co, Cu, Fe, Mn, Cr, and Ni	Graphene enhanced polylactic acid filament (PLA)	Air samples	Colorimetric	8.16, 45.84, 1.86×10^2, 10.08, 1.52×10^2 and 80.40 ng	Sun et al. (2018a)
Cr^{2+}	Diphenylcarbazide	Natural water	Colorimetric	$3\,\mu g/L$	Alahmad et al. (2018)
Cu^{2+}, Ni^{2+}, Hg^{2+}, Pb^{2+}, Cr^{2+}, Fe^{2+}	TCPP, amine of (3-aminopropyl) triethoxysilane ($-NH_2$)	Aqueous	Colorimetric	15, 0.5, 0.1, 0.3, and 3.58 μM	Idros and Chu (2018)
Cu^{2+}	Polyethyleneimine	Aqueous	Colorimetric	30 pM, 0.1−1 mM	Liu et al. (2018)
Cd^{2+}, Zn^{2+}	Sputtered Sn film electrode	Aqueous	Electrochemical	0.9 and 1.13 $\mu g/L$	Kokkinos et al. (2018)
Ni^{2+}, Cu^{2+}, Cr^{2+}	Dimethylglyoxime	Aqueous	Colorimetric	4.8,1.6, and 0.18 mg/L	Sun et al. (2018b)
$Cu2+$	Cationic porphyrin, 5,10,15,20-tetrakis(1-methyl-4-pyridinio) porphyrin tetraiodide	Aqueous	Fluorometric	0.16 ppm	Prabphal et al. (2018)
Cr-VI	1,5-diphenylcarbazide	Aqueous	Colorimetric	30 ppm	Asano and Shiraishi (2018)
Zn^{2+}	Bismuth-modified SPE	Biological fluids	Electrochemical	25 ng/L	Cinti et al. (2017)
Cu^{2+}, Hg^{2+}	CdTe QDs	Aqueous	Fluorometric	0.035 and 0.056 $\mu g/L$	Qi et al. (2017)
Cu^{2+}	Meso-tetrakis(1,2-dimethylpyrazolium-4-yl) porphyrin sulfonate	Aqueous	Colorimetric	1 ppm	Pratiwi et al. (2017)
Cd^{2+}	Pristine graphite foil free of any surface modifier	Aqueous	Electrochemical	1.2 and 1.8 $\mu g/L$	Shen et al. (2017)
Hg^{2+}	Dithizone in NaOH	Aqueous	Colorimetric		Cai et al. (2017)
Pb^{2+}, Cd^{2+}	Boron-doped diamond paste electrode	Aqueous	Electrochemical	1 and 25 ppb	Nantaphol et al. (2017)

(Continued)

TABLE 19.2   (Continued)

Analyte	Detection platform	Sample matrix	Transduction	Detection limits, range	References
Zn^{2+}	Zincon indicator	Tap water, river water	Colorimetric	0.53 μM	Kudo et al. (2017)
Mn^{2+}, Co^{2+}	4-(2-pyridylazo)resorcinol	Aqueous	Colorimetric	–	Meredith et al. (2017)
Pb^{2+}, Cd^{2+}, Cu^{2+}	Bismuth-modified boron-doped diamond electrode	Aqueous	Electrochemical and colorimetric	0.1 ng/mL	Chaiyo et al. (2016)
Cr^{3+}	Luminol oxidation by hydrogen peroxide	Natural water	Chemiluminescence	0.02 ppm	Alahmad et al. (2016)
Fe^{3+}	1,10-phenanthroline	Natural hot spring water	Colorimetric	100−1000 ppm	Ogawa and Kaneta (2016)
Hg^{2+}, Ag^+	Fluorescence labeled ssDNA functionalized graphene oxide	Food	Fluorometric	121 and 47 nM	Zhang et al. (2015)

HCFI, 3-(5-Hydroxy-4-carboxyphenylimino)-5-fluoroindol-2(H)one; *SPE*, screen-printed electrode; *ssDNA*, single-stranded DNA; *TCPP*, tetrakis(4-carboxyphenyl)porphyrin.

Gold nanoparticles are costly but shows rapid detection characteristics due its affinity with the analyte. Therefore gold nanoparticles were employed on the μPADs prepared by wax printing for the detection of Hg^{2+} ions from river and pond water. However, Hossain et al. used inkjet printing on 180 μm thick filter paper for developing μPAD for the detection of Hg^{2+}, Cu^{2+}, Cr^{6+}, and Ni^{2+} ions. Multiplexed β-galactosidase (B-GAL) modified μPADs substrate was employed for the colorimetric detection of mixture of heavy metals. Detection zone of μPAD has coating of chlorophenol red β-galactopyranoside and red-magenta color is formed after adding B-GAL. Each detection zone is preloaded with chromogenic reagents for detection of heavy metals. The detection limits of μPAD for Hg^{2+} and Cu^{2+} are 1 and 20 ppb, respectively, whereas these were 0.15 and 0.23 ppm for Cr^{6+} and Ni^{2+}, respectively (Hossain and Brennan, 2011). Though colorimetric detection of heavy metal is a popular method, its main drawback is the release of toxic gases during analysis and quantification of analyte concentration.

Electrochemical techniques are more popular in detection of heavy metals due to its high sensitivity and accuracy. For the first time, Dungchai reported that electrochemistry can be coupled with paper-based analytical devices. The authors used various electrode materials, for example, carbon, metals, micro wires, and nanoparticles and fabrication techniques, for example, screen/stencil printing, pencil/pen drawing, inkjet printing, and wire placement (Dungchai et al., 2009). Screen-printed electrodes have been extensively employed in electrochemical PADs due to their easy fabrication, low cost, and potential for large-scale production and achieving better quantitative results. Nie have demonstrated paper-based microfluidic devices for the electrochemical sensing of lead (Pb) in aqueous samples. The author fabricated three electrode sensors on paper by screen printing of carbon ink and microfluidic channels were prepared by patterning chromatography and detection was done using Anodic stripping voltammetry. The authors concluded that hydrodynamic microfluidic paper–based electrochemical devices have comparatively higher sensitivity and lower limit of detection (Nie et al., 2010). Determination of heavy-metal ions (e.g., Hg^{2+}, Ag^+, and Cr^{3+} ions) in groundwater samples has been demonstrated by Lee et al. recently. In this study, polypyrrole/cellulose composite paper was applied. Polypyrrole provides favorable electrochemical properties while the cellulose paper offers sufficient mechanical strength. The study demonstrated that this composite paper–based device serves as a multiplex detector for various heavy metals by unique signatures under analysis of principal component analysis (Lee et al., 2014). Another application has been published by Ruecha et al. wherein three electrode system was fabricated on filter paper by screen printing method. The working electrode was modified by nanocomposite of graphene–polyaniline, which was then used for trace level detection of Zn^{2+}, Cd^{2+}, and Pb^{2+} ions by square wave anodic stripping voltammetry (Ruecha et al., 2015). The introduction of nanostructured materials has been of great significance to improve detection limits, sensibility, selectivity, reproducibility, and miniaturization of sensing devices. Real-time injection system was developed by Xiao-Chen Dong et al., for the determination of heavy-metal ions by integrating the metal organic framework (MOF)-derived Mn_2O_3 modified SPE, portable 3D printed microfluidic cell, and universal serial bus interfaces (Hong et al., 2016).

19.5 Conclusion and future prospects

Paper-based microfluidic sensors signify emerging yet highly useful approach focusing on a variety of applications encompassing disease diagnostics, detection of environmental contaminants, industrial automation, forensics, etc. The major advantages of μPADs are their low cost, portability, low sample requirements, ease of mass production, user–friendliness, and above all their high sensitivity to detect analytes. Considering their miniature dimensions these sensor devices are not only beneficial in laboratories but also in field analysis, particularly in remote locations with limited resources and harsh environments. Another major advantage of these devices are the possibility to fabricate multiple sensing nodes on a single chip through multipatterning and thus, allowing simultaneous detection of multiple analytes (multiplexed analysis). All these features help in realizing true lab-on-chip systems for monitoring of a variety of parameters using single devices. These devices are based on simple transductions, namely, optical (colorimetric), electrochemical, chemiluminescence, and electrochemiluminescence, which also enable us to develop low-cost and reliable devices. These devices provide an opportunity for developing simple devices, which eliminates the need of highly skilled manpower required for operation of in-lab sophisticated instruments.

The μPADs have been extensively explored for environmental applications, particularly for detection of metals based on colorimetric and electrochemical techniques. The detection of elements, namely, Pb, As, Cr, and Hg has been reported with high sensitivity and selectivity. The performance in terms of sensitivity, selectivity, and cost of these devices is likely to increase with advent of recent materials, including graphene-based composites and 2D nanomaterials. The fabrication techniques have also advanced in past few years with more sophisticated printing options enabling high-end multipatterning at affordable costs. A combination of these approaches of highly efficient receptors and fabrication techniques is likely to offer new and exciting sensor devices capable of replacing the sophisticated instrumentation with the miniature devices for environmental monitoring, including real-time/in-line analysis.

Although microfluidic sensors have many advantages over conventional methods for measurement of water pollutants, they do have disadvantages as well. For example, microfluidic electrochemical sensor generally uses small amount of samples, therefore there are always chances of not getting accurate measurement, especially if the measurand is affected by any physical/chemical alteration during measurement and samples are not well mixed. Therefore unlike conventional techniques, microfluidic sensors are usually microfluidic integrated electrochemical biosensors and have less shelf life, lack of selectivity. More researches focusing on the improvement in sensitivity are needed in the future.

References

Alahmad, W., et al., 2016. A miniaturized chemiluminescence detection system for a microfluidic paper-based analytical device and its application to the determination of chromium(III). Anal. Methods 8 (27), 5414–5420. Available from: https://doi.org/10.1039/c6ay00954a.

Alahmad, W., et al., 2018. A colorimetric paper-based analytical device coupled with hollow fiber membrane liquid phase microextraction (HF-LPME) for highly sensitive detection of hexavalent chromium in water samples. Talanta 190, 78–84. Available from: https://doi.org/10.1016/j.talanta.2018.07.056.

Asano, H., Shiraishi, Y., 2018. Microfluidic paper-based analytical device for the determination of hexavalent chromium by photolithographic fabrication using a photomask printed with 3D printer. Anal. Sci. 34 (1), 71–74. Available from: https://doi.org/10.2116/analsci.34.71.

Cai, L., et al., 2017. Visual quantification of Hg on a microfluidic paper-based analytical device using distance-based detection technique. AIP Adv. 7 (8), 085214. Available from: https://doi.org/10.1063/1.4999784.

Chaiyo, S., et al., 2015. Highly selective and sensitive paper-based colorimetric sensor using thiosulfate catalytic etching of silver nanoplates for trace determination of copper ions. Anal. Chim. Acta 866, 75–83. Available from: https://doi.org/10.1016/j.aca.2015.01.042.

Chaiyo, S., et al., 2016. High sensitivity and specificity simultaneous determination of lead, cadmium and copper using μPAD with dual electrochemical and colorimetric detection. Sens. Actuators, B: Chem. 233, 540–549. Available from: https://doi.org/10.1016/j.snb.2016.04.109.

Cinti, S., et al., 2017. Sustainable monitoring of Zn(II) in biological fluids using office paper. Sens. Actuators, B: Chem. 253 (Ii), 1199–1206. Available from: https://doi.org/10.1016/j.snb.2017.07.161.

Cleary, J., Maher, D., Slater, C., Diamond, D., 2010. In-situ monitoring of environmental water quality using an autonomous microfluidic sensor, In: IEEE Sensors Applications Symposium (SAS). pp. 36–40. Available from: https://doi.org/10.1109/SAS.2010.5439385.

Lu, Y., Shi, W., Jiang, L., Qin, J., Lin, B., 2009. Rapid prototyping of paper-based microfluidics with wax for low-cost, portable bioassay. Electrophoresis 30, 1497–1500. <https://doi.org/10.1002/elps.200800563>.

Dieterich, K., 1902. Testing-paper and method of making same. U.S. patent 691,249.

Dungchai, W., Chailapakul, O., Henry, C.S., 2009. Electrochemical detection for paper-based microfluidics. Anal. Chem. 81 (14), 5821–5826. <https://doi.org/10.1021/ac9007573>.

Feng, L., et al., 2013. Enhancement of sensitivity of paper-based sensor array for the identification of heavy-metal ions. Anal. Chim. Acta 780, 74–80. Available from: https://doi.org/10.1016/j.aca.2013.03.046.

De Gans, B.J., Hoeppener, S., Schubert, U.S., 2007. Polymer relief microstructures by inkjet etching. J. Mater. Chem. 17 (29), 3045–3050. Available from: https://doi.org/10.1039/b701947e.

Hong, Y., et al., 2016. 3D printed microfluidic device with microporous Mn_2O_3-modified screen printed electrode for real-time determination of heavy metal ions. ACS Appl. Mater. Interfaces 8 (48), 32940–32947. Available from: https://doi.org/10.1021/acsami.6b10464.

Hossain, S.M.Z., Brennan, J.D., 2011. β-Galactosidase-based colorimetric paper sensor for determination of heavy metals. Anal. Chem. 83 (22), 8772–8778. Available from: https://doi.org/10.1021/ac202290d.

Idros, N., Chu, D., 2018. Triple-indicator-based multidimensional colorimetric sensing platform for heavy metal ion detections. ACS Sens. 3 (9), 1756–1764. Available from: https://doi.org/10.1021/acssensors.8b00490.

Jang, A., et al., 2011. State-of-the-art lab chip sensors for environmental water monitoring. Meas. Sci. Technol. 22 (3), 32001. <https://doi.org/10.1088/0957-0233/22/3/032001>.

Jiang, J., et al., 2014. Smartphone based portable bacteria pre-concentrating microfluidic sensor and impedance sensing system. Sens. Actuators, B: Chem. 193, 653–659. Available from: https://doi.org/10.1016/j.snb.2013.11.103.

Kim, U., et al., 2013. Rapid, affordable, and point-of-care water monitoring via a microfluidic DNA sensor and a mobile interface for global health. IEEE J. Transl. Eng. Health Med. 1, 3700207. Available from: https://doi.org/10.1109/JTEHM.2013.2281819.

Kokkinos, C., Economou, A., Giokas, D., 2018. Paper-based device with a sputtered tin-film electrode for the voltammetric determination of Cd(II) and Zn(II). Sens. Actuators, B: Chem. 260 (Ii), 223–226. Available from: https://doi.org/10.1016/j.snb.2017.12.182.

Kudo, H., et al., 2017. Paper-based analytical device for zinc ion quantification in water samples with power-free analyte concentration. Micromachines 8 (4). Available from: https://doi.org/10.3390/mi8040127.

Lee, J.E., et al., 2014. Real-time detection of metal ions using conjugated polymer composite papers. Analyst 139 (18), 4466–4475. Available from: https://doi.org/10.1039/c4an00804a.

Liu, B., et al., 2014. Paper-based electrochemical biosensors: from test strips to paper-based microfluidics. Electroanalysis 26 (6), 1214–1223. Available from: https://doi.org/10.1002/elan.201400036.

Liu, L., et al., 2018. Sensitive colorimetric detection of Cu^{2+} by simultaneous reaction and electrokinetic stacking on a paper-based analytical device. Microchem. J. 139, 357–362. Available from: https://doi.org/10.1016/j.microc.2018.03.021.

Maejima, K., et al., 2013. Inkjet printing: an integrated and green chemical approach to microfluidic paper-based analytical devices. RSC Adv. 3, 9258–9263. <https://doi.org/10.1039/c3ra40828k>.

Martinez, A.W., et al., 2007. Patterned paper as a platform for inexpensive, low-volume, portable bioassays. Angew. Chem. Int. Ed. 46 (8), 1318–1320. Available from: https://doi.org/10.1002/anie.200603817.

Martinez, A.W., et al., 2008. Simple telemedicine for developing regions: camera phones and paper-based microfluidic devices for real-time, off-site diagnosis. Anal. Chem. 80 (10), 3699–3707. <https://doi.org/10.1021/ac800112r>.

Meredith, N.A., Volckens, J., Henry, C.S., 2017. Paper-based microfluidics for experimental design: screening masking agents for simultaneous determination of Mn(II) and Co(II). Anal. Methods 9 (3), 534–540. Available from: https://doi.org/10.1039/c6ay02798a.

Miao, W., 2008. Electrogenerated chemiluminescence and its biorelated applications. Chem. Rev. 108 (7), 2506–2553. Available from: https://doi.org/10.1021/cr068083a.

Mujawar, L.H., El-Shahawi, M.S., 2019. Rapid and sensitive microassay for trace determination and speciation of Cu^{2+} on commercial book-paper printed with nanolitre arrays of novel chromogenic reagent. Microchem. J. 146, 434–443. Available from: https://doi.org/10.1016/j.microc.2019.01.025.

Muller, R.H., Clegg, A.N.D.D.L., York, N., 1951. Automatic paper chromatography. Ann. N. Y. Acad. Sci. 1123–1125. Available from: https://doi.org/10.1021/ac60033a032.

Nantaphol, S., et al., 2017. Boron doped diamond paste electrodes for microfluidic paper-based analytical devices. Anal. Chem. 89 (7), 4100–4107. Available from: https://doi.org/10.1021/acs.analchem.6b05042.

Nie, Z., et al., 2010. Electrochemical sensing in paper-based microfluidic devices. Lab Chip 10 (4), 477–483. Available from: https://doi.org/10.1039/b917150a.

Ogawa, K., Kaneta, T., 2016. Determination of iron ion in the water of a natural hot spring using microfluidic paper-based analytical devices. Anal. Sci. 32 (1), 31–34. Available from: https://doi.org/10.2116/analsci.32.31.

Olkkonen, J., Lehtinen, K., Erho, T., 2010. Flexographically printed fluidic structures in paper. Anal. Chem. 82, 10246–10250. <https://doi.org/10.1021/ac1027066>.

Prabphal, J., Vilaivan, T., Praneenararat, T., 2018. Fabrication of a paper-based turn-off fluorescence sensor for Cu^{2+} ion from a pyridinium porphyrin. ChemistrySelect 3 (3), 894–899. Available from: https://doi.org/10.1002/slct.201702382.

Pratiwi, R., et al., 2017. A selective distance-based paper analytical device for copper(II) determination using a porphyrin derivative. Talanta 174 (Ii), 493—499. Available from: https://doi.org/10.1016/j.talanta.2017.06.041.

Qi, J., et al., 2017. Three-dimensional paper-based microfluidic chip device for multiplexed fluorescence detection of Cu^{2+} and Hg^{2+} ions based on ion imprinting technology. Sens. Actuators, B: Chem. 251, 224—233. Available from: https://doi.org/10.1016/j.snb.2017.05.052.

Ratnarathorn, N., et al., 2012. Simple silver nanoparticle colorimetric sensing for copper by paper-based devices. Talanta 99, 552—557. Available from: https://doi.org/10.1016/j.talanta.2012.06.033.

Ruecha, N., et al., 2015. Sensitive electrochemical sensor using a graphene-polyaniline nanocomposite for simultaneous detection of Zn(II), Cd(II), and Pb(II). Anal. Chim. Acta 874, 40—48. Available from: https://doi.org/10.1016/j.aca.2015.02.064.

Shen, L.L., et al., 2017. Modifier-free microfluidic electrochemical sensor for heavy-metal detection. ACS Omega 2 (8), 4593—4603. Available from: https://doi.org/10.1021/acsomega.7b00611.

Songjaroen, T., et al., 2011. Novel, simple and low-cost alternative method for fabrication of paper-based microfluidics by wax dipping. Talanta 85 (5), 2587—2593. Available from: https://doi.org/10.1016/j.talanta.2011.08.024.

Sun, H., et al., 2018a. Multiplex quantification of metals in airborne particulate matter via smartphone and paper-based microfluidics. Anal. Chim. Acta 1044, 110—118. Available from: https://doi.org/10.1016/j.aca.2018.07.053.

Sun, X., et al., 2018b. Improved assessment of accuracy and performance using a rotational paper-based device for multiplexed detection of heavy metals. Talanta 178, 426—431. Available from: https://doi.org/10.1016/j.talanta.2017.09.059.

Wang, Joseph, 2006. Analytical Electrochemistry, first ed. John Wiley & Sons, Inc, Hoboken, NJ, <https://doi.org/10.1002/0471790303>.

Wu, X.J., et al., 2019. Ratiometric fluorescent nanosensors for ultra-sensitive detection of mercury ions based on AuNCs/MOFs. Analyst 144 (8), 2523—2530. Available from: https://doi.org/10.1039/c8an02414f.

Zhang, Y., Zuo, P., Ye, B.C., 2015. A low-cost and simple paper-based microfluidic device for simultaneous multiplex determination of different types of chemical contaminants in food. Biosens. Bioelectron. 68, 14—19. Available from: https://doi.org/10.1016/j.bios.2014.12.042.

Index

Note: Page numbers followed by "*f*" and "*t*" refer to figures and tables, respectively.